Ecological Studies, Vol. 179

Analysis and Synthesis

Edited by

M.M. Caldwell, Logan, USA
G. Heldmaier, Marburg, Germany
R.B. Jackson, Durham, USA
O.L. Lange, Würzburg, Germany
H.A. Mooney, Stanford, USA
E.-D. Schulze, Jena, Germany
U. Sommer, Kiel, Germany

Ecological Studies

Volumes published since 2001 are listed at the end of this book.

Luc Abbadie   Jacques Gignoux
Xavier Le Roux   Michel Lepage
Editors

Lamto
Structure, Functioning, and Dynamics
of a Savanna Ecosystem

With 158 Illustrations

Luc Abbadie
Jacques Gignoux
Biogéochimie et écologie des
  milieux continentaux
Ecole Normale Supérieure
75230 Paris cedex 05
France

Xavier Le Roux
Laboratoire d'écologie microbienne
CNRS–Université Lyon 1
Bâtiment Gregor Mendel
69622 Villeurbanne cedex
France

Michel Lepage
IRD
01 BP 182
Ouagadougou 01
Burkina Faso

Cover illustration: Typical view of Lamto savanna showing the mixture of grasses, shrubs/trees, and palm trees (photography by Jacques Gignoux).

Library of Congress Cataloging-in-Publication Data
Lamto: structure, functioning, and dynamics of a savanna ecosystem/Luc Abbadie ... [et al.].—1st ed.
  p. cm. — (Ecological studies; v. 179)
  Includes bibliographical references and index.
  ISBN 0-387-94844-9
  1. Savanna ecology—Côte d'Ivoire—Lamto. 2. Lamto (Côte d'Ivoire) I. Abbadie, Luc.
II. Series.
QH195.I9L26 2005
577.4′8′096668—dc22   2004056572

ISSN 0070-8356
ISBN-10: 0-387-94844-9
ISBN-13: 978-0387-94844-7

Printed on acid-free paper.

© 2006 Springer Science+Business Media, Inc.
All rights reserved. This work may not be translated or copied in whole or in part without the written permission of the publisher (Springer Science+Business Media, Inc., 233 Spring Street, New York, NY 10013, USA), except for brief excerpts in connection with reviews or scholarly analysis. Use in connection with any form of information storage and retrieval, electronic adaptation, computer software, or by similar or dissimilar methodology now known or hereafter developed is forbidden.
The use in this publication of trade names, trademarks, service marks, and similar terms, even if they are not identified as such, is not to be taken as an expression of opinion as to whether or not they are subject to proprietary rights.

Printed in the United States of America.   (SPI/EB)

9 8 7 6 5 4 3 2 1

springer.com

Preface

Lamto is a dream place for scientists. The first time one arrives there, one has the feeling of being in an ideal place to achieve scientific work. After a few days walking in the savanna, sampling soil, cutting grass, or carrying helium bottles or respiration chambers all the day long, learning botany and entomology with the local technicians, taming the apparently eternal old cars, searching for keys, augers, or screws during hours, learning at the library all the background on Lamto savannas needed for one's particular study, enjoying the food at "the popotte" thrice a day, canoeing on the Bandama River at dusk, and slowly sipping a Flag (the local beer and the best beer in the world) at night, the work is completed without difficulty. "When shall I come back?" is most often the question asked before leaving Lamto's Eden. This is a particularly open question in the difficult days Côte d'Ivoire is undergoing.

Because of its particular location at the edge of the rain forest, of the facilities offered to scientists, of the recurrent funding by the Ivorian and French research agencies, Lamto is one of the very few savanna sites in the world where ecological research has been going on for more than forty years. This makes it one of the best-known ecosystems in the world. This long and fascinating scientific adventure had to be told. We therefore attempted a synthesis, maybe just a feeling of the treasure, showing the top of the iceberg. We choose to consider ecosystem functioning as the main goal of this book, stressing the importance of studying the sometimes subtle relationships between ecosystem structure, functioning, and dynamics, a key issue of modern ecology. It was therefore not our purpose to list for each scientific field all the (often considerable) work that has been performed: this synthesis is therefore plant- and soil-biased, but we tried to give the reader a touch of the school of thought that emerged from obstinate research on the same system. We want to believe that this long and exacting work has promoted Lamto as the drosophila of savanna ecosystem science.

Lamto constitutes an invaluable model for studying the functioning and dynamics of ecosystems. It has now become a daily teaching tool for our colleagues from the universities of Côte d'Ivoire who currently implement original projects linking ecology and human development besides more strictly academic projects. Since September 2003, Côte d'Ivoire has experienced a dramatic crisis that it did not deserve and that mortgages its future. More than ever, we feel very close to its inhabitants, particularly to our colleagues, technicians, researchers, and university teachers, to whom we would like to ensure our brotherly support.

We would like to dedicate this book to all those who made Lamto, especially to those whose part is least visible in the scientific literature: first, to Professor Maxime Lamotte and the late Dr. Jean-Luc Tournier, the founders; then, to Professor Roger Vuattoux, the late Jean-Louis Tireford, and Souleymane Konaté, the directors of the ecology and geophysics stations, who carried Lamto on their shoulders for a long time; finally, to Kouassi Konan Germain, Konan N'Dri Alexis, Kouassi Etienne, N'Guessan François, Kouassi Guillaume, Loukou Martin, Raphaël Zou, and Sawadogo Prosper (technicians of the stations), and the traditional chiefs of the villages of Ahiérémou II, Zougoussi, and Kotiéssou, who made the adventure possible. Our hope is that, in spite of the difficult times Côte d'Ivoire is enduring, the adventure will continue.

Paris, France

Luc Abbadie
Jacques Gignoux
Xavier Le Roux
Michel Lepage

Contents

Preface V

Contributors... XVII

**1 History of the Lamto Ecology Station and
Ecological Studies at Lamto**
Roger Vuattoux, Souleymane Konaté, Luc Abbadie, Sébastien Barot,
Jacques Gignoux, and Gaëlle Lahoreau 1

1.1 Origin and installation of the station 1
1.2 The scientific programs that sustained the station from 1962 ... 6
    1.2.1 The IBP program: 1968 to 1978 6
    1.2.2 The SALT program: 1988 to 1998 7
1.3 The current programs: 2000 to present 8
1.4 Conclusion: Forty years of scientific production 9

Part I The Environment

2 Geology, Landform, and Soils
Luc Abbadie and Jean-Claude Menaut........................... 15

2.1 Geology 15
2.2 Landform 15
2.3 Soils 17
    2.3.1 Pedogenesis 17
    2.3.2 Diversity of soil profiles 18
    2.3.3 Typical soil profiles 18

3 Climate
Xavier Le Roux... 25

3.1 Introduction 25

3.2	The Lamto climate in the context of the West African climates	25
3.3	The Lamto Geophysical Station: Forty years of routine climatic observations	27
	3.3.1 Parameters monitored at Lamto	27
	3.3.2 Data quality assessment	29
3.4	Seasonal course of climatic parameters	29
	3.4.1 Precipitation and dew	30
	3.4.2 Radiation	31
	3.4.3 Air temperature	33
	3.4.4 Air water vapor pressure and saturation deficit	33
	3.4.5 Horizontal wind speed	34
	3.4.6 Evaporation	36
	3.4.7 Rainwater and aerosol chemistry	36
3.5	Interannual variability and temporal trends	37
3.6	Conclusion	40

4 Environmental Constraints on Living Organisms
Luc Abbadie, Jacques Gignoux, Michel Lepage,
and Xavier Le Roux 45

4.1	Introduction	45
4.2	Soil water	46
	4.2.1 Climatic influences	46
	4.2.2 Soil influences	46
4.3	Soil nutrients	48
4.4	Light	50
4.5	Fire	51
	4.5.1 Specificity of savanna fires	51
	4.5.2 Fire in Lamto: A driving force of the ecosystem	52
	4.5.3 Fire severity	53
4.6	Herbivory	56
4.7	Conclusion	57

5 Vegetation
Jean-Claude Menaut and Luc Abbadie 63

5.1	Introduction	63
5.2	Main savanna types	63
5.3	Structure of the vegetation	64
	5.3.1 The grass layer	64
	5.3.2 The tree and shrub layer	69
5.4	Life-Forms	69
5.5	Phenological cycles	71
5.6	Conclusion	72

Part II Structure and Functioning of Plant Cover

6 Soil-Plant-Atmosphere Exchanges
Xavier Le Roux and Bruno Monteny 77

6.1	Introduction	77
6.2	Overview of the 1991 to 1994 field campaign	78
6.3	Savanna radiation budget and spectral signatures	80
	6.3.1 Radiation budget above the savanna	80
	6.3.2 Radiation budget of the grass and shrub layers	83
	6.3.3 Surface spectral signatures	83
	6.3.4 Scientific gains and gaps in the knowledge	86
6.4	Energy budget	86
	6.4.1 Seasonal variations in the components of the savanna energy budget	86
	6.4.2 Aerodynamic and surface resistances, and savanna-atmosphere coupling	90
	6.4.3 Scientific gains and gaps in the knowledge	91
6.5	Water balance and plant water status	92
	6.5.1 Interception loss	92
	6.5.2 Runoff and infiltration	93
	6.5.3 Soil moisture	94
	6.5.4 Drainage	95
	6.5.5 Soil water uptake by roots	96
	6.5.6 Plant water status	97
	6.5.7 Scientific gains and gaps in the knowledge	98
6.6	CO_2 exchanges and leaf conductance	99
	6.6.1 At the leaf level	99
	6.6.2 At the canopy level	102
	6.6.3 Scientific gains and gaps in the knowledge	103
6.7	NO, NO_2, and O_3 exchanges	104
	6.7.1 NO emission from soils	104
	6.7.2 Behavior of the NO-NO_2-O_3 triade	105
	6.7.3 Scientific gains and gaps in the knowledge	106
6.8	Conclusion	106

7 Biomass Cycle and Primary Production
Jacques Gignoux, Patrick Mordelet, and Jean-Claude Menaut 115

7.1	Introduction	115
7.2	The aboveground phytomass cycle	115
	7.2.1 Methods for studying aboveground phytomass	119
	7.2.2 Tree phytomass	121
	7.2.3 The grass phytomass cycle	122

X   Contents

7.3 The belowground phytomass cycle 124
     7.3.1 Methods for studying belowground phytomass 124
     7.3.2 The root phytomass cycle 127
7.4 Primary production of Lamto savannas...................... 127
     7.4.1 Estimating primary production 127
     7.4.2 The primary production of Lamto savannas............ 130
     7.4.3 Relation between primary production and
            climatic indices 133
7.5 Plant allocation strategies: What can be inferred from phytomass
     measurements?... 134
7.6 Discussion: Toward an integrative approach of primary
     production and allocation 135

8 Tree/Grass Interactions
Patrick Mordelet and Xavier Le Roux 139

8.1 Introduction 139
8.2 Trees alter the understory grass environment.............. 140
     8.2.1 Aboveground microclimate and light availablity 140
     8.2.2 Soil physical and chemical characteristics, and nutrient
            availability 142
     8.2.3 Soil water availability 144
     8.2.4 A particular case: Tree clumps associated with termite
            mounds 145
     8.2.5 Summary: Relative importance of changes in resources
            availability under shrub clumps for grasses 145
8.3 Trees and grasses share the same soil resources 146
     8.3.1 Both trees and grasses are shallow-rooted 146
     8.3.2 Both trees and grasses uptake most of their water from
            upper soil layers 147
     8.3.3 A particular case: Palm tree roots associated with
            tree clumps... 149
     8.3.4 Lack of time partitioning of soil resources between trees
            and grasses... 149
     8.3.5 Summary: Likely importance of competition for
            soil resources between trees and grasses 150
8.4 Trees alter grass functioning and production 151
     8.4.1 Tree effect on grass leaf photosynthesis
            and water status 151
     8.4.2 Tree effect on grass aboveground biomass and primary
            production 152
     8.4.3 Tree effect on grass shoot/root ratio 153
8.5 Conclusion 154

9 Modeling the Relationships between Vegetation Structure and Functioning, and Modeling Savanna Functioning from Plot to Region
Xavier Le Roux, Jacques Gignoux, and Guillaume Simioni 163

- 9.1 Introduction 163
- 9.2 Models previously developed for predicting the functioning of Lamto savannas 164
- 9.3 TREEGRASS: A 3D model for simulating structure-functioning relationships in savanna ecosystems 165
  - 9.3.1 Overview of the model TREEGRASS 165
  - 9.3.2 Ability of TREEGRASS to predict the temporal and spatial variations in savanna functioning 168
  - 9.3.3 Analyzing the factors driving the outcome of tree/grass interactions.. 171
  - 9.3.4 Studying structure-functioning relationships in Lamto savanna 171
- 9.4 Modeling the functioning of savannas at large scales 173
  - 9.4.1 Modeling primary production and water balance in the West African savanna zone: A regional-scale approach using satellite data assimilation 175
  - 9.4.2 Modeling the primary production and phenology of the savanna biome: A global-scale approach in the context of General Circulation Models......................... 178
  - 9.4.3 Current limitations of savanna models operating at regional or global scale............................. 178
- 9.5 Conclusion 180
  - 9.5.1 State of the art of the modeling of the Lamto savanna .. 180
  - 9.5.2 Perspectives 181

10 Modification of the Savanna Functioning by Herbivores
Xavier Le Roux, Luc Abbadie, Hervé Fritz, and Hélène Leriche 185

- 10.1 Introduction 185
- 10.2 Herbivore densities, biomasses, and green grass consumption rate in Lamto savannas 185
  - 10.2.1 Invertebrate herbivores 185
  - 10.2.2 Large mammal herbivores 187
- 10.3 Field studies of grazing effect on the savanna functioning 189
  - 10.3.1 Response of grass production to grazing............... 189
  - 10.3.2 Response of soil microbial activities following a grazing event 192
- 10.4 Modeling approaches for understanding grazing effect on the savanna functioning 192
- 10.5 Conclusion 194

Part III Carbon Cycle and Soil Organic Matter Dynamics

11 Origin, Distribution, and Composition of Soil Organic Matter
Luc Abbadie and Hassan Bismarck Nacro 201

11.1	Introduction	201
11.2	The inputs of organic matter to soil	202
11.3	Soil organic matter distribution	203
	11.3.1 Variations of soil organic matter distribution at landscape scale	203
	11.3.2 Variations of soil organic matter distribution at organomineral particle scale	205
11.4	Chemical composition of soil organic matter	207
	11.4.1 Variations of the chemical composition of soil organic matter at landscape scale	207
	11.4.2 Variations of the chemical composition of soil organic matter at organomineral particles scale	212
11.5	Conclusion	214

12 Soil Carbon and Organic Matter Dynamics
Luc Abbadie, Hassan Bismarck Nacro, and Jacques Gignoux 219

12.1	Soil micro-organisms	219
12.2	The limitation of soil microbial activity by the supply of organic and mineral compounds	220
12.3	Plant litter decomposition	223
12.4	Soil organic matter mineralization and turnover	225
12.5	Modeling organic matter dynamic in Lamto soils	228
12.6	Conclusion	231

13 Perturbations of Soil Carbon Dynamics by Soil Fauna
Michel Lepage, Luc Abbadie, Guy Josens, Souleymane Konaté, and Patrick Lavelle 235

13.1	Introduction	235
13.2	Earthworms and termites: Abundances and spheres of influence	236
	13.2.1 Densities and biomasses	236
	13.2.2 Biogenic structures	236
13.3	Carbon distribution and storage	239
	13.3.1 Anecic species	240
	13.3.2 Geophageous species	241
13.4	Carbon mineralization	241
	13.4.1 Biological systems of regulation	241

	13.4.2 Termites and CO_2 release	243
	13.4.3 Scaling	245
13.5	Conclusion	245

Part IV The Nitrogen Cycle

14 Nitrogen Inputs to and Outputs from the Soil-Plant System
Luc Abbadie 255

14.1	Introduction	255
14.2	Dry and wet depositions	255
	14.2.1 Nitrate and ammonium concentrations	255
	14.2.2 Inputs of nitrogen to the ecosystem	257
14.3	Biological fixation of atmospheric dinitrogen	258
	14.3.1 The dinitrogen fixation by Cyanobacteria	258
	14.3.2 The symbiotic fixation of molecular nitrogen	259
	14.3.3 The non-symbiotic fixation of molecular nitrogen	262
14.4	Grass cover leaching	263
14.5	Soil leaching	264
14.6	Denitrification	264
14.7	Nitrogen monoxide emission	266
14.8	Impact of fire on the nitrogen cycle	267
14.9	N fluxes associated to grass consumption by animals	270
14.10	Conclusion: The input-output balance of nitrogen	271

15 Nitrogen Dynamics in the Soil-Plant System
Luc Abbadie and Jean-Christophe Lata 277

15.1	Introduction	277
15.2	Nitrogen dynamics in the shrub-tree layer	277
15.3	Nitrogen dynamics in the grass layer	278
	15.3.1 Nitrogen concentrations in the aboveground parts of herbaceous plants	278
	15.3.2 Nitrogen concentrations in the roots of herbaceous plants	280
	15.3.3 Nitrogen concentrations and pools in the grasses of the yearly burned savannas	280
	15.3.4 Nitrogen concentrations and pools in the grasses of unburned savanna	281
15.4	Annual nitrogen requirements of grasses	282
15.5	Origin of grass nitrogen	283
15.6	The transformations of nitrogen in soil	286
	15.6.1 The accumulation of organic nitrogen in soil	286

	15.6.2 The production of mineral nitrogen in soil	287
	15.6.3 Nitrification	290
15.7	Conclusion: The savanna, a system that retains nitrogen and mineral nutrients	294

16 Role of Soil Fauna in Nitrogen Cycling
Michel Lepage, Luc Abbadie, Guy Josens, and Patrick Lavelle 299

16.1	Introduction	299
16.2	Nitrogen storage and throughput in soil macrofauna and associated structures	299
	16.2.1 Nitrogen in anecic species	299
	16.2.2 Nitrogen in endogeic species	301
	16.2.3 Importance of the soil macrofauna in the N cycle	301
16.3	Impact of soil macrofauna on nitrogen dynamics and mineralization	303
	16.3.1 Termites	303
	16.3.2 Earthworms	305
	16.3.3 Impact on ecosystem dynamics	306
16.4	Stimulation of plant growth by soil macrofauna	306
16.5	Conclusion	308

Part V Plant Community Dynamics

17 Spatial Pattern, Dynamics, and Reproductive Biology of the Grass Community
Jacques Gignoux, Isabelle Dajoz, Jacques Durand, Lisa Garnier, and Michel Veuille 315

17.1	Introduction	315
17.2	Structure of the grass layer	315
	17.2.1 Grass population structure	315
	17.2.2 Spatial distribution of grass species	318
	17.2.3 Discussion	322
17.3	Dynamics of the grass layer	323
	17.3.1 Tuft dynamics	323
	17.3.2 Toward demographic studies of the grass layer	323
17.4	Reproduction system of *Hyparrhenia diplandra* and its population genetic structure as revealed by microsatellites	324
	17.4.1 Isolation and characterization of microsatellites	325
	17.4.2 Discussion	331
17.5	Conclusion	332

18 Structure, Long-Term Dynamics, and Demography of the Tree Community
Jacques Gignoux, Sébastien Barot, Jean-Claude Menaut, and
Roger Vuattoux... 335

18.1	Introduction	335
18.2	Factors influencing tree population dynamics.................	335
	18.2.1 Competition for resources	335
	18.2.2 Fire and the definition of demographic stages	337
18.3	Spatial patterns of tree species	339
	18.3.1 Spatial distribution of tree species....................	339
	18.3.2 Association to environment heterogeneities	343
	18.3.3 Case study: *Borassus aethiopum*	344
	18.3.4 Conclusion: Vital attributes of savanna trees inferred from their spatial patterns	345
18.4	Tree population dynamics	346
	18.4.1 Long-term dynamics.................................	346
	18.4.2 Size structure of tree populations.....................	350
	18.4.3 Demographic parameters	354
18.5	Discussion: The interaction of demography and spatial patterns and its effect on savanna stability	358

19 Modeling Tree and Grass Dynamics: From Demographic to Spatially Explicit Models
Jacques Gignoux and Sébastien Barot 365

19.1	Introduction	365
19.2	Persistence of savanna species under annual burning: Analysis through matrix population models...................	366
	19.2.1 Effect of fire on grass demography and persistence......	366
	19.2.2 Tree persistence and reproductive strategy: The case study of *Borassus aethiopum*	367
	19.2.3 Population persistence, fire, and demographic strategies of savanna plants....................................	369
19.3	Spatialized demographic models	370
	19.3.1 The role of spatial pattern and fire in savanna dynamics	370
	19.3.2 Continuous spatial model	370
	19.3.3 Cellular automaton model..........................	373
	19.3.4 Fire and the stability of Guinea savannas as mixed life-form systems	374
19.4	Conclusion: The dynamics of plant populations and spatial patterns	376

Part VI Toward an Integration of Savanna Structure, Functioning, and Dynamics

20 A Synthetic Overview of Lamto Savanna Ecology: Importance of Structure-Functioning-Dynamics Relationships
Xavier Le Roux, Luc Abbadie, and Jacques Gignoux 381

20.1 Introduction 381
20.2 Rationales for the "structure-functioning-dynamics relationships" approach 382
20.3 Structure-functioning relationships as a key to understanding the Lamto productivity paradox 384
20.4 Structure-dynamics relationships as a key to understanding changes in tree/grass balance 386
20.5 Current approaches for studying tree functioning-dynamics relationships in Lamto savanna 386
20.6 Modeling as a synthesis tool for studying structure-functioning-dynamics relationships 387
20.7 Conclusion 389

21 Perspectives: From the Lamto Experience to Critical Issues for Savanna Ecology Research
Jacques Gignoux, Xavier Le Roux, and Luc Abbadie 393

21.1 Introduction 393
21.2 Scaling across time 395
    21.2.1 Linking physiology and demography 395
    21.2.2 Linking vegetation dynamics to soil organic matter decomposition/sequestration 396
21.3 Scaling across space: From plot to landscape and region 397
    21.3.1 From plot to landscape 397
    21.3.2 From landscape to region 397
21.4 Scaling across system complexity 399
    21.4.1 Savanna biodiversity and functioning 399
    21.4.2 Savanna as a trophic web 399
    21.4.3 Savanna as a managed system 400
21.5 The whole picture: Modeling a spatially organized trophic web and its physical environment 401
21.6 Conclusion 403

Index 409

Contributors

Luc Abbadie
Biogéochimie et écologie des milieux continentaux (UMR 7618)
Ecole Normale Supérieure
46 rue d'Ulm
75230 Paris cedex 05, France
abbadie@biologie.ens.fr

Sébastien Barot
Laboratoire d'écologie des Sols Tropicaux (UMR 137)
32 Avenue H. Varagnat
93143 Bondy cedex, France
sebastien.barot@bondy.ird.fr

Isabelle Dajoz
Biogéochimie et écologie des milieux continentaux (UMR 7618)
Ecole Normale Supérieure
46 rue d'Ulm
75230 Paris cedex 05, France
isabelle.dajoz@biologie.ens.fr

Jacques Durand
Fonctionnement et évolution des systèmes écologiques (UMR 7625)
Université Pierre et Marie Curie
Bâtiment A - 7ème étage
7 quai Saint Bernard
75252 Paris cedex 05, France
jacques.durand@snv.jussieu.fr

Hervé Fritz
Centre d'études biologiques de Chizé (UPR 1934), BP 14
79360 Beauvoir-sur-Niort, France
fritzh@cebc.cnrs.fr

Lisa Garnier
Fonctionnement et évolution des systèmes écologiques (UMR 7625)
Ecole Normale Supérieure
46 rue d'Ulm
75230 Paris cedex 05, France

Jacques Gignoux
Biogéochimie et écologie des milieux continentaux (UMR 7618)
Ecole Normale Supérieure
46 rue d'Ulm
75230 Paris cedex 05, France
gignoux@biologie.ens.fr

Guy Josens
Systématique et écologie animales
ULB cp 160/13
Avenue Roosevelt, 50
1050 Bruxelles, Belgique
gjosens@ulb.ac.be

Contributors

Souleymane Konaté
Station d'écologie de Lamto
Centre de Recherches en Ecologie
Université d'Abobo-Adjamé
BP 28, N'Douci, Côte d'Ivoire
skonate@caramail.com

Gaëlle Lahoreau
Biogéochimie et écologie des milieux
continentaux (UMR 7618)
Ecole Normale Supérieure
46 rue d'Ulm
75230 Paris cedex 05, France
lahoreau@biologie.ens.fr

Jean-Christophe Lata
Laboratoire d'écologie, systématique
et évolution (UMR 8079)
Université de Paris-Sud
Bâtiment 362
91405 Orsay cedex, France
jean-christophe.lata@ese.
u-psud.fr

Patrick Lavelle
Laboratoire d'écologie des sols
tropicaux (UMR 137)
32 Avenue H. Varagnat
93143 Bondy cedex, France
patrick.lavelle@bondy.ird.fr

Michel Lepage
IRD, 01 BP 182
Ouagadougou 01
Burkina Faso
lepage@ird.bf

Hélène Leriche
Biogéochimie et écologie des milieux
continentaux (UMR 7618)
Ecole Normale Supérieure
46 rue d'Ulm
75230 Paris cedex 05, France

Xavier Le Roux
Laboratoire d'écologie microbienne
CNRS–Université Lyon 1
(UMR 5557–USC INRA 1193)
Bâtiment Gregor Mendel
43, bd du 11 novembre 1918
69622 Villeurbanne cedex, France
leroux@biomserv.univ-lyon1.fr

Jean-Claude Menaut
Centre d'études spatiales de la
biosphère (UMR 5126)
18, Avenue Edouard Belin
BPI 2801
31401 Toulouse cedex 4, France
mjc@cesbio.cnes.fr

Bruno Monteny
Centre IRD de Montpellier
bd Agropolis
34000 Montpellier, France

Patrick Mordelet
Centre d'études spatiales de la
biosphère (UMR 5126)
18, Avenue Edouard Belin
BPI 2801
31401 Toulouse cedex 4, France
patrick.mordelet@cesbio.cnes.fr

Hassan Bismarck Nacro
I.D.R./U.P.B., 01 BP 1091
Bobo-Dioulasso
Burkina Faso
nacrohb@hotmail.com

Guillaume Simioni
CRC Greenhouse Accounting,
Forest Products Commission
Locked Bag 888
Perth Business Centre
WA 6849, Australia
guillaume.simioni@csiro.au

Michel Veuille
Fonctionnement et évolution des
systèmes écologiques (UMR 7625)
Université Pierre et Marie Curie
Bâtiment A - 7ème étage
7 quai Saint Bernard
75252 Paris cedex 05, France
michel.veuille@snv.jussieu.fr

Roger Vuattoux
Station d'écologie de Lamto
Université d'Abobo-Adjamé
BP 28, N'Douci
Côte d'Ivoire

1

History of the Lamto Ecology Station and Ecological Studies at Lamto

Roger Vuattoux, Souleymane Konaté, Luc Abbadie, Sébastien Barot, Jacques Gignoux, and Gaëlle Lahoreau

1.1 Origin and installation of the station

The ecology station of Lamto was created after the abandonment of the Mt Nimba field station in 1958, caused by the arrival of president Sékou Touré in Guinea. At this time, Professor M. Lamotte was looking for a site to study the structure and the functioning of a complex ecosystem. In 1961, Professor M. Lamotte and M. J. L. Tournier, the director of the Institut Français d'Afrique Noire (IFAN), office of Abidjan, explored extensively the grass-based ecosystems of North and Central Côte d'Ivoire and chose the site of Lamto. The station was named from the first three and two letters of its founders (*Lamo*tte and *To*urnier).

The site of Lamto has several practical and scientific advantages. It is located 180 km north of Abidjan, the largest city of Côte d'Ivoire with the main harbor and airport. The station is only 15 km off the main North-South trunk road, which enables to reach Abidjan in two hours. This allows the quick transportation of persons, goods, and scientific materials to and out of the station. This has been critical to facilitate the maintenance of the station.

Lamto savannas are located at the southern edge of the "V Baoulé" (Fig. 1.1), an edge of savanna entering well within the rainforest area, with no apparent climatic or geologic explanation. They represent the wettest end of the Guinea savanna domain, characterized by a four-season cycle and the presence of the palm tree *Borassus aethiopum*. The density of the local inhabitants, the Baoulé ethnic group, was relatively low in the area and the Baoulé used to cultivate mainly forest soils. This diminished the risk of territorial conflict and ensured that the ecosystem was nearly deprived of human influence.

The field station was established in November 1961 with the authorization of the President of Côte d'Ivoire, M. F. Houphouët-Boigny. A 2500 ha plot was delimited on the traditional grounds of two villages: Ahiérémou II and Zougoussi. A first wooden building was quickly built in December 1961 (Fig. 1.2), and a fire protection experiment was set up immediately, with the secondary effect of protecting the building. Three French organizations funded the first researches and participated actively to the early development of Lamto: the Centre National de la Recherche Scientifique (CNRS),

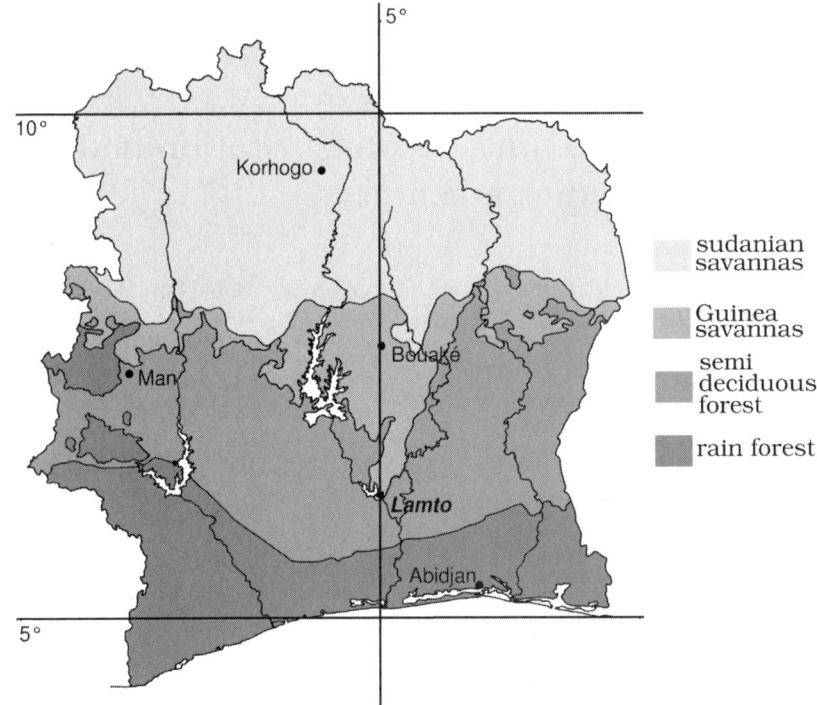

Fig. 1.1. Vegetation map of Ivory Coast and location of the Lamto station.

the zoology department of the Ecole Normale Supérieure (ENS Paris), and the Institut de Recherche pour le Développement (IRD), formerly known as the Office de Recherche Scientifique et Technique d'Outre Mer (ORSTOM). Efficient technical help and logicstic suppport were provided by the Abidjan IFAN center. A constant flow of students and researchers from the ENS, CNRS, and ORSTOM was established, and first scientific results were produced in the ecology of this savanna ecosystem (Fig. 1.3).

Quickly, a second scientific field arose in Lamto, i.e., geophysics. The first seismograph was installed in 1964, after a meteorological station was established in 1962 (Chap. 3). In 1965 J. L. Tournier became the director of an independent geophysical station, which was set up 800 m north of the first building of the ecology station. Since then, the two stations had separated finances and directors. Yet, the destinies of the two stations have remained linked until now; they share the track to the main road, the power supply, and the food is managed by a common cantina ("la popotte"), the heart of the social life of the stations.

In 1976, the CNRS, ENS, and ORSTOM stopped the direct and recurrent funding of the ecology station. The University of Abidjan became the only official institution in charge of the station. Funding from French institutions

Fig. 1.2. The first building of the ecology station (SCAF-1) in 1965 (top) and 1999 (bottom): Note the change from savanna in 1965 to forest in 1999 due to fire protection set up in 1962 (photography by Y. Gillon and G. Simioni, with permission).

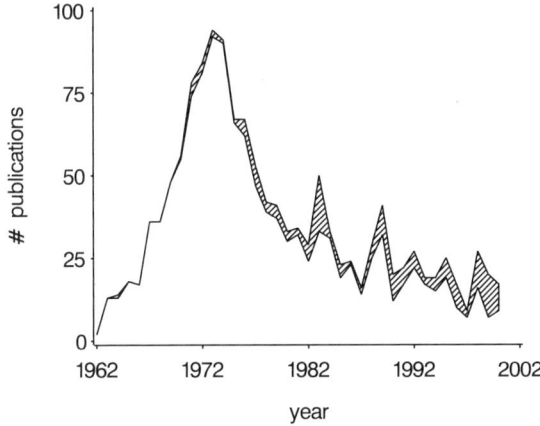

Fig. 1.3. The scientific production of the Lamto ecology research station from 1962 to 2000: Top line is the total production; the hatched area corresponds to publications in international peer-reviewed journals.

went on, but only selectively through the acquisition of scientific equipment, payment of accommodation, and wages of local technicians during long stays in Lamto, according to funding opportunities and contracts. The University of Abidjan provided a more modest but more regular funding from 1976 on that enabled to maintain and improve the work conditions, and affected a few University technicians permanently to the station. In 1996 the University of Abidjan established the Centre de Recherches en Ecologie (CRE), grouping the two field stations of Lamto and Taï (evergreen forest ecology station, North-West Côte d'Ivoire) and the former Institut d'Ecologie tropicale, and it was decided to develop the field stations as key components of the research strategy in ecology of Côte d'Ivoire.

Through the regular and obstinate activity of its directors, Roger Vuattoux (1962 to 1998) and Souleymane Konaté (since 1998), Lamto has progressively become the most practical field station in West Africa, with good housing conditions and the possibility to achieve quite easily field observations, samplings, and experiments. Facilities include (basic) accomodation for up to 30 people, a library, computers, vehicles, and an impressive amount of scientific equipment accumulated by all the people who worked in Lamto and left it at their departure. Technicians constitute a keystone of the station since they endure the continuity in the field research.

Yet, due to the irregularity of the funding, the research facilities have remained basic, and the life of the station has even been threatened periodically, for example when there was no longer any good vehicle to link the station to Abidjan. The number of technicians employed by the station has also decreased since the 1970s. During all these years (from 1976 to 2000), the durability of the station has been secured by its French director, R. Vuattoux (hired by the French ministry of Education as a professor at the University of Abidjan) and the unfailing support of the head of the University of Cocody (1976 to 1996), then the head of the University of Abobo-Adjamé (since 1996).

Lamto has resulted in scientific exchanges between Côte d'Ivoire and other countries worldwide, particularly thanks to a constant flux of students and researchers who have worked together at the station on many different subjects (see below). Lamto was also meant right from the start to be locally settled. An efficient way to build links between local populations and the station was to hire inhabitants of the neighboring villages as scientific technicians. This has also allowed scientists to take advantage of local knowledge about the savanna ecosystem and its species. This has at least resulted in the training of a pool of local villagers who now know very well the flora and fauna of the area, both in taxonomic and local names. They are familiar with all basic sampling techniques (soil sampling, plant biomass, soil humidity, plant census,...) and are partially responsible for the success of the station. Thanks to them, scientists who arrive for the first time at the station benefit quickly from the knowledge accumulated on this savanna ecosystem.

The human population density around the station has strongly increased since its foundation. This results in some small conflicts: poaching, illegal forest tree cutting and cattle grazing. Although the 2500 ha of the station now belong to the University of Abobo-Adjamé and have been set up as an integral reserve in 1998, it was never possible to employ full time guards on a regular basis. To secure the durability of Lamto savannas and to allow its long term study, it is now attempted to put the station under the UNESCO label of biosphere reserve. This would also allow the creation of a peripheral buffer zone in which development projects would be undertaken to enable local populations to benefit more from the station than through the employment of a few technicians and to foster the development of ecological engineering experiments.

A new impetus has been recently given to Lamto. First, in 1996 an international meeting was organized in Taabo (a small town near the ecology station) to summarize the past researches achieved at Lamto and to find a way to perpetuate the station. Second, thanks to S. Konaté and L. Abbadie, the housing facilities of the station have been renewed and a large classroom has been built. Consequently, it became possible to hold two-week training courses for African and European students and young researchers. Already two training courses, ecompassing theoretical and field training, have been organized, in 2001 on "ecology at the service of the durable use of soils" and in 2002 on "the dynamics of biodiversity and populations: towards new tools for the durable management of living resources".

1.2 The scientific programs that sustained the station from 1962

1.2.1 The IBP program: 1968 to 1978

The initial goal of the researches conducted in Lamto was the analysis of the composition, structure, and functioning of a complex biocenosis, the palm savanna of central Ivory Coast, with the objective of establishing its energy budget [25, 26, 32, 30, 31]. The research program was funded by the CNRS and the International Biosphere Program (IBP), and comprised three main steps:

1. identifying the flora and fauna, and characterizing the physical environment (climate and soils); this resulted in many systematic studies (e.g., [13, 10, 21, 5, 16, 17, 24, 33, 36, 45, 49] and many others) and general studies on environmental factors (e.g., [43, 44, 14]).
2. quantifying the different parts of the biocenosis, and their seasonal variations; almost all taxonomic groups have been studied in this way: grasses [1, 15, 18], trees [40, 50], invertebrates [3, 4, 20, 23, 37], specially insects, mammals [9, 12], birds [47, 48], and reptiles [7]. The favorite technique of this period was the work-consuming biomass sampling (Fig. 1.4).
3. studying the trophic interactions and the productivity of the components of the ecosystem (e.g., [11, 41]), enabling one to compute the energy budget of the ecosystem [27, 28, 29].

Fig. 1.4. Biomass sampling in the 1970s (from [26], with permission).

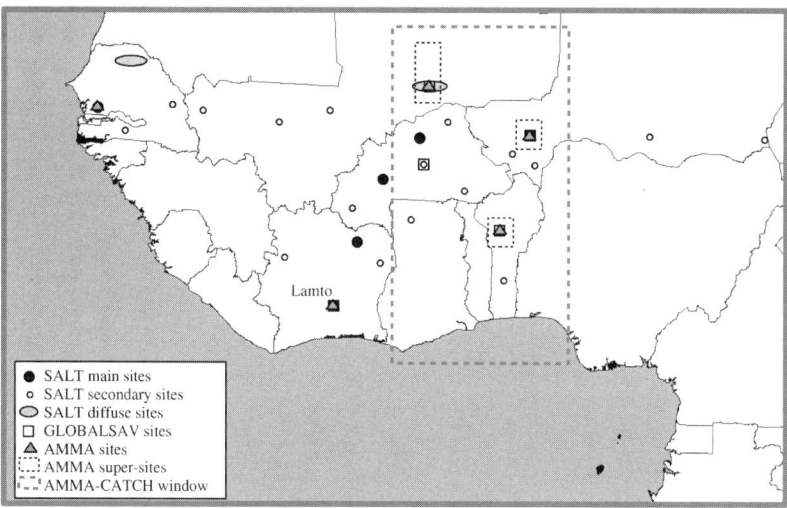

Fig. 1.5. The West African SALT, GLOBALSAV, and AMMA site networks.

This framework led to the publication of many papers in international peer-reviewed journals and books but also in French journals and in the so-called "gray literature" (Fig. 1.3). All these publications have provided an impressive amount of naturalist and quantitative information on the Lamto ecosystems.

1.2.2 The SALT program: 1988 to 1998

The IBP approach was relayed in 1988 by the Savannas on the Long Term (SALT) program, funded by the French ministry of research and by the CNRS, further labeled as a core project of International Geosphere Biosphere - Global Change and Terrestrial Ecosystem Program (IGBP-GCTE). SALT was actually the first transect of the GCTE research setup (Fig. 1.5). The aim of the SALT project was to identify and quantify the mechanisms regulating the equilibrium of savanna ecosystems, their response to natural and human-induced disturbances, and their long term dynamics. Two main questions had to be answered:

1. What are the structural and functional properties which promote their stability/resilience to stresses and seasonal disturbances?
2. Are there irreversible thresholds beyond which properties of savanna ecosystems are irreversibly altered, even when the disturbance factor has been removed?

This program funded many studies, not only in Lamto but also in several West African countries, on ecosystem functioning and on the relationships among ecosystem structure, functioning, and dynamics. In this period started the change to a more internationally visible publication strategy (Fig. 1.3) and to a more technological scientific culture in the field, with important micrometeorological and laboratory investments and the initiation of theoretical and simulation modeling. Key results obtained during this period deal with nitrogen cycling in the Lamto savanna [2], control of landscape variations in N oxide emission from savanna soils [34], water uptake patterns from dominant savanna tree and grass species [35], the role of the structure of the tree layer for savanna functioning [46], the effects of grazers on grass functioning [38, 39], and determinants of tree and grass dynamics [42, 22, 8, 19] among other issues. The SALT program ended in 1998.

1.3 The current programs: 2000 to present

Lamto is now at the heart of new research programs:

1. The French "Zones atelier de recherche sur l'environnement" program aims at setting up networks of sites for long term studies on environmental issues, with a strong implication of human sciences summarized by the concept of "anthroposystem", i.e., where man is a major driving force of the ecosystem, and not an external agent responsible for disturbances. Lamto and other sites have been labeled in 2001 as the "West African savannas research zone". First actions of this program include an archeological prospection of the savanna to document human influences over the last centuries in the Lamto area [6] (Fig. 1.6) and the set up of the Lamto thematic field courses (see http//www.Lamto.org).
2. The international "Multidisciplinary Analysis of the West African Monsoon" (AMMA) project aims at understanding the mechanisms responsible for the interannual variability of the West African monsoon; among these mechanisms, "surface" processes play an important role, and Lamto belongs to a set of West African sites where long term assessment of vegetation interannual variability is planned (see http://medias.obs-mip.fr/amma/index.en.html).

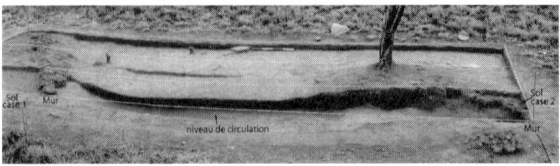

Fig. 1.6. Archeological remains of past human occupation in the Lamto reserve: Walls of traditional habitations (from [6]).

3. The international group on "Interactions in mixed tree/grass systems" aims at understanding the mechanisms of savanna functioning, dynamics, and productivity. It groups teams specialized in field and modeling studies all around the world (see https://www2.nceas.ucsb.edu). This activity has promoted coordinated studies and model comparison exercises that allowed ecological knowledge gained from studies performed at Lamto to be compared to knowledge gained in other savanna systems (e.g., House et al. submitted).

1.4 Conclusion: Forty years of scientific production

Up to now, more than 1250 scientific publications have been produced at Lamto based on Lamto data, among which 170 papers appear in international peer-reviewed journals (Fig. 1.3). This makes Lamto savannas one of the best known ecosystems in the world. All background fields needed to conduct modern ecology studies have been documented: systematics (many new species have been described in Lamto); trophic interactions; quantification of biomasses and productions; nutrient fluxes; ... . This background knowledge together with the real comfort of work in the field brings Lamto to the status of a "model ecosystem".

Should ecological work go on in Lamto? Apart from this background, many reasons call for a continuity of ecological studies in Lamto. First, Lamto is in an area of important demographic change (human population multiplied by 4 in the area), and the human pressure on the ecosystems is increasing). This promoted the development of environmental research. Second, the installation of the Taabo Dam 10 km upstream of Lamto in 1967 profoundly affected the water regime of the river and of the functioning of the riparian forest, causing changes in bird and mammal communities. Third, the density of large herbivores will strongly increase in Lamto if efficient laws to prevent hunting are applied: such an increase could have profound impacts on the Lamto savanna. Fourth, the fire exclusion experiment set up in 1962 initiated a succession toward rainforest, but stability of the new ecosystem is not yet reached. Fifth, long term plots where tree dynamics has been surveyed for decades still provide new data every year, documenting a significant encroachment of the Lamto savanna. Sixth, there is a need for improving the quality of life of the local populations through a better use of natural resources, based on an understanding of the ecological processes that make savnnas productive and sustainable environments. These six fields require further work in fundamental and applied ecology.

References

1. L. Abbadie. Contribution à l'étude de la production primaire et du cycle de l'azote dans les savanes de Lamto (Côte d'Ivoire). *Travaux des Chercheurs de Lamto (Côte d'Ivoire)*, 1:1–135, 1983.

2. L. Abbadie, A. Mariotti, and J.C. Menaut. Independence of savanna grasses from soil organic matter for their nitrogen supply. *Ecology*, 73:608–613, 1992.
3. J. Arbeille. *Recherches biologiques et écologiques sur les blattes de la région de Lamto (Côte d'Ivoire)*. Ph. D. thesis, Université Pierre et Marie Curie, Paris, 1986.
4. F. Athias. *Etude quantitative du peuplement en Microarthropodes du sol d'une savane de Côte d'Ivoire*. Ph.D. thesis, Faculté des Sciences, Université Pierre et Marie Curie, Paris, 1973.
5. F. Athias. *Scutacaridae* de la savane de Lamto (Côte d'Ivoire) (Acariens Tarsonemida). 2. *Imparipedinae*, avec description d'une espèce nouvelle, 1973.
6. S. Badey. Activités humaines passées et évolution du paysage: approche méthodologique de l'interaction homme-milieu dans une zone de contact forêt-savane au sud du V-Baoulé (Côte d'Ivoire centrale). DEA, Université Paris 1, Paris, 2002.
7. R. Barbault. *Structure et dynamique d'un peuplement de Lézards: Les Scincidés de la savane de Lamto (Côte d'Ivoire)*. Ph.D. thesis, Université Pierre et Marie Curie, Paris, 1973.
8. S. Barot, J. Gignoux, and J.-C. Menaut. Demography of a savanna palm tree: Predictions from comprehensive spatial pattern analyses. *Ecology*, 80(6):1987–2005, 1999.
9. L. Bellier. *Application de l'analyse des données à l'écologie des Rongeurs de la savane de Lamto (RCI)*. Ph.D. thesis, Université Pierre et Marie Curie, Paris, 1974.
10. P. Blandin, R. Vuattoux, and J. Plantrou. Les Lépidoptères *Charaxinae* (*Nymphalidae*) récoltés à la Station d'écologie tropicale de Lamto (Côte d'Ivoire). 1. Inventaire systématique et remarques faunistiques. *Bulletin de l'I.F.A.N.*, A, 37(4):840–858, 1975.
11. F. Bourlière. *Tropical savannas*, volume 13. Elsevier, Amsterdam, 1983.
12. F. Bourlière, E. Minner, and R. Vuattoux. Les grands mammifères de la région de Lamto, Côte d'Ivoire. *Mammalia*, 38(3):433–447, 1974.
13. J. Brunel and J.M. Thiollay. Liste préliminaire des oiseaux de Côte d'Ivoire. *Alauda*, 37:230–254, 1969.
14. D. Clément. *Modélisation des flux hydriques et thermiques dans la savane de Lamto (Côte d'Ivoire)*. Ph. D. thesis, Université Pierre et Marie Curie, Paris, 1982.
15. J. César. *Etude quantitative de la strate herbacée de la savane de Lamto (Moyenne Côte d'Ivoire)*. Thèse de 3ème cycle, Université de Paris 6, Paris, 1971.
16. J. Daget and P. Planquette. Sur quelques poissons de Côte d'Ivoire avec la description d'une espèce nouvelle, *Clarias lamottei* n. sp. (Pisces, Siluriformes, Clariidae). *Bulletin du Muséum National d'Histoire Naturelle, 2ème Série*, 39:278–281, 1967.
17. R. Darchen. Le nid de deux nouvelles espèces d'abeilles de la Côte d'Ivoire, *Trigona (Axestotrigona) sawadogoi* Darchen et *Trigona (Axestotrigona) eburnensis* Darchen (Hymen., *Apidae*). *Biologica gabonica*, 6:139–150, 1970.
18. A. Fournier. *Phénologie, croissance et production végétales dans quelques savanes d'Afrique de l'Ouest. Variations selon un gradient de sécheresse*. Ph. D. thesis, Université Pierre et Marie Curie, Paris, 1990.
19. L.K.M. Garnier and I. Dajoz. Evolutionary significance of awn length variation in a clonal grass of fire-prone savannas. *Ecology*, 82(6):1720–1733, 2001.

20. D. Gillon. *Recherches biologiques et écologiques sur les Hémiptères Pentatomides d'un milieu herbacé tropical.* Ph.D. thesis, Université Pierre et Marie Curie, Paris, 1973.
21. W.H. Gotwald and J. Lévieux. Systematics and zoogeography of African ants belonging to the genus *Amblyopone* (Hym. Form.). *Annual Meeting Entomology Society America*, pages 34–36, 1970.
22. M.E. Hochberg, J.C. Menaut, and J. Gignoux. The influences of tree biology and fire in the spatial structure of the West African savanna. *Journal of Ecology*, 82(2):217–226, 1994.
23. G. Josens. *Etudes biologiques et écologiques des Termites (Isoptera) de la savane de Lamto-Pakobo (Côte d'Ivoire).* Ph.D. thesis, Faculté des Sciences de Bruxelles, 1972.
24. D. Lachaise. Les drosophiles de la savane de Lamto. *Bulletin de Liaison des Chercheurs de Lamto*, pages 30–35, 1973.
25. M. Lamotte. Recherches écologiques dans la savane de Lamto (Côte d'Ivoire) : Présentation du milieu et du programme de travail. *La Terre et la Vie*, 21:197–215, 1967.
26. M. Lamotte. La participation au P.B.I. de la station d'écologie tropicale de Lamto (Côte d'Ivoire). *Bulletin de la Société d'Ecologie*, 1(2):58–65, 1970.
27. M. Lamotte. The structure and function of a tropical savannah ecosystem. In Golley, editor, *Tropical ecological systems: trends in terrestrial and aquatic research*, pages 179–222. Springer-Verlag, New York, 1975.
28. M. Lamotte. Première approche du bilan énergétique d'un écosystème herbacé tropical (Lamto, Côte d'Ivoire): production primaire et consommation animale. *Comptes-rendus de l'Académie des Sciences de Paris*, 284:1449–1452, 1977.
29. M. Lamotte. Structure and functioning of the savanna ecosystems of Lamto (Ivory Coast). Natural resources research. In *Tropical grazing land ecosystems*, pages 511–561. UNESCO, Paris, 1979.
30. M. Lamotte. Consumption and decomposition in tropical grassland ecosystems at Lamto, Ivory Coast. In B.J. Huntley and B.H. Walker, editors, *Ecology of Tropical Savannas, Ecological Studies*, pages 415–429. Springer-Verlag, Berlin, 1982.
31. M. Lamotte and F. Bourlière. Energy flow and nutrient cycling in tropical savannas. In F. Bourlière, editor, *Tropical Savannas, Ecosystems of the world*, pages 583–603. Elsevier, Amsterdam, 1983.
32. M. Lamotte, F. Bourlière, P. Kaiser, R. Barbault, P. Lavelle, and J.M. Thiollay. Secondary production : consumption and decomposition. Natural resources research. In *Tropical grazing land ecosystems*, pages 146–206. UNESCO, Paris, 1979.
33. B. Laporte. Descriptions de nouvelles espèces africaines de noctuelles (Insecta Lepidoptera). *Annales de la Faciulté des Sciences du Cameroun*, 14:199–226, 1973.
34. X. Le Roux, L. Abbadie, R. Lensi, and D. Serça. Emission of nitrogen monoxyde from African tropical ecosystems: control of emission by soil characteristics in humid and dry savannas of West Africa. *Journal of Geophysical Research*, 100:23,133–23,142, 1995.
35. X. Le Roux, J. Bariac, and A. Mariotti. Spatial partitioning of the soil water resource between grass and shrub components in a West African humid savanna. *Oecologia*, 104:147–155, 1995.

36. C. Lecordier. Deux espèces nouvelles de carabiques de la Côte d'Ivoire (Col. Carabidae). *Bulletin de la Société d'Entomologie française*, 73:218–221, 1968.
37. C. Lecordier. *Les peuplements de Carabiques (Coléoptères) dans la savane de Lamto (Côte d'Ivoire)*. Ph.D. thesis, Université Pierre et Marie Curie, Paris, 1975.
38. H. Leriche, X. Le Roux, J. Gignoux, A. Tuzet, H. Fritz, L. Abbadie, and M. Loreau. Which functional processes control the short-term effect of grazing on net primary production in grassland? *Oecologia*, 129:114–124, 2001.
39. C. De Mazancourt, M. Loreau, and L. Abbadie. Grazing optimization and nutrient cycling: Potential impact of large herbivores in a savanna system. *Ecological Applications*, 9(3):784–797, 1999.
40. J.C. Menaut. *Etude de quelques peuplements ligneux d'une savane guinéenne de Côte d'Ivoire*. Ph.D. thesis, Faculté des Sciences de Paris, Paris, 1971.
41. J.C. Menaut. Primary production: natural resources research. In *Tropical grazing land ecosystems*, pages 122–145. UNESCO, Paris, 1979.
42. J.C. Menaut, J. Gignoux, C. Prado, and J. Clobert. Tree community dynamics in a humid savanna of Côte d'Ivoire: Modelling the effects of fire and competition with grass and neighbours. *Journal of Biogeography*, 17:471–481, 1990.
43. Y. Monnier. *Les effets des feux de brousse sur une savane préforestière de Côte d'Ivoire*, volume 9 of *Etudes Eburnéennes*. Ministère de l'Education nationale de Côte d'Ivoire, Abidjan, 1968.
44. G. Riou. Les sols de la savane de Lamto. *Bulletin de Liaison des Chercheurs de Lamto*, Numéro spécial:3–43, 1974.
45. R. Roy. Deux nouvelles espèces d'*Amorphoscelis* en Côte d'Ivoire (Mantodea Amorphoscelidae). *Journal de l'I.F.A.N., A*, 27(4):1250–1258, 1965.
46. G. Simioni, J. Gignoux, and X. Le Roux. Tree layer spatial structure can affect savanna production and water budget: Results of a 3D model. *Ecology*, 84(7):1879–1894, 2003.
47. J.M. Thiollay. Le peuplement avien d'une savane préforestière (Lamto, Côte d'Ivoire). Thèse de spécialité, Université d'Abidjan, 1970.
48. J.M. Thiollay. *Les Rapaces diurnes dans l'Ouest africain: analyse d'un peuplement de savane préforestière et recherches sur les migrations saisonnières*. Ph.D. thesis, Université Pierre et Marie Curie, Paris, 1976.
49. L. Tsacas. *Drosophila teissieri*, nouvelle espèce africaine du groupe *melanogaster* et note sur deux autres espèces nouvelles pour l'Afrique (Dipt. Drosophilidae). *Bulletin de la Société d'Entomologie française*, 76:35–45, 1971.
50. R. Vuattoux. Le peuplement du Palmier Rônier (*Borassus aethiopum*) d'une savane de Côte d'Ivoire. *Annales de l'Université d'Abidjan, série E*, 1:1–138, 1968.

Part I

The Environment

2

Geology, Landform, and Soils

Luc Abbadie and Jean-Claude Menaut

2.1 Geology

Côte d'Ivoire is underlain by the crystalline Precambrian base of West Africa. The geological substratum is made up of two major types of rock: granite, dominant in the North and West, and metamorphic rock of sedimentary origin, dominant in the East (Fig. 2.1). The Antebirrimian material outcrops are small or large granite massives or metamorphic rock. The latter are (i) schists and micaschists organized in long NNE-SSW chains from the lower Birrimian period belonging to the "green rocks" group and outcropping inside the schistous chains, and (iii) rare granite and metamorphic rocks from the Tarkwanian period [2]. Lamto is located at the junction of three main geological formations: the large granitic Baoulé batholite in the North, the Dimbokro schistous massif in the East and the Hiré metamorphic chain in the West (Fig. 2.1). Granites are the dominant bedrock in the Lamto area and they frequently form ridges and blocks in the landscape. There are also amphibolites and green rocks belonging to the Hiré and Dimbokro formations in some places.

2.2 Landform

Two main geological features control the West African geomorphology: (i) the lack of significant orogenesis since the end of the Precambrian period and (ii) the slow continental warping toward the Atlantic Ocean following the breakup of Gondwanaland. Most African landscapes have been deeply marked by a long erosion period and the landform is roughly tabular [7]. Thus, Côte d'Ivoire is roughly a peneplain, whose mean altitude is 400 m in the North and below 50 m in the South.

Lamto is located in a complex geomorphological area belonging to a transition region between the plateaus in the North of the country and the lowlands in the South. The whole area shows features typical of an old erosion

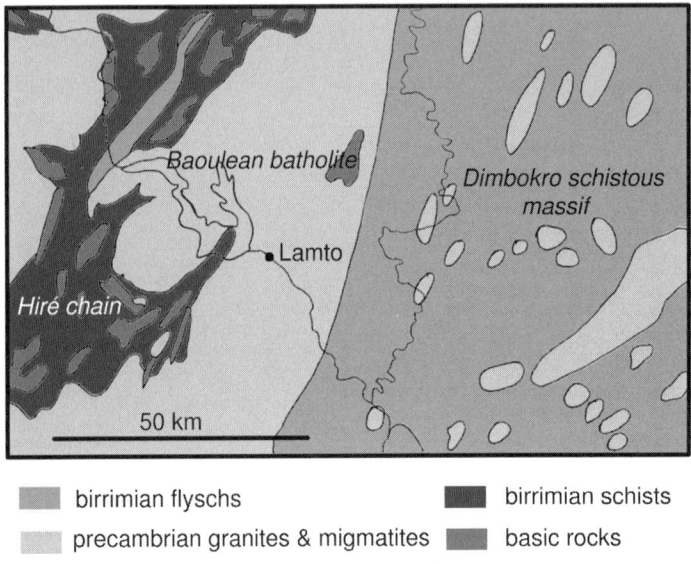

Fig. 2.1. Geological map of the Lamto area (simplified and redrawn from [8], with permission of the Institut de Recherche pour le Développement).

surface [2]. The uplift of the nearby Guinean mountain chain in addition to the effects of continental warping have produced successive erosion phases, removing several hundred meters of material. A large number of remnants of old erosion surfaces can be seen around Lamto, such as Mount Taabo (401 m above sea level), the closest to the station. Most are still covered by a bauxite and ferruginous shell, giving them a typical tabular form. These more or less disrupted hard-pans were very important sources of iron forming indurated surfaces on the slopes surrounding the residual reliefs, probably during the Pliocene period [6, 7].

The destruction of erosion surfaces in Lamto was mainly due to competition between the Bandama and N'Zi rivers and their tributaries. This competition was particularly intense during the Quaternary period, inducing severe erosion throughout the region. However, while the river base levels dropped, some portions of the indurated surface at 150 m altitude stayed in the same position. The "Plateau du Grand Nord" is probably one of these remnants, which was more or less restructured during a recent erosion phase into the long tabular inter-river areas in the northern territory of the station. There are also some river terraces that still persist along the Bandama: one located a few meters above the present river and the other 30 to 40 m away. A sharp change in the slope near the buildings of the station is the remnant of a former concave bank of the Bandama river [7].

The altitude in the Lamto reserve ranges from 75 to 125 m (Fig. 2.2). The North and East of the area belong to the N'zi river basin that is 20 km from

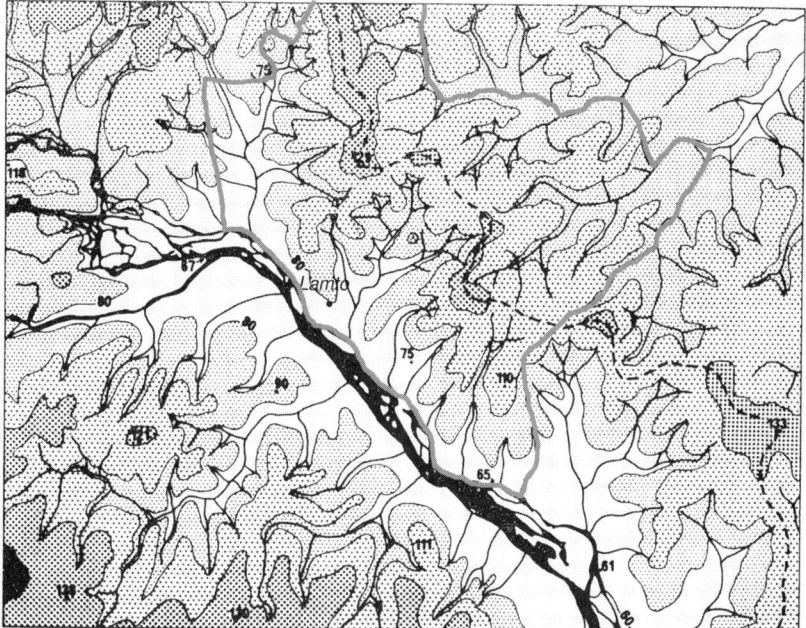

Fig. 2.2. Topographical map of the Lamto area: Elevations in m, equidistance between curves 20 m. Dotted line, watershed limit; thick gray line, limit of the Lamto reserve; fine solid lines, rivers and temporary water courses (after [7]).

Lamto. The topography in this area is fairly hilly. Average slopes are generally about 1%, except in the thalwegs where they reach 7% to 8%. The South and the West areas of the Lamto reserve belong to the Bandama river basin. The topography varies more rapidly and the inter-river areas are narrower. Slopes range from 10% to 15% on average although they can exceed 20% to 25% in the thalwegs. Some granite outcrops as subhorizontal flagstones or more or less spherical rocks which can sometimes be very large. Macrotermitinae termites mounds are found in some areas and can be locally very common [1]. They can be up to 3 m high and 10 m in diameter, thus affecting topography at a fine scale.

2.3 Soils

2.3.1 Pedogenesis

Present climatic conditions have induced ferralitic pedogenesis under tree vegetation and ferruginous pedogenesis under grass vegetation. Leucocratic granites have produced midly saturated ferralitic soils in the forest and ferruginous tropical soils in the savanna, with illite and kaolinite as the dominant clay minerals. Granite with oligoclase and amphibole have produced midly weathered ferralitic soils. Amphibolites and green rocks have produced brown eutrophic soils, with montmorillonite as the dominant clay [3, 7].

A large number of ancient geomorphological and climatic events have interacted with current climatic conditions and lithological constraints to produce different soil patterns. Erosion initially carried away large amounts of material, which depleted upper soil layers in some profiles and produced a large amount of colluvium topping lower layers in other profiles. In addition, ancient climatic conditions, either wetter or dryer than present ones, have left a large number of marks in the upper soil layers. These marks have been displaced from their original location by intense erosion during the recent Quaternary period. The wettest phases have left highly unsaturated ferralitic layers, while drier or semi arid phases have left some ferruginous gravel layers, pieces of shell and quartz blocks. Tropical ferruginous soil may sometimes top ferralitic layers [7].

2.3.2 Diversity of soil profiles

The importance of topography and drainage in soil evolution is demonstrated by the characteristic catena that can be seen in Lamto (Fig. 2.3). Four major profiles can be distinguished under grass vegetation [7]:

1. Leached tropical ferruginous soils are found on plateaus and upper sections of the slopes. These soils are shallow and gravelly (sometimes from 30 cm) and reworked (FAO-UNESCO: Acrisols).
2. Tropical ferruginous soils with a deep vertic horizon are found on the lowest inter-river areas, with bedrock made up of biotite granite.
3. The slopes have very sandy tropical ferruginous soils with a very low mineral nutrient content. They stay waterlogged for several months.
4. Vertisols are found where the bedrock is made up of amphibolites [7, 3].

In addition to these four main soil profile types, dominant in the savanna, other less common soils can be seen such as brown eutrophic soils on basic rocks, rare lithosols and regosols, or ferralitic soils in forests (FAO-UNESCO: Ferralsols). Some transition and complex soils may also be observed between deep sandy soils and hydromorphic soils downslope. Soil diversity is high in Lamto, as shown by the map produced by Riou [7] (Fig. 2.4).

2.3.3 Typical soil profiles

Complex tropical ferrugineous soil on plateaus

0–7 cm: grey, fine sand, some coarser quartz particles, simple grain structure, high porosity, many roots.
7–20 cm: greyish beige, fine sand, massive structure, high porosity, many grass roots.

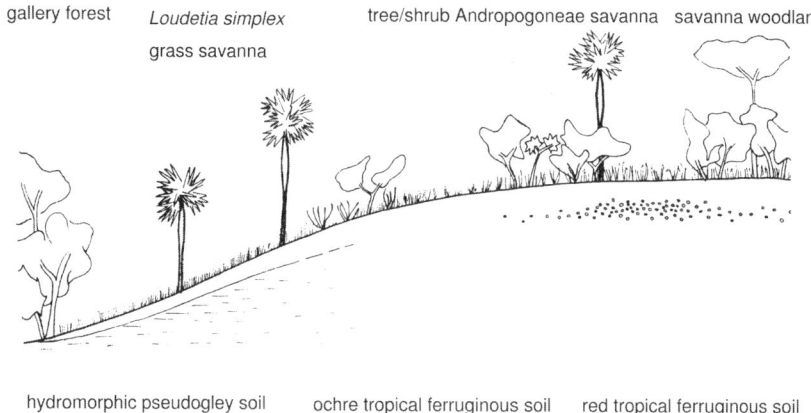

Fig. 2.3. Typical distribution of soil types along a catena at Lamto (reprinted from [5], with permission of the Ecological Society of America).

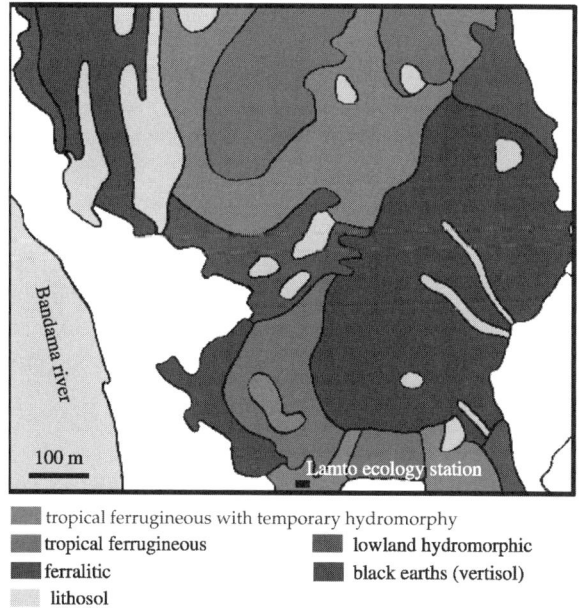

Fig. 2.4. Spatial distribution of soil types over the Lamto ecology station area (after [7]).

20–34 cm: greyish beige, fine sand, strong cohesion, less grass roots.
34–49 cm: beige, fine sand, slightly silty, high porosity, medium density of grass roots.
49–105 cm: yellowish beige, very firm, very gravelly, some quartz pieces less than 2 cm in diameter, irregular ferruginous gravels less than 5 cm in diameter, scarce roots.
105–137 cm: less firm, less gravels, many quartz pieces, high porosity.
137–240 cm: red at the top, with clear spots at the bottom, sandy clay texture with coarse sands at the bottom, pieces highly weathered granite, very large blocks of weathered feldspathic granite, shallow quartz veins at the bottom.
240–250 cm: highly weathered granite bedrock.

This is one of the commonest types of soil in Lamto savanna. It is well drained and fairly stable at the surface, but less stable in deep horizons. The gravel horizon is a major level of drainage, inducing an external drainage at the bottom of the sandy section of the profile (evolution of an A2 horizon). A weathered horizon is found at a depth of 100 to 110 cm, where the total amount of exchangeable bases, S, is 1.1 cmol kg^{-1} and the saturation ratio V is 61%. V exceeds 75% in deep horizons (Table 2.1). Clays are made of poorly crystallized illite and kaolinite. The profile is that of a tropical ferruginous soil that underwent ferralitization a long time ago.

Tropical ferruginous soil with a vertic horizon

0–6 cm: grey, sandy texture with low clay content, simple grain structure, high porosity, very abundant grass roots.
6–15 cm: dark grey, sandy texture with a low clay content, massive structure, high cohesion, many roots.
15–30 cm: greyish beige, sandy texture with a low clay content, massive structure, medium porosity.
30–85 cm: dark beige, sandy clay texture, massive structure, medium porosity. At the bottom of the horizon, clay texture, low porosity, some more or less ferruginous coarse pieces of quartz and pieces of weathered feldspaths.
85–130 cm: greenish grey, sandy clay texture (coarse ferruginous sands), massive structure with large cracks during dry season, strong cohesion, very firm, numerous small reddish, greyish or yellowish spots (a few millimeters in size).
130–148 cm: greenish grey, sandy clay (coarse ferruginous sands), massive structure, numerous blocks of white feldspath, no porosity, some small black spots and concretions.
148–185 cm: green, sandy clay, massive structure, complex network of cracks during dry season.
185 cm: irregularly weathered bedrock.

In this profile, the upper layers share physical and chemical characteristics with the upper layers of tropical ferruginous soil on plateaus. But significant differences are found in the lower horizons: the texture becomes sandy clay and some weathered minerals occur below 70 cm depth. Clays, belonging to the 2:1 type, induce a bad drainage. The total amount of exchangeable bases S reaches 14.03 cmol kg^{-1} and the saturation ratio V 95% at 1.7-1.8 m depth (Table 2.1).

Table 2.1. Major physical and chemical characteristics of typical soils in Lamto. S: total amount of exchangeable bases; V: saturation ratio (from [6], with permission of the Editions Universitaires de Côte d'Ivoire).

Soil type	Ferruginous soil on plateau						
Depth (cm)	0-10	35-40	100-110	140-150	150-165	190-200	240-260
Clay (%)	7.5	7.8	9.3	13.8	12.0	14.5	8.5
Fine silt (%)	5.8	8.8	7.8	11.0	11.0	16.5	5.8
Coarse silt (%)	8.2	7.1	6.7	6.2	7.0	4.7	2.4
Fine sand (%)	29.4	29.1	18.9	16.7	16.5	17.3	13.8
Coarse sand (%)	46.0	47.3	57.5	52.3	53.4	46.6	67.0
Organic matter (%)	2.0				1.1		
P2O5 (%)	0.18	0.16			0.10		
CaO (cmol kg^{-1})	2.34	0.81	0.78	1.80	1.56	2.40	6.84
MgO (cmol kg^{-1})	0.69	1.98	0.27	0.67	0.48	1.05	2.76
K2O (cmol kg^{-1})	0.12	0.06	0.06	0.12	0.10	0.10	0.14
Na2O (cmol kg^{-1})	0.05	0.00	0.00	0.00	0.00	0.02	0.00
S (cmol kg^{-1})	3.20	2.85	1.11	2.59	2.14	3.57	9.74
V (%)	79.5	72.4	61.0	75.4	72.9	80.0	90.9
pH (water)	6.2	5.4	5.8	5.8	5.8	5.9	5.9

Ferruginous soil w. vertic horizon				Deep sandy soil on slope				Black earth		
0-15	60-70	110-120	170-180	0-15	25-30	70-80	100-110	0-15	40-50	100-110
11.0	21.3	25.5	32.3	6.5	7.5	7.5	8.5	17.0	38.0	39.0
9.0	8.5	2.8	5.3	7.3	7.5	7.8	7.3	13.8	7.5	9.0
7.5	7.2	7.2	3.0	34.8	35.2	32.5	31.9	5.1	5.2	4.8
28.7	20.6	17.6	8.7	48.6	48.7	51.9	52.2	43.3	23.1	23.9
40.7	41.0	44.6	49.3	2.8	1.1	0.3	0.1	20.8	26.2	23.3
2.9					1.1					
		0.10			0.55				0.75	
2.60	3.39	5.85	11.55	1.64	0.74	0.98	0.74	7.45	6.18	6.92
1.90	0.14	0.17	0.15	0.93	0.60	0.42	0.55	8.19	14.78	15.06
0.20	0.14	0.17	0.15	0.11	0.07	0.07	0.07	0.57	0.13	0.10
0.04	0.00	0.14	0.38	0.05	0.03	0.03	0.05	0.02	0.05	0.08
4.80	4.67	6.55	14.03	2.73	1.44	1.50	1.41	16.23	21.14	22.16
78.0	77.0	88.0	95.3	68.1	49.3	61.5	70.5	89.9	94.5	95.5
6.3	5.6	6.3	7.2	6.3	6.1	6.1	6.7	6.4	6.7	6.9

Deep sandy soil on slopes

0–5 cm: dark grey, very sandy texture, medium or fine grain structure, weak or very weak cohesion, high porosity, high grass roots density.

5–21 cm: grey, sandy, massive structure, medium microporosity, high density of roots.

21–45 cm: greyish beige, sandy, massive structure, very weak cohesion, high microporosity, some small reddish spots, some gravel under 1.5 cm in diameter, some palm roots.

45–87 cm: beige, sandy, massive structure, very weak cohesion, some small reddish or beige veining.

87–130 cm: beige to reddish, texture a little richer in clay, coarse sand, some wider veining, two types of quartz gravel: transparent with slicing aristas or yellow, rounded, very worn, from river origin, sometimes ferruginous.

130–200 cm: light beige with reddish spots, sandy clay texture with coarse sand, some more or less ferruginous quartz gravel, numerous round red or black ferruginous stones 1 cm in diameter.

200 cm: very weathered granite bedrock.

This profile is characterized by its high sand content (around 85%). It is well drained, particularly in the upper horizons. In the lower horizons, some clays (below 10%) can decrease permeability thereby producing temporary hydromorphy and the evolution of yellowish or reddish veins and ferruginous stones. Drainage is very fast in the upper layers, very variable in the gravel horizons and slow in the weathering layers where the water table is located. Gravel horizons can appear below 50 cm depth. The total amount of exchangeable bases S is very low and that of phosphorus and potassium is extremely low. However, in downslope soils, the amount of exchangeable bases can increase up to 4.9 cmol kg^{-1}. The saturation ratio V is lower than in the plateau soils (Table 2.1). This profile is typical of tropical ferruginous soils although it shares some similarities with midly unsaturated ferralitic soils.

Black earth on amphibolitic rocks

0–25 cm: very black, sandy clay texture with fine sand, fine grain structure in the first centimeters, high macroporosity, very numerous small roots.

25–40 cm: black, sandy clay with fine sand, good cohesion, prismatic structure (massive at the bottom of the horizon), numerous palm roots.

40–55 cm: black or dark brown, sandy clay with fine sand, massive structure, very white large unaltered quartz stones, numerous pieces of feldspath, pieces of green rock.

55–80 cm: green, with brown or black veins, clay texture with fine and medium sands, small unaltered pieces of quartz, feldspath and, mainly, amphibole, massive (in dry season) and plastic (in wet season) structure.

80–110 cm: green, sandy clay, pieces of amplibolites of various sizes, weathered in the first millimeters.

110 cm: irregularly weathered bedrock.

Black earth extends on amphibolitic bedrock if slopes are gentle and drainage difficult. Such soils belong to the vertisols group. However, the Lamto region is not included in the area of true vertisols (i.e., North of Côte d'Ivoire and South of Burkina Faso). Rainfall is too high, the dry season too short and vertical and lateral drainages are relatively rapid. Not all clays belong to the 2:1 type. A large amount of montmorillonite is found in deep horizons, while chlorite and vermiculite are present in surface horizons. Although the total amount of exchangeable bases (S = 16 to 22 cmol kg^{-1}) and saturation ratio (V = 90 to 95%) are very high as compared to values obtained for other Lamto soils (Table 2.1), they are lower than in true vertisols. Black earth is less permeable and stays waterlogged for several months every year. Large shrinkage cracks appear during the wet season, while the earth becomes plastic and sticky during the wet season.

To summarize, most soils in the Lamto region are of the sandy weathered tropical ferruginous type (UNESCO-FAO: Acrisols). The upper layers, down to 50 to 200 cm depth, are influenced by recent and present climatic conditions. Weathered horizons and accumulation horizons with iron hydroxide and ferruginous concretions are formed in the savanna, while ferralitic soils are formed in the forest (UNESCO-FAO: Ferralsols). Very old soils can be found in deep horizons whose upper layers has been eroded under very different climatic and morphogenetic conditions. The transition between the upper and lower layers is generally marked by a stone line made of coarse pieces of quartz or ferruginous materials accumulated during periods of very intense erosion [6]. In addition to the large landscape-scale diversity of soils, Macrotermitineae termite mounds can substantially affect soil characteristics locally [1, 4], which further adds to soil diversity at Lamto and has strong ecological implications.

References

1. L. Abbadie, M. Lepage, and X. Le Roux. Soil fauna at the forest-savanna boundary: Role of termite mounds in nutrient cycling. In P.A. Furley, J. Proctor, and J.A. Ratter, editors, *Nature and dynamics of forest-savanna boundaries*, pages 473–484. Chapman & Hall, London, 1992.
2. J.M. Avenard. Aspects de la géomorphologie. In ORSTOM, editor, *Le milieu naturel de la Côte d'Ivoire*, pages 11–72. ORSTOM, Paris, 1971.
3. J. Delmas. Recherches écologiques dans la savane de Lamto, premier aperçu sur les sols et leur valeur agronomique. *La Terre et la Vie*, 3:216–227, 1967.
4. S. Konaté, X. Le Roux, D. Tessier, and M. Lepage. Influence of large termitaria on soil characteristics, soil water regime, and tree leaf shedding pattern in a West African savanna. *Plant and Soil*, 206:47–60, 1999.

5. J.C. Menaut and J. César. Structure and primary productivity of Lamto savannas, Ivory Coast. *Ecology*, 60(6):1197–1210, 1979.
6. G. Riou. Notes sur les sols complexes des savanes préforestières en Côte d'Ivoire. *Annales de l'Université d'Abidjan, série Lettres*, 1:17–36, 1965.
7. G. Riou. Les sols de la savane de Lamto. *Bulletin de Liaison des Chercheurs de Lamto*, Numéro Spécial(1):3–45, 1974.
8. B. Tagini. *Atlas de la Côte d'Ivoire, planche A2, Géologie, 1:2000000*. Institut de Géographie d'Abidjan, ORSTOM, 1979.

3
Climate

Xavier Le Roux

This chapter is dedicated to the late Jean Louis Tireford, Director of the Lamto Geophysical Station from 1985 to 1997.

3.1 Introduction

Climate exerts an overwhelming influence on the composition and activity of plant and animal communities [44]. In particular, a marked climatic seasonality (i.e., alternating wet and dry seasons) is recognized to be a major determinant of savanna structure and functioning [30, 11]. This chapter presents an overview of the Lamto climate attributes in the context of the West African climates. Climatic data that have been routinely recorded at the Lamto Geophysical Station since 1962 are presented, and seasonal and interannual trends are discussed. The consequences of climate on living organisms are presented in Chapter 4.

3.2 The Lamto climate in the context of the West African climates

The Lamto climate is strongly determined by the regional scale monsoon circulation of West Africa. However, a comprehensive review of the West African climatology is beyond the scope of this chapter. Only general characteristics are presented here, while further information can be found in Ojo [31], Leroux [21], Wauthy [43], and Janicot [13] among many others.

The location of the wind confluence known as the surface Inter-Tropical Discontinuity (ITD) fundamentally determines the climates experienced along a longitudinal gradient in West Africa, while its sun-controlled migration determines the seasonality of these climates. However, identification and hierarchization of the driving variables and processes that influence the West African precipitation regime remain highly controversial. Simple theories of rainfall generation through recycling by land surfaces were proposed [25, 38]. On the contrary, Polcher [36] assumed that moisture convergence is primarily sustained by the regional scale South West monsoon. In a deforestation experiment performed with a global circulation model (GCM), this author pointed out a potential feedback loop between land surfaces and climate and reformulated the Charney hypothesis. Finally, the low frequency variability in sea surface temperatures (SST), particularly in the tropical Atlantic Ocean, were found to be correlated to the ITD migration and precipitation trends over West Africa [14, 9, 22, 12]. However, no consensus exists on the relative importance of all these processes, and the discrimination between causal and only statistical relationships sometimes remains to be made.

The relationship between the different weather zones and the location of the surface ITD is shown in an idealized atmospheric cross-section given in Fig. 3.1. In short, the rainless and cloudless zone A corresponds to the southward penetration of the continental dry "hamattan" air at the surface. Zone B is a transitional weather zone where conditions are mainly dry and where fog often develops at night. These two zones are thus associated with dry season attributes. Zone C is periodically traversed by synoptic disturbances, the most important being squall lines, which are associated with short, highly intensive periods of rain. Convective thunderstorms giving widespread, variable and sporadic rainfalls also occur. Winds are generally light southwesterlies most of the time. At the end of this zone, frequent rainfalls are more prolonged and less intense. The sky is generally overcast. Zones C thus correspond to the rainy season. Zone D is associated with an overcast and rainless weather characteristic of the short dry season.

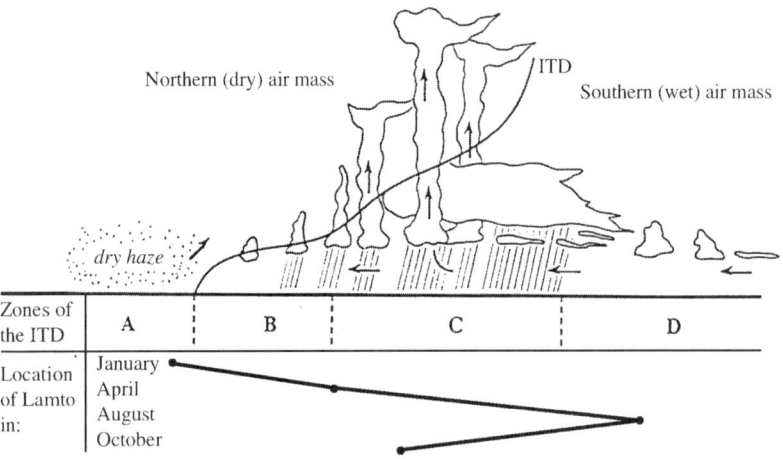

Fig. 3.1. Idealized atmospheric cross-section showing the relationship between the different weather zones and the location of the surface Inter-Tropical Discontinuity (ITD): The location of Lamto at four dates along the year is presented. Width of zones B, C, and D are typically around 200, 650, and 500 km, respectively. Adapted from [24].

At a given location, the longitudinal movement of the surface ITD determines the seasonal course of local climatic parameters such as precipitation, air temperature and water vapor pressure. However, the convection intensity may significantly influence precipitation for a given ITD position [29]. In January, the mean position of the surface ITD is around 6°N, close to the Lamto Station latitude (6°13 N). The occurrence of typical continental dry air mass conditions (zone A) at Lamto is thus rare and exhibits a high interannual variability. Generally, the harmattan dry continental wind reaches the Lamto region during around two weeks per year. With the northward movement of the ITD, the Lamto region lies in zone C and experiences rainy season conditions. The high humidity is usually accompanied by high temperatures. A cloud cover of less than 30% is rare. In July and August, when the surface ITD reaches its northernmost position, weather zone D sometimes prevails at Lamto, thus determining a so-called "short dry season". The ITD migrates southward from September onward and reaches its southernmost position in January.

To summarize, the Lamto climate (including both seasonal trends and interannual variability) and the associated vegetation functioning are essentially determined by the regional scale climatology.

3.3 The Lamto Geophysical Station: Forty years of routine climatic observations

3.3.1 Parameters monitored at Lamto

Since 1962, climatic data have been routinely acquired at the Lamto Geophysical Station. A history and overview of the station was given by Lamotte and Tireford [15]. Observed parameters and instruments used are reported in Table 3.1. Most of the standard climatic parameters have been recorded since January 1962, although solar radiation and mean wind speed have only been recorded since June 1977 and January 1971, respectively. A CIMEL automatic weather station was set up in 1995 in an area of shrubby savanna 300 m away from the geophysical station.

Recently, an effort has been made to characterize the atmospheric pollution in this environment. This includes studies of atmospheric turbidity, aerosol content and characteristics, and rainfall acidity and chemistry (Table 3.1). For instance, aerosol and rainwater chemistry was monitored during a two-year period (from January 1987 to December 1988) as part of the DECAFE (Dynamique Et Chimie de l'Atmosphère en Forêt Tropicale) IGAC/BIBEX program. Lamto was also chosen as a test site for long term monitoring of rainwater and aerosol chemistry, and atmospheric SO_2, NO_2, NH_3 and organic acids deposition fluxes by the DEBITS (DEposition of Biogeochemically Important Trace Species) IGBP/IGAC program. This survey began in 1995 in

Table 3.1. Climatic parameters observed at the Lamto Geophysical Station (after [15], completed).

Parameter	Observation time	Instrument	Period of monitoring
Precipitation	6:00, 18:00	Two 400 cm^2 pluviometers	Jan 1962 –
		1000 cm^2 daily pluviograph	Jan 1962 –
		400 cm^2 weekly pluviograph	Jan 1962 – Dec 1999
Insolation	6:00	Campbell heliograph	1963 –
Radiation	7:00	Gun Bellani pyranometer	June 1977–
Air temperature	6:00, 9:00, 12:00, 15:00, 18:00	Mercury thermometer + weekly thermograph	Jan 1962 –
Soil temperature (bare soil, −50 cm)		Thermograph	June 1969 –
Water vapor pressure	6:00, 9:00, 12:00, 15:00, 18:00	Ventilated psychrometer + weekly hydrograph	Jan 1962 –
Potential evaporation	6:00, 9:00, 12:00	Piche evaporimeter	Jan 1962 –
	7:00	1 m^2 ORSTOM pan	Jan 1971 – Mar 1982
Wind	7:00, 18:00	Cumulative anemometer	Jan 1971 –
Atmospheric pollution[1]	24 hours each Sunday	OMM aerosol collector	Sept 1982 –
	9:00, 12:00, 15:00	OMM photometer	Sept 1982 –
	each rainfall	OMM rain collector	April 1983 –
Atmospheric pollution[2]	Each rainfall	Rain samples	1995 –
	One sample per week	Aerosol samples	1995 –
	One sample per month	Gas passive samples	1996 –
Atmospheric turbidity[3]	6:00 to 18:00	CIMEL CE 318 photometer	1995 –
Automatic weather station[4]	20 min. mean, 24/24 hours	CIMEL CE 411 station	1995 –

[1] As part of the Background Air Pollution Monitoring Network (BAPMON).
[2] As part of the IDAF network.
[3] As part of the PHOtométrie pour le Traitement Opérationnel de Normalisation Satellitaire (PHOTONS) program.
[4] As part of the SAvannas in the Long Term (SALT) IGBP-GCTE program.

the framework of the IDAF (IGAC DEBITS AFRICA) network. At least, as part of the PHOTONS (PHOtométrie pour le Traitement Opérationnel de Normalisation Satellitaire) project, the optical properties of the atmosphere have been monitored routinely since 1995. This program has provided a field database of aerosol optical thickness, water content and ozone concentration that will be applied to check the parameters used in algorithms of atmospheric corrections of satellite data.

3.3.2 Data quality assessment

At Lamto, routine observations have been made without interruption since 1962 by a skilled staff. No change of design, exposure and sitting of instruments has occurred since the beginning of the observation for each parameter. A major drawback is that the reference climatic screen was chosen according to climatic standards for the tropics of the early 1960s. At this time, a belief in the inadequacy of standard Stevenson screens in the tropics resulted in the use of cages beneath thatched shelters in parts of southern Asia, Africa and Australia [34]. This type of shelter has been used to measure air temperature and water vapor pressure at 2 m over a short-cut sward since 1962 at Lamto. Thus, the absolute values of these two parameters cannot be compared to data acquired in Stevenson screens without caution. Comparisons of the Stevenson screen and tropical shed exposures are reviewed by Parker [34] and suggest a typical systematic overheating of about 0.4°C.

Direct measurements of solar radiation are scarce in West African weather networks. Since 1977, daily solar radiation has been measured routinely with a Gun Bellani pyranometer. The daily solar radiation R_{sgun} measured by this pyranometer was compared from January 1991 to May 1992 with the daily integrated solar radiation R_s measured by a Skye Instr. photovoltaic sensor calibrated against a Kipp and Zonen reference pyranometer. The following relationship was obtained [18]:

$$R_s - 0.952\, R_{sgun} + 8 \quad \left(\mathrm{W\,m^{-2}};\ R^2 = 0.95; d.f. = 412\right). \tag{3.1}$$

In conclusion, the Lamto Weather Station has been providing high quality data since 1962, which allows a reliable analysis of the temporal trends of the climatic parameters.

3.4 Seasonal course of climatic parameters

The major characteristics of the Lamto climate were reviewed by several authors, and attempts were made to define this climate according to existing classifications [3, 19, 10, 33, 37]. Ombrothermic diagrams are generally used to characterize climates by the seasonal courses of precipitation and temperature (Fig. 3.2). However, such representation is more adequate for temperate

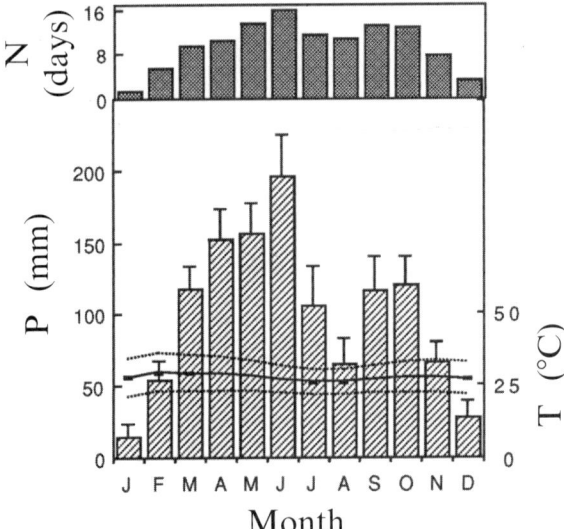

Fig. 3.2. Ombrothermic diagram for the 1962 to 1995 period: Bars are standard errors. The seasonal courses of mean, minimum and maximum air temperatures, T, monthly precipitation, P, and monthly number of rainy days, N, are presented (data from the Lamto Geophysical Station).

than humid tropical climates. Indeed, air temperature is not a major discriminant characteristic of humid tropical climates. Principal Components Analysis applied to all the climatic parameters recorded at Lamto [19, 10] showed that precipitation and atmospheric saturation deficit are the major discriminant factors, whereas temperature is of minor importance in this context.

3.4.1 Precipitation and dew

Annual mean precipitation recorded for the 1962 to 1995 period equals 1192 mm with 116 rainy days per year ($P > 0.1$ mm). However, this total value provides little information to understand the Lamto savanna functioning. The seasonal distribution of precipitation and the contribution of different precipitation size classes to the annual total are much more helpful in this context.

The Lamto climate is generally described as a transitional climate between the four-season subequatorial climate and the two-season tropical climate. Well-defined precipitation periods occur (Fig. 3.2): a long rainy season from March to November, partly interrupted by a so-called "short dry season" in August, and a long dry season generally from December to February. The rainfall maximum generally occurs in June and is associated with zone D, i.e., the deeper parts of the southwesterly current. However, the term "short dry season" is definitely false and misleading: August exhibits several attributes (low saturation deficits, a high number of rainy days and a low insolation)

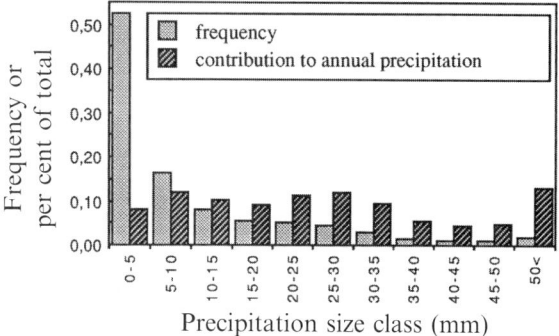

Fig. 3.3. Frequency and contribution to annual precipitation of different precipitation size classes: Values were calculated from individual observations performed during the 1962 to 1980 period at the Lamto Geophysical Station.

which are atypical of a dry month from a functional point of view. This led several authors [19, 10, 33] to consider that the Lamto climate can be regarded as a two-season climate.

At Lamto, small rainfalls account for a relatively small part of total precipitation (Fig. 3.3). Events of less than 10 mm account for 69% of the number of events but only 20% of the total annual rainfall. Dew often occurs at night, particularly during the dry season (26 days per month in December and January) and less frequently in September (14 days). During the year, 240 days experience dew occurrence. The mean amount of dew was estimated to be equal to 0.08 mm d^{-1} over a period of 148 days [41].

3.4.2 Radiation

Incoming solar radiation R_s was studied by Bony [5] and Le Roux et al. [18]. The seasonal trends in incoming solar radiation and sunshine hours are presented in Fig. 3.4. From November to May, cloudiness is relatively low and the average time course of R_s is partly determined by variations of solar radiation at the top of the atmosphere. The atmospheric transmission is approximately 0.45 to 0.48 at this time. Atmospheric transmission values around 0.6 were reported under clear sky conditions at the beginning of the rainy season [18]. The relatively low (0.45) atmospheric transmission recorded during the dry season despite cloudless conditions are due to high aerosol contents in the atmosphere that increase the attenuation of direct solar beams. From June to October, cloudiness, total precipitable water and solar radiation absorption by the atmosphere increase. Atmospheric transmission reaches 0.33 in August. Monthly mean downward shortwave radiation ranges from 12.5 MJ m^{-2} d^{-1} in August to 17.7 MJ m^{-2} d^{-1} in April, with an annual mean of 15.2 MJ m^{-2} d^{-1} for the 1978-1995 period. This value is similar to values typical at ca. 45°N in Western Europe [7] and lower than values recorded in a dry savanna

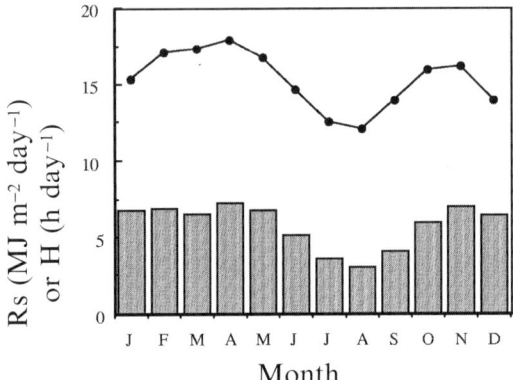

Fig. 3.4. Seasonal course of incoming solar radiation at the surface R_s (line) and sunshine hours H (bars) for the 1978 to 1995 period (data from the Lamto Geophysical Station).

environment (20 MJ m^{-2} d^{-1} at Nylsvley in South Africa, according to Scholes and Walker [39]). Thus, despite the low latitude of Lamto, solar radiation is potentially a limiting resource for vegetation growth, especially at the end of the fast growing season in July and August.

Sunshine hours at Lamto range from 106 in August to 220 in May with an annual total of 2140 h. A relationship between the (corrected) incident solar radiation at the surface R_s to solar radiation at the top of the atmosphere R_0 ratio $\left(\frac{R_s}{R_0}\right)$ and the actual to maximum insolation ratio $\left(\frac{n}{N}\right)$ was developed (X. Le Roux, unpublished):

$$\frac{R_s}{R_0} = 0.25 + 0.364 \frac{n}{N} \quad \left(R^2 = 0.80; \, d.f. = 669\right). \tag{3.2}$$

Apart from quantitative aspects, the spectral composition of downward solar radiation is a potentially important parameter in ecological studies. In particular, incoming photosynthetically active radiation (PAR) [23] provides relevant information for many purposes including plant photosynthesis and growth modeling. The spectral composition of incoming solar radiation at Lamto was first characterized by Leguesdron and Baudet [20]. However, these measurements must be regarded with caution due to inadequate UV and infrared filters used in this study. The annual mean value of the incident photosynthetic active radiation to incident solar radiation ratio ε_s is 0.47 [17]. A weak linear relationship was obtained between the daily ε_s value and the daily fraction of diffuse solar radiation $\frac{R_d}{R_s}$:

$$\varepsilon_s = 0.433 + 0.071 \left(\frac{R_d}{R_s}\right) \quad \left(R^2 = 0.193; \, d.f. = 334\right). \tag{3.3}$$

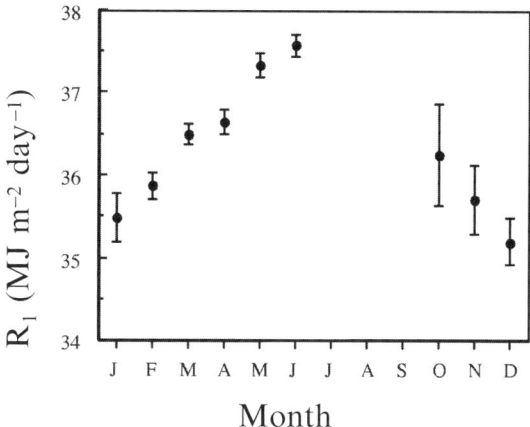

Fig. 3.5. Seasonal course of downward longwave radiation at the surface $R_{l\downarrow}$ calculated from data acquired from January 1991 to May 1992 (redrawn after [18], copyright 1994 American Geophysical Union, modified with permission of American Geophysical Union): Bars are confidence intervals ($P = 0.05$).

Despite its importance in surface energy balance, downward longwave radiation $R_{l\downarrow}$ is not routinely measured either by standard weather networks or at the Lamto Station. However, radiation observations carried out from January 1991 to May 1992 [18] provided an example of the seasonal course of $R_{l\downarrow}$ (Fig. 3.5). Monthly $R_{l\downarrow}$ ranges from 35.4 $MJ\,m^{-2}\,d^{-1}$ to 37.6 $MJ\,m^{-2}\,d^{-1}$. Minima are observed in December and January due to the low water vapor content of the continental air masses. Maxima are observed during the rainy season when the cloud cover and total precipitable water are high.

3.4.3 Air temperature

A uniformly high air temperature with little seasonal variation is typical of the Lamto climate (Fig. 3.2). Annual mean air temperature equals 27.8°C. The coldest month is August (mean temperature 26.2°C) and the warmest is February (29.4°C). The diurnal amplitude is relatively low, with an annual mean minimum air temperature of 22.4°C and an annual mean maximum air temperature of 32.3°C. The absolute recorded temperature range is between 13.9°C (22 Jan. 1975) and 41°C (25 Feb. 1995).

3.4.4 Air water vapor pressure and saturation deficit

The atmospheric water content is mostly controlled by the large scale movement of air masses. Thus, variations of air water vapor pressure are small on a daily basis and only slightly influenced by evapotranspiration from the surface. Daily variations of the saturation deficit are essentially determined by

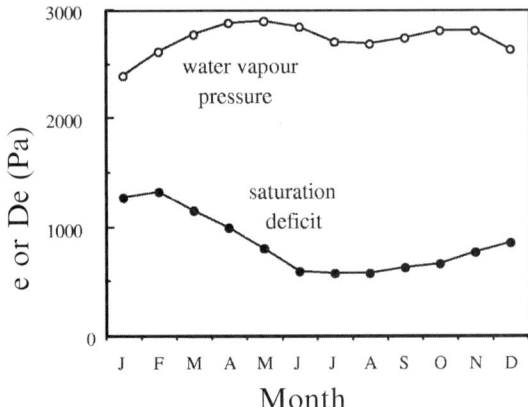

Fig. 3.6. Seasonal course of the diurnal mean air water vapor pressure e and saturation deficit D_e for the 1962 to 1995 period: Diurnal means were computed from values recorded at 6:00, 12:00, and 18:00 at the Lamto geophysical station.

variations of air temperature. The deficit at the height of 2 m is close to zero during the night when dew frequently occurs, and increases rapidly during the diurnal period.

During the rainy season, the daily mean air water vapor pressure ranges from 27 to 29 hPa (Fig. 3.6). Continental air mass conditions associated to the dry season explain the lower vapor pressure encountered in December and February (26 hPa) and in January (24 hPa). Daily mean saturation deficit ranges from 6 hPa in the middle of the rainy season when air temperature reaches its lowest values, to 14 hPa in February and March when air temperature is maximum.

3.4.5 Horizontal wind speed

The daily mean wind speed measured by a propeller anemometer at the height of 2 m in a clearing is low and around 0.6 m s^{-1}. Measurements of the horizontal wind speed made with propeller anemometers are subject to stalling at low wind speed. Thus, wind speed values are probably underestimated. At the seasonal scale, two wind speed maxima are observed during the early and late rainy season (Fig. 3.7). Wind speeds are minimum at the end of the year. A second minimum is observed in June.

Only the daily mean wind speed is measured at the Weather Station. However, continuous measurements providing means at 20 min intervals were acquired with two orthogonal Gill propeller anemometers for one year at a savanna site [16]. On a daily basis, wind speeds are higher during the diurnal period than during the nocturnal period (Fig. 3.8).

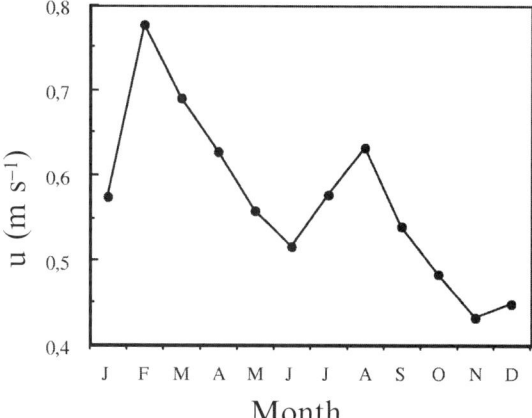

Fig. 3.7. Seasonal course of the daily mean horizontal wind speed, u, during the 1987 to 1995 period (data from the Lamto geophysical station).

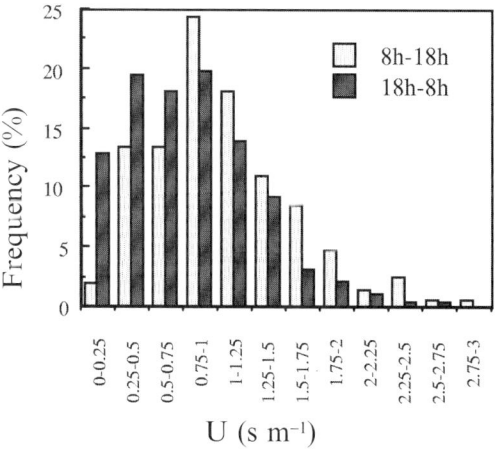

Fig. 3.8. Frequency distribution of horizontal wind speed, U, typical of the diurnal and nocturnal periods during the rainy season: Measurements were performed with two orthogonal propeller Gill anemometers at 2 m above the savanna grass canopy during one week in June 1991. Individual data are 20 min means obtained from instantaneous measurements acquired at 30 s intervals (X. Le Roux, unpublished).

3.4.6 Evaporation

Annual Piche evaporation equals 750 mm, with maximum values in March (87 mm) and minimum values from June to August (45 mm). These values reflect the seasonal trend of air temperature and of saturation deficit, but are not directly influenced by solar radiation. Turc potential evaporation, which depends on air temperature and solar radiation, was computed for the 1979-1986 period by Pagney [33]. Potential evaporation equals 1294 mm yr^{-1}, with maximum values in April (125 mm) and minimum values in August (89 mm). Low mean air saturation deficit, wind speed and incoming solar radiation explain these relatively low evaporation values.

3.4.7 Rainwater and aerosol chemistry

Rainwater and aerosol chemistry was studied at Lamto in the framework of the DECAFE program. The ionic composition of the aerosol was studied by Yoboué [45]. Aerosol exhibits high Ca^{2+}, K^+, Mg^{2+}, NO_2^- and SO_4^{2-} concentrations during the dry season, while Na^+ and Cl^- concentrations are higher during the rainy season [45]. The influence of biomass burning emissions and continental air masses during the dry season, and the influence of oceanic air masses during the rainy season explain these trends. According to Yoboué [45], total aerosol concentration at the surface equals 145 and 9 g m^{-3} during the dry and wet seasons, respectively. The aerosol derived from biomass burning is mainly composed of carbon (total carbon C_t, including black carbon C_b and organic carbon C_o) and was studied at Lamto by H. Cachier, M.H. Pertuisot, and C. Liousse. The temporal trend in carboneous aerosol (Fig. 3.9)

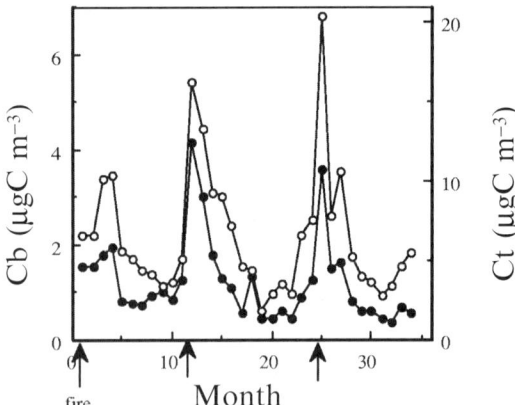

Fig. 3.9. Mean concentrations for total carbon C_t (open symbols) and black carbon C_b (solid symbols) in aerosols from January 1991 (month 1) to October 1993 (month 34) (M.H. Pertuisot, personal communication).

is determined by the regional source seasonality [8]: savanna fires explain maximum values recorded during the dry season, while anthropogenic sources (charcoal, agricultural and domestic fires) explain the significant background throughout the year. Reported concentrations are high and are similar to those obtained in temperate suburban regions.

As underlined by Yoboué [45], both the source of atmospheric water and rainfall intensity influence rainwater chemistry. All major ions are found in significant concentrations in rainwater. The most abundant ions are H^+, Na^+, K^+, NH_4^+, Ca^{2+}, Mg^{2+}, NO_3^-, Cl^- and SO_4^{2-}. Annual mean ionic concentrations are in the range 4.5 - 9.5 eq l^{-1} but notably higher for ammonium (39 eq l^{-1}), Ca^{2+} (26.6 eq l^{-1}) and chloride (47 eq l^{-1}). Particulate black carbon concentrations are also significantly high and range from 4 to 60 eq l^{-1}. During the dry season, concentrations are higher, except for Na^+ and K^+. During this season, rainwater pH is 5.02 on average, with recorded values as low as 4.58 due to the influence of biomass burning [45].

These results highlight the importance of the aerosol load in the atmosphere and the strong acidity of rainwater experienced at Lamto, both related to regional pollution sources.

3.5 Interannual variability and temporal trends

The temporal variability of climatic parameters at Lamto was analyzed by several authors: the interannual variations of precipitation are high, while variations of temperature or atmospheric humidity are low [10, 33]. Annual precipitation ranges from 797 mm (1983) to 1689 mm (1968) (Fig. 3.10) with a relatively high coefficient of variation (0.20). Monthly precipitations also exhibit a high interannual variability (e.g., coefficient of variation of 0.70 for

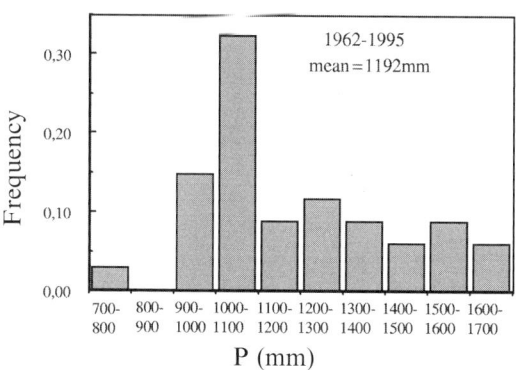

Fig. 3.10. Frequency distribution of annual precipitation P for the 1962 to 1995 period (data from the Lamto geophysical Station).

September). The maximum monthly precipitation recorded at Lamto was 481 mm in June 1979.

The temporal changes of rainfall over the last decades have been studied for humid West Africa by several authors. Conditions associated with peculiar rainfall patterns are discussed by Wagner [42], among others. A significant negative trend in annual rainfall was observed for the 1961-1987 period over southern Nigeria [2] and for the whole Guinea zone [26]. Moron [27] observed generalized dry rainfall anomalies from 1970 to 1990 for the Guinea zone. Olivry et al. [32] also suggested that annual rainfall significantly decreased during the 1950-1990 period. Recently, a decrease in annual precipitation was identified in the Ivory Coast for the 1950-1979 period [35]. Adejuwon et al. [1] obtained similar results in Nigeria and observed a delay in the onset of the rainy season and an earlier beginning of the dry season, concurrently to the decrease of annual rainfall.

At Lamto, the decreasing trend of annual precipitation with time for the 1964-2000 period is not significant, but becomes significant (linear model: $P = 0.01$) when moving averages over 3 years are considered (Fig. 3.11). Furthermore, the number of rainy days exhibited a highly significant (linear model: $P = 0.0001$) decreasing trend over the 1964-2000 period (Fig. 3.11). The annual number of rainy days ranged from 115 to 152 in the 1960s, to 83 to

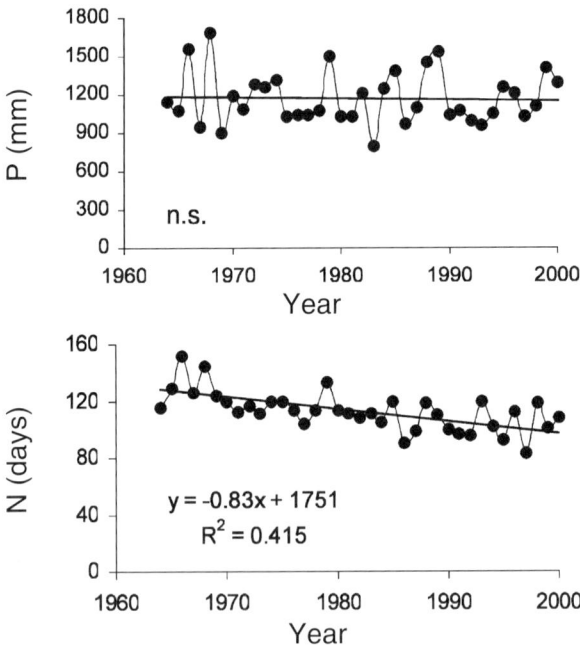

Fig. 3.11. Time course of annual precipitation, P, and annual number of rainy days, N, during the 1964 to 2000 period (data from the Lamto Geophysical Station).

119 during the 1990-2000 decade. This trend was particularly marked at the beginning of the rainy season (February; $P = 0.01$), when the ITD reached its northernmost position (June-July; $P = 0.001$ and 0.05, respectively), and at the beginning of the dry season (November; $P = 0.004$). Trends were not significant for the other months. These results are consistent with those reported by Adejuwon et al. [1] for Nigeria and by Servat [40] for the whole Ivory Coast.

Among others, the variation in sea surface temperature of the tropical Atlantic Ocean is generally statistically correlated to temporal changes of rainfall over West Africa. A very weak relationship ($P = 0.055$; $R^2 = 0.12$) is observed between the annual precipitation recorded at Lamto and the sea surface temperature anomaly of the Tropical Atlantic Ocean (30°N-30°S; not shown). A significant ($P = 0.005$), but weak relationship is observed between the annual precipitation recorded at Lamto and the sea surface temperature anomaly over the Guinea Gulf (Fig. 3.12). This relationship only briefly illustrates the complex and controversial relation between SSTs and precipitation (e.g., [9, 22, 12]). However, it is consistent with previous results reporting that positive rainfall anomalies are frequently observed in West Africa south of 10°N for positive SST anomalies in the Guinea Gulf. Nevertheless, SST anomalies only partly explain year to year variability of annual precipitation and do not explain the decreasing temporal trend observed over the 1962 to 1995 period, since SSTs exhibit no significant seasonal trend ($P = 0.49$) during the same period.

Annual mean, maximum and minimum air temperatures have been increasing significantly ($P = 0.0001$ for the three parameters) at Lamto since

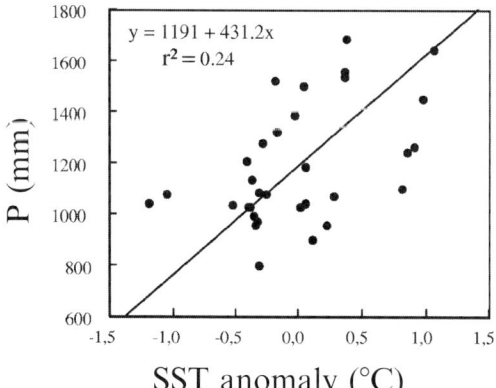

Fig. 3.12. Relationship between the annual precipitation recorded at Lamto, P, and the sea surface temperature (SST) anomaly over the Guinea Gulf (10°W-5°E and 10°N-5°S) from 1962 to 1994 (X. Le Roux, unpublished; SST data from the UKMO data base [6]): Anomalies are depatures from the 1951 to 1980 mean.

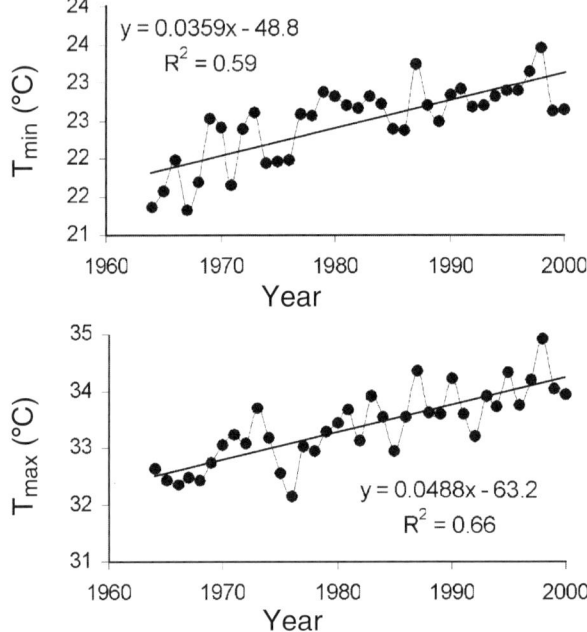

Fig. 3.13. Time course of the annual mean of maximum and minimum air temperatures during the 1964 to 2000 period (data from the Lamto geophysical station): Lines correspond to statistically significant linear regressions.

1964 (Fig. 3.13). This trend (+1.36°C for mean temperature over 34 years) is particularly important. The absence of a seasonal trend of SST cannot explain the increase in air temperature. However, the annual number of rainy days is highly correlated ($P = 0.0001$) with the annual mean air temperature at Lamto. The marked decrease of the annual number of rainy days at Lamto and the ensuing potential increase in surface heating could partly explain the increase in air temperature observed.

The decrease in annual precipitation observed over the last three decades at the Guinea region scale and in the Ivory Coast by different authors lends some credibility to the results obtained at Lamto. However, a causal relationship between the number of rainfall events and air temperature cannot fully be taken for granted.

3.6 Conclusion

Ecologists and other scientists working at Lamto have a reliable long term monitoring of climatic data at their disposal. A computerised database is now available. This is of primary importance when models of ecosystem functioning are to be supplied with high quality input variables. However, ecologists should keep in mind that, unlike incoming solar radiation, many climatic parameters are microclimatic parameters influenced by local topography and vegetation. Significant differences of microclimate between vegetation types were observed at Lamto. Over a one year period (March 1966 to March 1967), Ménager [28] compared climatic data measured at savanna, gallery forest and savanna-gallery forest boundary sites. Mean air temperature was 2°C higher at savanna sites than at forest sites. Nocturnal temperature differences between sites were low, while maximum temperatures were strongly affected by the location (e.g., 30.8°C at forest sites, 34.9°C at boundary sites and 35.2°C at savanna sites in March). Differences of air temperature and atmospheric water vapor pressure were also reported between *Loudetia* grasslands and Andropogoneae savanna sites [4], but comprehensive information on microclimate variability between savanna types is still needed. However, climatic data recorded at the weather station most often fulfill ecologists' needs.

To conclude, this chapter did not aim at classifying the Lamto climate but rather at underlining its major attributes as far as ecological studies are concerned. With a mean annual rainfall around 1200 mm, a 2 to 3 month dry season, a low annual incoming solar radiation of 15 $MJ\,m^{-2}\,d^{-1}$, low diurnal mean saturation deficits ranging from 6 to 15 hPa and a roughly constant mean air temperature of 27.8°C, Lamto is situated in the extreme southern and moist end of the West African savanna zone. This savanna ecosystem undergoes a high atmospheric pollution (high aerosol load, atmospheric oxidants levels and rainwater acidity mainly due to the desert-originating aerosol and to the biogenic emissions from savanna, agricultural or domestic fires. Among other characteristics, climatic traits exert considerable constraints on living organisms (see chap. 4). Furthermore, monitoring climatic data at the Lamto weather station suggests that the climate may have become significantly drier and warmer during the last decades. Thus, the potentially strong consequences of a negative long term trend in rainfall and soil water availability on the savanna functioning and dynamics should be assessed.

Acknowledgments

The author is greatly indebted to V. Moron (Centre de Recherche de Climatologie Tropicale, Université de Bourgogne, Dijon, France) and J. Polcher (Laboratoire de Météorologie Dynamique, CNRS, Paris, France) for useful comments on the manuscript. The author is also grateful to H. Cachier (Centre des Faibles Radioactivités, CNRS/CEA, Gif/Yvette, France) for helpful comments and data providing, and to J.P. Lacaux (Observatoire Midi-Pyrénées, Toulouse, France) for valuable suggestions.

References

1. J.O. Adejuwon, E.E. Balogun, and S.A. Adejuwon. On the annual and seasonal patterns of rainfall fluctuations in sub-saharan West Africa. *International Journal of Climatology*, 10:839–848, 1993.

2. R.N.C. Anyadike. Seasonal and annual rainfall variations over Nigeria. *International Journal of Climatology*, 13:567–580, 1993.
3. L. Bellier and J.L. Tournier. Climatologie de Lamto. Typologie des mois climatiques. Internal note, Lamto Station, 1971.
4. J. Bonvallot. Quelques mesures microclimatologiques dans les savanes de Lamto. In *Bulletin de liaison des chercheurs de Lamto*, pages 11–15, Paris, November 1970. Ecole Normale Supérieure.
5. J.P. Bony. Bilan radiatif du rayonnement solaire au dessus d'une savane de moyenne Côte d'Ivoire (Lamto). *Bulletin de liaison des chercheurs de Lamto*, S2:1–144, 1977.
6. M. Bottomley, C.K. Folland, J. Hsiung, R.E. Newell, and D.E. Parker. *Global ocean surface temperature atlas (GOSTA)*. HMSO, London, 1990.
7. M. Budyko. *Climate and life*. Academic Press, London, 1974.
8. H. Cachier, S. Ducret, M.P. Brémond, A. Gaudichet, J.-P. Lacaux, V. Yoboué, and J. Baudet. Characterization of biomass burining aerosol in a savanna region of the Ivory Coast. In J.S Levine, editor, *Global biomass burning*, pages 174–180. MIT Press, Cambridge, 1991.
9. C.K. Folland, T.N. Palmer, and D.E. Parker. Sahel rainfall and worldwide sea temperatures, 1901-85. *Nature*, 320(17):602–607, 1986.
10. P. Forge. Nouvelle approche de la définition des saisons climatiques de la région de Lamto (Côte d'Ivoire). *Annales de l'Université d'Abidjan*, XV:7–25, 1982.
11. P. Frost, E. Medina, J.-C. Menaut, O. Solbrig, M. Swift, and B. Walker. Responses of savannas to stress and disturbance. *Biology International*, S10:1–82, 1986.
12. S. Janicot. Spatiotemporal variability of West African rainfall. Part II: Associated surface and airmass characteristics. *Journal of Climatology*, 5:499–511, 1992.
13. S. Janicot and B. Fontaine. L'évolution des idées sur la variabilité interannuelle récente des précipitations en Afrique de l'ouest. *La Météorologie*, 8(1):28–53, 1993.
14. P.J. Lamb. Large-scale Tropical Atlantic surface circulation patterns associated with Subsaharan weather anomalies. *Tellus*, 30:240–251, 1978.
15. M. Lamotte and J.L. Tireford. Le climat de la savane de Lamto (Côte d'Ivoire) et sa place dans les climats de l'Ouest africain. *Travaux des chercheurs de Lamto*, 8:1–146, 1988.
16. X Le Roux. *Survey and modelling of water and energy exchanges between soil, vegetation and atmosphere in a Guinea savanna*. Ph.D. thesis, Université de Paris 6, Paris, 1995 (in French).
17. X. Le Roux, H. Gauthier, A. Bégué, and H. Sinoquet. Radiation absorption and use by humid savanna grassland, assessment using remote sensing and modelling. *Agricultural and Forest Meteorology*, 85:117–132, 1997.
18. X. Le Roux, J. Polcher, G. Dedieu, J.C. Menaut, and B. Monteny. Radiation exchanges above West African moist savannas: seasonal patterns and comparison with a GCM simulation. *Journal of Geophysical Research*, 99(D12):25,857–25,868, 1994.
19. C. Lecordier. Analyse d'un écosystème tropical humide: la savane de Lamto (RCI). Les facteurs physiques du milieu. I. Le climat. *Bulletin de liaison des chercheurs de Lamto*, S1:45–103, 1974.

20. H. Leguesdron and J. Baudet. Bilan énergétique en rayonnement ultra-violet, visible, infra-rouge au niveau du sol à Abidjan. *Annales de la Faculté des Sciences d'Abidjan*, 5:5–22, 1969.
21. M. Leroux. *Le climat de l'Afrique tropicale*. H. Champion & M. Slatkine, Paris, 1983.
22. J.M. Lough. Tropical Atlantic seas surface temperatures and rainfall variations in subsaharan Africa. *Monthly Weather Review*, 114:561–570, 1986.
23. K. McCree. Test of current definitions of photosynthetically active radiation against leaf photosynthesis data. *Agricultural Meteorology*, 10:442–453, 1972.
24. B. Monteny. *Contribution à l'étude des interactions végétation-atmosphère en milieu tropical humide*. Ph.D. thesis, Université de Paris XI-Orsay, 1987.
25. B. Monteny and A. Casenave. The forest contribution to the hydrological budget in tropical West Africa. *Annales Geophysicae*, 7(4):427–436, 1989.
26. V. Moron. Variabilité spatio-temporelle des précipitations en Afrique sahélienne et guinéenne (1933-1990). *La Météorologie*, 43-44:24–30, 1992.
27. V. Moron. Guinean and sahelian rainfall anomaly indices at annual and monthly scales (1933-1990). *International Journal of Climatology*, 14:325–341, 1994.
28. M.T. Ménager. Etude climatologique au contact de la forêt et de la savane (Lamto). In *Bulletin de liaison des chercheurs de Lamto*, pages 5–10, Paris, July 1970. Ecole Normale Supérieure.
29. S. Nicholson. Rainfall and atmospheric circulation during drought periods and wetter years in West Africa. *Monthly Weather Review*, 109:2191–2208, 1981.
30. H.A. Nix. Climate of tropical savannas. In F. Bourlière, editor, *Tropical savannas*, pages 37–62. Elsevier, Amsterdam, 1983.
31. O. Ojo. *The climate of West Africa*. Heinemann, London, 1977.
32. J.C. Olivry, J.P. Bricquet, and G. Mahé. Vers un appauvrissement durable des ressources en eau de l'Afrique humide? In *Hydrology of warm humid regions*, volume 216, pages 67–78, Yokohama, 1993. IAHS publications.
33. P. Pagney. Le climat de Lamto (Côte d'Ivoire). In M. Lamotte and J.L. Tireford, editors, *Le climat de la savane de Lamto (RCI) et sa place dans les climats de l'Ouest africain*, volume 8 of *Travaux des chercheurs de Lamto*, pages 31–79. Ecole Normale Supérieure, Paris, 1988.
34. D.E. Parker. Effects of changing exposure of thermometers at land stations. *International Journal of Climatology*, 14·1–31, 1994.
35. J.-E. Paturel, E. Servat, B. Kouame, and J.-F. Boyer. Manifestations de la sécheresse en Afrique de l'ouest non sahélienne. cas de la Côte d'Ivoire, du Togo et du Bénin. *Sécheresse*, 6:95–102, 1995.
36. J. Polcher. Sensitivity of tropical convection to land surface processes. *Journal of Atmospheric Sciences*, 52(17):3143–3161, 1995.
37. G. Riou. Proposition pour une géographie des climats en Côte d'Ivoire et au Burkina Faso. In J.L. Tireford and M. Lamotte, editors, *Le climat de la savane de Lamto (Côte d'Ivoire) et sa place dans les climats de l'Ouest africain*, volume 8 of *Travaux des chercheurs de Lamto*, pages 81–115. Ecole Normale Supérieure, Paris, 1988.
38. H. Savenije. New definition for moisture recycling and the relationship with land-use changes in the Sahel. *Journal of Hydrology*, 167:57–78, 1995.
39. R. Scholes and B. Walker. *An African savanna. Synthesis of the Nylsvley study*. Cambridge University Press, Cambridge, 1993.

40. E. Servat, J.-E. Paturel, H. Lubès, B. Kouamé, M. Ouedraogo, and J.M. Masson. Climatic variability in humid Africa along the Gulf of Guinea. Part I. Detailed analysis of the phenomenon in Côte d'Ivoire. *Journal of Hydrology*, 191:1–15, 1995.
41. J.L. Tournier. Fiches climatologiques année 1971. In *Bulletin de liaison des chercheurs de Lamto*, pages 9–12, Paris, mars 1972. Ecole Normale Supérieure.
42. R. Wagner. Surface conditions associated with anomalous rainfall in the Guinea coastal region. *International Journal of Climatology*, 14:179–199, 1994.
43. B. Wauthy. Introduction à la climatologie du golfe de Guinée. *Océanographie Tropicale*, 18(2):103–138, 1983.
44. F.I. Woodward. *Climate and plant distribution*. Cambridge University Press, Cambridge, 1987.
45. V. Yoboué. *Caractéristiques physiques et chimiques des aérosols et des pluies collectées dans la savane humide de Côte d'Ivoire*. Ph.D. thesis, Université de Toulouse, 1991.

4
Environmental Constraints on Living Organisms

Luc Abbadie, Jacques Gignoux, Michel Lepage, and Xavier Le Roux

4.1 Introduction

Water, light and nutrient availabilities, and fire and herbivory are the five factors constraining savanna structure, functioning and dynamics. In some other ecosystems of the temperate, boreal or subtropical zones, air temperature is also a major physical determinant of ecosystem functioning. In the humid savanna biome, primary productivity is generally not significantly related to temperature, except when savannas are located at elevations greater than 1000 m and exposed to low temperatures. In Lamto, annual mean air temperature (27.8°C) and annual minimum air temperature (22.4°C) are high (Chap. 3) and cannot be considered as major limiting factors.

In Africa, savannas occur across a rainfall gradient from 300 $mm\,yr^{-1}$ to 1500 $mm\,yr^{-1}$ [64]. Annual rainfall allows one to distinguish between wet and dry savannas. In dry savannas, with rainfall below ca. 600 $mm\,yr^{-1}$, grass production is positively related to rainfall. In wet savannas, primary production is both related to annual precipitations and nutrient availability. McVicar (1977, quoted in [29]) pointed out that savannas also occur across a fertility gradient and distinguished eutrophic savannas, where nutrient availability is high, from dystrophic savannas, where nutrient availability is low. The combination of water and soil nutrient availabilities results in the classification proposed by Huntley [29], where savannas vary from arid eutrophic to moist dystrophic savannas. The Lamto savanna belongs to the latter type.

In addition to soil and climate constraints, fire and herbivory also strongly influence savanna structural and functional features. Both fire and herbivores control the specific composition of plant cover, plant growth rate, plant reproductive performances and spatial distribution. They also remove some organic materials, deeply modifying the organic matter cycling and accelerating the mineralization process. The intensity of herbivory seems to be linked to the nutrient status of the savanna [17]. In Africa, there is a continuum from eutrophic areas, with relatively high animal biomass and low plant biomass, to dystrophic areas with low animal biomass and high plant biomass [4]. Lamto savanna obviously belongs to the latter type.

Fire is generally restricted to the herbaceous layer: the young trees can be burned, but not adult trees. Early fires (in the beginning of the dry season), are less violent than late fires and have a lower impact on tree regeneration. In the absence of regular fires, many of the grass open savannas in the humid regions, such as in Côte d'Ivoire, develop into woodlands or forests [64]. The effect of fire on organic matter and nutrient cycling is still in discussion, especially for the long term [39].

4.2 Soil water

4.2.1 Climatic influences

Climate seasonality exerts an influence of paramount importance on the savanna biome structure and functioning [18]. Despite the high annual precipitation, precipitation seasonality is marked at Lamto (Sect. 3.5). However, soil water availability also depends on other climatological, biological and soil parameters. As proposed by Woodward [65], an ideal analysis of constraints that water regime exerts on vegetation should be based on a water balance approach taking into account effective rooting depth, extractable water content in the rooting zone and seasonal patterns of leaf area index, evaporative demand and precipitation. A simplified water balance approach was developed for Lamto by Pagney [50]. The author compared the seasonal trends of observed precipitation P and Turc potential evaporation EP computed as a function of air temperature and solar radiation. During an 8 year period, annual mean precipitation and potential evaporation are 1134 mm and 1294 mm, respectively (Fig. 4.1). Water deficit $P - EP$ is negative in August and from November to March. However, the interannual variability of water balance is high (Fig. 4.1). Precipitation was higher than potential evaporation in 1979, while water deficit reached -484 mm in 1983. During this dry year, precipitation exceeded potential evaporation only in June. Thus, although Lamto lies in the humid savanna zone, water appears as a potentially major limiting factor for plant growth and fauna activity.

Water deficit $P - EP$ provides a useful index of soil water availability for plants. However, plants do not experience EP as such, but respond to the interaction of the seasonal courses of EP, leaf area index (LAI) and soil moisture in the rooting zone. A water balance/primary production model developed at Lamto (Chap. 9) was used to simulate actual length and severity of drought experienced by vegetation [54].

4.2.2 Soil influences

Deep, loamy soils can carry over moisture from the wet season to the dry season more efficiently than shallow, sandy soils. Thus, soil depth and texture

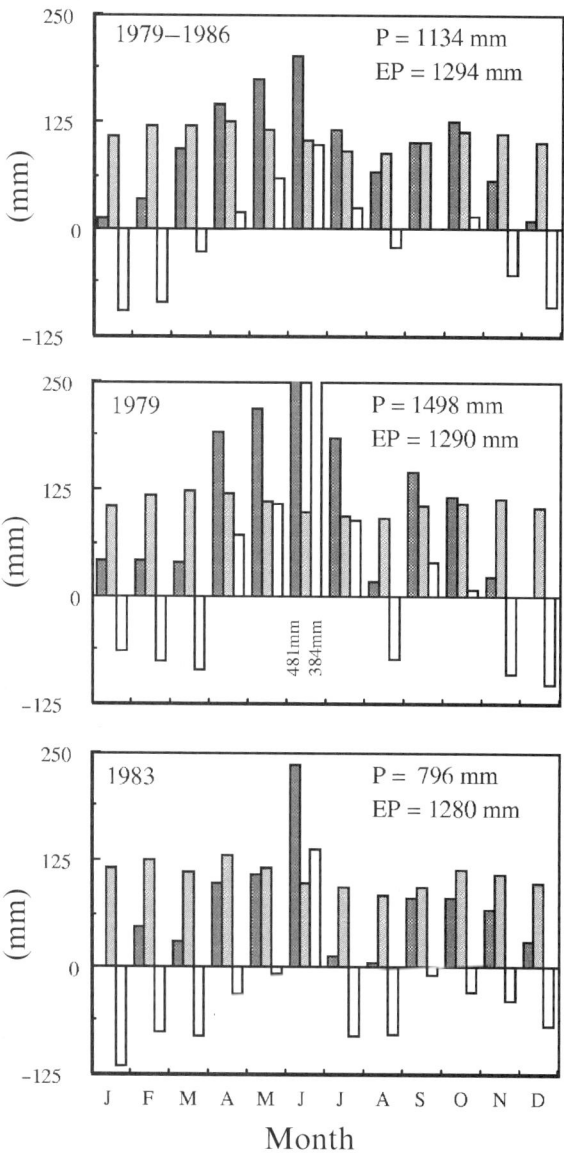

Fig. 4.1. Seasonal course of precipitation P (*dark grey*), Turc potential evaporation EP (*light grey*) and water deficit $P - EP$ (*white*) during the 1979 to 1986 period (top), the wet year 1979 (middle) and the dry year 1983 (bottom) (redrawn after [50]).

could strongly influence soil water availability in the seasonal Lamto climate. In this context, purely sandy soil layers encountered downslope at Lamto are less favorable than loamy soils upslope (Sect. 2.3 and Fig. 2.3). However, topography strongly affects precipitation redistribution and the ensuing soil water availability. The drought period can last up to 3 months on well-drained soils upslope. In contrast, soil saturation occurs frequently downslope during the rainy season and water table remains near the surface most of the time.

Within a given savanna type, soil fauna activity can significantly alter soil water retention ability (e.g., clay enrichment of soil by termites; influence of soil fauna on macroporosity). For instance, soil water content at holding capacity is around four times higher on eroded termite mounds than in the soil around in shrubby savannas [2, 32]. Maximum water content available for plants and length of periods with favorable topsoil water conditions were found to be higher on eroded termite mounds than in the surrounding soil [32]. In addition, isolated trees or tree clumps can significantly influence soil moisture dynamic (Sect. 8.2).

In conclusion, soil water availability is a major constraint for living organisms at Lamto and exhibits a high spatial and temporal variability. Particularly, patches of sufficient or limiting water availability can be observed at a same time, both at the catena scale and at smaller scales. This could have important implications for the savanna functioning.

4.3 Soil nutrients

The paper by Delmas [13] was the first one devoted to soil mineral nutrients at Lamto. It deals with the agricultural value of soils and focuses on upper layers, where most of the roots are located. Two major types of soil were studied (Sect. 2.3): the tropical ferruginous soils above granitic parent (FAO-UNESCO: Acrisols) and the black earths (FAO-UNESCO: Vertisols), located in small areas with amphibolites. Despite a complex pedogenesis (Sect. 2.3), most of the ferruginous soils of Lamto show a common feature in their superficial layers: they are made mostly of fine sands and are poor or very poor in clay (Table 2.1). Downslope, superficial layers are a little richer in clay, which induces a slightly higher content in organic matter and mineral nutrients. The clay is mainly made of poorly crystallized illite and kaolinite with a low adsorption capacity. Lamto ferruginous soils are always poor in organic matter (ca. 0.8% to 1.5% of total organic carbon and below 0.1% of total organic nitrogen), due to the rapid decomposition of dead plant matter and, likely, to the yearly grass biomass burning. This paucity in mineral and organic colloids of ferruginous soils have major detrimental consequences: the soils show a low structural stability and can be eroded quite easily in the upper layers, and they have a low content in extractable nutrients. In addition, they show a low level of microaggregation, also due to the lack of limestone. Their pH is neutral to acid, varying from 6.5 to 5.5. They are very poor in calcium (ca. 2 cmol kg^{-1}), potassium (below 0.1 cmol kg^{-1}), magnesium (ca. 1 cmol kg^{-1}) and phosphorus (below 0.03 cmol kg^{-1} of assimilable P$_2$O$_5$). All these nutrients are very easily leached and sucked down in the deep horizons, where only tree roots can uptake them. In the first 50 cm, the total amount of exchangeable cations rarely exceeds 5 cmol kg^{-1}, especially on the slopes where the deficiency in mineral nutrients is higher than on the plateaus. The concentration in mineral nitrogen rarely exceeds 2 mg of N per gram of dry soil [12] and adds to the limitation of primary production by soil nutrient status (Chap. 15).

The main physical feature of black earth soils is their high concentration in silt (20-25%) and clay (20-35%). Montmorillonite is the dominant form of clay. Consequently, in contrast to ferruginous soils, black earths show hydromorphic properties and have a high exchange capacity. Their total amount of exchangeable cations reaches 40 cmol kg^{-1}. Their contents in calcium and magnesium are high (ca. 10 cmol kg^{-1}), but their contents in potassium and assimilable P$_2$O$_5$ remain very low (below 0.3 cmol kg^{-1} and 0.01% respectively). pH varies from 6.0 to 6.9. However, despite their high organic matter concentration (between 2.0% and 2.5% of total organic carbon and ca. 0.15% of total organic nitrogen), black earth soils cannot be considered as fertile soils because they remain too depleted in potassium and phosphorus.

Most of Lamto soils have a very low nutritional value for plants. That could strongly limit the primary production, except during a short period after grass burning. Indeed, fire results in the ash deposit on the soil surface. Depending on climatic conditions (wind, rainfall, soil humidity), a part of these ashes is incorporated into the first 5 cm of the soil (Table 4.1). A slight increase of pH, cation exchange capacity (not shown) and calcium concentration is observed, whereas a strong increase is observed for potassium, magnesium and P$_2$O$_5$ concentrations (5- to 10-fold increase for the latter). This transient improvement of soil nutrient status likely boosts the regrowth of vegetation if it occurs just after fire.

This default in soil nutrients can also be reduced during short periods by precipitations, notably at the end of the dry season when rains are concentrated in nutrients, or during the heart of the humid season when rains are abundant. Annual precipitation inputs are very important for nitrogen

Table 4.1. Effect of fire on the chemical characteristics of savanna topsoil (0-5 cm depth): All mineral measurements in mg g^{-1} (after [13], with permission of Société Nationale de Protection de la Nature).

	Before fire	5 days after fire	10 days after fire
pH (water)	5.55	5.95	5.65
K	0.045	0.120	0.040
Na	0.002	0.003	0.005
Ca	0.423	0.470	0.282
Mg	0.098	0.134	0.068
P$_2$O$_5$	0.006	0.050	0.014

Table 4.2. Nutrient concentrations of rainfall and grass savanna leachates as mg l^{-1} from September to December 1971. ND: Not determined (reprinted from [59], with permission of Elsevier).

Source	Date	Ca	Mg	K	Na	PO$_4$	SiO$_2$
Rainwater	September	2.92	ND	ND	ND	0.15	ND
	October	2.30	0.01	0.50	0.92	0.20	ND
	November	2.42	0.20	0.25	0.60	1.20	0.40
	December	1.26	0.25	0.25	0.60	0.42	0.65
Leachate	September	2.10	0.30	1.40	0.50	0.50	1.40
	October	2.42	1.00	2.05	0.75	1.00	ND
	November	2.70	1.00	2.60	0.55	1.25	1.00
	December	2.30	0.80	2.45	0.40	1.35	ND

(16 kg ha^{-1} of organic nitrogen and 3 kg ha^{-1} of ammonium), calcium (27 kg ha^{-1}), sodium (11 kg ha^{-1}) and phosphorus (11 kg ha^{-1}). The supplies for magnesium, potassium and silica range between 3 and 5 kg ha^{-1} yr^{-1} [59]. An additional input comes from the washing of grasses and tree leaves. Table 4.2 gives nutrient concentrations in rainfall and leachates under Andropogoneae savanna for the last 4 months of the year 1971 [59]. It shows a strong effect of leaf leaching on the supply of magnesium, potassium and phosphorus to soil.

The contributions of rainfall and plant washing to the pool of available mineral nutrients in soil is not negligible for plant growth, especially for calcium and phosphorus, which are at very low concentrations in savanna soils. Calcium and phosphorus concentrations in Lamto grasses were estimated at 0.5% and 0.05-0.15%, respectively [60, 9]. The annual needs of grass cover (aboveground and belowground) are thus ca. 75 kg [Ca] ha^{-1} and 8-22 kg [P] ha^{-1}. With an annual supply of ca. 25 kg [Ca] ha^{-1} and 20 kg [P] ha^{-1}, rainwater and grass leachates thus potentially contribute to 33% and 90% of grass requirements in calcium and phosphorus, respectively.

4.4 Light

Temporal variations in incoming solar radiation are high throughout the year. Solar radiation input is maximum during the early rainy season and minimum in the middle of the year, when a very high cloud cover strongly reduces incident solar radiation at the surface (Chap. 3). In July and August, sunshine hours are as low as 3.4 h d^{-1} and atmospheric transmission of total shortwave radiation reaches 0.33 (i.e., 12.5 MJ m^{-2} d^{-1}). During the early growing season, incident PAR at midday frequently reaches around 2000 μmol m^{-2} s^{-1} while mean values for midday hours are ca. 1400 μmol m^{-2} s^{-1} (Fig. 4.2). These values are close to the saturating light values exhibited by canopy photosynthesis-radiation curves at this time [34] (see Fig. 6.19). In contrast, mean values for midday hours are ca. 800 μmol m^{-2} s^{-1}, and values as low as 400 μmol m^{-2} s^{-1} are recorded around midday on very cloudy days in August and

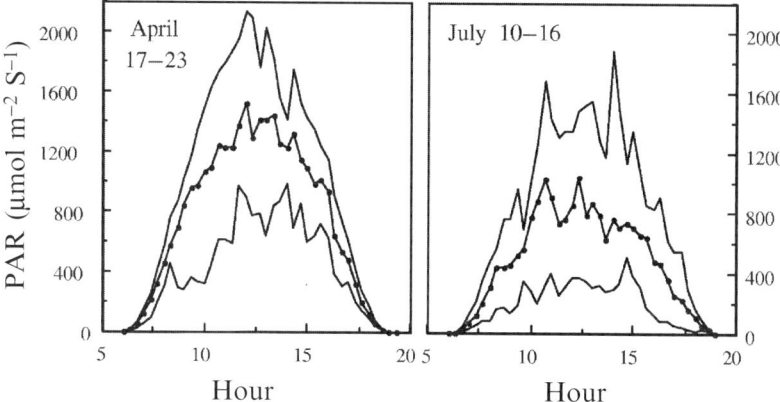

Fig. 4.2. Daily variation of the downward photosynthetically active radiation PAR at Lamto during a one-week period in the early rainy season (April 17 to 23, 1993) (left) and during a one-week period in middle of the year (July 10 to 16, 1993) (right). 20 min mean, maximum and minimum values are presented (after H. Gauthier and X. Le Roux, unpublished).

July (Fig. 4.2). Such values are much lower than saturating light values exhibited by canopy photosynthesis-radiation curves in the middle of the year (see Fig. 6.19). Thus, radiation is a potentially limiting resource for plant growth, essentially in July and August. Spatial variations of incoming solar radiation within a given savanna type have also to be taken into account. Indeed, isolated trees or tree clumps significantly influence the radiation regime of the grass layer below the tree canopy [47], which is a major constraint for grass growth (Sect. 8.2).

4.5 Fire

4.5.1 Specificity of savanna fires

Savannas are characterized by the coexistence of grasses and trees and the regular occurrence of surface fires which destroy the herbaceous layer [6]. Fire is therefore considered a very important feature of savannas, justifying its inclusion in their definition.

Savanna fires are set by man for various purposes (clearing, protection against uncontrolled fires, hunting, grazing management; see Monnier [44] for a discussion on fire causes in Guinea savannas). As a result, fires are frequent, usually occurring every 1-5 years in wet savannas [19]. They can be considered as relatively mild compared to forest fires [5]. They burn the grass layer and the young trees included in it, leaving adult trees alive, affecting tree recruitment but not adult survival. Fuel load typically ranges between 2.0 and 10.0 Mg ha^{-1} [55, 46]. Generally, the fire front is narrow and moves quickly: temperature measurements using thermocouples show that a given point is usually exposed to temperatures above 50°C for less than 2 min [55, 44]. Flame height is usually 2-3 m high [19], although very variable. During fire, maximum temperatures are usually encountered between 0 and 50 cm above the ground [51, 42]. In the soil, temperature rise quickly becomes negligible with depth, with no significant rise below 5 cm [7, 8].

4.5.2 Fire in Lamto: A driving force of the ecosystem

Lamto represents the wettest end of the West African aridity gradient: it is located at the bottom end of the V Baoulé, a region of Guinea savannas surrounded by rainforests. Fire has been proposed as the main explanation for this unexpected presence of savannas where climate is able to sustain rainforests [44]: the high water availability enables high fuel (grass phytomass) production. As a result, fire is severe and occurs every year. Under this hypothesis, fire would have a stabilizing role by preventing tree invasion on long time scales, freezing the forest-savanna boundary in a historical position [22].

The hypothesis has been tested in Lamto by experiments of fire exclusion: 30 ha of savanna have been protected from fire since 1962 (Fig. 4.3). In these areas, plots have been set and followed for up to 30 years [62, 63, 37, 15]. Results show an invasion by savanna trees, mainly by sprouting; after 6 years, forest trees start to invade the plots and start to outcompete savanna species after 12-15 years [15]. Similar results have been found in other wet savannas in West Africa and South America [53, 56], clearly demonstrating the stabilizing role of fire in these areas.

Although fire clearly has a negative effect on savanna trees, it is not yet clear whether it actually prevents their invasion. Results from simulation models [41, 26] and evidence of a doubling in tree density over 20 years under a regime of yearly fires in Lamto [11, 20] tend to prove that fire only slows down tree invasion, but does not prevent it. Monnier [44] reports from a long-term experiment of the CTFT (Centre Technique Forestier Tropical) in the northern end of the Guinea savanna zone of Ivory Coast. In this area, tree invasion is promoted by fire protection and is prevented only in a late fire regime; the early fire regime has almost no effect compared to complete fire exclusion. Because fires in Lamto occur regularly during the dry season, as middle fires, they may not be sufficient to prevent tree invasion; a late fire from time to time might be needed to reduce tree density.

These results are confirmed by other studies: paleoecological data on Gabon wet savannas very similar to Lamto have demonstrated that these savannas were unstable under a natural fire regime on a 3000 year time scale and were maintained only if fire occurred yearly through human action [48]: tree invasion was possible only in periods where humans were absent.

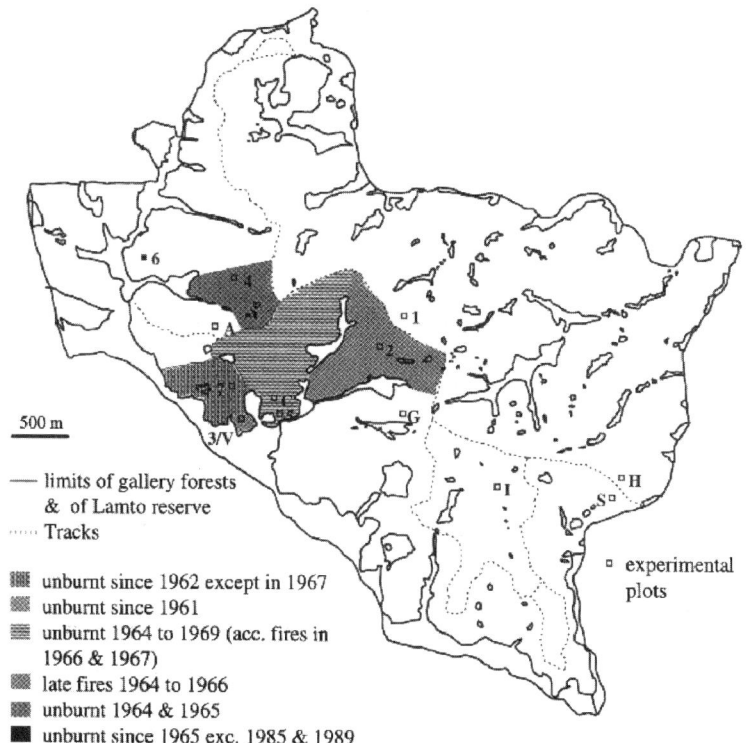

Fig. 4.3. Map of the experiments of fire protection in Lamto savanna: Experimental plots 1-7 and S have been set up in 1962 and all trees have been censused every 3 years until 1999. Plots A, C, G, H, I and V have been set up in 1969, and all trees have been mapped and measured in 1970, 1973, 1975, 1989, and every year between 1991 and 1995; all seedlings have been tagged and mapped from 1991 to 2002. (J. Gignoux, R. Vuattoux and G. Lahoreau, unpublished)

Conclusion of these findings is that fire can prevent tree invasion in Guinea savannas, but:

1. Only very frequent (yearly), human-induced, fires can prevent tree invasion.
2. Only late fires can reduce tree density, so that under an artificially regular regime of middle-season fires, tree invasion is not prevented but only slowed down.

4.5.3 Fire severity

The easiest way to measure fire severity is through temperature measurements [25], although fire behavior is better described by energy fluxes and many

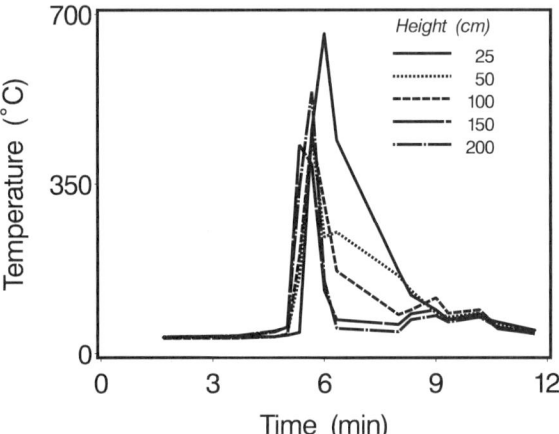

Fig. 4.4. Thermocouple measurement of fire temperatures at different heights above soil surface during January 1993 fires in Lamto. Data kindly provided by J.M. Brustet (unpublished).

other variables [5]. This can be done with thermocouples or, more easily, with thermal paints and pencils (paints that show a color change at a fixed temperature). Thermocouples measure the instantaneous temperature (Fig. 4.4), while thermosensitive paints include a time effect in their response and theoretically measure the maximal temperature over the time they have been exposed to heat: this makes the correlation between these two types of measurements difficult. We found a good correlation between temperature measured by thermosensitive paints and the time/temperature sums computed from thermocouple measurements ($T_{tag} = 1.066 \cdot \sum T_{thermocouple} - 581$; $F = 38.08$ with 1 and 3 d.f.; $P = 0.0086$; $R^2 = 0.93$), where T_{tag} is the time/temperature sum above 50°C of thermocouple measurements of Fig. 4.4). This latter variable is well correlated to fire energy [55]. We can therefore consider the temperatures measured with thermosensitive paints as a measure of fire energy rather than maximal temperature, as is frequently proposed [25]. Furthermore, the lethal temperature of plant tissues is usually fitted to a semi-logarithmic line ($time = a - b \, \mathrm{Log(Temperature)}$; [35]), and the temperature at which a color change of thermal paints occurs can also be fitted to a semi-logarithmic curve (Gignoux, unpublished). As thermosensitive paints respond to exposure to high temperature in a similar way as living tissues do, they are a very good tool for measuring fire severity in a biological perspective.

Fire temperatures in Lamto have been measured with thermosensitive paints in 1966 [23], 1982 and 1983 [43] and 1992 and 1993 [21]. Thermocouple measurements have been made in 1972 ([57] reproduced in [45]) and 1993 (Fig. 4.4). All measurements show a maximum temperature between 10 and 80 cm, usually around 20 cm, a decrease at ground level and toward the top of the profile, but there is a substantial variation in the temperature profiles [21].

Fire behavior is influenced by many factors [61], among which the most important are fuel load [36], fuel quality [27], wind [3], slope [58], air humidity and air temperature [5]. All these factors may vary in space and time, adding variability to the intrinsic variability of fire as a turbulent phenomenon [3]. Variations in fire severity have been noticed for a long time in Lamto, with records of exceptionally intense fires (1967) occurring under especially favorable conditions (long dry periods, high fuel load, high wind speed).

In Lamto, fuel load (grass phytomass at the end of the rainy season) is low compared to those recorded for forest fires and is among the highest recorded for surface fires (6.61 ± 1.94 (s.e.) $Mg\,ha^{-1}$; [46]). Grass phytomass is substantially reduced in dense tree clumps (down to zero) or small tree clumps (4.47 ± 1.37 $Mg\,ha^{-1}$; [46]) and on shallow soils, either hydromorphic or with rock outcrops [10, 16, 40, 52]. This results in significant variation of fire temperatures in space (Fig. 4.5) and the existence of fire-safe areas. Fire exclusion results in the accumulation of grass necromass until an equilibrium between decomposition and production is reached (and before trees outcompete grass). César [10] reports values of more than 13.0 $Mg\,ha^{-1}$ of grassy fuel in a plot unburned for 5 years in Lamto. Accidental fires in such previously unburned areas are much more severe than the usual savanna fires and behave like forest fires (with crown fires and 10-15 m high flame walls, never observed in usual savanna fires).

Fuel relative water content also has a major effect on fire severity [5]. It depends on the phenological stage of the grass layer and on weather conditions at the date of the fire. In Lamto, relative water content of the whole grass layer decreases from ca. 50% at the end of the wet season to ca. 30% at the heart of the dry season [38, 14]. As a result, early fires destroy only 20 to 25% of grass biomass, while late fires can destroy up to 100% of biomass.

Fire severity is also affected by very local weather conditions, like air humidity and wind speed at the time of the fire: local people always light fire in the early morning, before any wind is present and when air is still wet, so that is remains possible to control the fire; they also tend to avoid lighting fire during dry spells with northerly desert winds. Wind speed affects the rate of spread of the fire, thus indirectly affecting burning efficiency, as demonstrated at a very local scale [28]: fluctuating wind tends to produce waves leaving unburned areas of grasses in the field.

In conclusion, spatial and temporal variations in fire severity are important, due to many causes acting at different scales (effects of tree cover locally reducing fuel load, year-to-year variability, plot effect, plus intrinsic variability; [21]). Such variations have possible effects on vegetation dynamics (Chaps. 17 to 19).

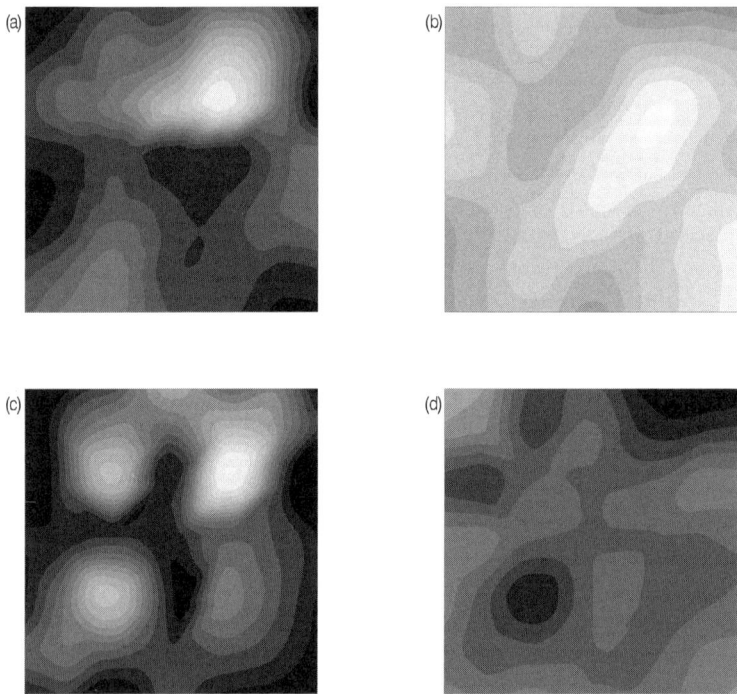

Fig. 4.5. Map of fire temperatures on 50×50 m savanna plots: Temperatures were measured with thermosensitive paints on aluminium tags, set on 2 m high poles. Twenty-five poles were set on three different savanna plots (Fig. 4.3), with six measurements at different heights. Results are shown here for (a) the A-plot at 25 cm high in January 1992, (b) the A plot at 2 m high in January 1992, (c) the A-plot at 25 cm high in January 1993, and (d) the C-plot at 25 cm in January 1992. Contour colors vary from 0-50°C (*white*) to 750-800°C (*black*) in steps of 50°C (J. Gignoux, unpublished data).

4.6 Herbivory

Herbivory is not a main determinant of Lamto savannas, because herbivores—either arthropods or Mammals—have a very low biodiversity and biomass (Chap. 10). On the reserve area, large grazers are mainly Kob antelopes (0.025 ha^{-1}, *Kobus kob*) and buffalos (0.024 ha^{-1}, *Syncerus caffer nanus*) [24]. Such low densities of large herbivores is quite typical of protected West African savannas and contrasts with high herbivore loads characterizing East or South African savannas [17] (see Chap. 10). This is mainly due to the poor nutrient concentration of mature grasses. In the Serengenti, the ratio P/C (protein over carbohydrates) is much higher, with crude protein concentration varying from 13.5% to 8% of the dry matter and herbivores remove over 40% of the annual grass production [49]. In contrast, several authors have underlined the low forage value of the Lamto savannas: according to Hédin [30], P/C during the first month of the growing season is around 0.19 in Andropogoneae savannas, decreasing down to 0.3% at the maturation period, which corresponds to protein concentration values ranging from 11.9% to 1.9% [1]. These values should be compared to the minimum 5% crude protein concentration that is the critical level necessary for the maintenance of cattle during the dry season.

Therefore, the low numbers and diversity of mammals in Lamto could be explained both by an historical trend of stocking (wild and domestic mammals) in this zone and the low rates of herbivory, due to the poverty of the grass layer.

Including non-mammal herbivores, the biomasses of the main herbivore populations in Lamto amounted to 1.14 $g\,m^{-2}$ [33], including mostly foraging termites (0.6 $g\,m^{-2}$ [31]) and granivorous ants (0.4 $g\,m^{-2}$). This low value could be compared with the 10-12 $g\,m^{-2}$ of total termite populations and 30 $g\,m^{-2}$ of earthworm populations. According to [33], of a total of 47,600 $kJ\,m^{-2}\,yr^{-1}$ produced, only 420 $kJ\,m^{-2}\,yr^{-1}$ (0.9%) are used by herbivores.

4.7 Conclusion

The high rainfall occurring in Lamto savannas directly reduces the role of water as a limiting factor. It also promotes fire as the key driving factor, especially for tree dynamics. As a result of both of these factors, nutrient losses are potentially high and thus the apparent soil nutrient poverty is reinforced. It is therefore not surprising that nutrients become a major limiting factor, both for grass production and for herbivore diversity. Lamto savannas could thus be qualified as oligotrophic savannas.

References

1. L. Abbadie. *Aspects fonctionnels du cycle de l'azote dans la strate herbacée de la savane de Lamto.* Ph.D. thesis, Université de Paris 6, Paris, 1990.
2. L. Abbadie, M. Lepage, and X. Le Roux. Soil fauna at the forest-savanna boundary: role of the termite mounds in nutrient cycling. In J. Proctor, editor, *Nature and dynamics of forest-savanna boundaries,* pages 473–484. Chapman & Hall, London, 1992.
3. T. Beer. The interaction of wind and fire. *Boundary-Layer Meteorology,* 54:287–308, 1991.
4. R.V.H. Bell. The effect of soil nutrient availability on community structure in african ecosystems. In B.J. Huntley and B.H. Walker, editors, *Ecology of tropical savannas,* pages 193–216. Springer-Verlag, Berlin, 1982.

5. W.C. Bessie and E.A. Johnson. The relative importance of fuels and weather on fire behavior in subalpine forests. *Ecology*, 76(3):747–762, 1995.
6. F. Bourlière and M. Hadley. Present-day savannas: an overview. In F. Bourlière, editor, *Tropical savannas*, volume 13 of *Ecosystems of the world*, pages 1–17. Elsevier, Amsterdam, 1983.
7. R.A. Bradstock and T.D. Auld. Soil temperatures during experimental bushfires in relation to fire intensity: Consequences for legume germination and fire management in south-eastern Australia. *Journal of Applied Ecology*, 32(1):76–84, 1995.
8. L.M. Coutinho. Ecological effects of fire in Brazilian cerrado. In B.J. Huntley and B.H. Walker, editors, *Ecology of tropical savannas*, volume 42 of *Ecological Studies*, pages 273–291. Springer-Verlag, Berlin, 1982.
9. J. César. Cycles de la biomasse et des repousses après coupe en savane de Côte d'Ivoire. *Revue d'élevage et de médecine vétérinaire des pays tropicaux*, 34(1):73–81, 1981.
10. J. César. *La production biologique des savanes de Côte d'Ivoire et son utilisation par l'homme*. CIRAD - Institut d'élevage et de médecine vétérinaire des pays tropicaux, Maisons-Alfort, 1992.
11. J.M. Dauget and J.C. Menaut. Evolution sur vingt ans d'une parcelle de savane boisée non protégée du feu dans la réserve de Lamto (Côte d'Ivoire). *Candollea*, 47:621–630, 1992.
12. P. de Rham. Recherches sur la minéralisation de l'azote dans les sols des savanes de Lamto. *Revue d'Ecologie et de Biologie du Sol*, 10(2):169–196, 1973.
13. J. Delmas. Recherches écologiques dans la savane de Lamto (Côte d'Ivoire): Premier aperçu sur les sols et leur valeur agronomique. *La Terre et la Vie*, 21(3):216–227, 1967.
14. R. Delmas, J.P. Lacaux, J.C. Menaut, L. Abbadie, X. Le, Roux, G. Helas, and J. Lobert. Nitrogen compound emission from biomass burning in tropical African savanna, FOS/DECAFE 1991 experiment (Lamto, Ivory Coast). *Journal of Atmospheric Chemistry*, 22:175–193, 1995.
15. J.L. Devineau, C. Lecordier, and R. Vuattoux. Evolution de la diversité spécifique du peuplement ligneux dans une succession préforestière de colonisation d'une savane protégée des feux (Lamto, Côte d'Ivoire). *Candollea*, 39:103–134, 1984.
16. A. Fournier. *Phénologie, croissance et production végétales dans quelques savanes d'Afrique de l'Ouest*. ORSTOM, Paris, 1991.
17. H. Fritz. Low ungulate biomass in West African savannas: primary production or missing megaherbivores or predator species? *Ecography*, 20:417–421, 1997.
18. P.G.H. Frost, E. Medina, J.C. Menaut, O. Solbrig, M. Swift, and B.H. Walker. Responses of savannas to stress and disturbance. *Biology International*, S10:1–82, 1986.
19. P.G.H. Frost and F. Robertson. The ecological effects of fire in savannas. In B.H. Walker, editor, *Determinants of tropical savannas*, volume 3 of *Monograph series*, pages 93–140. International Council of Scientific Unions Press, Miami, FL, 1985.
20. L. Gautier. Contact forêt-savane en Côte d'Ivoire centrale: Evolution du recouvrement ligneux des savanes de la réserve de Lamto (sud du V baoulé). *Candollea*, 45:627–641, 1990.
21. J. Gignoux. *Modélisation de la coexistence herbes/arbres en savane*. Ph.D. thesis, Institut National Agronomique Paris-Grignon, Paris, 1994.

22. D. Gillon. The fire problem in tropical savannas. In F. Bourlière, editor, *Tropical savannas*, volume 13 of *Ecosystems of the world*, pages 617–642. Elsevier, Amsterdam, 1983.
23. D. Gillon and J. Pernès. Etude de l'effet du feu de brousse sur certains groupes d'arthropodes dans une savane préforestière de Côte d'Ivoire. *Annales de l'Université d'Abidjan, Série E*, 1(2):113–197, 1968.
24. S. Glémin. *Mise au point d'une méthode de recensement de grands herbivores dans une savane de type mosaïque*. M.Sci. thesis, Université de Paris 6, Paris, 1997.
25. R.J. Hobbs, J.E.P. Currall, and C.H. Gimingham. The use of 'Thermocolor' pyrometers in the study of heath fire behaviour. *Journal of Ecology*, 72:241–250, 1984.
26. M.E. Hochberg, J.C. Menaut, and J. Gignoux. The influences of tree biology and fire in the spatial structure of the West African savanna. *Journal of Ecology*, 82(2):217–226, 1994.
27. J.C. Hogenbirk and C.L. Sarrazin-Delay. Using fuel characteristics to estimate plant ignitability for fire hazard reduction. *Water, Air and Soil Pollution*, 82(1-2):161–170, 1995.
28. B. Hopkins. Observations on savanna reserve burning in the Olokemeji forest reserve, Nigeria. *Journal of Applied Ecology*, 2:367–381, 1965.
29. B.J. Huntley. Southern African savannas. In B.J. Huntley and B.H. Walker, editors, *Ecology of tropical savannas*, pages 101–119. Springer-Verlag, Berlin, 1982.
30. L. Hédin. Recherches écologiques dans la savane de Lamto (Côte d'Ivoire): la valeur fourragère de la savane. *La Terre et la Vie*, 3:249–261, 1967.
31. G. Josens. *Etudes biologique et écologique des termites (Isoptera) de la savane de Lamto-Pakobo (Côte d'Ivoire)*. Ph.D. thesis, Université libre de Bruxelles, Bruxelles, 1972.
32. S. Konaté, X. Le Roux, D. Tessier, and M. Lepage. Influence of large termitaria on soil characteristics, soil water regime, and tree leaf shedding pattern in a West African savanna. *Plant and Soil*, 206:47–60, 1999.
33. M. Lamotte. Structure and functioning of the savanna ecosystems of Lamto (Ivory Coast). In UNESCO, editor, *Tropical grazing land ecosystems*, pages 511–561. UNESCO, Paris, 1979.
34. X. Le Roux and P. Mordelet. Leaf and canopy CO_2 assimilation in a West African humid savanna during the early growing season. *Journal of Tropical Ecology*, 11(4):529–545, 1995.
35. R.E. Martin. A basic approach to fire injury of tree stems. In *Tall Timbers Fire Ecology Conference*, pages 151–162, Tallahassee, FL, 1963.
36. R.S. McAlpine. Testing the effect of fuel consumption on fire spread rate. *International Journal of Wildland Fire*, 5(3):143–152, 1995.
37. J.C. Menaut. Evolution of plots protected from fire since 13 years in a Guinea savanna of Ivory coast. In *Actas Del IV Symposium Internacional De Ecologia Tropical*, pages 541–558, Panama, 1977.
38. J.C. Menaut, L. Abbadie, F. Lavenu, P. Loudjani, and A. Podaire. Biomass burning in West Africa. In J.S. Levine, editor, *Global biomass burning - Atmospheric, climatic and biospheric implications*, pages 133–142. MIT Press, Cambridge, MA, 1991.

39. J.C. Menaut, L. Abbadie, and P. Vitousek. Nutrient and organic matter dynamics in tropical ecosystems. In P.J. Crutzen and J.G. Goldammer, editors, *Fire in the environment: the ecological, atmospheric and climatic importance of vegetation fires*, pages 215–231. John Wiley & Sons, London, 1993.
40. J.C. Menaut and J. César. Structure and primary productivity of Lamto savannas, Ivory Coast. *Ecology*, 60(6):1197–1210, 1979.
41. J.C. Menaut, J. Gignoux, C. Prado, and J. Clobert. Tree community dynamics in a humid savanna of the Côte d'Ivoire: Modelling the effects of fire and competition with grass and neighbours. *Journal of Biogeography*, 17:471–481, 1990.
42. A.C. Miranda, H.S. Miranda, I. de Fatima Oliveira Dias, and B. Ferreira De Souza Dias. Soil and air temperatures during prescribed cerrado fires in Central Brazil. *Journal of Tropical Ecology*, 9:313–320, 1993.
43. B. Monfort. *Dynamique du renouvellement des populations de deux Papilionoidaea herbacées d'une savane brûlée de Basse Côte d'Ivoire (Lamto)*. Ph.D. thesis, Université des Sciences et Techniques du Languedoc, Montpellier, 1985.
44. Y. Monnier. *Les effets des feux de brousse sur une savane préforestière de Côte d'Ivoire*, volume 9 of *Etudes Eburnéennes*. Ministère de l'Education Nationale de Côte d'Ivoire, Abidjan, 1968.
45. Y. Monnier. *La poussière et la cendre*. Agence de Coopération Culturelle et Technique, Paris, 1981.
46. P. Mordelet. *Influence des arbres sur la strate herbacée d'une savane humide (Lamto, Côte d'Ivoire)*. Ph.D. thesis, Université de Paris 6, Paris, 1993.
47. P. Mordelet. Influence of tree shading on carbon assimilation of grass leaves in Lamto savanna, Ivory coast. *Acta Oecologica*, 14(1):119–127, 1993.
48. R. Oslisly and L. White. La relation homme/milieu dans la réserve de la Lopé (Gabon) au cours de l'holocène: les implications sur l'environnement. In *Dynamique à long terme des écosystèmes forestiers intertropicaux*, pages 163–165, Bondy, 1996. CNRS-ORSTOM.
49. R.N. Owen-Smith. Factors influencing the consumption of plant products by large herbivores. In B.J. Huntley and B.H. Walker, editors, *The ecology of tropical savannas*, pages 359–404. Springer-Verlag, Berlin, 1982.
50. P. Pagney. Le climat de Lamto (Côte d'Ivoire). In M. Lamotte and J.L. Tireford, editors, *Le climat de la savane de Lamto (RCI) et sa place dans les climats de l'Ouest africain.*, volume 8 of *Travaux des chercheurs de Lamto*, pages 31–79. Ecole Normale Supérieure, Paris, 1988.
51. A. Pitot and H. Masson. Quelques données sur la température au cours des feux de brousse aux environs de Dakar. *Bulletin de l'I.F.A.N.*, 13(3):711–732, 1951.
52. J.P. Puyravaud. *Processus de la production primaire en savanes de Côte d'Ivoire. Mesures de terrain et approche satellitaire*. Ph.D. thesis, Université de Paris 6, Paris, 1990.
53. J.J. San José and M.R. Fariñas. Temporal changes in the structure of a trachypogon savanna proected for 25 years. *Acta Oecologica*, 12(2):237–247, 1991.
54. G. Simioni. *Importance de la structure spatiale de la strate arborée sur les fonctionnements carboné et hydrique des écosystèmes herbes-arbres*. Ph.D. thesis, Université de Paris 6, Paris, 2001.
55. N.R.H. Stronach and S.J. MacNaughton. Grassland fire dynamics in the Serengeti ecosystem, and a potential method of retrospectively estimating fire energy. *Journal of Applied Ecology*, 26:1025–1033, 1989.

56. M.D. Swaine, W.D. Hawthorne, and T.K. Orgle. The effects of fire exclusion on savanna vegetation at Kpong, Ghana. *Biotropica*, 24(2):166–172, 1992.
57. R. Viani, J. Baudet, and J. Marchant. Réalisation d'un appareil portatif d'enregistrement magnétique de mesures. application à l'étude de l'évolution de la température lors du passage d'un feu de brousse. *Annales de l'Université d'Abidjan, Série E*, 6(2):295–304, 1973.
58. D.X. Viegas. Some thoughts on the wind and slope effects on fire propagation. *International Journal of Wildland Fire*, 4(2):63–64, 1994.
59. P. Villecourt and E. Roose. Charge en azote et en éléments minéraux divers des eaux de pluie, de pluviolessivage et de drainage dans la savane de Lamto. *Revue d'Ecologie et de Biologie du Sol*, 15(1):1–20, 1978.
60. P. Villecourt, W. Schmidt, and J. César. Recherches sur la composition chimique de la strate herbacée de la savane de Lamto. *Revue d'Ecologie et de Biologie du Sol*, 16(1):9–15, 1979.
61. R.G. Vines. Physics and chemistry of rural fires. In A.M. Gill, R.H. Groves, and I.R. Noble, editors, *Fire and the Australian biota*, pages 129–150. Australian Academy of Sciences, Canberra, 1981.
62. R. Vuattoux. Observations sur l'évolution des strates arborée et arbustive dans la savane de Lamto (Côte d'Ivoire). *Annales de l'Université d'Abidjan, Série E*, 3(1):285–315, 1970.
63. R. Vuattoux. Contribution à l'étude de l'évolution des strates arborée et arbustive dans la savane de Lamto (Côte d'Ivoire). Deuxième note. *Annales de l'Université d'Abidjan, Série C*, 7(1):35–63, 1976.
64. B.H. Walker. Structure and function of savannas: an overview. In J.C. Tothill and J.J. Mott, editors, *Ecology and management of the world's savannas*, pages 83–91. Australian Academy of Sciences, Canberra, 1985.
65. F.I. Woodward. *Climate and plant distribution*. Cambridge University Press, Cambridge, 1987.

5
Vegetation

Jean-Claude Menaut and Luc Abbadie

5.1 Introduction

The landscape of Lamto is a forest-savanna mosaic (Figs. 5.1 and 5.2). Evergreen forests are located in the thalwegs, along intermittent rivers (gallery forests), or along the permanent Bandama river (riparian forest), while semi-deciduous forests are located on plateaus. Savannas occur between these two types of forest, with shrubs and trees in variable densities. In savannas, the tallest and most characteristic tree is the palm *Borassus aethiopum* (Fig. 5.3). The flora contains ca. 400 savanna species (pyrophytes) and a number of woody species originating from forest or secondary thickets (Table 5.1) [2, 14, 22, 9]. The herbaceous layer is made of perennial grasses which contribute up to 95% of the total aboveground grass biomass. Legumes are scarce and often absent. Among the woody species, ca. 30 are fire tolerant and belong to the Sudan flora. The others come from the surrounding forests and occur in savanna areas where fire intensity is low (in savanna tree clumps for example). Adjanohoun [1] classified the Lamto savannas in the *Brachiaria brachylopha* association, *Loudetia simplex* sub association.

5.2 Main savanna types

Different types of savanna vegetation are organized according to the topographical gradient along a catena [19]. Grass savannas, dominated by *Loudetia simplex*, occur in the lower part of the slopes or on hydromorphic plateaus (Fig. 2.3). Trees and shrubs are rare and small and occur mostly on termite mounds, likely due to soil water regime (soils are waterlogged several months during the rainy season, except on mounds), with the exception of the palm *Borassus aethiopum* that is present everywhere in the savanna. On well drained soils, the dominant species are *Andropogon schirensis*, *Hyparrhenia diplandra*, *H. smithiana*, and *Imperata cylindrica*, all belonging to the Andropogoneae tribe. These grass species show variable densities in relation

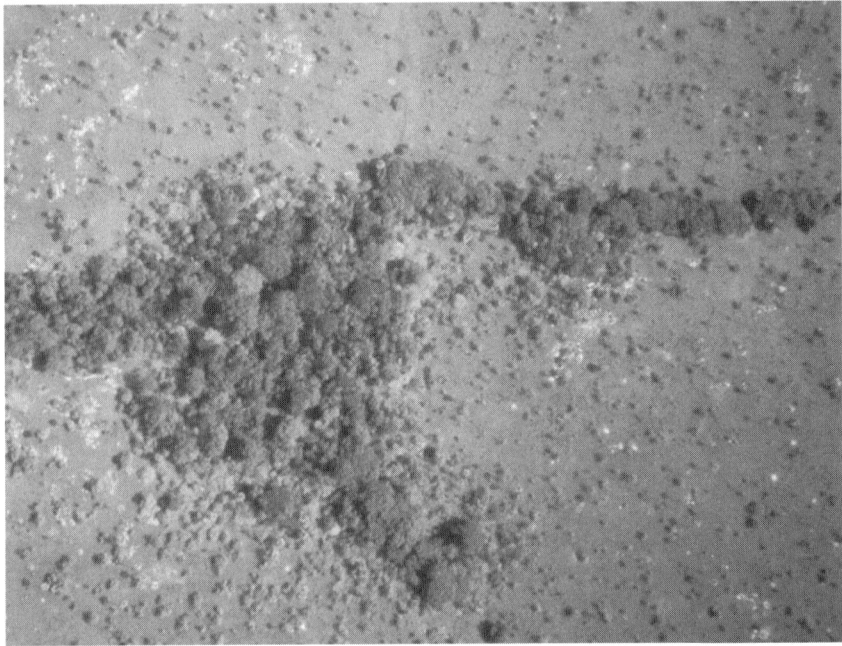

Fig. 5.1. Aerial photography of the Lamto area (photography by L. Gautier, copyright Conservatoire et Jardin botaniques de la ville de Genève, with permission).

to the tree cover density. More than 90% of the woody layer is composed of four species: *Bridelia ferruginea*, *Crossopteryx febrifuga*, *Cussonia arborea*, *Piliostigma thonningii*. An open shrubland grows at the bottom of the slopes, with a woody cover lower than 5%. This woody cover increases with elevation and vegetation progressively changes from open shrub savanna to dense shrub savanna, until savanna woodland on plateaus (Figs. 5.2 and 2.3). In the areas with high tree density, the vegetation is often a mosaic of open and closed environments, the open environments being generally linked to accumulations of big granitic rocks.

5.3 Structure of the vegetation

The Lamto savanna vegetation is made of three strata: the grass stratum with some low woody species (0-2 m), the shrub stratum (2-8 m), and the tree stratum (over 8 m).

5.3.1 The grass layer

The herbaceous stratum is continuous, except in the most dense woodland where it is generally discontinuous and invaded by forbs (Fabaceae,

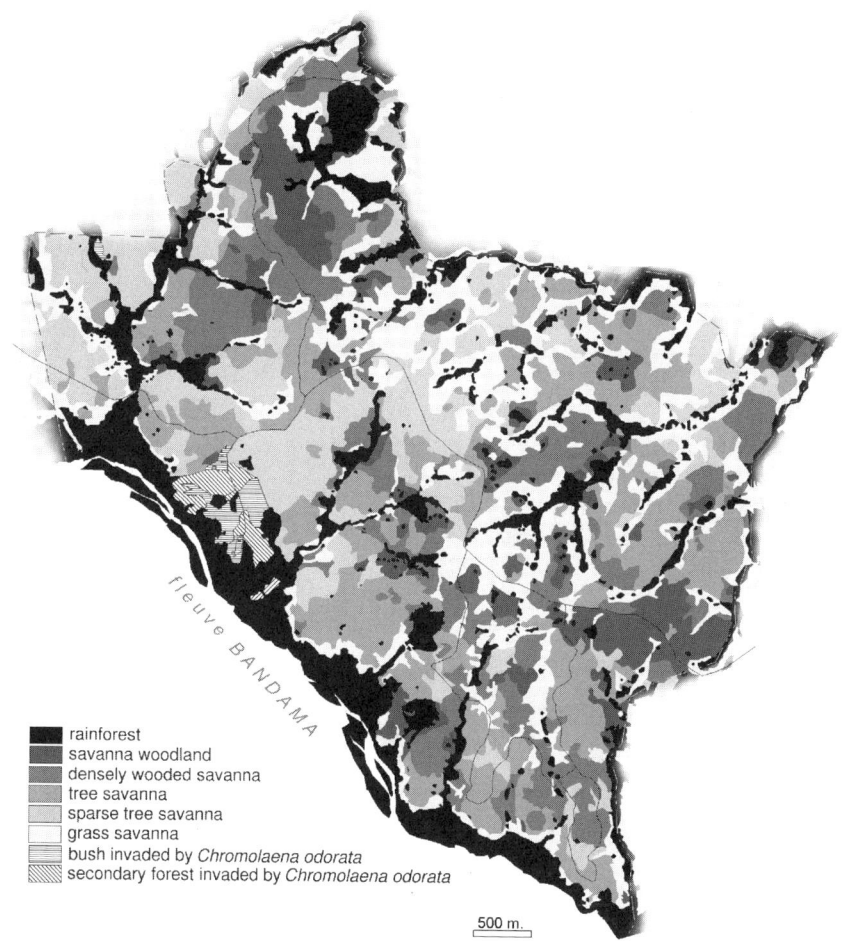

Fig. 5.2. Vegetation map of the Lamto reserve. Rainforest is 100% tree cover, savanna woodland is over 62% tree cover, densely wooded savanna is between 37% and 61% tree cover, tree savanna is between 19% and 36% tree cover, sparse tree savanna is between 7% and 18% tree cover, and grass savanna is below 7% tree cover (after [7], kindly reworked by C. Châtelain, copyright Conservatoire et Jardin botaniques de la ville de Genève, with permission).

Rubiaceae). Ninety percent of the herbaceous stratum is made of perennial grasses. Grass cover of the main savanna facies (Andropogoneae savannas) consists in 7 main species (Table 5.1). When leaves are fully developed, cover reaches 100% at the top of the grass canopy, but is low at ground level (between 10% and 15% for basal area [4, 16, 15]) because all species grow in compact tufts (Fig. 5.4).

Fig. 5.3. Typical view of Lamto savanna showing the mixture of grasses, shrubs/trees, and palm trees, with a gallery forest in the background (photography by J. Gignoux).

Table 5.1. Dominant grass and tree species of Lamto savanna: Nomenclature follows the Flora of West Tropical Africa [10], with revisions [8].

Species	Biological type	Family
Andropogon ascinodis C. B. Cl.	Grass	Poaceae (Andropogoneae)
Andropogon canaliculatus Schumach.	Grass	Poaceae (Andropogoneae)
Andropogon schirensis Hochst. ex A. Rich.	Grass	Poaceae (Andropogoneae)
Annona senegalensis Pers.	Shrub/tree	Annonaceae
Borassus aethiopum Mart.	Palm tree	Arecaceae
Brachiaria brachylopha Stapf	Grass	Poaceae(Paniceae)
Bridelia ferruginea Benth.	Tree	Euphorbiaceae
Crossopteryx febrifuga (Afzel. ex G. Don) Benth.	Tree	Rubiaceae
Cussonia arborea A. Rich.	Tree	Araliaceae
Hyparrhenia diplandra (Hack.) Stapf	Grass	Poaceae (Andropogoneae)
Hyparrhenia smithiana (Hook. f.) Stapf	Grass	Poaceae (Andropogoneae)
Imperata cylindrica (Linn.) P. Beauv.	Grass	Poaceae (Andropogoneae)
Loudetia simplex (Nees) C.E. Hubbard	Grass	Poaceae (Arundinelleae)
Piliostigma thonningii (Schum.) Milne-Redhead	Tree	Cesalpiniaceae

Fig. 5.4. Typical aspect of the grass layer: Top, 2 weeks after the fire; bottom, 3 months after the fire (photographs by X. Le Roux).

The distribution of leaf angles is erectophilous [17]. The maximum height reached by grass leaves is ca. 100 cm [6, 15], but inflorescences can reach up to 250 cm. Most tufts are monospecific, but 10% of the total basal cover consists in plurispecific tufts [16]. The most frequent association of species within tufts is *Andropogon schirensis* + *Andropogon ascinodis*. The mean size of the tufts varies with environment: it is smaller in savannas dominated by the grass *Loudetia simplex* than in those dominated by Andropogoneae. Shoots of woody species, such as *Bridelia ferruginea*, *Annona senegalensis*, *Piliostigma thonningii* and *Ficus sur* grow in the grass layer, including those which appear as annual suckers from rootstocks.

The distribution of herbaceous species is strongly controlled by that of woody plants (Fig. 5.5). This mainly results from the variation in the amount

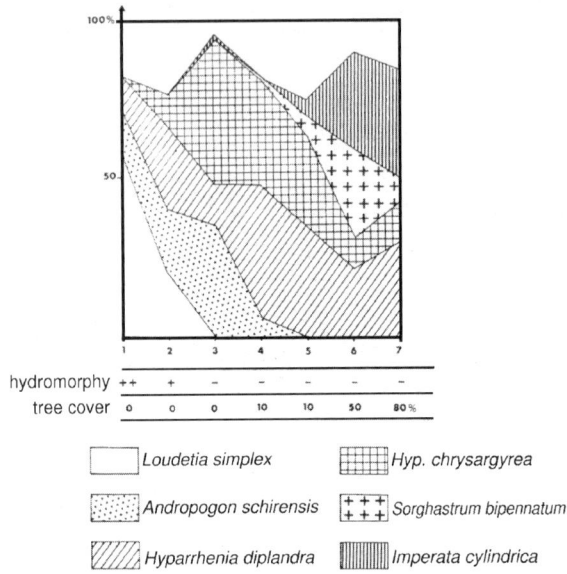

Fig. 5.5. Influence of tree cover and hydromorphy on the species composition (% of total biomass) of the grass layer (from [3]).

of light (see Sect. 8.2) and, secondary, from that of fire intensity, the latter being related to litter accumulation (see Sect. 4.5). For example, the transition zone from an open grassland to more closed woody savanna shows the decrease of the density of grasses such as *Andropogon* spp. and *Hyparrhenia* spp. to the benefit of the grass *Imperata cylindrica* and *Aframomum latifolium*. In addition, in the areas dominated by *Andropogon* spp., *Andropogon schirensis* is gradually replaced by *Andropogon canaliculatus* as trees and shrubs appear. The same pattern is observed for the density of tree seedlings and woody hemicryptophytes that are highly correlated to the density of adult shrubs and trees.

Belowground, the distribution of grass roots closely matches that of tufts (Fig. 5.6). This results in a highly heterogeneous spatial distribution of grass roots within the soil, with 7-fold variations in root density in 10 cm distance. A semi-variogram analysis (Abbadie et al. unpublished) indicates a correlation distance of 0.3 to 0.4 m for the 0 to 5 cm soil stratum and 0.6 to 0.7 m for the 5 to 10 cm soil stratum, values quite compatible with the average tuft size and spacing. This pattern has important consequences for soil organic matter. The belowground heterogeneity of roots observed in lamto has been reported in other systems: soil characteristics vary at small scales correlated to the tufted structure of the grass layer in North American steppes [11, 12]. Such tufted vegetation structure has important effects on ecosystem functioning (Chap. 15).

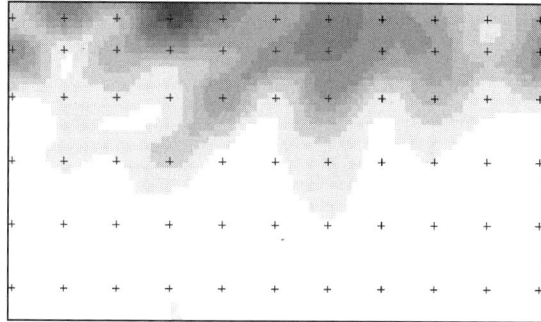

Fig. 5.6. Belowground distribution of root density in a vertical plan: + represent the location of sampling points, every 10 cm horizontally and vertically. Contours vary from 1 kg m^{-3} (white) to 13 kg m^{-3} (black) by steps of 1 kg m^{-3} (after [16]).

5.3.2 The tree and shrub layer

The density of woody plants varies up to ca. 1500 stems per hectare, with a mean density of 250 per hectare. Shrubs may grow all over the savanna, but they are mostly gathered into patches, making the average tree cover as low as 15-20%. In the dense shrub savanna and savanna woodland, the upper stratum is rich in mesophanerophytes such as *Erythrophleum guineense* and *Terminalia shimperiana*. The shrubs such as *Crossopteryx febrifuga* and *Ficus sur* grow their full height up to 10-12 m and connect the two woody strata. Trees are generally isolated and dominate a shrub cluster. Most often, the tree stratum consists only of palms, quite regularly distributed in the savanna. They bear waterlogged soils and are the only trees in the *Loudetia* savanna. In marshy sites, they frequently occur on termite mounds with the other palm *Phoenix reclinata*.

5.4 Life-Forms

The classification of Raunkiaer has been used in Lamto to describe the growth forms of plants [19]. A major feature is that therophytes are poorly represented. Only the grass *Sorghastrum bipennatum*, the Asteraceae *Aspilia bussei* and a few legumes (*Indigofera* spp. and *Tephrosia* spp.) belong to this type. The seeds germinate during the first rains, generally in March. The seedlings compete severely with the adult perennial grasses and grow slowly until August.

Geophytes benefit from ideal protection against fire. The root system accumulates reserves that allow rapid growth, flowering and fruiting soon after burning, i.e., before other herbaceous plants grow. Examples are given by a number of monocotyledons with bulbs, such as *Urginea indica*, tubers with *Eulophia* spp. and rhizomes with the grass *Imperata cylindrica*.

Hemicryptophytes are the main contributors to the herbaceous layer's primary production. Grasses have strong fasciculate roots. They start growing just after fire and develop rapidly their aerial parts until seed formation, from September to November for most of them. They tolerate fire due to the dense base of tillers and leaf sheaves that protect buds against heating. They are tufted: Their bases cover 10% of the soil surface on average [19]. A large number of dicotyledons is scattered in the grass layer. Their stems are often prostrate and twining, as for *Galactia tenuiflora*, *Rhynchosi sublobata*, *Sphenostylis holosericea* or *Vigna* spp. Some species are erect such as *Indigofera polysphaera*. The root system of these dicotyledons is often as developed as that of small shrubs. *Cochlospermum planchoni*, *Cissus doeringii*, *Eriosema psoraleoides* and *Lippia rugosa* are bushy species with aerial systems intolerant to fire. Their development is basitonic (ramifications on the lower part of the axes) at the beginning of the wet seasons, then acrotonic (ramifications on the upper part of the axes). Most trees and shrubs in Lamto behave as hemicryptophytes during their first developmental stages, but look like phanerophytes after.

Chamaephytes do not survive long in burnt savannas. The only representative is *Crossopteryx febrifuga* during the first years of their life [5]. Other species grow directly from hemicryptophyte to phanerophyte form.

The development of the phanerophytes is perturbed by fire and the variability of their response is high between species and even individuals. For *Crossopteryx febrifuga*, fire only slows growth. When it is young, it behaves as a chamaephyte with annual shoots springing from pollards. The root system of *Bridelia ferruginea* allows it to survive as a hemicryptophyte for many years, producing annual and fruiting suckers. It becomes a phanerophyte when a long thick shoot appears, tolerating fire and ramifying. During the first years of its life, *Cussonia arborea* also behaves as an hemicryptophyte producing annual nonfruiting shoots from a fast growing taproot, until a single fire tolerant shoot develops while the others disappear. Many individuals of *Piliostigma thonningii* are true hemicryptophytes: their belowground system produces a strong taproot with perennial plagiotropic belowground shoots whose buds develop annual suckers with a complete reproductive cycle. This type of individual is stable and occurs in the same ecological conditions as the phanerophytes.

The shoots of the young trees are periodically destroyed by fire and most species take a bushy form during the first years of their development. Their shoots are first annual, then perennial, the latter branching during growth and giving the plant its final architecture. In the experimental areas protected from fire (Sect. 4.5), the behavior of trees is not greatly modified. Their growth is faster than in burnt savanna, but their transitory forms still occur. Particularly, many individuals of *Piliostigma thonningii* are found as hemicryptophytes and some species such as *Ficus sur*, *Ficus vallis-choudae* and *Nauclea latifolia* often show bushy forms although they can become trees in both burnt and unburnt savanna.

5.5 Phenological cycles

The annual cycle of vegetation is basically controlled by the seasonality of rainfall (Chap. 3). During the dry season, in December and January, most of perennial plants wither and lose their leaves while annual species die. Fire generally occurs in January, i.e., in the heart of the dry season, and destroys most of the biomass and dead matter of the herbaceous layer. In unburnt areas, tree leaf fall spreads over several weeks. A rapid regrowth of herbaceous species (Fig. 17.1) and the flowering of some of them can occur before the beginning of the rainy season if soil humidity is not too low. During the long wet season, the biomass of all species considerably increases, the short dry season (August) having only a small impact of the primary production rate. Most species flower in October and November, then growth stops and withering starts.

Six types of herbaceous producer have been defined by César [3], considering the variability of emergence date, flowering date and length of vegetative state (Fig. 5.7). Species with a forward cycle, such as *Cyperus tenuiculmis*, flower in the first weeks following fire and spread their seeds before being covered by the grasses, sometimes even before the spring of any vegetative shoot. They often have a short development cycle and their aerial parts have

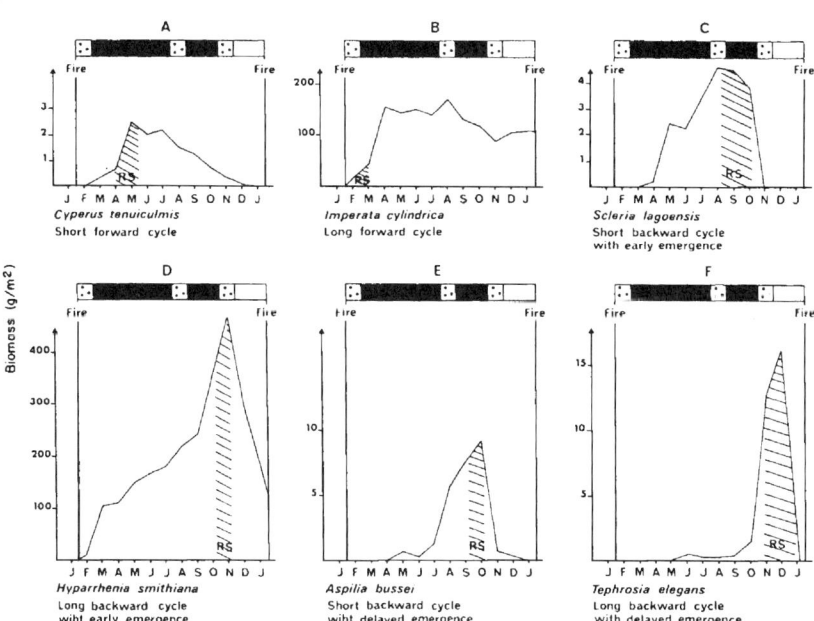

Fig. 5.7. Phenological cycles of six characteristic herb and grass species at Lamto: Hatches indicate the reproductive stage (RS) of the species. The upper part of the graph represents monthly rainfall: <50 mm, white; 50-100 mm, dotted; >100 mm, black. Forward cycle = flowering soon after emergence; backward cycle = flowering during the wet season or at beginning of the dry season (after [19], with permission of the Ecological Society of America).

withered when the wet season occurs. However, some of them, such as *Imperata cylindrica*, have a long development cycle and grow until the next burning. Many species with a forward cycle are geophytes. Most plants with a backward cycle flower during the short wet season, generally when biomass peaks. The grasses contributing to most of primary production, such as *Hyparrhenia* spp. and *Andropogon* spp., belong to this group. They have a strong root system that supplies a rapid regrowth at the beginning of the wet season.

The phenological behavior of the woody species is generally more homogenous, particularly for leaf outbreak and leaf fall [18, 21], but not for flowering dates. For example, *Annona senegalensis* flowers and fruits before being covered by grasses, while *Bridelia ferruginea* flowers when the first leaves appear. In contrast, *Piliostigma thonningii* flowers later, until the beginning of the dry season. The leaf fall generally occurs within a short period for all the species, but with strong interannual variations for both date and duration [18]. Leaf fall may occur after fire at 90% but may also begin in July, with only 25% of the leaves falling after fire. Although tropical tree leaf shedding patterns are often controlled by atmospheric factors [24] and plant water status [20, 23], the phenological variations observed for trees at Lamto have rarely been related to climatic or soil water cycles. However, Konaté et al. [13] observed that the leaf shedding pattern of *Crossopteryx febrifuga* was delayed and less pronounced on eroded termite mounds than in control areas. Given that soil water status in the 0.20-0.60 cm soil layer was improved on mounds during the beginning of the dry season, these results suggest that soil water status substantially modulates the leaf shedding pattern of *Crossopteryx febrifuga*. The location of individual trees also affects tree phenology for some species. For instance, budburst and leaf shedding occur earlier in isolated as compared to clump trees in *Crossopteryx febrifuga* [21], although the factors driving this difference are still debated.

5.6 Conclusion

Lamto vegetation encompasses savanna areas with contrasting tree densities, mixed with moist and dry forest areas. The different savanna types are roughly organized according to topographical transects. The grass layer is dominated by grass tussock species from the Andropogonaea family, whereas four species dominate the tree layer. Herbaceous species exhibit different phenological patterns. In contrast, trees are all deciduous but exhibit a marked time variability of leaf fall. Such variability has important implications for carbon cycle since most of the leaves falling before fire are burned and have no chance to contribute to soil organic matter accumulation.

Acknowledgments

We thank Laurent Gautier and Cyrille Châtelain for specially redigitizing and reworking the Lamto vegetation map of Fig. 5.2 for us.

References

1. E. Adjanohoun. *Végétation des savanes et des rochers découverts en Côte d'Ivoire*, volume 7 of *Mémoire*. ORSTOM, Paris, 1964.
2. V. Bänninger. Inventaire floristique des dicotylédones de la réserve de Lamto (V-baoulé) en Côte d'Ivoire centrale. Technical report, Département de botanique et de biologie végétale, Université de Genève, Genève, 1995.
3. J. César. *Etude quantitative de la strate herbacée de la savane de Lamto*. Ph.D. thesis, Université de Paris, Paris, 1971.
4. J. César. *La production biologique des savanes de Côte d'Ivoire et son utilisation par l'homme*. CIRAD - Institut d'élevage et de médecine vétérinaire des pays tropicaux, Maisons-Alfort, 1992.
5. J. César and J.C. Menaut. Analyse d'un écosystème tropical humide: La savane de Lamto (Côte d'Ivoire). II. Le peuplement végétal. *Bulletin de Liaison des Chercheurs de Lamto*, S2:1–161, 1974.
6. H. Gauthier. *Echanges radiatifs et production primaire dans une savane humide d'Afrique de l'Ouest (Lamto - Côte d'Ivoire)*. M.Sci. thesis, Université Paul Sabatier, Toulouse, 1993.
7. L. Gautier. *Carte du recouvrement ligneux de la réserve de Lamto*. Conservatoire et Jardin Botanique de la ville de Genève, 1988.
8. L. Gautier. *Contact forêt-savane en Côte d'Ivoire - Rôle de Chromolaena odorata (L) dans la dynamique de la végétation*. Doctorat, Université de Genève, Genève, 1992.
9. L. Gautier and R. Spichiger. The forest-savanna transition in west africa. In L. Poorter, F. Bongers, F. N. G. Kouamé, and W. D. Hawthorne, editors, *Biodiversity of West African Forests*, pages 33–40. CABI Publishing, Wallingford, 2004.
10. J. Hutchinson and J.M. Dalziel. *Flora of West tropical Africa*. Millbank, London, 1954.
11. R.B. Jackson and M.M. Caldwell. Geostatistical patterns of soil heterogeneity around individual perennial plants. *Journal of Ecology*, 81:683–692, 1993.
12. R.H. Kelly and I.C. Burke. Heterogeneity of soil organic matter following death of individual plants in shortgrass steppe. *Ecology*, 78(4):1256–1261, 1997.
13. S. Konaté, X. Le Roux, D. Tessier, and M. Lepage. Influence of large termitaria on soil characteristics, soil water regime, and tree leaf shedding pattern in a West African savanna. *Plant and Soil*, 206:47–60, 1999.
14. F. N. G. Kouamé. Contribution au recensement des Monocotyledonae de la Réserve de Lamto (Côte d'Ivoire centrale) et à la connaissance de leur place dans les différents faciès savaniens. Technical report, Faculté des Sciences et Techniques, Université Nationale de Côte d'Ivoire, Abidjan, 1993.
15. J.C. Lata. *Interactions entre processus microbiens, cycle des nutriments et fonctionnement du couvert herbacé: Cas de la nitrification dans les sols d'une savane humide de Côte d'Ivoire sous couvert à Hyparrhenia diplandra*. Ph.D. thesis, Université de Paris 6, Paris, 1999.
16. E. Le Provost. *Structure et fonctionnement de la strate herbacée d'une savane humide (Lamto, Côte d'Ivoire)*. M.Sci. thesis, Universités de Paris 6 & 11, Institut National Agronomique Paris-Grignon, Paris, 1993.
17. X. Le Roux, H. Gauthier, A. Bégué, and H. Sinoquet. Radiation absorption and use by humid savanna grassland, assessment using remote sensing and modelling. *Agricultural and Forest Meteorology*, 85:117–132, 1997.

18. J.C. Menaut. Chutes de feuilles et apport au sol de litière par les ligneux dans une savane préforestière de Côte d'Ivoire. *Bulletin d'Ecologie*, 5:27–39, 1974.
19. J.C. Menaut and J. César. Structure and primary productivity of Lamto savannas, Ivory Coast. *Ecology*, 60(6):1197–1210, 1979.
20. P.B. Reich and R. Borchert. Water stress and tree phenology in a tropical dry forest in the lowlands of Costa Rica. *Journal of Ecology*, 72:61–74, 1984.
21. G. Simioni, J. Gignoux, X. Le Roux, R. Appé, and Benest D. Spatial and temporal variation in leaf area index, specific leaf area, and leaf nitrogen of two co-occuring savanna tree species. *Tree Physiology*, 24(2):205–216, 2003.
22. R. Spichiger. Contribution à l'étude du contact entre flores sèche et humide sur les lisières des formations forestières humides semi-décidues du V-baoulé et de son extension nord-ouest (Côte d'Ivoire centrale). Technical report, Département de Biologie végétale, Université de Genève, Genève, 1975.
23. R.J. Williams, B.A. Myers, W.J. Muller, G.A. Duff, and D. Eamus. Leaf phenology of woody species in a north Australian tropical savanna. *Ecology*, 78:2542–2558, 1997.
24. S.J. Wright and F.H. Cornejo. Seasonal drought and leaf fall in a tropical forest. *Ecology*, 71:1165–1175, 1990.

Part II

Structure and Functioning of Plant Cover

6
Soil-Plant-Atmosphere Exchanges

Xavier Le Roux and Bruno Monteny

6.1 Introduction

Interactions between plant structure and functioning can be summarized as follows: at a given time, plant structure is the result of carbon allocation to the formation of the different plant parts that has occurred in the past, and the resulting new structure has an impact on the local environments experienced by plant parts and the ability of plants to conduct their metabolic functioning (resource acquisition and storage) in the future [42]. While Chap. 7 focuses on plant biomass, production and allocation at time scales ranging from one month to one year, this chapter focuses on the local environments experienced by plant parts and their sub-instantaneous metabolic functioning. Plant-atmosphere (or plant-soil) exchanges are controlled by surface exchanges, namely leaves (or roots). Thus, a key factor linking plant biomass and its functioning is specific leaf area (or specific root area) that determines the biomass investment per unit leaf (or root) surface. Values of specific leaf area reported for the main tree and grass species found in Lamto savannas are given in Table 6.1.

This chapter presents information acquired during the last three decades in the study of exchanges of radiation, energy, water, carbon and other trace gases between the soil, plants and the atmosphere in the Lamto savanna. The chapter focuses more particularly on recent advances because this study has developed rapidly in recent years mainly as a result of concern over global climate change. The next section presents an overview of the coordinated, cross-disciplinary field campaign that has been held from 1991 to 1994 at Lamto and that is a milestone investigation in the studies of soil-plant-atmosphere interactions for this ecosystem. The next sections summarize some of the most significant findings in this context and outline some gaps in the scientific knowledge and desirable future research directions.

Table 6.1. Values of specific leaf area ($cm^2\ g^{-1}$) for the major grass and tree species in Lamto. *: range of mean values obtained at three or four dates (Nov., Dec., Feb., and Apr.). **: values from an unpublished dataset by Menaut and Gignoux, obtained in late April.

Group	Species	Specific leaf area	Source
Grasses	*Andropogon* spp. (green)	165 (Feb.) to 80 (Oct.)	[37]
	Hyparrhenia spp. (green)	130 (Feb.) to 60 (Oct.)	[37]
	whole grass layer (green)	$SLA = 128 - 62\left(1 - e^{-0.0102\,biomass}\right)$	[37]
	grass spp. (dead)	144	[37]
Trees	*Bridelia ferruginea*	77	**
		85 to 111*	[23]
	Crossopteryx febrifuga	70	**
		82 to 103*	[23]
	Cussonia arborea	91	**
		98 to 144*	[23]
	Ficus sur	71	**
	Nauclea latifolia	76	**
	Piliostigma thonningii	70	**
		86 to 106*	[23]
	Pterocarpus erinaceus	118	**
	Terminalia shimperiana	69	**
	Vitex doniana	81	**

6.2 Overview of the 1991 to 1994 field campaign

As part of the monitoring effort of the SAvannas on the Long Term (SALT) program (see Chap. 1), soil-plant-atmosphere exchanges were surveyed on a 500 m × 500 m shrubby savanna test site, which represents the major vegetation type at Lamto (see Sect. 5.2). Field experiments benefited from the cooperation of different disciplines and research institutes. The test site became the focus of an extended monitoring program for acquiring biological, biophysical, biogeochemical and micrometeorological data over a four-year period (Table 6.2). In situ and satellite remotely sensed data were acquired concurrently on this site or over the Lamto region. Most experiments were extensive and designed to obtain comprehensive information on the radiation, water, carbon and energy balances throughout the annual vegetation cycle. The major objectives were (1) to derive a better understanding of the role of biological and micrometeorological controls on savanna-atmosphere interactions, (2) to provide relevant information on how seasonality affects this ecosystem and its radiation, water and energy budgets, and (3) to provide a useful database to develop and test models of savanna-atmosphere interactions.

For obvious reasons, the objectives of this program did not include bridging the gap between the understanding of small-scale processes and the representation of processes operating at larger scales (in particular, assessment of the

Table 6.2. Overview of the experiments dealing with soil-plant-atmosphere exchanges carried out at a same test site at Lamto from January 1991 to June 1994 as part of the SALT program. ×continuous measurements; — punctual measurement.

Focus	1991	1992	1993	1994	Parameters measured	Main references
Lamto weather station[1]	× × ×	× × ×	× × ×	×	T, e, u, R_s, P	Chap. 3
In situ automatic weather station	× ×	× ×	× ×		T, T_{soil}, e, u, R_s, PAR, P	id.
Biological parameters	× ×	× ×	× ×	×	Green and dead LAI, SLA, rooting patterns, fractional coverage	[34]
Throughfall		×	×	×	Throughfall under grass canopy	id.
Radiation budget	× ×	×	× ×	×	R_n, R_s, αR_s, T_s, R_a	[44]
Energy budget	× ×	×	× ×	×	R_n, G, H, λE	[44]
Soil moisture		× × ×			0-170 cm (neutron probe)	[40]
	× ×				0-60 cm (gravimetric)	[39]
		— —	×	×	0-2 cm (gravimetric)	[23]
Water uptake by roots			×	×	δ^2H and δ^{18}O of sap and soil water	[39]
Plant water status	× ×	× × ×			Diurnal cycles of Ψ_l	[37]
CO_2 exchanges	— —			— —	Leaf and canopy photosynthesis	[17, 43]
Radiation absorption and conversion			× ×	×	f_{PAR} and ε_n	[23, 41]
Trace gas emissions[2] from biomass burning	— —				CO_2, CO, CH_4, NO_2, NO, N_2O, NH_3, O_3,	[34, 15][3]
Trace gas emissions[2] from soils		— —	— —		NO_2, NO (all periods), O_3 (first period)	[74, 75, 38]
In situ reflectances			×	×	SPOT channels	[23, 41]
Satellite data			TM S	S	S = SPOT, TM = LANDSAT	-
NOAA data	× ×	× × ×	×	×		-

[1] 400 m from the site.
[2] DECAFE/SALT joint program.
[3] See J. Atmos. Chem. FOS/DECAFE special issue.

variability of surface parameters, and interactions with the atmosphere at landscape or regional scales). Progressing toward a better understanding and modeling of soil-plant-atmosphere exchanges in a major but in this context largely unstudied biome, namely humid savannas, was thought to be a sufficiently ambitious task. All the results and interpretations presented hereafter thus only refer to one-way coupling processes between savanna and the atmosphere at the site scale (i.e., atmospheric forcing of local surface conditions) and do not address two-way coupling processes.

Concurrently with this project, the Dynamique Et Chimie Atmosphériques en Forêt Equatoriale (DECAFE) program and the SALT program jointly developed a project on (i) the emission of aerosols and carbon- and nitrogen-compounds from savanna fires and (ii) O_3 and NO_x ($NO+NO_2$) fluxes and associated physical and biological controls. The corresponding experiments were held on the same test site (Table 6.2). The interpretation of the high heterogeneity of NO emission rates and ecological knowledge gained from the DECAFE-SALT joint project is presented in Sect. 14.7. Radiation, energy, water, carbon, NO_x and O_3 exchanges are discussed herein.

6.3 Savanna radiation budget and spectral signatures

Radiation exchanges, that are arguably among the most important feature of vegetation-atmosphere interactions, were studied by Bony [10], Gauthier [23], Le Roux et al. [44], Le Roux [37] and Le Roux et al. [41]. These studies encompass the seasonal variations of (i) the components of the savanna radiation budget, (ii) light transmission, interception and absorption by the grass and shrub canopies and (iii) surface spectral reflectances.

6.3.1 Radiation budget above the savanna

The radiation budget of the savanna can be written as

$$R_n = (1 - \alpha) R_s + R_{l\downarrow} - R_{l\uparrow}, \qquad (6.1)$$

where R_n is net radiation (energy available at the surface for heat transfer), R_s is incoming solar radiation, α is the savanna albedo for shortwave radiation, $R_{l\downarrow}$ is downward longwave radiation (atmospheric radiation) and $R_{l\uparrow}$ is upward longwave radiation (emitted or reflected by the surface). Data collected between January 1991 and May 1992 provided 258 complete daily cycles of the 5 components of the radiation budget in an open area, i.e., above a soil-grass-low shrub system (one daily cycle is presented in Fig. 6.1). The methodology used and a verification of the radiation balance closure are described by Le Roux et al. [44]. The albedo of shrubby areas (including shrubs 2 to 7 m high) or open areas (with only shrubs lower than 2 m) are similar (H. Gauthier and X. Le Roux, unpublished); thus, the observed albedo can

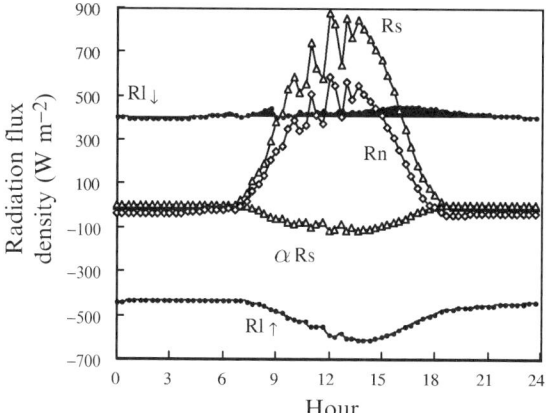

Fig. 6.1. Daily variation of the savanna radiation balance components on 26 February 1992: The 20 min mean values of the incoming (positive) and outgoing (negative) shortwave radiation R_s (triangles), longwave radiation R_l (dots) and net radiation R_n (circles) are shown (after [37]).

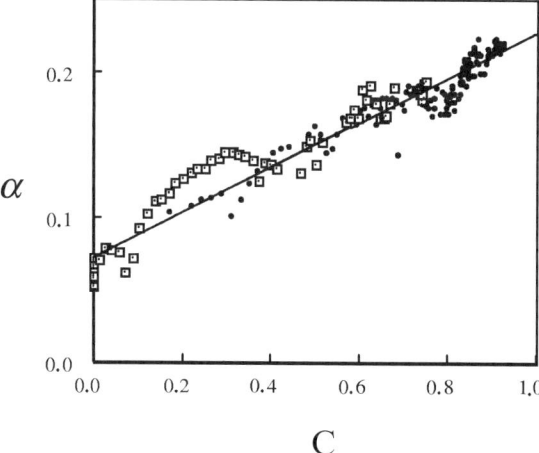

Fig. 6.2. Relationship between the daily savanna albedo α, and the canopy fractional coverage C, during the rainy season: dots, 1991; squares, 1992 (after [37]).

be assumed to be representative of the test site. The relationship between the daily albedo of the savanna canopy α and its fractional coverage by vegetation C (Fig. 6.2) can be written

$$\alpha = \alpha_s(1-C) + \alpha_c C \quad (r^2 = 0.95), \tag{6.2}$$

where the soil albedo α_s equals 0.070 and the closed canopy albedo α_c equals 0.224. The savanna monthly albedo ranges from 0.075 during the month after fire to 0.224 when the canopy is closed [44]. These values are consistent with those reported by Bony [10] for a similar savanna in Lamto: 0.080 to 0.234. The annual mean albedo reported by Bony [10] and by Le Roux et al. [44] are 0.188 and 0.194, respectively. These values are close to those reported by Oguntoyinbo [58] for Nigerian mesic savannas, but lower than values typical of Sahelian dry savannas [2] and higher than values reported for rainforests [78]. The daily net radiation to daily solar radiation ratio $\frac{R_n}{R_s}$ is linearly related to the vegetation fractional coverage:

$$\frac{R_n}{R_s} = 0.53\,(1 - C) + 0.73C \quad (r^2 = 0.79)\,. \tag{6.3}$$

The daily net radiation to daily solar radiation ratio observed at Lamto during the rainy season is close to the ratio observed on a rainforest site in Amazonia (Fig. 6.3). This is explained by a daily net longwave radiation loss slightly lower at Lamto than at the forest site. Thus, net radiation, i.e., the energy available at the surface, is particularly high for Lamto savanna during the rainy season. Due to a lower vegetation fractional coverage and a lower leaf area index (LAI), $\frac{R_n}{R_s}$ is lower in Sahelian dry savannas of Niger even during the dry season (Fig. 6.3).

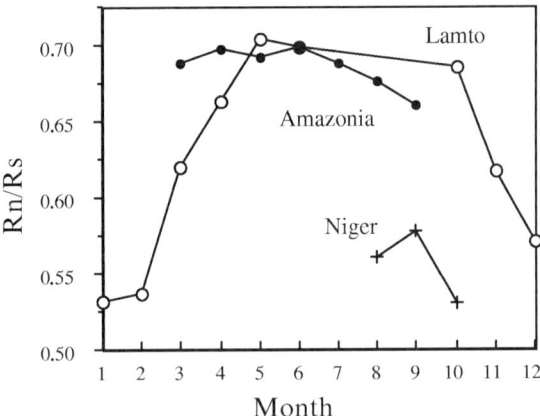

Fig. 6.3. Seasonal variations of the ratio between daily (24 h) net radiation R_n and daily incoming solar radiation R_s for the Lamto savanna (open circles), for an Amazonian rainforest (solid circles) and for a dry fallow savanna in Niger (crosses). Seasonal variations were computed from the 1983-1984 ARME database [76] and from the 1992 HAPEX-Sahel data of Monteny et al. for the forest and dry savanna, respectively (after [37]).

6.3.2 Radiation budget of the grass and shrub layers

The components of the canopy radiation budget (incoming and transmitted radiation, and canopy and bare soil albedo) for photosynthetically active radiation (PAR) and total solar wavebands were measured continuously from April 1993 to May 1994 in an open area and on a shrubby site. The methodology used is presented by Gauthier [23] and Le Roux et al. [41]. The fraction of the PAR absorbed by the canopy f_{aPAR} was calculated as

$$f_{aPAR} = 1 - \tau - \frac{\alpha}{\rho} + \tau \frac{\alpha_s}{\rho_s}, \qquad (6.4)$$

where τ is the ratio between transmitted and incident radiation, and $\frac{\alpha}{\rho}$ and $\frac{\alpha_s}{\rho_s}$ are the canopy and bare soil albedos for PAR, respectively. Simulations performed with a radiative transfer model were used to compute the fractional PAR absorption by the green vegetation from the observed total f_{aPAR} [41]. In open areas, f_{aPAR} is closely related to LAI (Fig. 6.4). At the shrubby site, two sets of sensors randomly located below the sparse shrub layer (one set above and one set below the grass layer) allowed the calculation of the fractional PAR interception by the shrub and grass canopies [23]. This was used to assess competition for light between the shrub and grass layers as discussed in Sect. 8.2.

6.3.3 Surface spectral signatures

Hemispherical spectral reflectance, transmittance, and absorptance factors of individual green leaves of a dominant grass species at Lamto, i.e., *Andropogon*

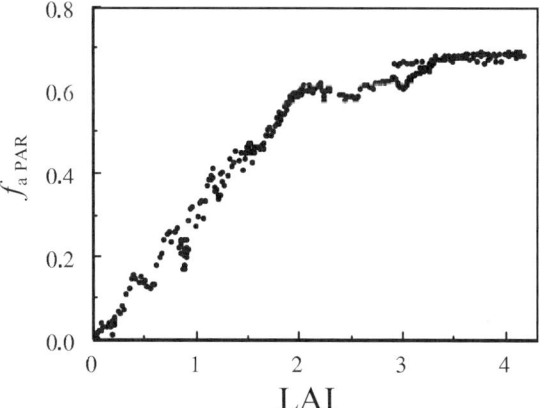

Fig. 6.4. Relationship between the estimated daily PAR absorption efficiency by green leaves f_{aPAR} and the green leaf area index (LAI) of the savanna grass canopy (adapted from [41], copyright (1997), with permission of Elsevier).

macrophyllus, were measured by Bony [10]. The spectral response of this grass resembles typical responses of healthy green leaves obtained for other plant species, with low and high visible transmittance and reflectance factors in the visible and near infrared regions, respectively (Fig. 6.5). During the rainy season, chlorophyll a and chlorophyll b concentrations of green grass leaves are 1.65±0.32 and 0.66±0.14 mg g^{-1} dry matter, respectively (H. Gauthier and X. Le Roux, unpublished; the chlorophyll concentration was determined according to Vernon [86]).

Surface spectral reflectances were measured at ground level throughout an annual cycle with a radiometer simulating the sensors on board SPOT satellites. The methodology used is described by Gauthier [23] and Le Roux et al. [41]. Measurements were performed above the bare soil, grass canopy and shrub canopy on the test site. Near infrared and red reflectance factors were used to compute the normalized difference vegetation index (NDVI). The bare soil spectral reflectance factors are typically around 11%, 15% and 22% for the 500-590, 620-680, and 790-890 nm wavebands, respectively. However, a significant seasonal evolution of soil reflectances is observed during the first months after fire, due to the removal of ash from the soil surface by rainwater (H. Gauthier and X. Le Roux, unpublished). Furthermore, even after the soil surface has been washed, soil reflectances significantly decrease with increasing moisture in the surface soil layer (Fig. 6.6). For soil moistures higher than 4%, two-order polynomials adequately describe this relationship:

$$\text{SRF}_{red} = 15.91 + 0.778 W_{0-2} - 0.093 W_{0-2}^2 \quad (r^2 = 0.75), \qquad (6.5)$$

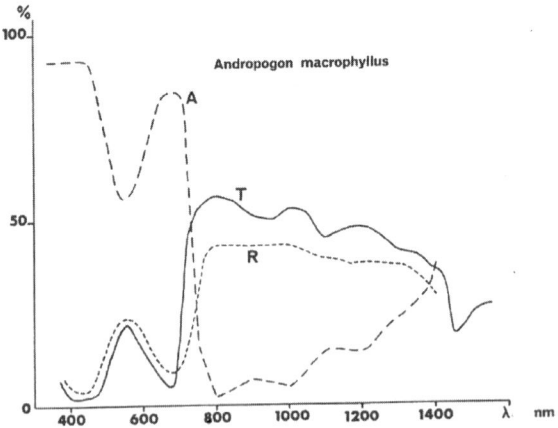

Fig. 6.5. Hemispherical spectral reflectance (R), transmittance (T) and absorptance (A) factors of a green leaf of *Andropogon macrophyllus* (after [10]).

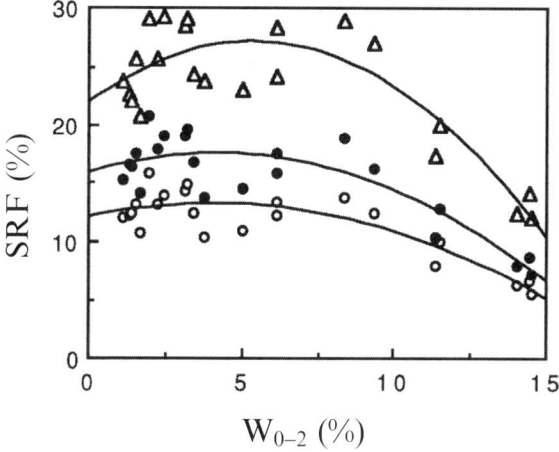

Fig. 6.6. Hemispherical spectral reflectance factor of bare soil (SRF) in the 500-590 nm (open circles), 620-680 nm (solid circles) and 790-890 nm (triangles) wavebands as a function of the gravimetric moisture of the upper 2 cm soil layer W_{0-2} (H. Gauthier and X. Le Roux, unpublished). Fitted lines for the last two channels correspond to equations 6.5 and 6.6. The equation for the first channel is SRF $= 12.03 + 0.575 W_{0-2} - 0.069 W_{0-2}^2$ $\left(r^2 = 0.75\right)$.

$$\mathrm{SRF}_{NIR} = 21.84 + 1.947 W_{0-2} - 0.18 W_{0-2}^2 \ \left(r^2 = 0.77\right), \qquad (6.6)$$

where W_{0-2} is the moisture in the upper 2 cm soil layer (% dry weight), and SRF_{red} and SRF_{NIR} are the soil reflectances (%) in the 620-680 and 790-890 nm wavebands, respectively. Soil reflectances can be assumed to be constant for soil moisture lower than 4%.

The bare soil NDVI is around 0.25 and relatively constant throughout the year (Fig. 6.7), exhibiting slight variations due to the influence of fire and surface soil moisture. The grass canopy NDVI strongly increases during the growing season from bare soil values to approximately 0.7. A decrease of the grass canopy NDVI is observed at the end of the cycle during the dry season and senescence. In contrast, shrubby canopies exhibit NDVI higher than 0.6 40 days after fire when most leaves have developed. Shrub NDVI values only slightly decrease during the dry season, although a high variability of the signal is expected due to the heterogeneity of the leaf fall pattern (see Sect. 5.5). The relationship between the grass canopy NDVI and its fractional coverage C is adequately described by the following equation:

$$\mathrm{NDVI} = C\,\mathrm{NDVI}_g + (1 - C)\mathrm{NDVI}_s \ \left(r^2 = 0.69;\, n = 23\right), \qquad (6.7)$$

where NDVI_g is the asymptotically limiting value of NDVI for an infinitely thick grass canopy (0.83), and NDVI_s is the bare soil NDVI (0.24).

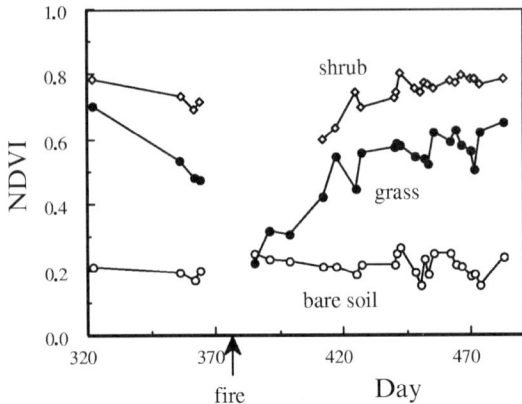

Fig. 6.7. Seasonal variations of the normalized difference vegetation index (NDVI) of the bare soil, and of the grass and shrub savanna canopies (H. Gauthier and X. Le Roux, unpublished). Day 366 is 1 January 1994.

6.3.4 Scientific gains and gaps in the knowledge

As part of the savanna-atmosphere interactions, radiation exchanges are probably the best-documented processes at Lamto. Comprehensive information acquired along with a good description of both the grass canopy structure and the shrub layer structure (see Chap. 5 and [23]), offers a unique opportunity to generalize to the heterogeneous savanna ecosystem using these point observations. Considerable modeling effort has recently been directed at this goal (see Chap. 9). In addition, information about the spectral signature of the bare soil, leaves and canopy, along with the comparison between satellite- and ground-based signatures, provides the basis on which satellite remotely sensed data can be used with some confidence at Lamto. However, the influence of aging and inter-specific variability on the spectral properties of leaves remains to be documented.

6.4 Energy budget

6.4.1 Seasonal variations in the components of the savanna energy budget

The energy budget of the savanna can be written as

$$R_n = G + H + \lambda E, \tag{6.8}$$

where R_n is the net radiation flux density, G is the soil heat flux density, H is the sensible heat flux density and λE is the latent heat flux density (Wm^{-2}).

Energy partitioning was measured on a 500 × 500 m open area savanna site by the Bowen ratio-energy budget method that uses robust sensors allowing continuous measurements in remote sites [22]. The methodology and instruments used are detailed by Le Roux [37]. The existence of a constant flux layer above the grass canopy height was documented [37]. Data collected between March 1991 and May 1992 provided 143 complete diurnal cycles of the energy budget components (e.g., Fig. 6.8). In addition, complete net radiation and soil heat flux data were obtained for 86 additional days.

Maximum values of net radiation are observed in April (Fig. 6.9) when the atmospheric transmission is maximal; R_n is low during the dry season due to low atmospheric transmission and low downward longwave radiation at this time (see Sect. 3.5). During the growing season, the soil heat flux to net radiation ratio $\frac{G}{R_n}$ is linearly related to the vegetation fractional coverage C:

$$\frac{G}{R_n} = 0.30 - 0.22C \quad (r^2 = 0.78). \tag{6.9}$$

The seasonal variations of actual evapotranspiration were parameterized by the equilibrium evaporation and the soil water availability. Equilibrium evaporation rate λE_{eq} is used by many micrometeorologists as an energy-driven reference evaporation rate and is defined as follows:

$$\lambda E_{eq} = \left(\frac{\Delta}{\Delta + \gamma}\right)(R_n - G), \tag{6.10}$$

where Δ is the derivative of the air saturation water vapor pressure with respect to air temperature $(\mathrm{Pa\,K^{-1}})$ and γ is the psychrometric constant $(\mathrm{Pa\,K^{-1}})$. The actual to equilibrium evaporation diurnal ratio $\frac{E}{E_{eq}}$ increases from 0.07 in January to 0.9 during the rainy season when the canopy is closed

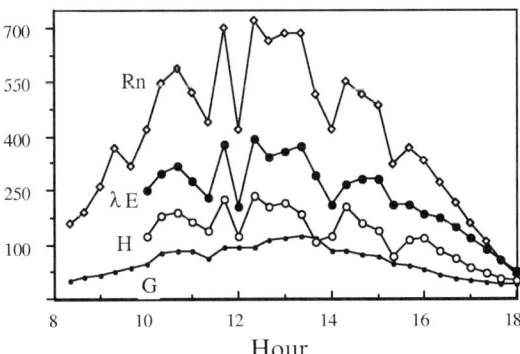

Fig. 6.8. Diurnal variations of the savanna energy budget components on 4 May 1992: The 20 min mean values of net radiation R_n, soil heat flux G, sensible heat flux H and latent heat flux λE are presented (from [37]).

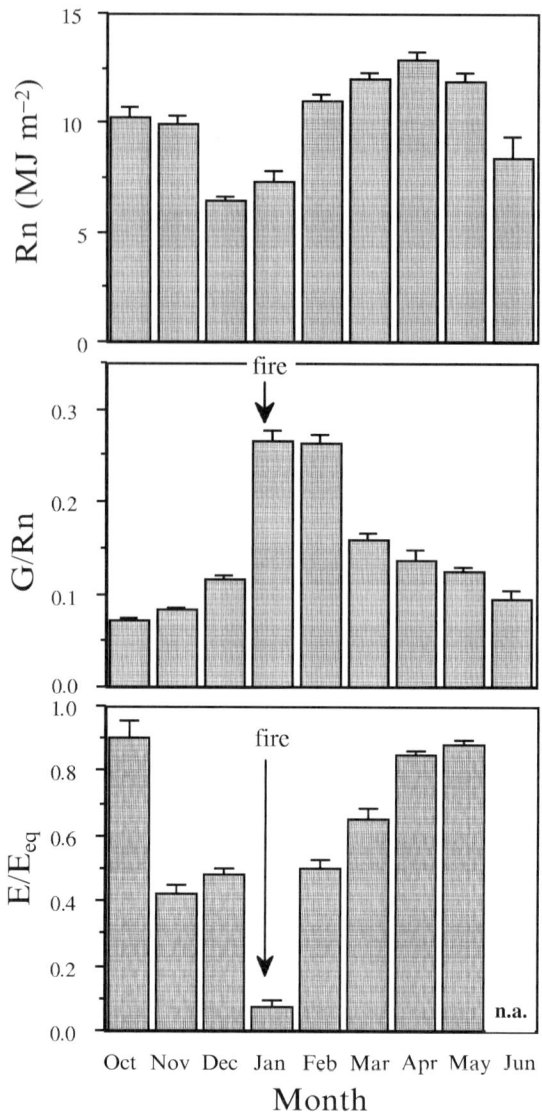

Fig. 6.9. Seasonal variations of the savanna diurnal energy budget: Monthly mean of net radiation R_n, soil heat flux to net radiation ratio $\frac{G}{R_n}$ and actual to equilibrium evaporation ratio $\frac{E}{E_{eq}}$ are presented. Bars are standard errors (after [37]).

(Fig. 6.9). Lower values are recorded at the end of the year due to water shortage. When the canopy is closed, the ratio of actual to equilibrium evapotranspiration is correlated to the water content W of the 0-60 cm soil layer (Fig. 6.10). With sufficient water availability ($W > 60$ mm), $\frac{E}{E_{eq}}$ is around

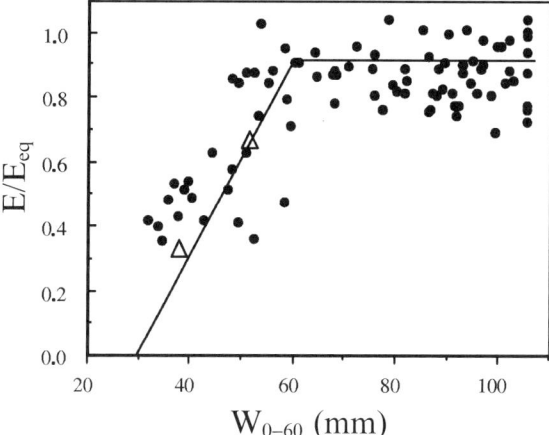

Fig. 6.10. Relationship between the savanna actual to equilibrium evaporation ratio $\frac{E}{E_{eq}}$ and the water content of the 0-60 cm soil layer W_{0-60} for a closed canopy (LAI>1.25). Values computed from micrometeorological data (dots) and from soil moisture survey by a neutron probe during a period without drainage (triangles) are presented (after [37]).

0.9. Under this threshold, $\frac{E}{E_{eq}}$ decreases linearly with W. Given the soil characteristics at the study site, the critical W value corresponds to a ratio of actual to maximum soil water content available for plants equal to 0.39 for the 0 to 60 cm layer [37].

The absence of isotopic ($\delta^{18}O$, $\delta^{2}H$) fractionation for leaf water during the night [39] suggests that the nocturnal transpiration flux is negligible. This could be explained by air water vapor deficits always close to zero, occasional dew occurrence and very low wind speeds at canopy level at this time. Le Roux [37] estimated that about 9% of the mean diurnal net radiation would be needed to evaporate dew.

The partitioning of net radiation into latent and sensible heat fluxes exhibits a marked seasonality, in contrast with results reported for an evergreen tropical forest that experiences neither severe drought nor leafless stages (Fig. 6.11). However, this seasonal trend is close to that reported for a deciduous forested canopy in the southern Ivory Coast by Monteny [54]. Despite a low green LAI at the Lamto savanna site (2 to 3) compared to the LAI at the deciduous and evergreen forest sites (4.5 according to Monteny [54] and 6.6 according to Roberts et al. [64]), the maximum $\frac{\lambda E}{R_n}$ ratio recorded at Lamto approaches values experienced by moist forests, i.e., around 0.7 (Fig. 6.11). This value corresponds to a $\frac{E}{E_{eq}}$ ratio around 0.92. This result sustains the emerging opinion that tropical forests and grass-dominated vegetation could exhibit similar maximum evapotranspiration rates, as anticipated for temperate coniferous forests and grasslands by Kelliher et al. [30].

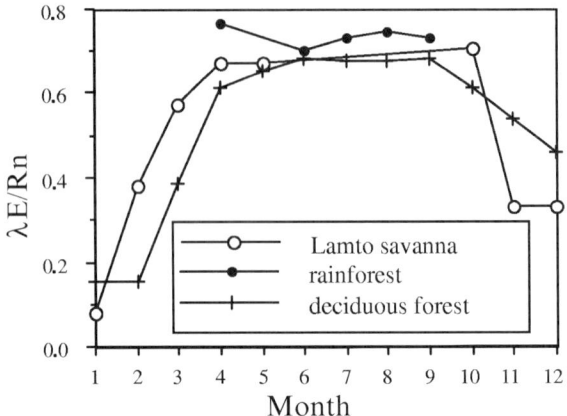

Fig. 6.11. Seasonal variations of the ratio between daily (24 h) latent heat flux and net radiation $\frac{\lambda E}{R_n}$ for the Lamto savanna, for an Amazonian rainforest and for a deciduous tropical forested canopy in the Ivory Coast. Seasonal variations were computed from the 1983-1984 ARME database [76] and from data of Monteny [54] for the rainforest and the deciduous forested canopy (after [37]).

6.4.2 Aerodynamic and surface resistances, and savanna-atmosphere coupling

Among the wide range of existing canopy evapotranspiration models [77], the most versatile one, i.e., the Penman-Monteith equation [53], was used to analyze the control of the savanna evapotranspiration rates by biological and micrometeorological parameters. This equation reduces the representation of the canopy to a big leaf and subsumes the evapotranspiration process into two driving variables (available energy $R_n - G$ and air water vapor deficit D_a) and two resistances (aerodynamic r_a and surface r_s resistances). Calculation of r_s is detailed by Le Roux [37]. The surface resistance was calculated by inversion of the Penman-Monteith equation:

$$\lambda E = \frac{\Delta (R_n - G) + \rho c_p \frac{D_a}{r_a}}{\Delta + \gamma \left(1 + \frac{r_s}{r_a}\right)}, \qquad (6.11)$$

where ρ is the air density (kg m^{-3}), c_p is the air specific heat at constant pressure (J kg^{-1} K^{-1}), D_a is the air water vapor pressure deficit at the reference level (Pa) and r_a and r_s are the aerodynamic and surface resistances (s m^{-1}), respectively. In order to assess the degree of coupling between the vegetation and the atmosphere, equation 6.11 can be re-expressed as [29]

$$\lambda E = \Omega \lambda E_{eq} + (1 - \Omega) \lambda E_{imp}, \qquad (6.12)$$

where Ω is the decoupling coefficient, and λE_{imp} is the evaporation rate imposed by the effects of air saturation deficit:

$$\Omega = \left(1 + \frac{\gamma}{\Delta + \gamma} \frac{r_s}{r_a}\right)^{-1}, \qquad (6.13)$$

$$\lambda E_{imp} = \frac{\rho c_p D_a}{\gamma r_s}. \qquad (6.14)$$

Over aerodynamically smooth canopies, Ω approaches 1, resulting in energy-driven λE rates close to λE_{eq}. Over aerodynamically rough canopies, Ω approaches 0, resulting in saturation deficit-driven λE rates close to λE_{imp}. Since the use of the Penman-Monteith equation is hampered, among other factors, by its inability to predict λE over sparse canopies, only the results obtained when the canopy was closed are presented. For the well-watered grass-dominated savanna canopy in May, surface resistance was typically around 130 s m^{-1} during the diurnal period (10:00-17:00), while aerodynamic resistance increased from 40 to 70 s m^{-1} when wind speed decreased from 2 to 1 m s^{-1}. Thus, the generally low reference wind speeds (see Sect. 3.5) and the possible shelter effect from trees upwind explain the high r_a values observed at Lamto. The decoupling factor Ω ranged from 0.59 to 0.73 for wind speed ranging from 2 to 1 m s^{-1}, suggesting a weak vegetation-atmosphere coupling and the prime importance of the energy availability on evapotranspiration rates in open areas. This is consistent with the fact that evapotranspiration from grass-dominated open areas within shrubby savanna sites occurs at a rate close to the equilibrium rate and exhibits little dependence on the saturation deficit. Such a decoupling was reported for a windbreak-sheltered kiwifruit orchard by McAneney et al. [48]. One consequence of this weak vegetation-atmosphere coupling is that transpiration from open areas should be relatively insensitive to stomatal regulation in conditions of sufficient water availability.

6.4.3 Scientific gains and gaps in the knowledge

To our knowledge, the only reliable studies on evapotranspiration in the humid savanna zone were carried out in neotropical savannas, on a Cerrado site in Brazil [47, 52] and on Llanos in Venezuela [68, 69]. Comprehensive information on evapotranspiration from dry savanna in Niger was obtained part of the HAPEX-Sahel experiment [55]. The results obtained at Lamto provide an extensive database useful for (1) understanding the role of microclimatic and biological controls on vegetation-atmosphere interactions in grassy areas of West African humid savannas and (2) assessing how seasonality affects their energy budget. Along with concurrent measurements of the radiation budget, soil moisture, phenology and patterns of water withdrawal by roots (Table 6.2), this database provides a unique opportunity to acquire a comprehensive view of the functioning of grass-dominated surfaces in this environment and to develop and/or test ensuing models.

However, the energy budget was studied only in an open grassy area. Although the soil water balance was surveyed in open and shrubby areas (see hereafter), transpiration rates of trees are not documented at Lamto. Thus, direct measurements of transpiration from the dominant woody species (e.g., using the sap-flow method) are warranted. Furthermore, the energy budget of a shrubby savanna (that is a heterogeneous ecosystem) cannot be properly squeezed out of few point flux data acquired for shrubby and open areas by a simple summation process. Indeed, one of the most important features of energy exchanges in savanna-like ecosystems is the dependence of the microclimatic driving variables on the momentum and energy exchanges between the vegetation and the atmosphere. This dependence implies spatially distributed feedbacks between the atmosphere and the vegetation function and ensuing fluxes. Thus, both the structural characteristics of the savanna (i.e., the architecture of the tree-grass association) and the physiological properties of the dominant tree and grass species should be taken into account to infer evapotranspiration rates and, moreover, the sensitivity of evapotranspiration to changes in stomatal resistance, environmental variables, and vegetation physiognomy. These requirements spur on current advances in data acquisition (a field campaign for documenting leaf gas exchanges and sap flow rates of major tree species in Lamto has been launched in 2000 and 2001) and the design of models representing savanna-atmosphere interactions (see Chap. 9).

6.5 Water balance and plant water status

The savanna water balance can be written as

$$P = E + R + D + \Delta W, \tag{6.15}$$

where P is precipitation (including dew fall), E is actual evapotranspiration (including interception loss), R is runoff, D is the drainage below the rooting zone and ΔW is the change of stored water in the soil profile (in mm). The sub-surface lateral flow is neglected. Along with soil characteristics, these processes control the temporal courses of the soil water availability and plant water status, which strongly determine the savanna functioning (i.e., periods favorable to growth and mineralization, plant germination and mortality rates, amount and dryness of above ground matter at the end of the dry season and, thus, fire intensity).

6.5.1 Interception loss

Interception loss includes the water evaporated during rainfall and the water held on the canopy that evaporates after rainfall. Remaining water reaches the soil as throughfall and stemflow. No stemflow data are available. However, throughfall was measured using 2 m × 0.15 m troughs on *Hyparrhenia* and *Loudetia* savanna sites from September 1971 to December 1971 [87]. During this period, throughfall accounted for 70% of precipitation at both sites.

Throughfall below the grass canopy (including standing dead matter) was also measured in an open area during a 18-month period by twelve 80 cm^2 collectors randomly distributed at the ground level [37]. For low rainfall depths and a high LAI, stemflow is assumed to be negligible. For a rainfall depth less than 14 mm, the difference between precipitation and throughfall P–T is related to the precipitation amount and the LAI (Fig. 6.12). For 12 mm of rainfall, P–T ranges from 2.5 to 6 mm for increasing LAI. The interception loss-precipitation relationship obtained for a LAI ranging from 3 to 6 is close to that obtained by Butler and Huband [11] for a wheat crop. These results show that interception loss is probably a significant component of the grass canopy water balance.

In the riparian forest, throughfalls measured by five 26 cm diameter gauges accounted for 76% of annual precipitation (1084 mm) in 1971 [87]. However, stemflow was not measured by the authors. Interception by savanna trees has not been studied at Lamto.

6.5.2 Runoff and infiltration

Along with precipitation and soil moisture, runoff R was studied for two soil types (a ferrugineous soil on a shrubby savanna site, and a sandy soil downslope on a *Loudetia* grassland site) at Lamto by De Jong [14] from February to April 1982. On each site, slope was approximately 5%, and measurements were carried out for areas covered by grasses, trees or the bare soil. Runoff plots of 4.7 m^2 were used. In the open area at the shrubby savanna site, surface runoff can be described by the following function:

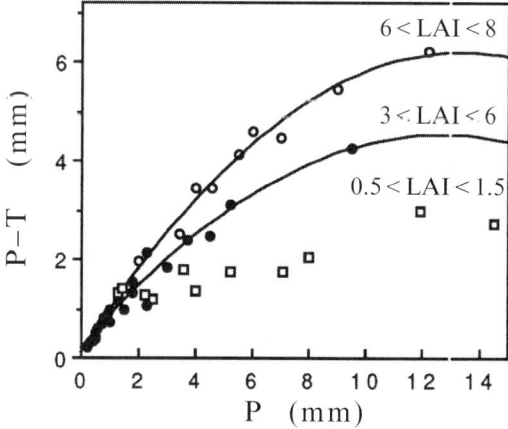

Fig. 6.12. Relationship between gross precipitation minus throughfall (P–T) and gross precipitation P above the grass canopy for increasing total (green plus dead) LAI (X. Le Roux, unpublished).

$$\begin{cases} P < P_0 : R = 0 \\ P \geq P_0 : R = a\,(P - P_0) \end{cases} \quad \left(r^2 = 0.66\right), \tag{6.16}$$

where the threshold precipitation value P_0 for surface runoff generation equals 22 mm, and α is equal to 0.139. Thus, surface runoff is a quantitatively minor term of the water balance in shrubby savanna areas upslope at Lamto. This is consistent with the absence of runoff on plateau reported for a Guinea-savanna dominated catchment near Lamto [35]. Higher runoff rates can be observed for hydromorphic soils downslope. During the rainy season, these normally porous soils become impermeable when the groundwater table rises to the surface.

Low runoff rates reported upslope are explained by high soil hydraulic conductivities at saturation (ca. 860 cm d^{-1} for the 0–40 cm layer according to De Jong [14]) and, in particular, by the influence of vegetation and soil fauna on soil macroporosity.

6.5.3 Soil moisture

The relationship between soil matrix potential and soil water content was established in the laboratory for different layers of the dominant soil types at Lamto by Athias [5], Lavelle [36], Vannier [84], Le Roux [37] and Konaté et al. [32]. Results reported for the 0–10 cm layer on shrubby savanna sites on plateau are illustrated in Fig. 6.13. Although the concepts of soil water content

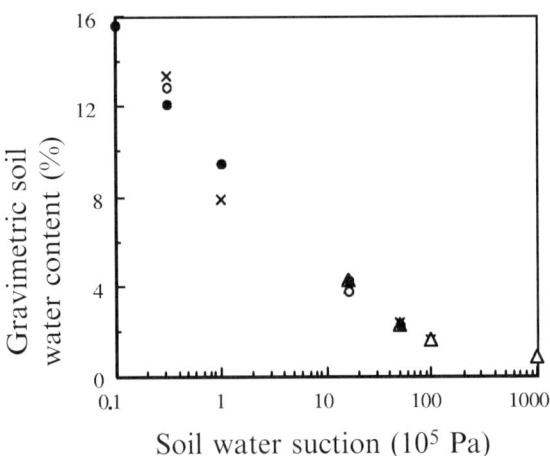

Fig. 6.13. Water retention curves reported for the 0–10 cm soil layer on shrubby savanna sites on plateau by Vannier [84] (triangles), Lavelle [36] (dots), Athias [5] (x's) and Le Roux [37] (circles). All the data are adequately described by Gardner's equation: $y = 635.7x - 2.970$ $\left(r^2 = 0.98\right)$.

Table 6.3. Soil water contents of the 0–60 and 60–170 cm soil layers at the field capacity (W_{fc}) and at the permanent wilting point (W_{wp}) calculated from in situ neutron probe observations or from laboratory measurements (in mm). The maximum available water content AW is calculated as the difference between these two values (according to Le Roux and Bariac [39]). *: the permanent wilting point may not be reached in the deeper layers, and AW may thus be underestimated.

	W_{fc}		W_{wp}		AW	
Depth	In situ	Lab	In situ	Lab	In situ	Lab
0–60 cm	104.6	109.4	30.9	31.1	73.7	78.3
60–170 cm	185.0	—	120.9*	—	64.1*	—

at the field capacity (i.e., the water content after excess water has drained from the soil profile) and the permanent wilting point (i.e., the residual water content unavailable for plants) are physically and biologically ill-defined, they are widely used in empirical water balance models. Values reported for a shrubby savanna site on a plateau are presented in Table 6.3. The maximum water content available for plants is ca. 75 mm in the 0-60 cm soil layer. This value is close to values reported for three locations in the Ivory Coast (64, 69, and 76 mm for the same soil layer according to Amadou and Yao [3], Roose [65], and Yao et al. [88], respectively).

The seasonal variations of soil moisture were investigated on 6 sites located along a catena from April 1967 to April 1968 by Bonvallot [9]. Soil moisture was surveyed on 5 of these sites from December 1968 to December 1969 by Lavelle [36]. During the first year, the maximum measurement depth ranged from 80 cm at a gallery forest boundary to 200 cm in a shrubby savanna and 210 cm in a *Loudetia* savanna. Based on measured values of soil moisture at field capacity and wilting point, the seasonal course of water availability was determined on each site [9]. No water was available at the depth of 20 cm during 90, 15 and 15 days on the shrubby savanna, gallery forest and *Loudetia* savanna sites, respectively. Extensive monitoring of soil moisture was carried out on burnt and unburnt savanna sites during an 18-month period by Athias [5]. Variations of soil moisture occurred more rapidly at the burnt savanna site. Recently, soil moisture was measured down to 30 cm deep [57], 170 cm deep [39], and 60 cm deep [79] in open areas and under tree clump canopies in shrubby savanna sites. The seasonal variations of soil moisture are generally not significantly different between open areas and beneath shrub clumps (see Sect. 8.2).

6.5.4 Drainage

Vertical drainage was studied from 1970 to 1972 by Villecourt and Roose [87] on a shrubby savanna site upslope. Three lysimeters (30 cm in diameter and 80 cm deep) with a vegetated ferrugineous soil were used. Only few drainage occurrences were observed. Annual drainage accounted for 5% of precipitation in 1971 (P=1089 mm) and 16% in 1972 (P=1324 mm). Based on a simple water balance approach, annual vertical drainage was estimated to range from 0 to 500 mm for the 1962 to 1971 period, with a mean value around 130 mm, i.e., 10% of precipitation [87]. Sub-surface lateral flow was found to be negligible (annual value of 0.034%) for a Guinea shrubby savanna site with a 4% slope by Avenard and Roose [7].

6.5.5 Soil water uptake by roots

The pattern of soil water withdrawal by grass or shrub roots was investigated by comparing the isotopic composition (^{18}O and ^{2}H) of soil water and grass or tree xylem sap water under rainy and dry conditions [40] (see Sect. 8.3) and by correlating the seasonal variations of soil moisture and grass or shrub water potential [39]. Both grasses and shrubs acquire most of their water in the upper soil layers during the rainy season, which is consistent with their shallow rooting patterns (Sect. 8.3). However, during dry periods, deciduous shrub species can exhibit different water uptake patterns: *Cussonia arborea* can extract water from deeper soil layers while *Crossopteryx febrifuga* has only limited access to deep soil layers [39]. Grasses essentially use water from surface soil layers, and a shift of their water uptake pattern toward deeper horizons is observed only during very dry conditions (Fig. 6.14). The functioning of tree and grass root systems is detailed and its ecological implication is discussed in Chap. 8.

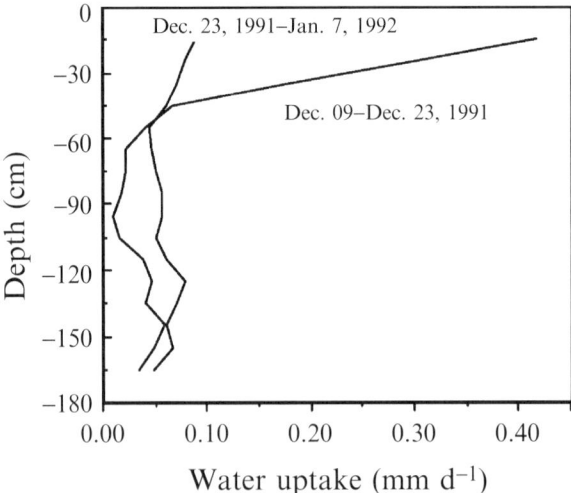

Fig. 6.14. Profile of soil water uptake by roots in open areas during two drying stages at the end of the dry season. Soil water content in the 0-60 cm layer was 61, 42 and 34 mm on 9 December, 23 December, and 7 January, respectively (after [40], with kind permission of Springer Science and Business media).

6.5.6 Plant water status

Leaf and twig water potentials were surveyed concurrently with moisture of the 0-60 cm soil layer during a two year period for the grass *Hyparrhenia diplandra* and the trees *Crossopteryx febrifuga* and *Cussonia arborea* [39]. The seasonal changes of predawn and minimum shoot water potentials were pronounced for the grass species and the shrub *Crossopteryx*, but weak for the shrub *Cussonia* (Fig. 6.15). Both the grass species and *Crossopteryx* did

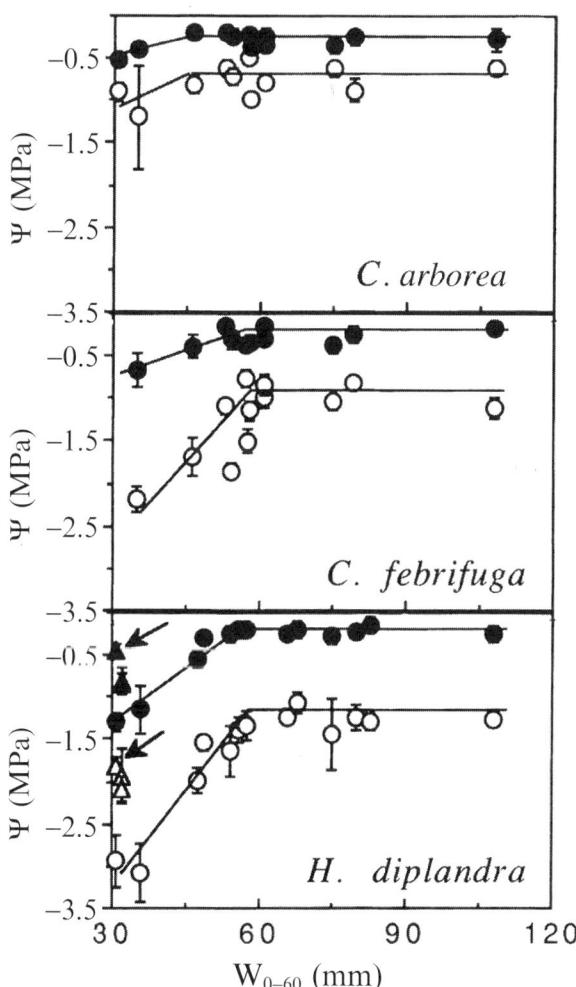

Fig. 6.15. Predawn and minimum shoot water potentials Ψ_l of the shrubs *Cussonia arborea* and *Crossopteryx febrifuga* and the grass *Hyparrhenia diplandra*, as a function of the water content W_{0-60} of the 0-60 cm soil layer. Bars are 95% confidence intervals (after [39], with kind permission of Springer Science and Business media).

not maintain high water potentials when soil moisture in the upper soil layers decreased. In contrast, *Cussonia* exhibited predawn water potential around −0.5 MPa when the 0-60 cm soil layer was at the permanent wilting point. This is due to the different water uptake patterns exhibited by these species (see above).

Plant water relations of these three species were characterized during rainy and dry periods with the pressure-volume curve technique [39]. Relatively low osmotic potential and tissue elasticity measured for *Crossopteryx* as compared to *Cussonia* were consistent with the use of water from upper soil layers by this species, since these characteristics could help water withdrawal and promote turgor maintenance at lower shoot water potentials. Different water uptake patterns could explain the contrasted leaf shedding patterns reported for the two shrub species [51]. These results emphasize that the water economy of deciduous shrub species of African humid savannas can differ significantly and that both grasses and some shrub species acquire water in the upper soil layers even during dry spells.

6.5.7 Scientific gains and gaps in the knowledge

Knowledge of water flows in the soil-plant-atmosphere continuum at Lamto allows the assessment of the relative importance of each component of the savanna water balance. Runoff appears as a negligible component on plateau, since only downslope hydromorphic soils can generate substantial runoff rates during the rainy season. Drainage below the rooting zone is also low on the plateau. Thus, evapotranspiration is expected to be a component of prime importance in the annual savanna water balance at least on the plateau and upslope. The components of the water balance of Lamto savanna, as estimated by a modeling approach, are presented in Chap. 9.

However, results dealing with interception loss, runoff and sub-surface water flows provide, at best, a basis for the empirical treatment of these processes. Such treatment can be criticized since major variables (e.g., rainfall intensity and duration, interval between rainfall events) are not taken into account. A further criticism is that extrapolation of purely empirical equations is a perilous exercise. Studies of the physical bases of the above-mentioned processes and associated parameters such as canopy water retention capacity, throughfall and stemflow fractions, soil hydraulic and thermal conductivities are thus warranted before physically based water balance models can be properly applied at Lamto.

Furthermore, information on processes associated with each water balance component acquired on plots ranging from 1 m^2 to a few hectares does not provide a sufficient basis to upscale these processes to the landscape or catchment scale, because, at this scale, water flow pathways critically depend on hillslope morphology and the spatial variability of rainfall inputs, soil characteristics and vegetation types. Water flows at the catchment scale have not been studied at Lamto. However, results obtained at the Sakassou catchment, that lies 30 km away from Lamto and shares the same morphology, soils and vegetation types [35], can be generalized to the Lamto region. This study reveals the key role of downslope hydromorphic soils in runoff generation at the catchment scale. Furthermore, the water table rising on plateau during the rainy season appears to be controlled by a deep lateral supply from waterlogged areas downslope rather than by a water supply by vertical percolation on plateau.

6.6 CO_2 exchanges and leaf conductance

Examining leaf and canopy CO_2 assimilation rates provides a basis for understanding the surprisingly high primary productivity of the Lamto savanna outlined in Chap. 7. A number of field and laboratory studies have been conducted on leaf photosynthesis of savanna grass or tree species. However, data on CO_2 assimilation rates of savanna canopies and their seasonal variations is very scarce [66], thus restricting the validation of savanna productivity process-based models. This section presents an overview of information dealing with soil-plant-atmosphere CO_2 exchanges acquired at Lamto, at the leaf and canopy levels.

6.6.1 At the leaf level

At Lamto, most savanna grass species possess the C_4 photosynthetic pathway. For well-watered plants, at high irradiance, moderate temperatures, moderate vapor pressure deficits, and ambient air CO_2 concentration (360 ppm), C_i is about 130 ppm. C_4 photosynthesis, among many other features [60], may allow for a higher photosynthetic efficiency in the moist, high temperature environment experienced in the West African humid savanna zone.

Leaf photosynthetic rates were reported for the grass *Hyparrhenia diplandra* by Mordelet [56], Dignac [17], Le Roux and Mordelet [43] and Sueur [83]. According to leaf photosynthesis-photosynthetic photon flux density(PPFD) relationships established in situ for the youngest fully expanded leaves (Fig. 6.16), quantum yield is 0.045 under full-light or shaded conditions [56]. This value is close to values reported for savanna grasses (see references in [43]; [4]) but relatively low for C_4 species [18]. The PPFD compensation point equals 60 μmol m^{-2} s^{-1} and 30 μmol m^{-2} s^{-1} under full-light and shaded conditions, respectively. Light saturation of leaf photosynthesis starts above 1000 μmol m^{-2} s^{-1}. Maximum photosynthesis rates ranged from 18 to 33 μmol CO_2 m^{-2} s^{-1}. These values are close to the maximum values reported for humid or mesic savanna grass species (see references in Le Roux and Mordelet [43]).

The variability of leaf photosynthetic capacity between species or induced by aging, microclimatic conditions or genetic background was assessed. Photosynthetic rates of the other major grass species *Andropogon schirensis* are

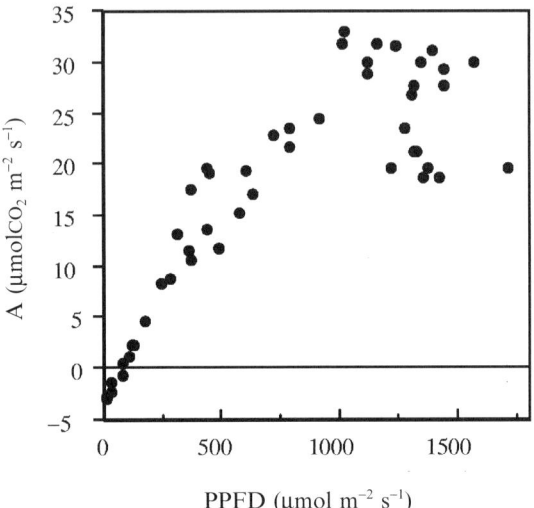

Fig. 6.16. Relationship between leaf net CO_2 assimilation rate A and photosynthetic photon flux density PPFD. Measurements were performed in situ on youngest fully expanded leaves of *Hyparrhenia diplandra* during the early growing season (redrawn after [43], with permission of Cambridge University Press).

not significantly different from those exhibited by *Hyparrhenia diplandra* (P. Mordelet, unpublished). Photosynthetic rates do not differ significantly between different pools of inbred full-siblings grown under greenhouse conditions [83]. However, a stratification of leaf photosynthesis with height is observed in situ within the canopy [43]. This could be due partly to aging, as documented by Sueur [83], and partly to light availability within the canopy according to the carbon gain optimization theory [28]. Furthermore, the heterogeneity of light availability imposed by the discontinuous shrub canopy leads to a subsequent high heterogeneity in CO_2 gain by grasses (see Chap. 8).

The major savanna grass species at Lamto may be considered as hypostomatous species. For *Hyparrhenia diplandra*, typical abaxial and adaxial stomatal densities are 325 and 50 stomata mm^{-2}, respectively [83]. This corresponds to a relatively high total stomatal density when compared to values reported for other Gramineae [31]. For well-watered plants and at high irradiance, the stomatal conductance to water vapor measured in situ ranges from 100 to 950 $mmol\,m^{-2}\,s^{-1}$ (i.e., stomatal resistance ranging from 40 to 400 $s\,m^{-1}$ [56]). Significant variations of stomatal conductance (135 to 270 $mmol\,m^{-2}\,s^{-1}$, i.e., resistance ranging from 150 to 300 $s\,m^{-1}$) were observed between greenhouse-grown family lines by Sueur [83]. The in situ assimilation rate is linearly related to stomatal conductance for conductance values below 600 $mmol\,m^{-2}\,s^{-1}$ [56]. For photosynthetic rates around 30 $\mu mol[CO_2]\,m^{-2}\,s^{-1}$ and air water vapor pressure deficit ranging from 3.3 to 4.3 kPa, the in situ water use efficiency ranges from 2.3 to 2.9 μmol $[CO_2]$ mmol^{-1} $[H_2O]$ (P. Mordelet, personal communication). For plants grown under greenhouse conditions, water use efficiency (measured at 32°C, 1250 μmol m^{-2} s^{-1}, and air water vapor pressure deficit around 2.5 kPa) was ca. 3 μmol $[CO_2]$ mmol^{-1} $[H_2O]$ [83].

Recently, Simioni et al. [81] parameterized two biochemically based photosynthesis models for C_3 species [20] and C_4 species [12]. Parameter values were derived for two major grass (C_4) species (*Hyparrhenia diplandra* and *Andropogon schirensis*) and two dominant tree (C_3) species (*Cussonia arborea* and *Crossopteryx febrifuga*). Relationships between photosynthetic capacity and leaf N per unit area, N_a were determined for each species (Fig. 6.17). The spatial and temporal changes in N_a were also reported for the two tree species [79]. The authors showed that maximum carboxylation rate was higher for *Cussonia arborea* than for *Crossopteryx febrifuga*, except at the end of the vegetation cycle [79].

One of the most striking features of leaf CO_2 exchanges at Lamto is that savanna grasses maintain high photosynthetic capacities despite their very low nitrogen concentration (typically 0.8% to 1.1% in situ, see Chap. 14). The widely recognized influence of leaf nitrogen on leaf photosynthesis [21] was quantified for greenhouse-grown individuals on a leaf mass basis (Fig. 6.18). The relationship obtained is consistent with that obtained for humid savanna grasses by Baruch et al. [8]. On a leaf area basis, the following relationship was obtained:

$$A_{max} = 0.521N - 6.97 \quad (r = 0.50), \tag{6.17}$$

where N and A_{max} are expressed in mmol m^{-2} and μmol m^{-2} s^{-1}, respectively. The corresponding nitrogen use efficiency values are consistent with

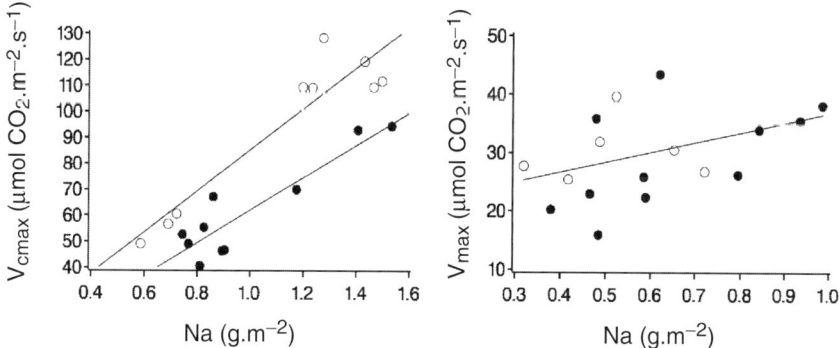

Fig. 6.17. Variations of the maximum carboxylation rate with leaf nitrogen content on an area basis N_a for the main tree and grass species at Lamto: Left, trees: *Crossopteryx febrifuga* (solid circles) and *Cussonia arborea* (open circles). Right, grasses: *Andropogon canaliculatus* (solid circles) and *Hyparrhenia diplandra* (open circles). Lines represent regression lines significant at the 0.05 level, per species for trees, and pooled for grasses (after [81], with kind permission of Springer Science and Business media).

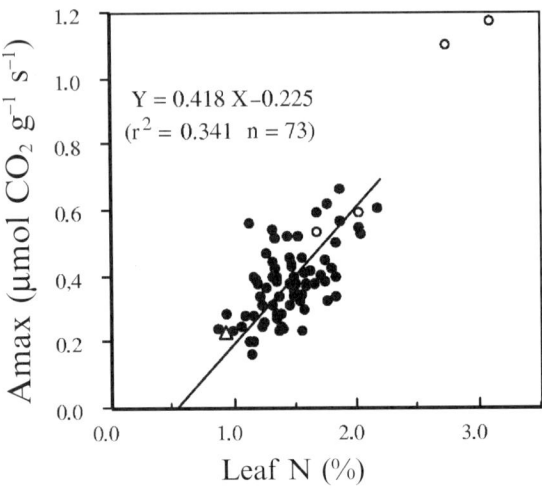

Fig. 6.18. Relationship between leaf net CO_2 assimilation rate at saturating light (1250 μmol m^{-2} s^{-1}) A_{max} and leaf nitrogen concentration N, both expressed per unit of leaf mass (X. Le Roux and J. Sueur, unpublished). Data (solid circles) were obtained for 73 leaves of different ages from 19 greenhouse-grown *Hyparrhenia diplandra* individuals. CO_2 assimilation rates were measured with a ADC-LCA2 system, leaf area with an optical leaf area meter (DT Devices) and leaf nitrogen concentration by gas chromatography (Fisons NA 1500N). Data reported for humid savanna C_4 grasses (open circles [8]) and for *Hyparrhenia diplandra* in situ at Lamto (triangle [43]) are presented.

those reported for C_4 grasses of seasonal savannas by Baruch et al. [8] and Anten et al. [4]. Such values are much higher than the values reported for C_3 species (reviewed by Evans [19]) and close to maximum values observed for C_4 species [61, 67, 46, among others]. The high sensitivity of leaf photosynthesis on leaf nitrogen content makes nitrogen a potentially major limiting factor for the primary productivity of this ecosystem (see Chap. 14).

6.6.2 At the canopy level

Total soil respiration (defined here as soil and root respiration plus respiration of the tussock base and standing dead matter) is very high throughout the rainy season. Values typically range from 5 to 9 μmol[CO_2] m^{-2} s^{-1} with no discernible daily trend [17, 43]. Lower values (4.0±0.6 μmol[CO_2] m^{-2} s^{-1}) are reported at the begining of the dry season [43]. These values are consistent with soil respiration rates reported in an oil palm plantation in southern Bénin (Lamade, personal communication): 3 to 5 μmol[CO_2] m^{-2} s^{-1} for relatively dry soils and 4 to 10 μmol[CO_2] m^{-2} s^{-1} for wet soils.

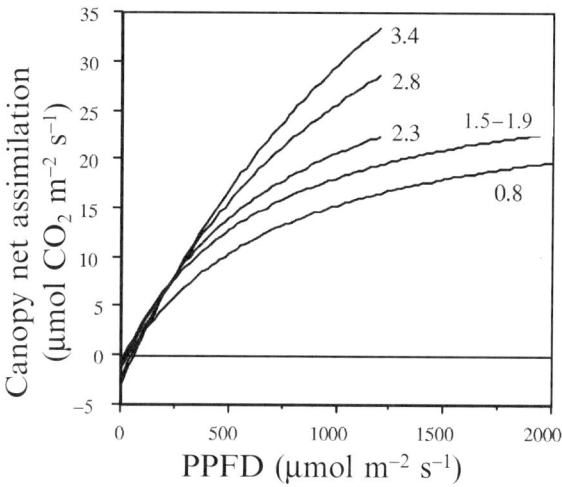

Fig. 6.19. Relationships between the savanna grass canopy net CO_2 assimilation (ground area basis) and photosynthetic photon flux density PPFD at five phenological stages indicated by green LAI values: 0.8, and 1.5 to 1.9 after [43]; 2.3, 2.8 and 3.4 after [17]. Measurements were made in conditions of sufficient soil water availability.

The grass canopy net photosynthesis measured by a chamber system (Fig. 6.19) is adequately described by a rectangular hyperbola for each phenological stage (for parameter values, see [17, 43]). Le Roux and Mordelet [43] compared the maximum canopy net photosynthesis reported for different biomes. CO_2 assimilation rates of the Lamto savanna grass layer are close to those of temperate C_4 grasslands and to those obtained for C_4 crop canopies whose nutrient status is generally higher. Moreover, the Lamto savanna exhibits relatively high photosynthetic rates when compared to tropical rainforests that share the same bioclimatic zone but exhibit a higher LAI. This supports the emerging opinion that the primary productivity of humid savanna ecosystems could be close to the primary productivity of tropical rainforests [6, 24].

6.6.3 Scientific gains and gaps in the knowledge

A better insight into the soil-plant-atmosphere CO_2 exchanges at Lamto has been gained in the past few years, therefore contributing to a better understanding of the savanna primary production. One of the most significant findings is the high photosynthetic nitrogen use efficiency exhibited by leaves of humid savanna grass species. This result sustains the opinion that tropical savannas could be highly productive despite the generally low nutrient status that they experience. Hitherto, ecophysiological information for savanna ecosystems was scarce, particularly for grass species, and neotropical savannas were better documented than African savannas [72, 25, 49, 71, 82]. A recent attempt to identify "typical" conductance values of different vegetation types [33] revealed this lack. Similarly, despite its importance in the ecosystem carbon budget, soil respiration remains poorly quantified for humid savannas, and data are only available for cerrados in Venezuela and Brazil [27, 62, 50]. In the same way, until recently, canopy CO_2 assimilation rates obtained at Lamto were the only available data for savanna canopies [66]. Thus, assessment of the carbon budget in the savanna biome (e.g., [73]) was not supported by direct data on CO_2 flux above savanna canopies. Comprehensive information on the seasonal course of canopy net assimilation is now available for a cerrado *sensu stricto* in central Brazil [52] and for dry fallow savannas of Niger (e.g., [85, 45, 26, 55]). Along with these data, information acquired at Lamto will contribute in providing key ecophysiological features and in understanding the carbon balance of the savanna biome.

However, major gaps in the knowledge of CO_2 exchanges at Lamto remain. Furthermore, models simulating CO_2 exchanges now generally represent photosynthesis using biochemically based modules rather than empirical relationships describing CO_2 assimilation rate as a function of irradiance. Such modules provide more mechanistic information useful to assess photosynthesis response to changing environmental variables and are widely used in regional and global scale models of biosphere. To our knowledge, Lamto is probably the first site of humid savanna where a comprehensive database allows accurate parameterization of those biochemically based modules for dominant tree and grass species. This information will be used to better predict grass/shrub competition and changes in species balance in future environmental scenarios. However, drought effects on CO_2 exchanges are not documented. Studies of leaf and canopy photosynthesis under water shortage conditions are thus warranted. In addition, source-sink regulations of photosynthetic performance at the whole-plant or canopy scale, and long-lasting effects of carbon partitioning and soil-plant interactions under field conditions are worth addressing since they may be of paramount importance in anticipating the response of the canopy to disturbances or changing environmental conditions. Indeed, photosynthesis is only one of the many characteristics that contribute to plant ecological success in a given environment.

6.7 NO, NO_2, and O_3 exchanges

6.7.1 NO emission from soils

NO emission rates from soils were measured during the dry and rainy seasons at shrubby savanna, grassland, unburned savanna and gallery forest sites by an open chamber technique [38, 75]. In the shrubby savanna, fluxes are always very low and range from 0.1 $ng[N-NO]\,m^{-2}\,s^{-1}$ at the end of the dry season to 0.6 $ng[N-NO]\,m^{-2}\,s^{-1}$ during the wet season. Higher emission rates are observed in the gallery forest, in the savanna protected from fire and on termite mounds which appear as hot spots characterized by high emissions relative to the surroundings [38]. The high heterogeneity of NO fluxes encountered at Lamto is explained by the heterogeneity of soil characteristics and microbial activities as discussed in Sect. 14.7. However, whatever the site and the season, NO emission rates at Lamto are low when compared to fluxes reported in other tropical ecosystems [75].

6.7.2 Behavior of the NO-NO$_2$-O$_3$ triade

Ozone concentration was measured at a height of 4 m from October 1990 to August 1992 as part of the DECAFE program. The daily range of ozone concentration is typically 5 to 50 ppb and 5 to 15 ppb during the dry and rainy seasons, respectively [1].

Vertical profiles of NO, NO$_2$ and O$_3$ concentrations were measured in the surface boundary layer at the end of the rainy season. A chemical correction system was applied to calculate trace gas net fluxes and deposition velocities [74]. Substantial flux divergences (i.e., height-dependent fluxes) are observed (Fig. 6.20). Most of the time, the NO flux is positive (emission) near the ground but negative above 4 m. Ozone fluxes are negative and absolute values slightly increase with height. A marked decrease of the NO$_2$ deposition flux is observed with height. During the few days studied in October 1991, the mean NO$_2$ deposition rate at 2.2 m was 3.3 ng[N − NO$_2$] m^{-2} s^{-1}. Due to low NO emission rates, the Lamto savanna is thus a net sink of nitrogen oxides, at least during the early dry season. At a height of 4.1 m, the O$_3$ dry deposition velocity is about 0.6 cm s^{-1} and NO$_2$ deposition velocity ranges from 0.46 to 1.51 cm s^{-1} [74].

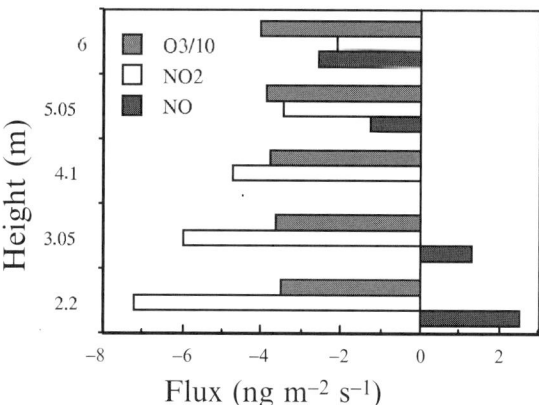

Fig. 6.20. Vertical profiles of NO, NO$_2$ and O$_3$ fluxes above the savanna at 14:30 on October, 21 1991. Fluxes were calculated at five levels from 2.2 to 6.0 m using a chemical correction system (D. Serça and X. Le Roux, unpublished data).

Interestingly, the relatively high NO_x concentrations encountered at Lamto are not explained by the very low NO emissions from savanna soils. Thus, NO_x deposition rates have to be explained by relatively high ambient atmospheric NO_x and O_3 concentrations imposed by remote biomass burning emissions at the regional scale [74].

6.7.3 Scientific gains and gaps in the knowledge

Apart from a previous study by Delmas and Servant [16], these pioneer investigations provided the only data available for West African savanna ecosystems concerning NO emissions as reviewed by Davidson and Kingerlee [13]. In particular, no other attempt was made to investigate the $NO-NO_2-O_3$ triad behavior in West African savannas. Together with results obtained in South American (e.g., [70, 62]) and South African (e.g., [59]) savannas (see the review of Davidson and Kingerlee [13]), these results should help quantify (i) annual NO emission rates in savanna environments and (ii) the atmospheric input of highly reactive oxidant compounds that represents a burden on the savanna-soil system and could contribute to irreparable damage to vegetation.

However, further studies of the mechanistic basis of NO emission rates throughout the year (in particular, soil moisture influence on nitrification and denitrification activities), characterization of the bulk sorption resistance for those chemical compounds and an important effort to model transfers and chemical reactions within the atmospheric boundary layer are warranted. This should help improving models of local, regional or global soil-biogenic nitrogen trace gas emissions (e.g., [63, 89]).

6.8 Conclusion

Both the continuous work carried out over the last 3 decades and the progress achieved in the past few years have paved the way to a comprehensive view of soil-vegetation-atmosphere interactions at Lamto. This is a crucial step toward a good understanding of this savanna ecosystem functioning and toward reliable predictions of its response to disturbances. Radiation exchanges are probably the best-documented processes in this context, both for grass- and shrub-dominated areas. The fund of knowledge acquired in this context fulfill our present needs to assess and model radiation exchanges for a wide range of vegetation types, from grass-dominated to densely woody savannas. Only marginal vegetation types (*Loudetia* savannas and gallery forests) have not been studied. Savanna-atmosphere exchanges of energy, water and carbon were accurately and extensively determined in grass-dominated areas on the plateau or upslope. However, at present, we lack important information on the spatial variability of momentum, heat, water and carbon exchanges at the ecosystem and landscape scales. Particularly, very little data are available for shrub-dominated areas, *Loudetia* savannas downslope and riparian or gallery forests. In contrast, nitrogen emission rates and driving soil characteristics were intensively studied for all the major vegetation types encountered at Lamto. This provided an insight into the source of spatial variability of NO emission rates at the landscape scale. Nevertheless, lack of flux monitoring prevented an accurate assessment of how seasonality affects these fluxes. This highlights the dilemma between (1) the need of long-term process monitoring in order to determine the savanna functioning in a wide range of environmental conditions and phenological stages and (2) the need to assess the spatial variability of processes in order to work out extrapolation strategies to determine the ecosystem functioning at the landscape scale. Since extensive monitoring in the different vegetation types encountered at Lamto would have be prohibitive, researchs have focused on the widest spread and presently best-documented vegetation type, i.e., shrubby savannas. In this way, one could begin with site-specific observations, simulations and identification of key parameters and processes, and pursue broader generalizations with considerable caution.

Although the widespread implications of the results already obtained are apparent, serious problems remain. There is much to be learned about the plant resistance to water shortage, the physiological characteristics of the dominant tree species and the regulations of plant performance at the individual and canopy scales, among other issues. However, the establishment of an accurate, relevant and predictive view of the savanna functioning requires to explore the link between the patchy tree distribution and the ensuing patchiness in the distribution of soil-plant-atmosphere interactions. A first step in this direction has already been made with the development of the spatially explicit TREEGRASS model [80] that greatly improves our ability to study the link between savanna structure and function (see Chap. 9). However, further development of the model will require field information about (i) the impact of the ecosystem structure (i.e., spatial distribution of spaced trees) on spatial patterns of momentum, heat and water exchanges in the field and (ii) associated feedback mechanisms that come into play at successively larger scales. Predicting savanna functioning without taking such feedbacks into account would be widely open to serious deviation from reality. Thus, as for most savanna functioning issues (see Chap. 20), understanding the link between the savanna structure and its functioning may be the major goal which researchers interested in savanna-atmosphere interactions should aspire to in the future.

Acknowledgments

The authors are indebted to a number of institutions and individuals whose cooperation and assistance have greatly contributed to the success of the field and greenhouse experiments during the 1991-1994 period, from which most of the material of this chapter was drawn: T. Bariac and A. Mariotti

(Laboratoire de Biogéochimie Isotopique, CNRS/INRA/University Paris 6, Paris), A. Bégué (Maison de la Télédétection, CIRAD, Montpellier), G. Cornic and J.-Y. Pontailler (Laboratoire d'Ecologie Végétale, CNRS/University Paris sud, Orsay), R. Delmas and D. Serça (Observatoire Midi Pyrénées, University Toulouse) and H. Gauthier and J. Sueur (Laboratoire d'Ecologie, CNRS/ENS/University Paris 6, Paris). Many thanks to A. Konan, E. Kouassi, F. N'Guessan, G. Kouassi, M. Loukou and P. Savadogo (Station d'Ecologie de Lamto, Côte d'Ivoire) for providing technical assistance in the field. The authors are also grateful to Dr. S.J. Allen and Dr. J.H.C. Gash (Institute of Hydrology, Wallingford, U.K.) for their critical reviews of the chapter.

References

1. B. Ahouha. Mesure de l'ozone en Afrique intertropicale. In *Proceedings of the IGAC-DEBITS-Africa workshop: dry and wet depositions in Africa*, pages 1–43, 1994.
2. S. Allen, J.S. Wallace, J. Gash, and M. Sivakumar. Measurements of albedo variation over natural vegetation in the Sahel. *International Journal of Climatology*, 14:625–636, 1994.
3. O. Amadou and N.R. Yao. Détermination in situ de la capacité au champ d'un sol ferrallitique au moyen de la sonde à neutrons. *Bull. G.F.H.N*, 23:11–24, 1988.
4. N.P.R. Anten, M.J.A. Werger, and E. Medina. Nitrogen distribution and leaf area indices in relation to photosynthetic nitrogen use efficiency in savanna grasses. *Plant Ecology*, 138:63–75, 1998.
5. F. Athias. *Etude quantitative du peuplement en microarthropodes du sol d'une savane de Côte d'Ivoire*. Ph.D. thesis, Université de Paris 6, Paris, 1973.
6. G. Atjay, P. Ketner, and P. Duvignaud. Terrestrial primary production and phytomass. In B. Bolin, E. Degens, S. Kempe, and P Ketner, editors, *The global carbon cycle. SCOPE 13*, pages 129–181. John Wiley & Sons, New York, 1979.
7. J.M. Avenard and E.J. Roose. Quelques aspects de la dynamique actuelle sur versants en Côte d'Ivoire. Internal note, ORSTOM, Adiopodoumé, 1972.
8. Z. Baruch, M.M. Ludlow, and R. Davis. Photosynthetic responses of native and introduced C_4 grasses from Venezuelan savannas. *Oecologia*, 67:388–393, 1985.
9. J. Bonvallot. Etude du régime hydrique de quelques sols de Lamto (Côte d'Ivoire). Internal note, ORSTOM, Adiopodoumé, 1968.
10. J.P. Bony. Bilan radiatif du rayonnement solaire au dessus d'une savane de moyenne Côte d'Ivoire (Lamto). *Bulletin de liaison des chercheurs de Lamto*, S2:144, 1977.
11. D. Butler and N. Huband. Throughfall and stem-flow in wheat. *Agricultural and Forest Meteorology*, 35:329–338, 1985.
12. G.J. Collatz, M. Ribas-Carbo, and J.A. Berry. Coupled photosynthesis-stomatal conductance model for leaves of C_4 plants. *Australian Journal of Plant Physiology*, 19:519–538, 1992.
13. E.A. Davidson and W. Kingerlee. A global inventory of nitric oxide emissions from soils. *Nutrient Cycling in Agroecosys*, 48:37–50, 1997.

14. K De Jong. *Research on the water balance in a savannah ecosystem. A study for two soil types at Lamto, Ivory Coast.* M.Sci. thesis, Agronomy University of Wageningen, Wageningen, 1983.
15. R. Delmas, J.P. Lacaux, J.C. Menaut, L. Abbadie, X. Le Roux, J. Lobert, and G Helas. Nitrogen compound emissions from biomass burning in tropical African savannas, FOS/DECAFE 1991 experiment. *Journal of Atmospheric Chemistry*, 22:175–193, 1995.
16. R. Delmas and J. Servant. Echanges biosphère-atmosphère d'azote et de soufre en zone intertropicale: transferts entre les écosystèmes forêt et savane en Afrique de l'Ouest. *Atmospheric Research*, 21:53–74, 1987.
17. J.M. Dignac. *Etude préliminaire de la production primaire d'une savane humide d'Afrique de l'Ouest.* M.Sci. thesis, University of Paris 6, Paris, 1994.
18. J. Ehleringer and R.W. Pearcy. Variation in quantum yield for CO_2 uptake among C_3 and C_4 plants. *Plant Physiology*, 73:555–559, 1983.
19. J.R. Evans. Photosynthesis and nitrogen relationships in leaves of C_3 plants. *Oecologia*, 78:9–19, 1989.
20. G.D. Farquhar, S. von Caemmerer, and J.A. Berry. A biochemical model of photosynthetic CO_2 assimilation in leaves of C_3 species. *Planta*, 149:78–90, 1980.
21. C. Field and H. Mooney. The photosynthesis-nitrogen relationship in wild plants. In T. Givnish, editor, *On the economy of plant form and function*, pages 25–55. Cambridge University Press, Cambridge, 1986.
22. L. Fritschen and J. Simpson. Surface energy and radiation balance systems: general description and improvements. *J. Appl. Meteorol*, 28:680–689, 1989.
23. H. Gauthier. *Echanges radiatifs et production primaire dans une savane humide d'Afrique de l'Ouest (Lamto - Côte d'Ivoire).* M.Sci. thesis, University of Toulouse, Toulouse, 1993.
24. R. Gifford. Carbon storage in the biosphere. In G Pearman, editor, *Carbon dioxide and climate*, pages 167–181. Australian Academy of Science, Canberra, 1980.
25. G. Goldstein, F. Rada, P. Rundel, A. Azocar, and A. Orozco. Gas exchange and water relations of evergreen and deciduous tropical savanna trees. *Annales des Sciences Forestières*, 46s:448–453, 1989.
26. N.P. Hanan, P. Kabat, A.J. Dolman, and J.A. Elbers. Photosynthesis and carbon balance of a Sahelian fallow savanna. *Global Change Biology*, 4:523–538, 1998.
27. W. Hao, D. Scharffe, P.J. Crutzen, and E. Sanhueza. Production of N_2O, CH_4, and CO_2 from soils in the tropical savanna during the dry season. *Journal of Atmospheric Chemistry*, 7:93–105, 1988.
28. T. Hirose and M.J.A. Werger. Maximizing daily canopy photosynthesis with respect to the leaf nitrogen allocation pattern in the canopy. *Oecologia*, 72:520–526, 1987.
29. P. Jarvis and K. McNaughton. Stomatal control of transpiration: scaling up from leaf to region. *Advances in Ecological Research*, 15:1–49, 1986.
30. F. Kelliher, R. Leuning, and E. Schulze. Evaporation and canopy characteristics of coniferous forests and grasslands. *Oecologia*, 95:153–163, 1993.
31. C. Kelly and D. Beerling. Plant life form, stomatal density and taxonomic relatedness: a reanalysis of Salisbury (1927). *Functional Ecology*, 9:422–431, 1995.

32. S. Konate, X. Le Roux, D. Tessier, and M. Lepage. Influence of large termitaria on soil characteristics, soil water regime, and tree leaf shedding pattern in a West African savanna. *Plant and Soil*, 206:47–60, 1999.
33. C. Körner. Leaf diffusive conductances in the major vegetation types of the globe. In E.-D. Schulze and M.M Caldwell, editors, *Ecophysiology of photosynthesis*, pages 463–490. Springer-Verlag, Berlin, 1995.
34. J.P. Lacaux, J.M. Brustet, R. Delmas, J.C. Menaut, L. Abbadie, B. Bonsang, H. Cachier, J. Baudet, M. Andreae, and G. Helas. Biomass burning in the tropical savannas of Ivory Coast. An overview of the field experiment Fire Of Savannas (FOS/DECAFE 91). *Journal of Atmospheric Chemistry*, 22:195–216, 1995.
35. A. Lafforgue. Etude hydrologique des bassins versants de Sakassou (Côte d'Ivoire, 1972-1977). Travaux et documents, ORSTOM, Adiopodoumé, 1982.
36. P. Lavelle. *Etude démographique et dynamique des populations de* Millsonia anomala *(Acanthodrilidae, oligochètes)*. Ph.D. thesis, University of Paris 6, Paris, 1971.
37. X Le Roux. *Survey and modelling of water and energy exchanges between soil, vegetation and atmosphere in a Guinea savanna (in French)*. Ph.D. thesis, University of Paris 6, Paris, 1995.
38. X. Le Roux, L. Abbadie, R. Lensi, and D. Serça. Emission of nitrogen monoxide from African tropical ecosystems: control of emission by soil characteristics in humid and dry savannas of West Africa. *Journal of Geophysical Research*, 100(D11):23,133–23,142, 1995.
39. X. Le Roux and T. Bariac. Seasonal variations in soil, grass and shrub water status in a West African humid savanna. *Oecologia*, 113:456–466, 1998.
40. X. Le Roux, T. Bariac, and A. Mariotti. Spatial partitioning of the soil water resource between grass and shrub components in a West African humid savanna. *Oecologia*, 104(2):147–155, 1995.
41. X. Le Roux, H. Gauthier, A. Bégué, and H. Sinoquet. Radiation absorption and use by a humid savanna grassland, and assessment by remote sensing and modeling. *Agricultural and Forest Meteorology*, 85:117–132, 1997.
42. X. Le Roux, A. Lacointe, A. Escobar-Gutierrez, and S. Le Dizes. Carbon-based models of individual tree growth: A critical appraisal. *Annals of Forest Science*, 58(5):469–506, 2001.
43. X. Le Roux and P. Mordelet. Leaf and canopy CO_2 assimilation in a West African humid savanna during the early growing season. *Journal of Tropical Ecology*, 11(4):529–545, 1995.
44. X. Le Roux, J. Polcher, G. Dedieu, J.C. Menaut, and B. Monteny. Radiation exchanges above West African moist savannas: seasonal patterns and comparison with a GCM simulation. *Journal of Geophysical Research*, 99(D12):25857–25868, 1994.
45. P.E. Levy and P.G. Jarvis. Stem CO_2 fluxes in two Sahelian shrub species (*Guiera senegalensis* and *Combretum micranthum*). *Funct. Ecol*, 12:107–116, 1998.
46. M. Li. Leaf photosynthetic nitrogen-use efficiency of C_3 and C_4 *Cyperus* species. *Photosynthetica*, 29(1):117–130, 1993.
47. G.T. Maitelli and A.C. Miranda. Evapotranspiracao e fluxos de energia no cerrado - estaçao chuvosa. *Anais da Academia Brasileira de Ciências*, 63(3):265–272, 1991.

48. K. McAneney, P. Prendergast, M. Judd, and A. Green. Observations of equilibrium evaporation from a windbreak-sheltered kiwifruit orchard. *Agricultural and Forest Meteorology*, 57:253–264, 1992.
49. E. Medina and N. Motta. Metabolism and distribution of grasses in tropical flooded savannas in Venezuela. *Journal of Tropical Ecology*, 6:77–89, 1990.
50. P. Meir, J. Grace, A. Miranda, and J. Lloyd. Soil respiration in a rainforest in Amazonia and in cerrado in central Brazil. In J.H.C. Gash, C.A. Nobre, J.M. Roberts, and R.L Victoria, editors, *Amazonian deforestation and climate*, pages 319–329. John Wiley & Sons, Chichester, 1996.
51. J.C. Menaut. Chute de feuilles et apport au sol de litière par les ligneux dans une savane préforestière de Côte d'Ivoire. *Bull. Ecol*, V(1):27–39, 1974.
52. A. C. Miranda, H.S. Miranda, J. Lloyd, J. Grace, J.A. McIntyre, P. Meir, P. Riggan, R. Lockwood, and J. Brass. Carbon dioxide fluxes over a cerrado sensu stricto in central Brazil. In J.H.C. Gash, C.A. Nobre, J.M. Roberts, and R.L Victoria, editors, *Amazonian deforestation and climate*, pages 353–363. John Wiley & Sons, Chichester, 1996.
53. J. Monteith. Evaporation and environment. In G.E Fogg, editor, *Society of Experimental Biology Symposium. The state and movement of water in living organisms*, pages 205–233. Cambridge University Press, Swansea, 1965.
54. B. Monteny. *Contribution à l'étude des interactions végétation-atmosphère en milieu tropical humide. Importance du rôle du système forestier dans le recyclage des eaux de pluies*. Ph.D. thesis, University of Paris 11, Paris, 1987.
55. B. Monteny, J. P. Lhomme, A. Chehbouni, D. Troufleau, M. Amadou, M. Sicot, A. Verhoef, S. Galle, F. Said, and C. R. Lloyd. The role of the Sahelian biosphere on the water and the CO_2 cycle during the HAPEX-Sahel experiment. *Journal of Hydrology*, 188:516–535, 1997.
56. P. Mordelet. Influence of tree shading on carbon assimilation of grass leaves in Lamto savanna, Côte d'Ivoire. *Acta Oecologica*, 14(1):119–127, 1993.
57. P. Mordelet, L. Abbadie, and J.C. Menaut. Effects of tree clumps on soil characteristics in a humid savanna of West Africa (Lamto, Côte d'Ivoire). *Plant and Soil*, 153:103–111, 1993.
58. J. Oguntoyinbo. Reflection coefficient of natural vegetation, crops and urban surfaces in Nigeria. *Quart. J. R. Meteorol. Soc*, 96:430–441, 1970.
59. D. Parsons, M. C. Scholes, R. Scholes, and J.S. Levine. Biogenic no emissions from savanna soils as a function of fire regime, soil type, soil nitrogen and water status. *Journal of Geophysical Research*, 101(D19):23,683–23,688, 1996.
60. R. W. Pearcy and J. Ehleringer. Comparative ecophysiology of C_3 and C_4 plants. *Plant, Cell Env*, 7(1):1–13, 1984.
61. R. W. Pearcy, K. Osteryoung, and D. Randall. Carbon dioxide exchange characteristics of C_4 Hawaiian *Euphorbia* species native to diverse habitats. *Oecologia*, 55:333–341, 1982.
62. M. Poth, I.C. Anderson, H.S. Miranda, A.C. Miranda, and P.J. Riggan. The magnitude and persistence of soil NO, N_2O, CH_4, and CO_2 fluxes from burned tropical savanna in Brazil. *Global Biogeochem. Cycles*, 9(4):503–513, 1995.
63. C.S. Potter, P.A. Matson, P.M. Vitousek, and E.A. Davidson. Process modeling of controls on nitrogen trace gas emissions from soils worldwide. *Journal of Geophysical Research*, D1:1361–1377, 1996.
64. J. Roberts, O. Cabral, G. Fisch, L. Molion, C. Moore, and W. Shuttleworth. Transpiration from an Amazonian rainforest calculated from stomatal conductance measurements. *Agricultural and Forest Meteorology*, 65:175–196, 1993.

65. E. Roose. Dynamique actuelle d'un sol ferrallitique gravillonaire issu de granite sous culture et sous savane arbustive soudanienne du nord de la Côte d'Ivoire. Korhogo: 1967-1975. mémoire, ORSTOM, Adiopodoumé, 1980.
66. A. Ruimy, P. Jarvis, D. Baldocchi, and B. Saugier. CO_2 fluxes over plant canopies and solar radiation: a review. *Adv. Ecol. Res*, 26:1–68, 1995.
67. R.F. Sage and R.W. Pearcy. The nitrogen use efficiency of C_3 and C_4 plants. II. Leaf nitrogen effects on the gas exchange characteristics of *Chenopodium album* (L.) and *Amaranthus retroflexus* (L.). *Plant Physiology*, 84:959–963, 1987.
68. J.J. San José, R. Bracho, and N. Nikonova. Comparison of water transfer as a component of the energy balance in a cultivated grass (*Braciara decumbens* Stapf.) field and a savanna during the wet season of the Orinoco Llanos. *Agricultural and Forest Meteorology*, 90:65–79, 1998.
69. J.J. San José, N. Nikonova, and R. Bracho. Comparison of factors affecting water transfer in a cultivated paleotropical grass (*Braciara decumbens* Stapf.) field and a neotropical savanna during the dry season of the Orinoco lowlands. *Journal of Applied Meteorology*, 37(5):509–522, 1998.
70. E. Sanhueza, W. Hao, D. Scharffe, L. Donoso, and P. Crutzen. N_2O and NO emissions from soils of the northern part of the Guyana shield, Venezuela. *Journal of Geophysical Research*, 95(D13):22,481–22,488, 1990.
71. G. Sarmiento. Adaptative strategies of perennial grasses in south American savannas. *Journal of Vegetation Science*, 3:325–336, 1992.
72. G. Sarmiento, G. Goldstein, and F. Meinzer. Adaptative strategies of woody species in neotropical savannas. *Biol. Rev*, 60:315–355, 1985.
73. R.J. Scholes and D.O. Hall. The carbon budget of tropical savannas, woodlands and grasslands. In A.I. Breymeyer, D.O. Hall, J.M. Melillo, and G.I Agren, editors, *Global change: effects on coniferous forests and grasslands. SCOPE*, pages 69–100. John Wiley & Sons, New York, 1996.
74. D. Serça. *Echanges biosphère-atmosphère de composés de l'azote en milieu tropical*. Ph.D. thesis, Université de Toulouse 3, Toulouse, 1995.
75. D. Serça, R. Delmas, X. Le Roux, D. A. B. Parsons, M. C. Scholes, L. Abbadie, R. Lensi, O. Ronce, and L. Labroue. Comparison of nitrogen monoxide emissions from several African tropical ecosystems and influence of season and fire. *Global Biogeochemical Cycles*, 12(4):637–651, 1998.
76. W. Shuttleworth. Evaporation from Amazonian rainforest. In *Proceedings of the Royal Society (London)*, volume 233, pages 321–346, 1988.
77. W. Shuttleworth. Evaporation models in hydrology. In J.C. Schmugge and T. André, editors, *Land surface evaporation. Measurement and parameterization*, pages 93–119. Springer-Verlag, New York, 1991.
78. W.J. Shuttleworth. Micrometeorology of temperate and tropical forest. *Philosophical Transactions of the Royal Society (London), Series B*, 324:299–334, 1989.
79. G. Simioni, J. Gignoux, X. Le Roux, R. Appé, and D. Benest. Spatial and temporal variation in leaf area index, specific leaf area, and leaf nitrogen of two co-occuring savanna tree species. *Tree Physiology*, 24:205–216, 2004.
80. G. Simioni, X. Le Roux, J. Gignoux, and H. Sinoquet. Treegrass: a 3D, process-based model for simulating plant interactions in tree-grass ecosystems. *Ecological Modelling*, 131:47–63, 2000.
81. G. Simioni, X. Le Roux, J. Gignoux, and A. S. Walcroft. Leaf gas exchange characteristics and water- and nitrogen- use efficiencies of dominant grass and tree species in a West African savanna. *Plant ecology*, 173:233–246, 2004.

82. M.A. Sobrado. The influence of leaf age and drought on photosynthesis of the tropical evergreen tree *Curatella americana* L. In P. Mathis, editor, *Photosynthesis: From light to biosphere*, volume IV, pages 549–552. Kluwer, Boston, 1995.
83. J. Sueur. *Comparaison des échanges gazeux foliaires de différents génotypes d'une graminée de savane (*Hyparrhenia diplandra*)*. M.Sci. thesis, Ecole Normale Supérieure de Lyon, Lyon, 1995.
84. G. Vannier. Mesures du pF dans les sols de 6 stations prospectées par P. Lavelle à Lamto. In *Bulletin de liaison des chercheurs de Lamto*, pages 21–28, Paris, November 1970. Ecole Normale Supérieure.
85. A. Verhoef, S.J. Allen, H.A.R. De Bruin, C.M.J. Jacobs, and B.G. Heusinkveld. Fluxes of carbon dioxide and water vapour from a Sahelian savanna. *Agricultural and Forest Meteorology*, 80:231–248, 1996.
86. L.P. Vernon. Spectrophotometric determination of chlorophylls and pheophytins in plant extracts. *Analytical Chemistry*, 32(9):1144–1150, 1960.
87. P. Villecourt and E. Roose. Charge en azote et en éléments minéraux majeurs des eaux de pluie, de pluviolessivage et de drainage dans la savane de Lamto (Côte d'Ivoire). *Revue d'Ecologie et de Biologie du Sol*, 15(1):1–20, 1978.
88. N. Yao, B. Goué, and J. Janeau. Consommation en eau d'une culture de manioc (*Manihot esculenta* Crantz) à l'échelle de la parcelle. *Bulletin du Groupe Francophone Humidimétrie et Transferts en Milieu Poreux*, 21:85–103, 1987.
89. J.J. Yienger and H Levy, II. Empirical model of global soil-biogenic NO_x emissions. *Journal of Geophysical Research*, 100(D6):11,447–11,464, 1995.

7

Biomass Cycle and Primary Production

Jacques Gignoux, Patrick Mordelet, and Jean-Claude Menaut

7.1 Introduction

Monitoring grass aboveground biomass has been one of the favorite field surveys done at Lamto since its foundation in December 1961, partly because most of the work done at Lamto during the 70s was funded by the IBP (see Chap. 1), and later because most ecological surveys needed these data as a preliminary input. Abundant data thus exist on grass biomass cycles, from 1962 to 2003, fully covering more than 25 different seasonal cycles; tree biomass cycles have also been studied, although less regularly. From the initial consideration of grass as a homogenous cover, studies progressively paid more and more attention to species composition and small scale spatial heterogeneity of the grass layer, both in modeling and field sampling. The main questions addressed in this chapter are as follows:

1. the average production level of savannas relative to other biomes, and in particular the relative contributions of the tree and grass components,
2. the interannual variability in biomass cycle and production, and
3. the spatial variability of these patterns along the topographic sequence.

We use the classical ecological terminology, with *biomass* for green plant parts, *necromass* for dead plant parts and *phytomass* for the total (biomass + necromass).

7.2 The aboveground phytomass cycle

Aboveground grass phytomass dynamics has been surveyed for various purposes in 1962 [22], 1963 [15], 1969 [3], 1970 to 1972, 1977 to 1980 [4], 1981-82 [1], 1984-86 (Fournier and Abbadie, unpublished data), 1986-87 [21], 1989 [17, 18], 1991 to 1993 [10] and 1999 (Simioni, unpublished data). Most of these surveys covered the whole cycle with usually a sampling point per month (Table 7.1), and some concerned a fairly large number of plots (Table 7.2).

Table 7.1. Variables and sampling methods of the phytomass surveys performed at Lamto since 1962. All weights are dry weights unless specified. Shortcuts: s. = sample, dp. = deep, p.i. = per individual plant, bl. = blocks, p.sp. = per species, u.d. = unpublished data, mth. = month, inc. = increment w./ = washed on, biom. = biomass, necrom. = necromass, phytom. = phytomass.

Survey	Year	Sampling methods for phytomass variables		Variables recorded		Frequency	Source
		Aboveground	Belowground[1]	Aboveground	Belowground		
Grass cycle	1962	16/100 m² s., fresh weight	—	Biomass, necromass	—	Monthly, 12 mth.	[22]
Grass regrowth after fire	1963	12 m² s., contact point sampling	Excavating 1 m² to 20 cm dp., sorting roots	Biomass, necromass, cover	Total phytomass	Monthly to Sept.	[15]
Tree allometry	1963	Cutting trees - fresh weight	Digging out main roots	Biomass of branches, trunk, leaves, fruits p.i.	Large root biomass p.i.	Once	[15]
Palm tree allometry	1965	Cutting trees - fresh weight	—	Biomass of trunk, leaves, fruits p.i.	—	Once	[27]
Productivity of grasslands	1969-72; 1977-80	4 m² s. × 4	Bl. 20 × 20 cm to 70 cm dp. by 10 cm strata, w./2mm sieve	Biomass, necromass, water content - p.sp.	Fine root (2 mm) phytomass	Monthly, 12 mths.	[3, 4]
Tree primary production	1969-73	Cutting trees ($n>160$), basal ∅ inc., leaf surface	Digging out main roots	Biomass of trunk, fruits, leaves, branches; crown maps, leaf area p.i.	Large root biomass, maps of main roots, rooting depth p.i.	Once (biomass); May 1969, 70, 73 (∅ inc.)	Menaut, u.d. and [14]
Grass nitrogen dynamics	1981-82	10 m² s., dry weight	Bl. 25 × 25 cm to 50 cm dp. × 4; cores 8 cm ∅ to 50 cm dp. × 10; dry sorting of roots on 1 mm sieves	Biomass, necromass, nitrogen content	Fine root (<2 mm) phytomass and nitrogen content, soil nitrogen content	Monthly, 12 mths.	[1]

7 Biomass Cycle and Primary Production

Topic	Years	Sampling	Method	Measurements	Root type	Frequency	Reference
Grass long term dynamics	1983	1 m² s. ×4	As above, but w./2 mm and 1 mm sieves	Biomass and necromass/phytomass	Fine root (<1 mm) phytomass, root humidity	Oct. and Nov.	Fournier, u.d.
	1984-86	4 m² s. ×4	—	Biomass, necromass	—	Jan. 84., monthly May–Dec. 85, Mar. 86	Abbadie, u.d.
	1986-87	4 m² s. ×4	Bl. 20×20 cm to 60 cm dp., dry sorting on 1 mm sieves	Biomass, necromass	Fine root (<2 mm) phytomass	Monthly, 11 mths.	[21]
Tree-grass interactions	1989	1 m² s. ×12	Bl. 20×20 cm to 60 cm dp., w./ 1 mm sieves, sorting tree/grass by $\delta^{13}C$	Biomass, necromass	Coarse root (>2 mm), fine root (<2 mm) tree/grass phytomass	monthly, 13 mths.	[17]
Primary production processes	1991-93	4 m² s. ×4	7 cores 4.5 cm ∅ to 180 cm dp., w./0.5mm sieves	Biomass, necromass	Fine root phytomass	Monthly	[10]
	1999-2000	1 m² s. ×4, contact point		Biomass, necromass		Monthly	[24]

[1] In all studies, large roots (i.e., >2 mm in diameter) were manually sorted out.

Table 7.2. Plots and vegetation types of the phytomass surveys of Table 7.1. A = Andropogoneae savanna, L = *Loudetia* savanna, A/L = transition savanna, G = grass savanna, S = shrub savanna, W = savanna woodland. Plot names appear as in original publications.

Survey	Year	Plot	Savanna type	Treatment
Grass cycle	1962	Plateau 1	A	Mid-season fire
		Unburned sav. 1	A	Unburned
Grass regrowth after fire	1963	P, M	A	Mid-season fire
		A, D	A/L	Mid-season fire
		B, C		Mid-season fire
		F, G	L	Late fire (March)
		R1, R2, Z	A	Very late fire (May)
		K	L	Very late fire (May)
Productivity of grasslands	1969-72; 1977-80	F / Lh		Mid-season fire
	1969-70	J1 & J2 / Lt	A/L, open	Mid-season fire
	1969-70; 1977-80	A / Lg	AG	Mid-season fire
	1969-72	I1 & I2 / La	AS open	Mid-season fire
	1969-72	H / Lb	AW	Mid-season fire
	1969	D / Lp	AS	Unburned
Grass nitrogen dynamics	1981-82	SA	A	Mid-season fire
		SL	L	Mid-season fire
		SNB	A	Unburned
Grass long-term dynamics	1983	JLB		Mid-season fire
		JM	A/L	Mid-season fire
		JAR	AW	Mid-season fire
		JLP	L	Mid-season fire
		SNB	A, open	Unburned
		LL / SL		Mid-season fire
		LA / SA	A	Mid-season fire
		GA	A	Mid-season fire
	1984-86	SA	A	Mid-season fire
		SL	L	Mid-season fire
	1986-87	SA	A	Mid-season fire
		SL	L	Mid-season fire
Tree grass interactions	1989	S1	A	Mid-season fire
		S2	AW	Mid-season fire
Primary production processes	1991-93	Plateau 2 / PL	A	Mid-season fire
	1999	Plateau 3	A	Mid-season fire

Tree phytomass was assessed through the periodic census of permanent plots (Chap. 18) and the establishment in 1969 [14] of size-biomass allometry relationships enabled one to infer yearly values for tree phytomass. Tree data are therefore less detailed than grass data: they do not cover the whole cycle, but give an average picture of every year. The aim of all these studies slowly moved from documenting the variability of primary production using a "black box" approach, to understanding the processes governing primary production and trying to simulate the observed variability.

The sampling designs were set up to average the spatial variability observed in aboveground phytomass, focusing on the interannual variability due to climate and fire.

7.2.1 Methods for studying aboveground phytomass

Sampling tree phytomass

Estimating tree phytomass in the field is a particularly difficult task, especially in Lamto where trees do not display clear growth rings (Menaut, unpublished data). Menaut and César [14] estimated tree phytomass and production by the following:

(1) sampling more than 150 trees after measurement of their height, basal diameter, crown surface, lateral root extent, cutting them into above and belowground components (trunks + sawdust, branches, large roots, twigs of the year, leaves, flowers), and weighting all this (fresh) material;
(2) estimating water content of the above parts by drying samples for each tree × component combination;
(3) measuring on 5 different plots (Table 7.1) the basal diameter of all trees in 1969, 1970, 1973, 1975 and 1989 to estimate the increment of the trunk.

Based on allometric relations deduced from (1) and (2) and the census data, this huge sampling effort enabled to produce reliable estimates of tree primary production for the Lamto savanna [14].

The palm tree *Borassus aethiopum* was not included in this study, but a few studies [15, 27] document the (fresh) phytomass of the various palm components for 7 individuals.

Non-destructive sampling of the grass layer

The grass layer has been quantified by contact-point sampling three times only, for estimating the relative abundances of grass species [20] and for quantifying cover during regrowth after fire [15, 10]. In both studies, the cover or frequency estimate was based on counting a large number (100-600) of contacts between the vegetation and a sampling line. In the last two studies [15, 10], destructive sampling has been used on the same plots as contact-point counting. The relationship between the two estimates is good until grass cover reaches 100%, i.e., around July (Fig. 7.1). The absence of a relation after that period oriented further work to destructive sampling, which has the drawback of confounding spatial and temporal variability.

Destructive sampling of the grass layer

The survey method that quickly became established in Lamto is based on the destructive sampling of relatively large surfaces (usually 4 to 16 m^2). All the standing crop in a square of the proper surface is clipped at ground level, sorted by state (alive or dead) and eventually species, oven dried until weight stabilization and weighted. Fresh weight was only used by Roland [22] because he preferred to focus on controlling spatial variability by using large

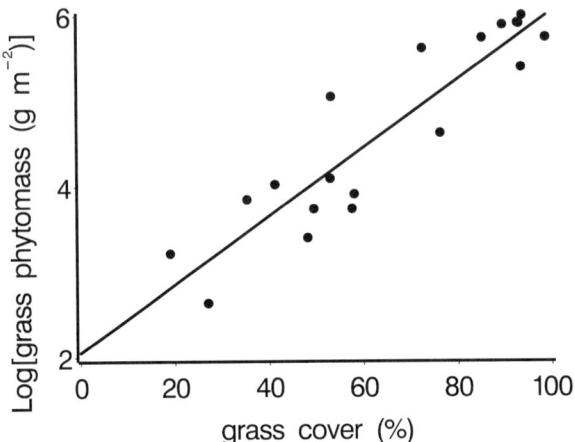

Fig. 7.1. Relationship between standing crop (phytomass) and grass cover measured by contact point sampling. Regression: $Y = 2.060 + 0.040\,X$; $F_{1,17} = 86.8$; $P < 0.0001$; $r^2 = 0.84$ (data from [15]).

samples (16 to 100 m²), which prevented drying all the cropped material. César [3] documented the seasonal course of grass relative water content, found significant variation (e.g., Fig. 7.2) and recommended to preferably use dry weight (see also [6]). All authors then conformed to this advice.

The major problem of this method is that temporal and spatial variability are confounded, and first measurements demonstrated that the spatial variability is high in Lamto, even in the open. Much effort has been put in estimating the measurement error and the optimal sampling design in order to assess its representativity [3, 21, 10]. César [3] estimated two components of measurement error:

(1) the error due to the positioning of the sampling frame in the field is ca. 2 cm for squares of 2×2 m, resulting in a 2% error on the surface sampled;
(2) the error due to the spatial variability in the grass cover is between 10 and 20%, depending on date.

He then determined the optimal sample size to ensure a sample estimate not beyond 10% of the mean of a large reference sample (Fig. 7.3). He recommended using samples of not less than 8 m². With usually 12 sampling dates per year, this requires ca. 100 m² of relatively "homogeneous" grass to perform a study of phytomass dynamics.

This makes this method representative of relatively coarse scale processes (e.g., vegetation dynamics at the scale of a 1 ha plot), but makes it less adapted to fine scaled issues like tree-grass interactions (Chap. 8). Remarking that César's tests of sampling size were done for the biomass/necromass of the whole grass layer, the method should (and usually does) fail for species-level estimations, except for the major dominant species. The key advantage of

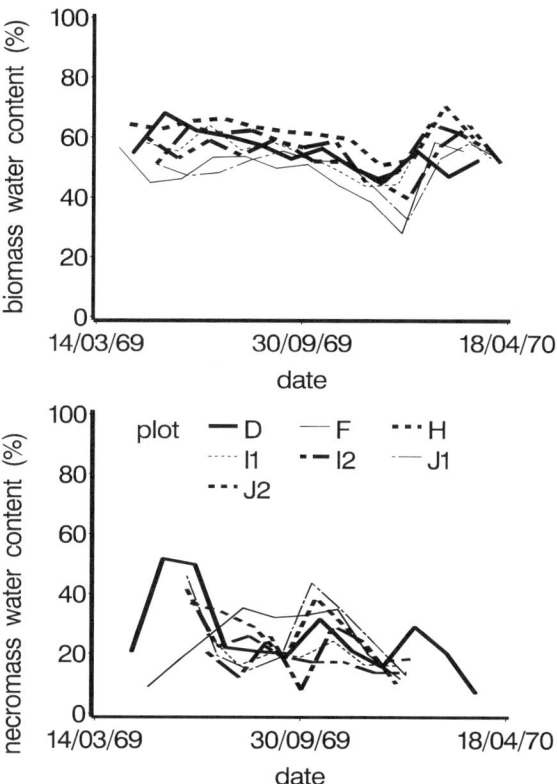

Fig. 7.2. Relative water content of grass biomass and necromass along the 1969 cycle for 7 different plots (redrawn from [3]; see Table 7.1).

this method is its low technical requirement in the field, which is particularly important for working in the bush.

7.2.2 Tree phytomass

The woody phytomass widely varied according to the facies, i.e., to the tree cover and density. Menaut and César [14] estimated aboveground phytomass values of 750 g m^{-2} for a *Loudetia*-Andropogoneae transition savanna with 120 trees ha^{-1}, and in Anropogoneae savannas: 2200 g m^{-2} for open shrub savanna (160 trees ha^{-1}), 3300 g m^{-2} for dense shrub savanna (300 trees ha^{-1}) and 5500 g m^{-2} for savanna woodland (800 trees ha^{-1}). The above-ground woody parts (branches + stems) made up nearly 2/3 of the total phytomass. Consequently, leaves contribute little to the total tree and shrub phytomass (from 2.2% to 4.4% of the total), most of it being immobilized into wood.

In the *Loudetia* savanna, tree aboveground phytomass was low (less than 10 g m^{-2}), but palm trees were not included in the sampling.

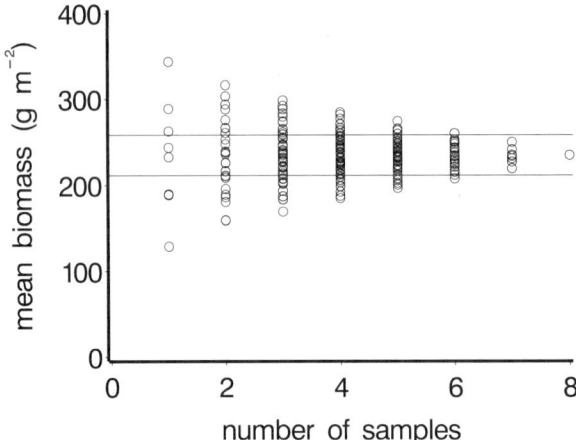

Fig. 7.3. Determination of the optimal sample area for the destructive sampling of biomass. Eight 1 m^2 samples were combined. Averages of 1 to 8 samples shown. Horizontal bars indicate a 10% deviation from the average of the 8 samples (data from [3]).

7.2.3 The grass phytomass cycle

Seasonality and interannual variability

Although the dataset on aboveground grass phytomass covers more than 20 full seasonal cycles (Table 7.1), its heterogeneity is high and reconstructing an apparently homogenous time series should be made only with caution (Chap. 9). However, even on shorter series, what is striking is the strong seasonality of the phytomass cycle and its high interannual variability (Fig. 7.4). Depending on climate, (1) the peak biomass will reach between 400 and 800 g m^{-2} and will occur between October and December, (2) necromass will appears after 1 to 4 months of growth, (3) a decrease in biomass will be observed or not at the onset of the short dry season and (4) necromass at the end of the cycle will reach 50% to 200% of biomass.

Regrowth after fire

Fire is a key factor in the functioning of Lamto savannas (Sect. 4.5), because it acts as a starter that clears off the aboveground phytomass every year, synchronizing the vegetation cycle. Monnier [15] designed an experiment to characterize the dynamics of regrowth after fire. He compared different fire treatments, from mid-season fires to late fires (Table 7.2). Regrowth after fire is the fastest for the May fires and the slowest for the January fires (Fig. 7.5). This is due to the better water availability when the rainy season has started [15]. The biomass reached for late fires is comparable to that reached in early fires: late fires have apparently no harmful effects on grass tufts (by contrast to trees: see Sect. 18.4).

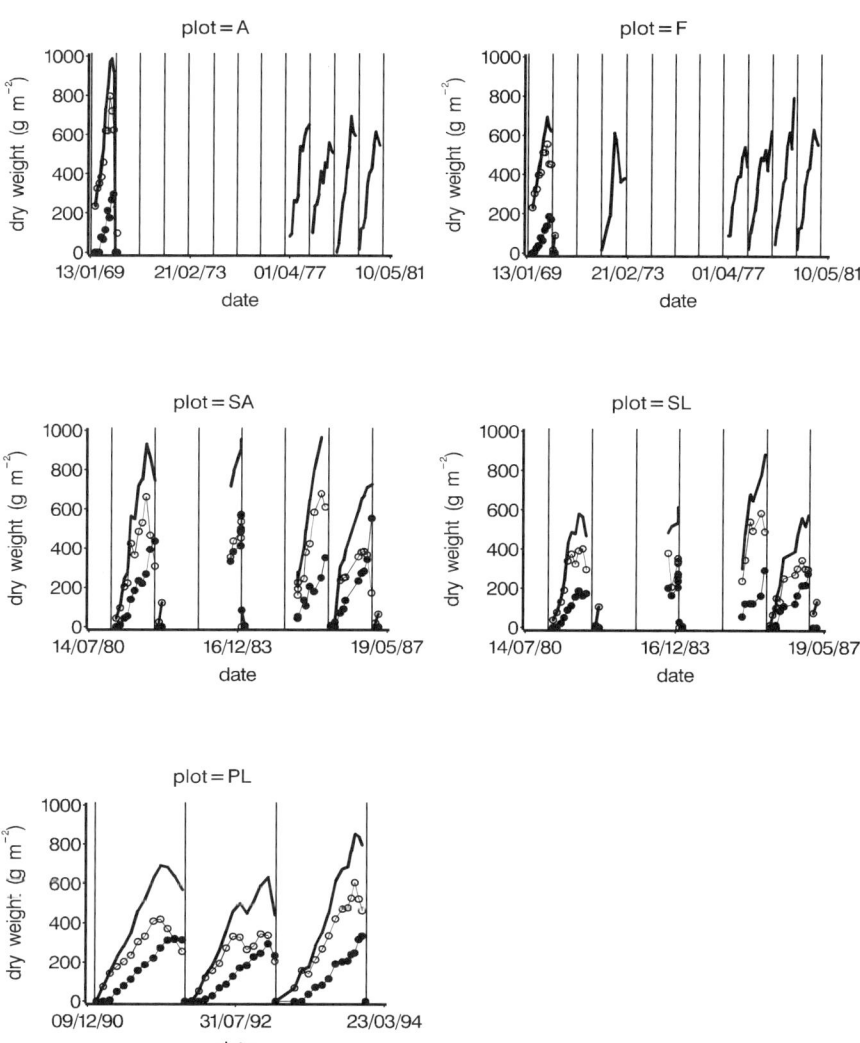

Fig. 7.4. Main homogeneous time series (i.e., measured on the same plots) of biomass (open circles), necromass (solid circles) and phytomass (thick line) measurements made at Lamto since 1962. Left: Andropogoneae savannas; Right: *Loudetia* savannas. Top: after [3, 4]; Middle: After [1, 21] and unpublished data by Abbadie (1986) and Fournier (1984); Bottom: After [10]. Vertical lines represent fire occurrences.

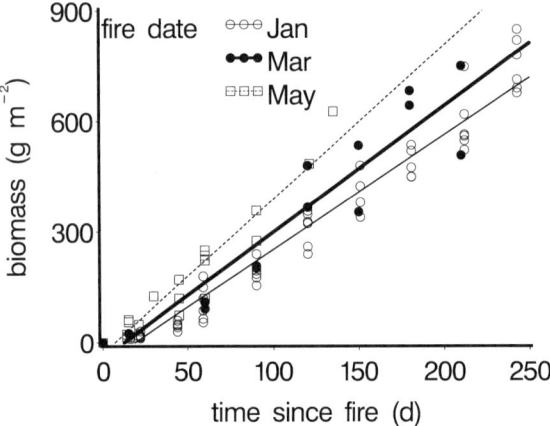

Fig. 7.5. Regrowth of grass biomass after fire, averaged over plots by fire treatment (data from [15]).

The effect of fire on the phytomass cycle

Although a continuous survey of grass in the unburned savanna has not been set up, the existing data covering the first years of the fire protection [15, 22, 1, 3] enable one to assess the effect of fire on the phytomass cycle (Fig. 7.6). In unburned savannas, the biomass cycle is similar to that of the burned savannas, but there is an important increase in necromass accumulation during the first years of fire protection: necromass can reach up to 1500 g m^{-2} the second year after protection. There is a significant decrease in necromass with time, with peaks at 1000 g m^{-2} seven years after protection and 500 g m^{-2} twenty years after protection. A regression line over time fitted to biomass is not significant ($F_{1,34} = 0.4$; $P = 0.543$) while a regression fitted to necromass is significant ($F_{1,34} = 82.4$; $P < 0.0001$). Therefore, the up to 100% increase in phytomass observed on the unburned plot is rather due to the accumulation of necromass than to a higher production. The decrease in necromass with time to levels similar to that of unburned plots may be due to the slow onset of the litter incorporation and decomposition process of aboveground litter otherwise prevented by fire. From the comparison of Fig. 7.6 and Fig. 7.5, fire appears as the principal cause of the phytomass fluctuations observed in burned plots, since phytomass shows little seasonal variation on unburned plots.

7.3 The belowground phytomass cycle

7.3.1 Methods for studying belowground phytomass

Sampling of belowground material raises many problems: sorting roots from the soil is time-consuming; species and state (alive or dead) cannot be deter-

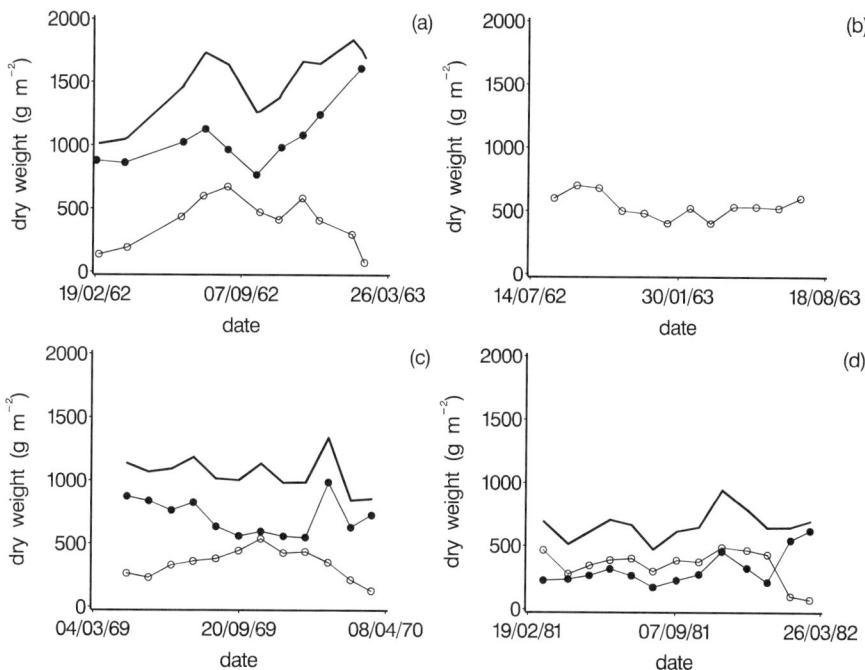

Fig. 7.6. Dynamics of aboveground biomass (open circles), necromass (solid circles) and phytomass (thick line) in the open of unburned savanna on red tropical soils. Data from references [22] (a); [15] (b); [3] (c); [1] (d). Dry weights for the Roland [22] dataset were computed on the basis of an average water content of 54% for biomass and 23% for necromass ([3] and Fig. 7.2).

mined through visual observation (except for the palm tree *Borassus aethiopum*); during the dry season, the soil is hard for digging or using an auger; gravels can prevent from reaching deep soil layers. Various methods have been used (Table 7.1):

- for sampling large tree roots and studying root system extension, complete excavation of the root system was performed down to 1 cm diameter of roots [14, 17];
- for sampling the grass root systems, Monnier [15] performed a complete excavation over 1 m² down to 20 cm, root extraction by hand, oven drying and weighting;
- for sampling fine roots, all other authors either extracted blocks of soil (20 × 20 cm) or soil cores using an auger. Except when chemical analyses were performed on samples, samples were washed on sieves under tap water to extract the roots, oven dried to constant weight and weighted. Abbadie and Lata (unpublished data) have found no significant difference in root phytomass between dry and wet extraction methods. The mesh size of sieves varied between studies (Table 7.1). Estimated phytomasses are heavily influenced by

Table 7.3. Percent of roots extracted by washing samples on sieves with different mesh sizes, and multiplicative correction factor used for root phytomass datasets (Fig. 7.8).

Mesh size (mm)	% of roots extracted on sieve (±s.e.)		Average cumulated root phytomass (%)	Correction factor
	César and Bigot 1984 [5], 1 sample	Mordelet 1993 [17], 5 samples		
2	37		37	2.70
1	29	50 (±0)	58	1.72
0.5	—	24 (±8)	74	1.35
Remaining	33	26 (±8)	100	1.00

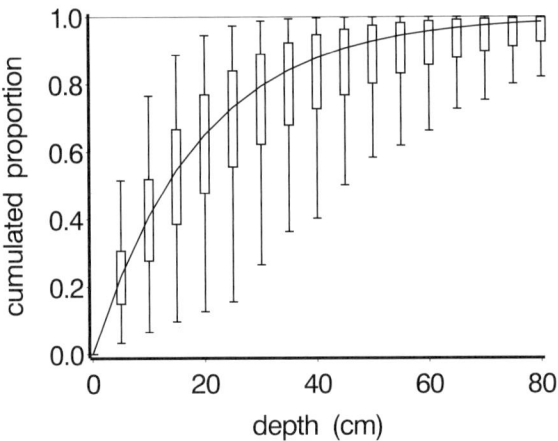

Fig. 7.7. The average root profile used to correct belowground phytomass data for different sampling depths. One hundred seventy profiles of different savanna types (from [3, 16]; Abbadie, unpublished data) with at least 5 sampling depths were used. A nonlinear model of the form $R = R_{max}(1 - e^{-\alpha d})$ (with R the cumulated root density, R_{max} the asymptotic total root density of the profile, d the depth and α an initial slope parameter) was fitted to each profile. The solid line represents the median profile, the box represents the 25% and 75% quantiles and the top and bottom ticks represent the extreme predicted root profiles.

mesh size (63% loss with 2 mm mesh [21]). However, two studies [17, 5] assessed this effect, spanning the different mesh sizes used (2, 1 and 0.5 mm). We used the results of these experiments to extrapolate comparable phytomasses from the raw data (Table 7.3), plus an average vertical root profile (Fig. 7.7 and Table 7.4) to estimate root phytomass over the whole profile.

The correction we used yielded comparable orders of magnitude for phytomass data (Fig. 7.8).

Table 7.4. The average root profile used to correct root phytomass data depending on profile depth, and the multiplicative correction factor used for root phytomass datasets (Fig. 7.8). Estimates deduced from Fig. 7.7.

Depth (cm)	Average %	95% Confidence interval	Correction factor
5	23.7	8.5-41.4	4.22
10	40.7	16.2-65.6	2.46
15	53.2	23.3-79.9	1.88
20	62.5	29.8-88.2	1.60
30	75.1	41.2-95.9	1.33
40	82.7	50.8-98.6	1.21
50	87.6	58.8-99.5	1.14
60	90.9	65.5-99.8	1.10
70	93.1	71.9-99.9	1.07
80	94.7	75.8-100	1.06

7.3.2 The root phytomass cycle

It is hard to see any consistent seasonal dynamics in fine root phytomass: the minimum apparently occurs more frequently during the wet season, although not always (e.g., the 1977-78 and 1986-87 cycles, Fig. 7.8). The peak phytomass is usually two times larger than the minimum phytomass. *Loudetia* savannas maintain a significantly larger belowground phytomass than Andropogoneae savannas. Fluctuations in numbers are rather erratic, and a study with a weekly sampling [17] suggests that, apart from spatial variability problems, fluctuations in belowground phytomass occur on a faster time scale than aboveground.

7.4 Primary production of Lamto savannas

7.4.1 Estimating primary production

Tree aboveground and coarse root production

For trees, production between two census years was estimated as the total leaf biomass plus the total woody biomass increment deduced from allometric relations between basal diameter and biomass [14].

Grass aboveground production

Estimating primary production with field data is a difficult exercise, as emphasized by the variety of methods proposed (see a review in Singh et al. [26]) - of which none is really satisfactory. The key problem in estimating

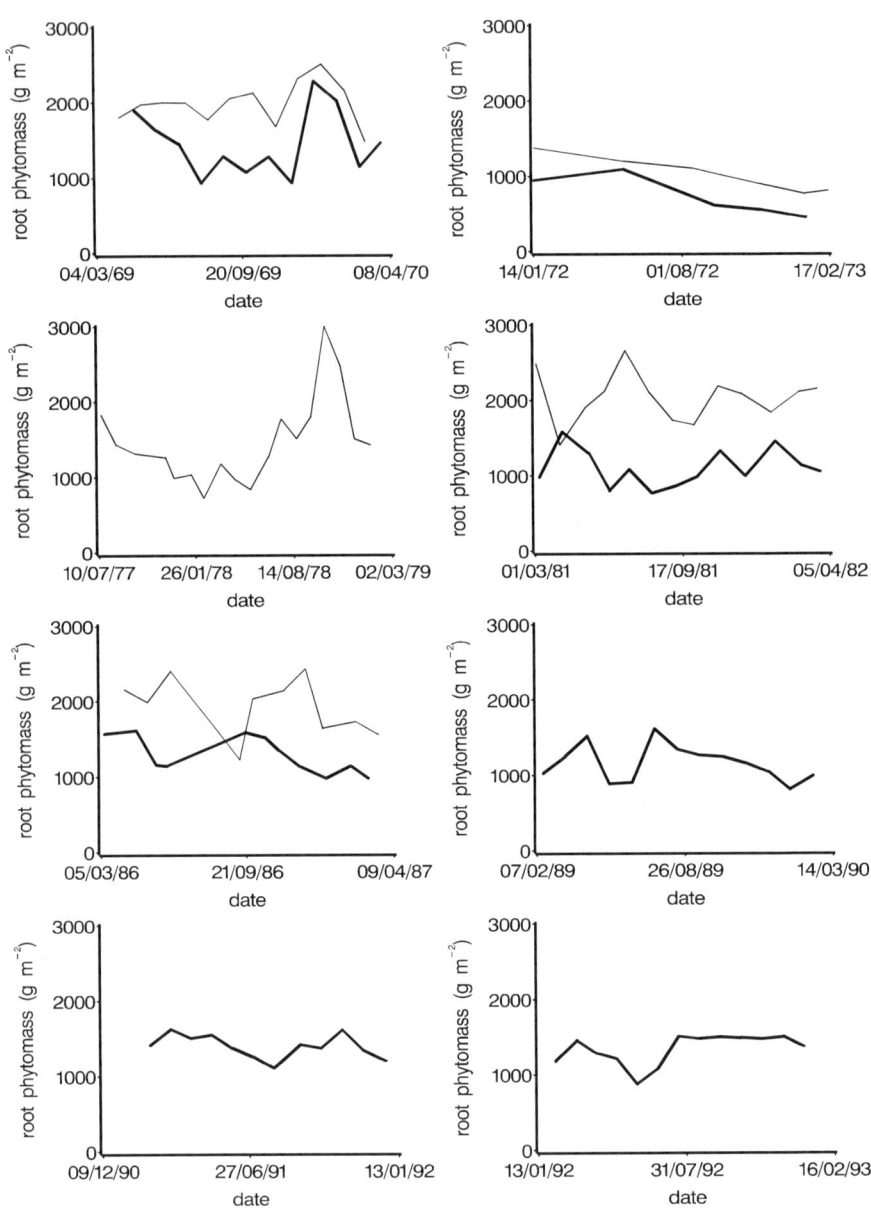

Fig. 7.8. Dynamics of the total (tree + grass) fine root (<2 mm diameter) phytomass over 8 seasonal cycles. Data corrected for sieve mesh size and profile depth effects (see text). Thin line, *Loudetia* savanna plots; thick lines, Andropogoneae grass savannas (After [3, 4, 1, 21, 17, 10] and unpublished data by Abbadie and Fournier).

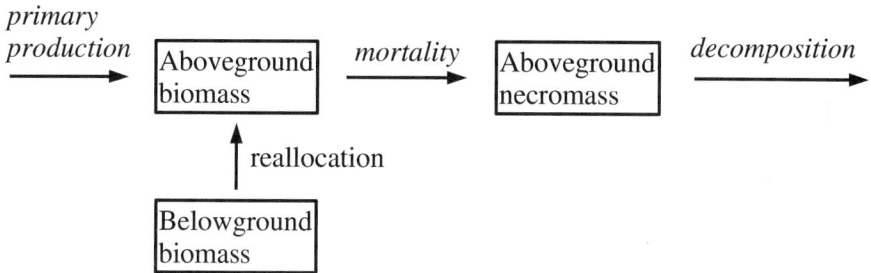

Fig. 7.9. The simple box model underlying our primary production estimates. All boxes have been measured. Unknown parameters are in italics.

primary production from phytomass measurements is that one parameter (i.e., either the necromass disparition rate or the mortality rate) is always missing (Fig. 7.9). We use here the following estimate:

$$\mathrm{NPP}(t) = \frac{\Delta B(t)}{\Delta t} + \frac{\Delta N(t)}{\Delta t} + \Gamma_D(t) N(t), \tag{7.1}$$

where NPP(t) is net primary production at time t, $\Delta B(t)$ and $\Delta N(t)$ are the variations in biomass and necromass observed at time t over the time interval Δt and $\Gamma_D(t)$ is the decomposition rate of necromass. Biomass and necromass variations are known, and the necromass decomposition rate has been estimated at 0.015 d^{-1} [10]. One of the interests of our method is that it enables one to estimate the biomass mortality rate $\Gamma_M(t)$. Given the paucity of data we have to estimate the decomposition rates, we had to assume a constant value of $\Gamma_D(t)$ over the year, which weakens our estimates. However, Le Roux [10] showed that whereas the primary production estimate was heavily influenced by taking into account decomposition, the sensitivity to the estimated decomposition value was relatively low: a change of 33% in $\Gamma_D(t)$ caused a variation smaller than 16% in the estimated primary production. At the beginning of the cycle, allocation from belowground parts to aboveground parts occurs [12], which explains the very high apparent relative growth rates of grasses [10] and apparent light use efficiencies (i.e., production to absorbed radiation ratio) [12].

Total fine root production

Even after correction for sampling methods (Sect. 7.3), belowground production cannot be accurately estimated because biomass cannot be distinguished from necromass: applying measured decomposition rates to phytomass will result in an overestimated production. On the contrary, ignoring dead root decomposition will result in an underestimation of root production. Furthermore, the monthly sampling rate of most studies might be too short given

130 Jacques Gignoux et al.

the high frequency of variations in root phytomass: the above equation will only work if variations are monotonous over Δt, otherwise production will be underestimated.

Estimates of belowground production are therefore very crude. We used different hypothetic values for the yearly average proportion of necromass in the belowground phytomass to yield estimates of production. We use a value of 0.030 d^{-1} for $\Gamma_D(t)$ based on the finding that the root decomposition rate was twice as fast as the leaf litter decomposition rate for Loudetia simplex [19].

7.4.2 The primary production of Lamto savannas

Tree aboveground and coarse root production

Production (ignoring palm trees) ranged between 60 and 670 g m^{-2} y^{-1} according to savanna facies [14]. Leaves and green shoots constitute 72% to 84% of this production; coarse roots constitute 5% to 9% (the contribution of fine roots is unknown). Those figures are consistent with litter harvesting which gave estimates ca. 20% lower [13]. This latter experiment also showed that interannual variability in tree production is probably high. Palm tree production has been crudely estimated at 200-300 g m^{-2} y^{-1} (over a total of 25-30) by Roland [22], i.e., represented 9% of the total production (while other trees represented a part comparable to the figures of Menaut and César).

Grass aboveground production

Estimates of cumulated yearly primary production vary between 1100 and 1900 g m^{-2} y^{-1} according to year for Andropogoneae grass savannas, which is higher than estimates computed ignoring necromass decomposition rate (Table 7.5). Grass production is significantly reduced in high-tree cover areas and even more in Loudetia savannas. Estimates of the necromass decomposition rate in Lamto are comparable to those observed in other West African wet savannas (discussion on this in Le Roux [10]), which range between 0.0025 and 0.021 d^{-1}. Better estimations of primary production therefore require better estimation of this parameter and its variation along the seasonal cycle.

Leaf demography

Leaf demography has been explored as an alternative to heavy biomass sampling to better estimate primary production in Lamto. Leaf demography has been studied for grasses by Fournier [8] in 1980, Puyra-vaud [21] in 1986, Abbadie in 1993 (unpublished data) and Le Roux [10]. These authors marked cohorts of leaves on a weekly, monthly or bimonthly periodicity and followed their fate until complete disparition over a complete vegetation cycle (Fig. 7.10).

Table 7.5. Aboveground grass primary production of Lamto savannas in g m^{-2} y^{-1}: Estimations based on different values of necromass disparition rate Γ_D. NPP, net primary production; CM, cumulated dead biomass production; CD, cumulated necromass disparition. Plot names after Table 7.2.

Plot	Year	$\Gamma_D = 0.000\,\text{d}^{-1}$			$\Gamma_D = 0.010\,\text{d}^{-1}$			$\Gamma_D = 0.015\,\text{d}^{-1}$			$\Gamma_D = 0.020\,\text{d}^{-1}$		
		NPP	CM	CD	NPP	CM	CD	NPP	CM	CD	NPP	CM	CD
Andropogoneae savannas, open or less than 5% tree cover													
A	1969	920	296	0	1290	666	370	1475	851	555	1660	1036	739
SA	1981	745	437	0	1408	1100	663	1740	1432	995	2072	1764	1327
SA	1985	965	353	0	1496	884	531	1761	1150	796	2026	1415	1062
SA	1986	727	554	0	1478	1305	751	1854	1681	1127	2230	2057	1503
S1	1989	655	470	0	1539	1354	884	1980	1796	1325	2422	2238	1767
PL	1991	569	314	0	1159	904	590	1454	1199	885	1749	1494	1180
PL	1992	439	234	0	911	706	472	1148	943	709	1384	1179	945
PL	1993	797	334	0	1194	731	397	1392	930	596	1591	1128	794
Andropogoneae savannas, tree clumps or more than 50% tree cover													
H	1969	672	425	0	1184	938	513	1441	1195	769	1697	1451	1026
S2	1989	404	366	0	1202	1164	799	1602	1564	1198	2001	1963	1597
Loudetia grass savannas													
F	1969	621	171	0	873	422	251	998	548	377	1124	674	502
SL	1981	468	173	0	757	462	289	901	606	433	1045	750	577
SL	1985	782	292	0	1179	689	397	1378	888	596	1576	1087	795
SL	1986	572	275	0	996	699	424	1208	911	636	1419	1122	847

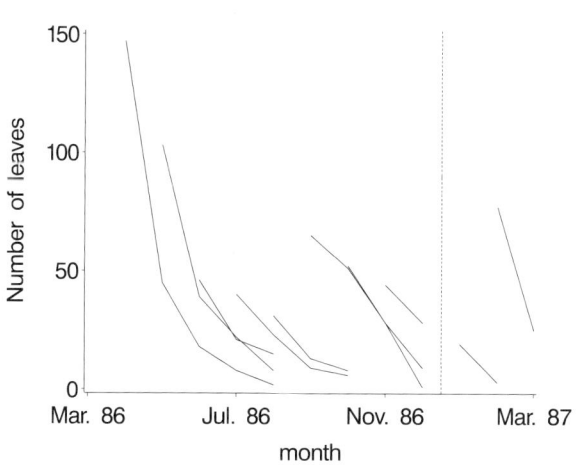

Fig. 7.10. Fate of cohorts of leaves of *Hyparrhenia smithiana* (redrawn from [21]): Numbers of leaves per monthly cohort in 1986 to 1987. Vertical line indicates fire date.

Leaf natality is concentrated in the first months after fire, just after the first rainfalls, and is relatively constant over the remaining of the cycle. The mortality pattern is less obvious: the mortality rate seems relatively constant over the year for *H. smithiana* and *H. subplumosa*, around 0.02 d^{-1}. Based on the unpublished data of Abbadie, Le Roux [10] estimated a maximal mortality rate of 0.031 d^{-1} for aboveground grass biomass. No conspicuous relation with rainfall or soil water was found.

Those results indicate a quick replacement of leaves, since 50% of leaves will be dead after one month. This probably strongly affects the estimation of primary production [10].

Fine root production

Belowground production figures should only be considered as an indication given the numerous sources of uncertainity discussed before. A large proportion of root necromass (e.g., 30%, Table 7.6) yields very high, and we think, unlikely, belowground productions. To keep the belowground/aboveground production ratio between 1 and 5, we need to assume a proportion of root necromass between 10% and 20% (Table 7.7). Given the probably quick decomposition of dead roots (Chap. 15), such a low proportion is likely. These root production estimates are fairly higher than all the previous

Table 7.6. Total belowground primary production of Lamto savannas in g m^{-2} y^{-1}: Estimations based on different values of necromass proportion p_N (from 10% to 30% of the total phytomass), computed with data corrected for mesh size effects (see text) and whole profile root distribution (Fig. 7.7), assuming a necromass disparition rate $\Gamma_D = 0.03\,\text{d}^{-1}$.

Plot	Year	Belowground net primary production			
		$p_N = 10\%$	$p_N = 15\%$	$p_N = 20\%$	$p_N = 30\%$
Andropogoneae savannas, open or less than 5% tree cover					
A	1969	1075	1839	2609	4150
SA	1981	1304	1760	2378	3614
SA	1986	878	2142	2882	4360
S1	1989	1243	1538	2178	3460
PL	1991	1289	3108	3876	5411
PL	1992	1699	2419	3164	4654
Andropogoneae savannas, tree clumps or more than 50% tree cover					
H	1969	1544	2278	3068	4647
S2	1989	1747	2609	3424	5054
Loudetia grass savannas					
F	1969	1853	4431	5535	7743
F	1977	653	2361	2986	4235
F	1978	1448	3524	4592	6730
SL	1981	1862	4390	5480	7660
SL	1986	1324	2316	3295	5254

Table 7.7. Ratio of (total) belowground to (grass only) aboveground primary production (computed from Tables 7.5 and 7.6).

Plot	Year	Belowground/aboveground production ratio		
		$p_N = 10\%$	$p_N = 15\%$	$p_N = 20\%$
Andropogoneae savannas, open or less than 5% tree cover				
A	1969	0.72	1.24	1.77
SA	1981	0.75	1.01	1.37
SA	1986	0.47	1.16	1.55
S1	1989	0.63	0.78	1.10
PL	1991	0.89	2.14	2.67
PL	1992	1.48	2.11	2.76
	Average	*0.82*	*1.41*	*1.87*
Andropogoneae savannas, tree clumps or more than 50% tree cover				
H	1969	1.07	1.58	2.13
S2	1989	1.07	1.60	2.10
	Average	*1.07*	*1.59*	*2.11*
Loudetia grass savannas				
F	1969	1.86	4.44	5.55
SL	1981	2.07	4.87	6.08
SL	1986	1.10	1.92	2.73
	Average	*1.67*	*3.74*	*4.79*

estimates [2, 14, 10], because we corrected raw phytomass data for sampling method biases.

Thus, Lamto appears as a very productive ecosystem, with a total (tree + grass and aboveground + belowground) production ranging between 3000 and 6000 g m^{-2} y^{-1} (ignoring palm trees), without possibility to further sharpen the confidence interval because of the uncertainity in root production estimation. Tree contribution to production ranges from 0% to 20% in the savanna woodland, although they consitute the largest part of the biomass in this facies.

7.4.3 Relation between primary production and climatic indices

Despite the marked seasonality of Lamto climate (see Chap. 3), neither monthly primary production (aboveground or belowground) nor mortality are correlated with monthly rainfall or drought duration (Table 7.8). There is a very weak correlation between yearly aboveground NPP and yearly rainfall (Fig. 7.11). Cleary, primary production processes are too complex to be inferred from regression on simple climatic indices. This is probably due to shifts across years between dominant limiting factors: water availability might be limiting for some years, but another factor like nitrogen availability might account for other years. These findings constituted the impetus of most of the research conducted at Lamto on the processes responsible for NPP variations (e.g., Chaps. 6, 9, 15, 14, 16 and 20).

134 Jacques Gignoux et al.

Table 7.8. Regression statistics of daily aboveground NPP, daily belowground NPP and daily mortality rate of aboveground parts to max drought duration (longest period without rainfall > 5 mm) and total rainfall over the sampled period. Periods between two samples varied between 25 and 90 days, with a mode at 30-31 days. Italics indicate significant regression models.

Y	X	$F_{1,101}$	P	r^2
Daily aboveground NPP	Max drought duration	2.147	0.1459	0.02
	Total rainfall	1.676	0.1984	0.02
Daily belowground NPP	Max drought duration	1.055	0.3068	0.01
	Total rainfall	1.842	0.1777	0.02
Daily mortality	Max drought duration	4.998	*0.0276*	0.05
	Total rainfall	3.304	0.0723	0.03

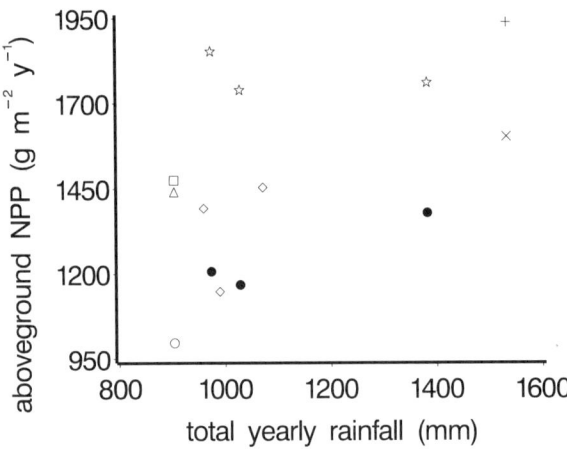

Fig. 7.11. Relationship between cumulated yearly aboveground NPP and total rainfall: Symbols represent different plots (after [3, 4, 1, 21, 10, 17] and unpublished data by Abbadie (1986) and Fournier (1984)).

7.5 Plant allocation strategies: What can be inferred from phytomass measurements?

The ratio of belowground to aboveground biomass for the whole savanna lies between 0.5 and 6.1 according to facies and proportion of necromass in root phytomass (Table 7.7). However, it is not a very biologically meaningful figure. For trees and ignoring fine roots, the ratio is ca. 0.5, but since the ratio of coarse to fine roots is unknown for trees this is of little help. For grasses and considering an open facies to avoid considering too many fine tree roots, the average ratio (assuming 10% of dead root phytomass) is 0.8 for the Andropogoneae savanna and 1.7 for the *Loudetia* savanna. This twofold variation according to facies is conserved if the proportion of dead roots is changed (Table 7.7). In some cases, the true root/shoot ratio of grasses was precisely estimated by using $\delta^{13}C$ measurements to sort tree and grass fine roots

[16, 18, 11]: see Chap. 8). In the whole dataset analyzed here, it is not possible to sort tree and grass fine roots. As a result, the allocation strategy of the two life-forms cannot be safely inferred, except for particular experiments designed to assess this point.

We presented (Sect. 7.4) results from an experiment [12] designed to estimate the allocation from belowground to aboveground parts during the beginning of the life cycle. This allocation is significant: about 60% of the regrowth observed just after fire is due to reallocation from belowground, the other half being primary production; reallocation can last up to 1.5 months. It is crucial for grasses to achieve their cycle, since after fire there are no photosynthetic tissues left and regrowth has to be due to allocation. It probably explains the advantage of perennial plants over annuals in Lamto: thanks to their large root system, perennials have a larger reserve to start with than annuals.

For trees, Gignoux [9] designed an experiment aiming at estimating the investment of tree seedlings into roots during their first growing season, as an indication of their fire resistance strategies. Results (Table 7.9) show a dramatic change in the aboveground/belowground ratio along the growing season: young savanna tree seedlings only grow belowground during their first year, in contrast to forest species. This was interpreted as an adaptation of trees to fire.

These two experiments show that belowground processes must be accurately accounted for to better estimate and predict Lamto savanna production. Two factors probably determining the high root/shoot ratios observed are: fire, which causes young trees to invest most of their production belowground and forces grasses to rely on their root system to start their cycle and lack of nutrients, which causes grasses to invest a lot of resources in their root system.

7.6 Discussion: Toward an integrative approach of primary production and allocation

The large phytomass dataset presented here makes Lamto one of the best documented savanna ecosystems in the world regarding interannual and seasonal variations in primary production (compare with the other major net

Table 7.9. Root/shoot ratios of seedlings of different tree species (from Gignoux [9]). Ratios were estimated as the slopes of regressions of belowground biomass over aboveground biomass for each species. *Ceiba pentandra* is the only forest species.

Species	Sample size	Belowground/aboveground biomass ratio	
		Initial	After 6 months growth
Bridelia ferruginea	10	0.29	1.92
Ceiba pentandra	7	0.21	0.73
Piliostigma thonningii	10	0.40	4.83
Pterocarpus erinaceus	7	0.35	3.57

primary production sites accessible at http://www-eosdis.ornl.gov/NPP/npp _home.html). The production of Lamto savannas is high compared to similar ecosystems [14]: 910 g m^{-2} y^{-1} in *Trachypogon* Venezuelian savannas [23]; in India, the *Cynodon* and *Dichantium* savannas are somewhat more productive than the Andropogoneae of Lamto [25]; in West Africa, the quantitative data given by Egunjobi [7] for Guinea savannas in Nigeria are very close to those of Lamto.

What transpires from the long series of results presented here is that primary production is not easily measured in savannas because of the very different habits of the two dominant life forms. Although total yearly rainfall explains some of the interannual variability in primary production, many other factors affect the process, so that for the same rainfall NPP can double (Fig. 7.11). These processes have started to be analyzed in the field (Chap. 6) and coupled into models of savanna functioning (Chap. 9). Such advances rely on the dataset presented here, which constitute invaluable tools for testing and validating models of ecosystem funtioning.

References

1. L. Abbadie. *Contribution à l'étude de la production primaire et du cycle de l'azote dans les savanes de Lamto (Côte d'Ivoire)*, volume 1 of *Travaux des chercheurs de Lamto*. Ecole Normale Supérieure, Paris, 1983.
2. L. Abbadie. Evolution saisonnière du stock d'azote dans la strate herbacée d'une savane soumise au feu en Côte d'Ivoire. *Acta Oecologica, Oecologia Plantarum*, 5:321–334, 1984.
3. J. César. *Etude quantitative de la strate herbacée de la savane de Lamto*. Ph.D. thesis, Université de Paris, Paris, 1971.
4. J. César. *La production biologique des savanes de Côte d'Ivoire et son utilisation par l'homme*. CIRAD - Institut d'élevage et de médecine vétérinaire des pays tropicaux, Maisons-Alfort, 1992.
5. J. César and A. Bigot. Une technique d'extraction des racines dans les échantillons de sol. Note technique 13/84, IDESSA, April 1984.
6. R. Delmas, J.P. Lacaux, J.C. Menaut, L. Abbadie, X. Le, Roux, G. Helas, and J. Lobert. Nitrogen compound emission from biomass burning in tropical African savanna, FOS/DECAFE 1991 experiment (Lamto, Ivory Coast). *Journal of Atmospheric Chemistry*, 22:175–193, 1995.
7. J.K. Egunjobi. Dry matter, nitrogen and mineral element distribution in an unburnt savanna during the year. *Oecologia Plantarum*, 9:1–10, 1974.
8. A. Fournier. *Phénologie, croissance et production végétales dans quelques savanes d'Afrique de l'Ouest*. ORSTOM, Paris, 1991.
9. J. Gignoux. *Modélisation de la coexistence herbes/arbres en savane*. Ph.D. thesis, Institut National Agronomique Paris-Grignon, Paris, 1994.
10. X. Le Roux. *Etude et modélisation des échanges d'eau et d'énergie sol-végétation-atmosphère dans une savane humide (Lamto, Côte d'Ivoire)*. Ph.D. thesis, Université de Paris 6, Paris, 1995.

11. X. Le Roux, T. Bariac, and A. Mariotti. Spatial partitioning of the soil water resource between grass and shrub components in a West African humid savanna. *Oecologia*, 104:147–155, 1995.
12. X. Le Roux, H. Gauthier, A. Bégué, and H. Sinoquet. Radiation absorption and use by humid savanna grassland, assessment using remote sensing and modelling. *Agricultural and Forest Meteorology*, 85:117–132, 1997.
13. J.C. Menaut. Chutes de feuilles et apport au sol de litière par les ligneux dans une savane préforestière de Côte d'Ivoire. *Bulletin d'Ecologie*, 5:27–39, 1974.
14. J.C. Menaut and J. César. Structure and primary productivity of Lamto savannas, Ivory Coast. *Ecology*, 60(6):1197–1210, 1979.
15. Y. Monnier. *Les effets des feux de brousse sur une savane préforestière de Côte d'Ivoire*, volume 9 of *Etudes Eburnéennes*. Ministère de l'Education Nationale de Côte d'Ivoire, Abidjan, 1968.
16. P. Mordelet. Root biomass dynamics and profiles considering tree/grass relationships and influence of soil water status. In *3rd ISRR Symposium*, pages 105–108, Wien, 1992.
17. P. Mordelet. *Influence des arbres sur la strate herbacée d'une savane humide (Lamto, Côte d'Ivoire)*. Ph.D. thesis, Université de Paris 6, Paris, 1993.
18. P. Mordelet and J.C. Menaut. Influence of trees on above-ground production dynamics of grasses in a humid savanna. *Journal of Vegetation Science*, 6:223–228, 1995.
19. H.B. Nacro. *Hétérogénéité de la matière organique dans un sol de savane humide (Lamto, Côte d'Ivoire): Caractérisation chimique, étude in vitro des activités microbiennes de minéralisation du carbone et de l'azote*. Ph.D. thesis, Université de Paris 6, Paris, 1997.
20. J. Poissonet and J. César. Structure spécifique de la strate herbacée dans la savane à palmier rônier de Lamto (Côte d'Ivoire). *Annales de l'Université d'Abidjan, Série E (Ecologie)*, 5(1):577–601, 1972.
21. J.P. Puyravaud. *Processus de la production primaire en savanes de Côte d'Ivoire. Mesures de terrain et approche satellitaire*. Ph.D. thesis, Université de Paris 6, Paris, 1990.
22. J.C. Roland. Recherches écologiques dans la savane de Lamto (côte d'ivoire) : données préliminaires sur le cycle annuel de la végétation herbacée. *La Terre et la Vie*, 21(3):228–248, 1967.
23. J.J. San José and E. Medina. Produccion de materia organica en la sabana de trachypogon, calabozo, venezuela. *Boletin de la sociedad Venezolana de Ciencias Naturales*, 33:75–100, 1977.
24. G. Simioni. *Importance de la structure spatiale de la strate arborée sur les fonctionnements carboné et hydrique des écosystèmes herbes-arbres*. Ph.D. thesis, Université Pierre et Marie Curie, Paris, 2001.
25. A.K. Singh. Structure and primary production and mineral contents of two grassland communities of chakia hills, varanasi, india. Dissertation, Banaras Hindu University, Varanasi, India, 1972.
26. J.S. Singh, W.K. Lauenroth, and R.K. Steinhorst. Review and assessement of various techniques for estimating net aerial primary production in grasslands from harvest data. *Botanical Review*, 41(2):181–232, 1975.
27. R. Vuattoux. Le peuplement du palmier rônier (*Borassus aethiopum*) d'une savane de Côte d'Ivoire. *Annales de l'Université d'Abidjan*, 1(1):1–138, 1968.

8

Tree/Grass Interactions

Patrick Mordelet and Xavier Le Roux

8.1 Introduction

The basic feature of savannas consists in the codominance of a continuous grass layer with a more or less discontinuous tree layer [49]. Both layers contribute to the water balance, nutrient cycle and primary production of the ecosystem. This coexistence of two codominant, strongly differing life-forms, contrasts with the usual dominance of a single life-form observed in other major biomes (i.e., grasslands or forests).

This feature has been taken into account in recent research programs. It is now widely accepted that the complex interactions between tree and grass individuals that compete for light, water and nutrient resources is a major characteristic of savanna functioning. Thus, being able to predict grass and tree functioning separately does not enable one to predict the functioning of the coupled tree/grass system [67, 29]. Furthermore, understanding how trees and grasses can coexist in savanna ecosystems requires a study on whether these two life-forms exploit the same soil resources or not [79, 64, 65].

The results presented in Chapters 6 and 7 were obtained either for the grass or woody component of the Lamto savanna, which already provides a broad knowledge on the ecosystem functioning. The aim of this chapter is to give a comprehensive view of what is known about tree/grass interactions in the Lamto savanna. The data available essentially focus on the influence of trees on the grass layer and allow answering four main questions: (i) to what extent trees may alter the grass environment, particularly the light, water and nutrient availability? (ii) do trees and grasses share the same soil resources or not? (iii) what is the outcome of the tree influence for the grass production? (iv) is it possible to assess the relative importance of the competition for light, water and nutrient for determining the outcome of the tree influence on grasses?

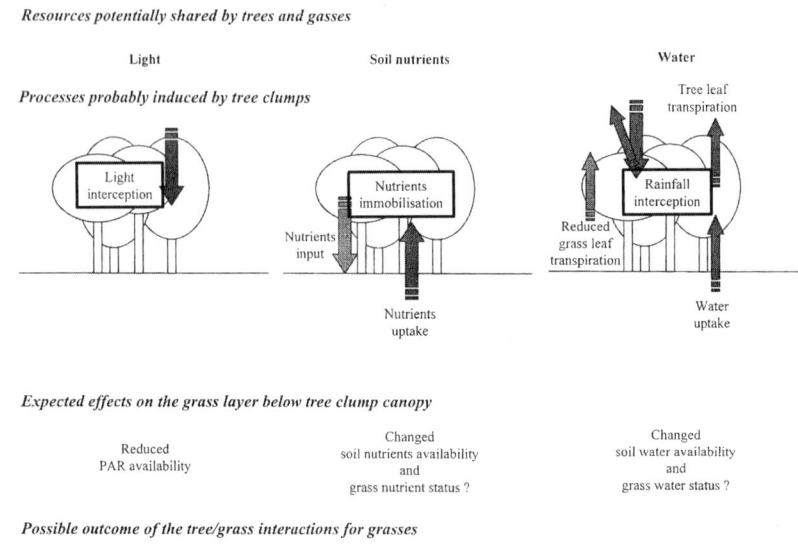

Fig. 8.1. Major processes determining the effect of the tree cover on the understory grass layer through changes in the availability of light, water and nutrient resources. The linkage between different processes (e.g., changes in the soil water availability influence changes in nutrient availability; light interception by trees contribute to the decrease in the understory evapotranspiration) is not represented.

8.2 Trees alter the understory grass environment

Tree/grass interactions can affect a lot of environmental variables, generally closely linked to each other, and driving grass primary production (Fig. 8.1). Among them are the primary determinants of savanna structure and functioning [22]: soil water and nutrient availability. Competition for light resource is another major aspect of these interactions.

8.2.1 Aboveground microclimate and light availablity

One of the most conspicuous tree/grass interaction is the competition between both layers for light. Woody plants are favored because they are taller and can thus intercept a large fraction of incident radiation [62, 59, 6, 7, 12]. This can strongly reduce grass photosynthetic activity. However, under tree canopy, air and soil temperatures, air humidity and CO_2 concentration can be different as compared to grassy areas [60, 6, 11], and all these variables can affect grass photosynthesis.

In Lamto, maximum tree leaf area index (LAI) computed on a projected area basis is around 4 for *Cussonia arborea* and *Crossopteryx febrifuga*

[70]. LAI is maximum from May to August and from July to October for *Crossopteryx febrifuga* and *Cussonia arborea*, respectively [70]. During the rainy season in Lamto, leaf irradiance was measured for leaves outside tree canopy and for shaded leaves under a tree canopy [53, 52]. The irradiance of shaded grass leaves beneath a tree canopy was strongly reduced compared to the open situation (Fig. 8.2) but remained higher than the compensation point for light (30 μmol m^{-2} s^{-1}). At a similar shrub savanna site, the amounts of radiation intercepted by the grass layer and by the woody layer were surveyed on a 144 m^2 shrubby plot ([25]; Gauthier and Le Roux, unpublished). The incoming radiation, the radiation intercepted by the tree layer (50 sensors below the tree layer), the radiation intercepted by both layers (50 sensors below the tree+grass layer) and the radiation reflected by the vegetation cover (4 sensors) or bare ground (2 sensors) were measured by light sensors, distinguishing photosynthetically active radiation (PAR) and total solar radiation. During the first part of the growing season, the amount of PAR intercepted (iPAR) is slightly higher for trees than for grasses (Fig. 8.3). Both iPAR by trees and grasses increase from January to May, with the development of leaf surfaces. At the end of the year, iPAR is lower for trees than for grasses (Fig. 8.3). In December, iPAR of grasses slightly decrease due to the reduction of grass phytomass, while iPAR of trees markedly decrease due to tree defoliation. All these results show that grasses below a shrub clump canopy experience a strong decrease in the availability of the light resource.

During the rainy season, air characteristics were measured under a tree canopy and outside a tree canopy [53, 52]. On average, carbon dioxide concentration in the air (50 cm above grasses) is slightly higher under a tree canopy than in the open (Fig. 8.4). Although many inputs and outputs can influence air CO_2 concentration, the difference between both situations is probably due

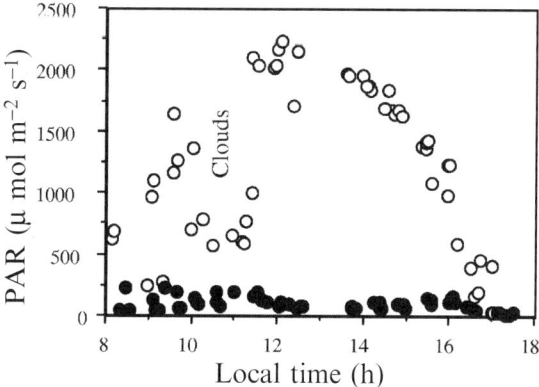

Fig. 8.2. Diurnal variations in the grass leaf irradiance during the rainy season, in the open (open symbols) and canopy (full symbols) situations (reprinted from [53], copyright (1993) with permission of Elsevier).

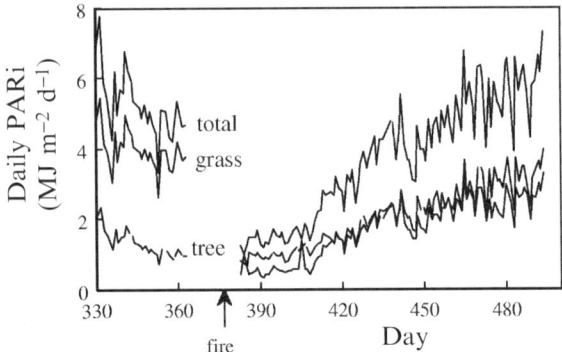

Fig. 8.3. Seasonal variations in the daily amount of PAR intercepted (PARi) by the tree and grass layers from November (beginning of the dry season) to May (maximum leaf area index) 1993 (X. Le Roux and H. Gauthier, unpublished).

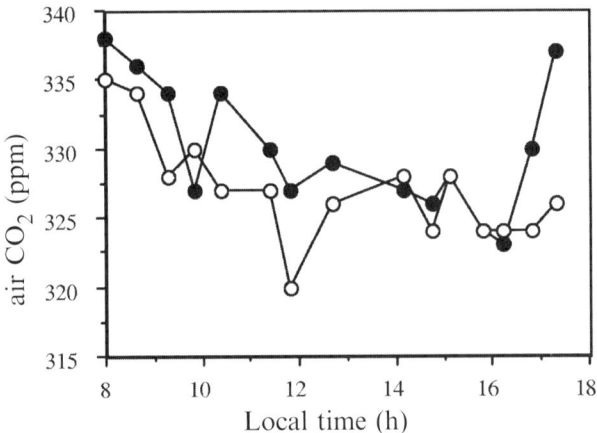

Fig. 8.4. Diurnal variations in the air carbon dioxide concentration during the rainy season in the open (open symbols) and under canopy (solid symbols) (after [52]).

to a higher soil organic matter mineralization rate (see below) and a lower photosynthetic activity of the grass layer beneath tree clumps. The released CO_2 could be trapped under the tree canopy where wind speed is reduced.

8.2.2 Soil physical and chemical characteristics, and nutrient availability

Woody plants can deeply influence soil nutrient status, but there is no rule of thumb to predict whether this influence will be positive or negative for soil fertility. Indeed, tree occurrence can be associated to changes in soil texture, soil structure, soil chemical characteristics and microbial activity.

The occurrence of woody plants in a savanna is sometimes associated to a higher soil clay content than in the surrounding grassy areas. Clay content difference is either due to or results in woody occurrence [8, 30]. In Lamto, the soil particle size distribution is similar in the open and canopy situations [54] except when tree clumps are located on termite mounds (see below). This result is in agreement with previous studies performed in *Faidherbia albida* savanna [14].

Soil bulk density is generally slightly lower beneath tree canopies than in the open situation (Table 8.1), but the difference fades away with depth. The higher macro-porosity observed under shrub clumps can be mainly attributed to a higher activity of the soil fauna, particularly termites and earthworms. In Lamto, indeed, termites eat tree leaf litter rather than grass leaf litter [45]. The higher macro-porosity under shrub clumps could enhance water and air circulation in the soil, as well as root penetration.

Under shrub clumps, soil chemical parameters showed significantly higher values than in grassy areas (Table 8.1). This is especially true for cation exchange capacity, pH, available phosphorus, organic carbon content and total nitrogen content. It has been shown that trees are able to deplete soil nutrients, at least provisionally, because trees fix part of these nutrients in their woody, perennial structures for a long period [9]. In contrast, other studies have revealed that trees can concentrate, in the shallow soil layers below their canopy, the nutrients that they uptake from the deep layers and outside their canopy [36, 35, 6, 17, 80]. This concentrating process, that seems to be dominant in Lamto, has generally been attributed to nutrient release in the topsoil layer due to leaf and root litter decomposition [14, 16, 80, 31, 17, 23, 30].

The C/N ratio in soil organic matter is lower beneath woody plants than in open areas. On one hand, tree leaf litter contributes to decrease the C/N ratio beneath the tree canopy because tree leaf C/N ratio is lower than grass leaf C/N ratio (see Sects. 15.2 and 15.3). On the other hand, the soil carbon and nitrogen mineralization potentials are higher for soil collected under woody plants than in the surrounding grassy areas (Fig. 8.5) The actual mineralization rate can therefore be assumed to be higher beneath the tree canopy. This activity releases carbon dioxide and immobilizes mineral nitrogen and

Table 8.1. Mean soil characteristics under tree clump canopy and in the open, between 0 and 30 cm depth (after [54], with kind permission of Springer Science and Business media).

	Canopy	Open
Bulk density (g m^{-2})	1.55	1.62
Cation exchange capacity (cmolkg^{-1})	3.4	3.0
pH	6.4	6.1
Available phosphorus (mg kg^{-1})	3.9	3.3
Total nitrogen (g kg^{-1})	0.5	0.4
Total carbon (g kg^{-1})	6.2	5.7
C/N ratio	12.6	15.2

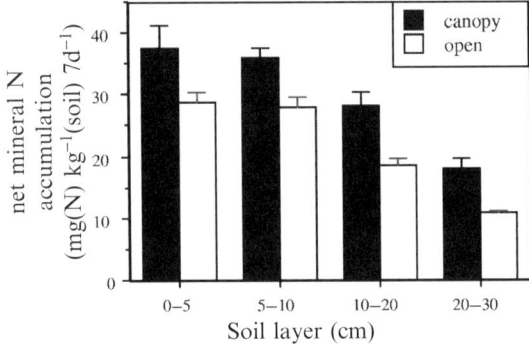

Fig. 8.5. Mean potential soil mineral nitrogen accumulation in the open and canopy situations, between 0 and 30 cm depth, after a 7 day incubation. Bars are standard errors ($n = 18$) (after [54], with kind permission of Springer Science and Business media).

then contributes to decrease the C/N ratio of the soil organic matter. Consequently, soil organic matter appears to be more available and to have a better quality beneath tree clumps than in the open.

All these features show the improvement of the soil macroporosity and nutrient availability beneath shrub clumps compared to open areas. Thus, shrub clumps can be viewed as more fertile spots for the grass layer than open areas.

8.2.3 Soil water availability

Since various processes acting in opposite ways determine the effect of a tree canopy on the soil water balance as compared to a grassy area, tree effect is not easy to forecast [37]. Tree leaves partly intercept the incoming precipitation, so that the water input is reduced [66, 6, 59, 12]. In addition, tree transpiration enhances water output, and consequently tends to exhaust more quickly the available soil water [39]. On the contrary, the higher macroporosity observed under the trees improves water infiltration [36, 6]. In addition, tree shading reduces evapotranspiration from the soil surface and the grass layer and can thus restrict water losses, and trees can generate water movements from deep to shallow soil layers through hydraulic lift [47]. This can improve the water status of the grass layer [40, 59, 33, 34, 73].

In Lamto, the different components of the soil water budget have been characterized in open areas (see Sect. 6.5) but not under tree clumps. However, the dynamics of the soil water content has been studied in both open and canopied situations [54, 43, 24]. The seasonal variations in the soil water content of the 0-30 cm layer ([54], Fig. 8.6) and the 0-60 cm layer [43] are similar under tree clumps and in the open most of the year. Mordelet et al. [54]

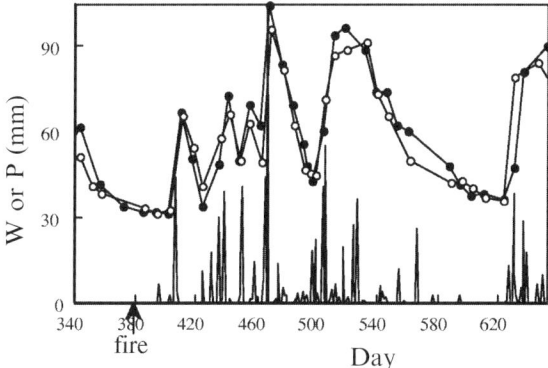

Fig. 8.6. Seasonal variations in soil water content under tree canopy (empty symbols) and in the open (solid symbols) situation. Water content W was computed for the 0-60 cm layer from neutron probe data. Daily precipitation P is indicated (redrawn after [43], with kind permission of Springer Science and Business media).

found that soil water content is slightly lower under tree clumps than in the open in the first 30 cm in July and August, but this trend was not observed by Le Roux and Bariac [43] at another site. These results show that soil water availability for plants does not strongly differ between open areas and shrub clumps (when significant, the slightly lower availability in July and August occurs when transpiration demand is low due to the reduced radiation and high air humidity).

8.2.4 A particular case: Tree clumps associated with termite mounds

In some restricted areas of the Lamto savanna, tree clumps are mainly located on old termite mounds [1]. In this case, soil clay concentration is higher under the tree clumps than in the surrounding open areas: on average, soil clay concentration ranges from 14% to 21% and from 6% to 8% on the top of termite mounds and off mounds, respectively [1, 41]. Such a higher soil clay concentration generally improves soil organic matter and nutrient concentrations [1] and soil water availability for plants [41]. Due to the soil fauna activity, termite mounds are thus fertile sites where plant water status is improved, which promotes tree settlement and growth. Indeed, tree density is 1630 ha^{-1} on mounds and only 620 ha^{-1} off mounds, and specific composition of the vegetation differs on and off mounds (see [1], Chaps. 13 and 16).

8.2.5 Summary: Relative importance of changes in resources availability under shrub clumps for grasses

In Lamto, as compared to the open situation, grasses beneath a tree canopy generally experience a strong reduction of light availability, a similar soil water availability and a slightly improved soil nutrient availability. In the special case of tree clumps located on old termite mounds, grasses beneath a tree canopy experience a strong reduction of light availability, and improved soil water and nutrient availability. However, in the context of tree/grass interactions, soil water and nutrient availability has been only characterized for the 0-60 cm layers. It could be argued that trees and grasses do not necessarily share these resources in the topsoil layers, because trees could use preferentially deeper soil layers.

8.3 Trees and grasses share the same soil resources

Walter [79] proposed the hypothesis of a spatial partitioning of soil resource, especially for soil water, between grasses and trees. According to this hypothesis, grass roots would be mostly located in the topsoil where they uptake water more efficiently than tree roots. On the contrary, tree roots would be mostly located in the subsoil where they are alone to uptake the water draining from the topsoil layers. This belowground niche separation could explain the stable coexistence between both life-forms thanks to competition avoidance, as discussed in several models [77, 18]. However, even in semi-arid savannas, field studies have indicated that trees and grasses can have widely overlapping rooting patterns [39, 65, 5]. Nevertheless, in dry environments, trees or shrubs generally use deeper soil water than grasses [39, 28, 63].

For humid savannas, this hypothesis is weakened because high precipitation should lessen competitive interactions for soil water between both vegetation components. In this case, it can thus be assumed that nutrient availability becomes a strong constraint that influences both plant structure and functioning. In humid West Africa, woody plants have generally shallow root systems [42, 48, 58]. In Lamto, detailed studies on tree and grass rooting patterns, soil water dynamics and stem and soil water isotopic signals have provided information to test whether trees and grasses exploit the same soil resources or not.

8.3.1 Both trees and grasses are shallow-rooted

Mordelet [52], Le Roux et al. [44] and Mordelet et al. [57] studied tree and grass root distribution along vertical profiles. Mordelet [52] also quantified root distribution along transects from the center of tree clumps to open areas. In all these studies, the measurement of the relative natural abundance of ^{13}C ($\delta^{13}C$) in each root sample was used to calculate the ratio of plant material originating either from C_3 or from C_4 photosynthetic pathway [72, 46, 74, 20]. Given that in Lamto savanna, trees are C_3 plants and grasses are C_4 plants, it is thus possible to assess the proportion of tree and grass roots in each sample (for isotopic signals of plant species in Lamto, see [45]).

Woody roots larger than 2 mm in diameter are essentially found in the upper soil layers, right under tree crown canopy [57]. This distribution is due to architectural constraints, but also corresponds to the nutrient rich soil volume. It contributes to enhance the soil organic matter content beneath tree clumps, and consequently to enhance the soil resource heterogeneity. Nevertheless, a few large tree roots are also observed more than 30 m away from the tree bole. This illustrates tree ability to canvass the environment looking for good soil conditions. Below 60 cm, large tree roots are very rare.

For both trees and grasses, fine roots (i.e., roots thinner than 2 mm in diameter) are mostly located in the upper soil layers (Fig. 8.7). These fine roots are concentrated in the top 30 cm and are rare below 60 cm for both life-forms [52, 44]. Fine grass root phytomass in the 0-60 cm layer is higher in the open than beneath tree clumps, while the reverse is observed for fine tree roots (Fig. 8.7). Mordelet [52] observed that total (grass plus tree) fine root phytomass decreases from the center to the edge of tree clumps, and increases with distance from the edge of tree clumps in grassy areas (Fig. 8.8). These variations can be attributed on one hand to a decrease of fine woody root phytomass away from the clump, and on the other hand to a simultaneous increase of grass root phytomass. These shallow rooting patterns suggest a potentially high competition between tree and grass roots for soil resources.

8.3.2 Both trees and grasses uptake most of their water from upper soil layers

In order to determine whether trees and grasses uptake water from the same soil layers or not, the deuterium and ^{18}O signals of grass and tree xylem water were compared to the signals of soil water extracted at different depths [44].

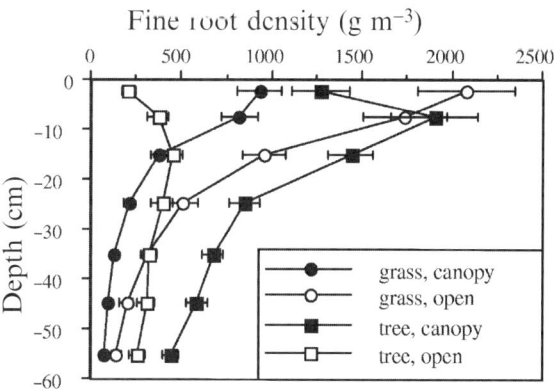

Fig. 8.7. Profiles of grass and fine tree root density in the open and canopy situations between 0 and 60 cm depth. Bars are standard errors ($n = 20$) (after [52]).

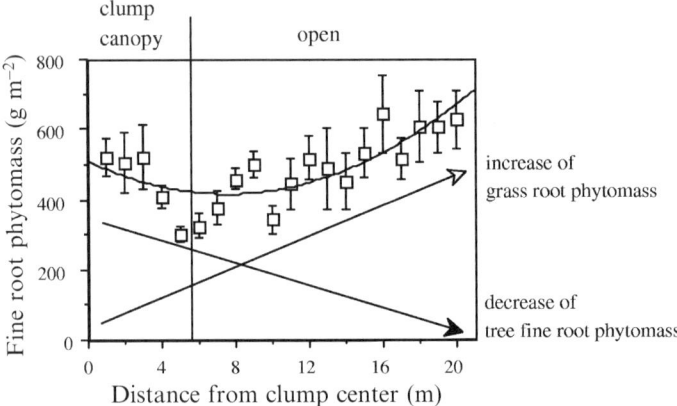

Fig. 8.8. Variations in the grass plus tree fine root phytomass in the 0–60 cm layer as a function of the distance to shrub clump center. Bars are standard errors ($n = 12$) (after [52]).

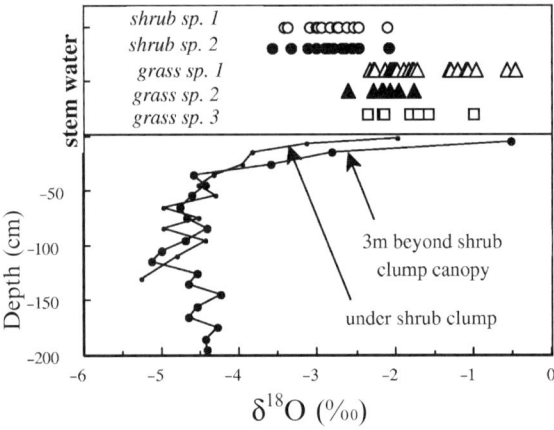

Fig. 8.9. Comparison between the isotopic signal of soil water at different depths, and the xylem sap water of tree and grass species during the rainy season (from [44], with kind permission of Springer Science and Business media, completed with unpublished data by X. Le Roux and T. Bariac).

This method allows qualitative assessment of the soil layers in which water is mostly uptaken by a plant species [81, 2, 78, 19].

During the rainy season, both trees and grasses uptake water in the upper soil layers, while uptake below 60 cm depth is negligible [44] (Fig. 8.9). During the dry season, grasses still essentially uptake water from the topsoil. At this time, the isotopic signals obtained for tree xylem water did not allow one to conclude whether trees still uptake water from the topsoil [44]. However, the relationships obtained between shoot water potential and water content of the 0-60 cm and 60-180 cm soil layers show that some tree species like *Crossopteryx febrifuga* actually rely on water of the topsoil even during the dry season, while other tree species like *Cussonia arborea* can use water from the deeper soil layers when the topsoil layers are dry [43]. Such spatial partitioning of soil water among tree species have been reported for other ecosystems (e.g., [76, 51, 32]).

Surveys of water content along 0-170 cm soil profiles showed that water depletion mainly occurs in the 0-60 cm horizon both in open areas and under shrub clumps [44], which is consistent with the conclusion of the isotopic approach.

To sum up, the soil water uptake patterns of the shrub *Crossopteryx febrifuga* is rather close to that of the grass species during both dry and rainy seasons, while *Cussonia arborea* takes up water from upper soil layers during the rainy season: this provides evidence for potentially significant competition between at least some shrub and grass species for the soil water resource. Thus, the two-layer hypothesis of water resource partitioning between grasses and shrubs in savanna ecosystems is a misleading simplification of actual water uptake patterns in Lamto savanna. In order to better understand interspecific interactions and species coexistence in this ecosystem, the diversity of strategies of water uptake exhibited by the different tree species, along with their concurrent different strategies of fire resistance (see Chap. 18), should be taken into account.

8.3.3 A particular case: Palm tree roots associated with tree clumps

The rooting pattern of palm trees (*Borassus aethiopum*) illustrates how soil nutrient heterogeneity can influence root distribution. Due to competition for light, palm tree seedlings do not grow when shaded by trees, so that adult palm trees rarely occur within a tree clump (see Barot et al. [3], who showed that no tree clump included palm trees within a 3 ha study area). However, palm tree root densities are significantly higher beneath a tree clump than in the open. This result was also emphasized by transect experiments [55]. Palm tree roots explore their environment and proliferate in nutrient rich patches, that is, beneath tree canopies. This strategy reinforces soil nutrient heterogeneity, but at the same time contributes to enhance nutrient use efficiency, especially for nitrogen.

8.3.4 Lack of time partitioning of soil resources between trees and grasses

Time partitioning of soil resources can reduce competitive interactions: different vegetation components using the same resources may have a differential exploitation of the resources by out-of-phase phenologies [10, 61, 50].

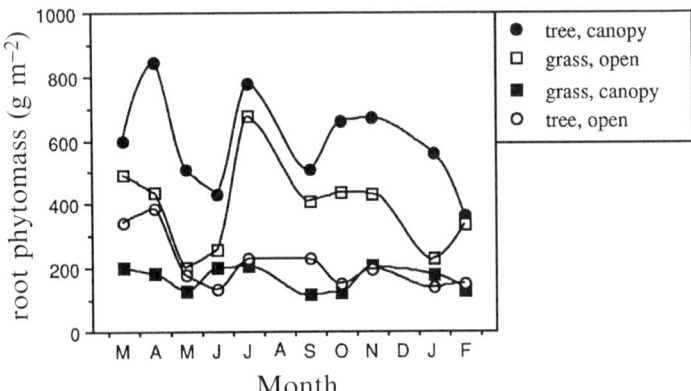

Fig. 8.10. Seasonal dynamics of the tree and grass fine root phytomass in the 0-60 cm soil layer, in the open and canopy situations (after [57], with kind permission of Springer Science and Business media).

This process was mentioned for the Nylsvley savanna, although it could hardly account for much of the stable coexistence between trees and grasses in this savanna [65].

In Lamto, tree and grass fine root phytomasses exhibit similar monthly variations beneath tree clumps and in the open (Fig. 8.10), which cannot allow a time partitioning of soil resources.

8.3.5 Summary: Likely importance of competition for soil resources between trees and grasses

According to the tree and grass rooting and water uptake patterns, strong competitive interactions for the use of soil resources should occur in the tree canopy areas of the Lamto savanna. However, in humid savannas, nitrogen availability, more than water availability, is likely to be the major limiting factor of primary production (see Chap. 4). Then, the observed shallow rooting patterns are probably mainly driven by the shallow nutrient resource distribution (i.e., existence of nutrient rich shallow soil layers). Similarly, the low density of tree fine roots in open areas, even a few meters from shrub clump edges, is probably driven by the patchy nutrient resource distribution (i.e., existence of nutrient rich plots beneath tree canopy). At the ecosystem scale, such a shallow and patchy root distribution reinforces the spatial heterogeneity of soil nutrient resource and could enhance competitive interactions between trees and grasses. However, it could also improve the soil mineral resources use efficiency, thanks to high nutrient concentrations in particular sites, as shown by Abbadie et al. [1] for nitrogen.

8.4 Trees alter grass functioning and production

Grass primary production can be considered as the target variable to assess the outcome of the beneficial and detrimental effects of woody plants on aboveground and belowground resource availability for the underneath grass layer. Some field studies conducted in the savanna concluded that grass primary production is higher beneath the trees than in the open [36, 6, 7, 80, 26]. On the contrary, other authors found a negative effect of trees on grass primary production [4, 27, 66].

In Lamto, grass ecophysiological characteristics that partly control grass primary production, such as leaf carbon assimilation rate and plant water status, were compared between shrub clump and open situations [52, 56].

8.4.1 Tree effect on grass leaf photosynthesis and water status

Leaf CO_2 assimilation was measured during the growth period (rainy season) on tufted grasses, either shaded by trees or in open sunny areas [52]. The tree canopy effect is potentially very important. Between 8 am and 6 pm, a leaf that remains sunlit uptakes 5 times more CO_2 than a leaf located under the tree canopy shade (Fig. 8.11). These measurements were performed in the two situations with the most possible contrast in order to assess the range of the difference. The actual difference between the two situations should be lessened by sunny areas and sunflecks under shrub clumps (they were not taken into account during this study).

During the fast growing period, grass leaf water potentials are lower in the open than beneath tree canopy, particularly in the early afternoon, but do

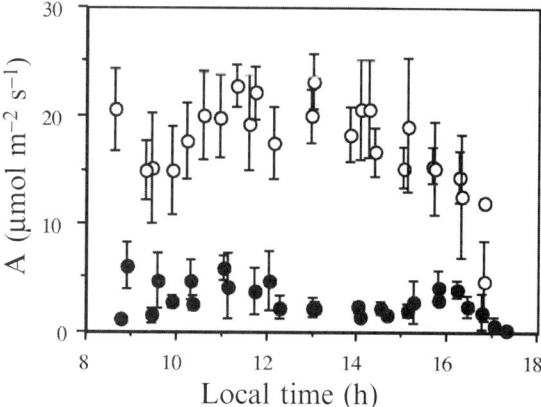

Fig. 8.11. Diurnal variations in the photosynthetic assimilation rate of grass leaves located in the open and canopy situations during the rainy season. Bars are standard errors ($n = 4$) (reprinted from [53], copyright (1993) with permission of Elsevier).

not reach the incipient plasmolysis, i.e., −1.2 MPa (Fig. 8.12). Given that soil water contents are similar between the two situations (see below), this result is explained by the higher transpiration demand in the open than beneath tree canopy, due to higher temperature and net radiation at the leaf surface in sunny areas. This is consistent with data reported for two grass species growing beneath savanna tree canopies in Kenya [38].

8.4.2 Tree effect on grass aboveground biomass and primary production

In Lamto, the major grass species contribute equally to the total grass biomass beneath a woody canopy and in the open [56]. The grass tuft cover is two-fold lower beneath tree canopy than in the open. This results from a lower tuft number and from a lower tuft width (Table 8.2). Maximum

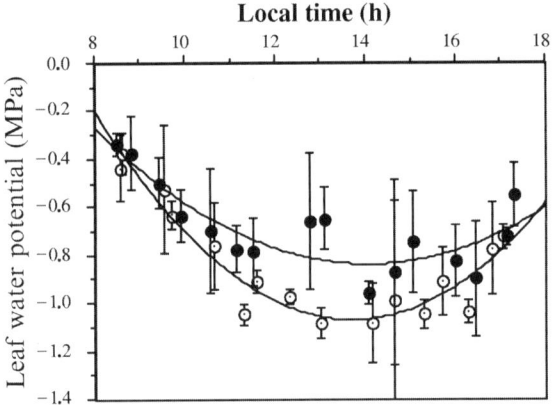

Fig. 8.12. Diurnal variations in the leaf water potential of grass leaves located in the open and canopy situations during the rainy season. Bars are standard errors ($n = 4$) (reprinted from [53], copyright (1993) with permission of Elsevier).

Table 8.2. Mean grass tuft characteristics and standard errors ($n = 144$) under tree clump canopy and in the open. Ground cover is expressed as linear meter covered by herbaceous vegetation per ground linear meter (after [56], with permission of Opulus Press).

	Canopy		Open	
	Mean	s.e.	Mean	s.e.
Basal ground cover (%)	6.23	0.45	12.03	0.60
Tuft number per linear ground m	1.32	0.07	2.06	0.08
Tuft width (cm)	4.54	0.30	6.23	0.33

aboveground grass biomass is two-fold lower beneath a woody canopy (about 300 g m^{-2}) than in the pure grass areas (more than 600 g m^{-2}) [56]. Grass phytomass dynamics is similar from mid-January (bush fire) to June, but from July onward, aboveground grass biomass remains stable or decreases beneath the tree canopy, whereas it keeps increasing until October in the open (Fig. 8.13). Thus, after June, grass production beneath the tree canopy is not high enough to compensate for leaf mortality and decomposition. The major factor explaining this result could be the limitation of grass photosynthesis due to the radiation interception by the tree canopy (Sect. 8.2). Simioni et al. [69] showed that grass phytomass at the end of the year is highly correlated to the sky gap fraction measured with fish-eye photographs taken skyward above grass plots (Fig. 8.14). This approach allows spatial generalization of the findings of the classical approach comparing open areas to dense shrub clumps situations. Temporal surveys of grass biomass and necromass were also made for different gap fraction (i.e., light level) classes [68].

Annual grass primary production was calculated by summing the positive increments of biomass, and the positive increments of necromass when the increment of biomass is positive and when the absolute value of the increment of biomass is smaller than the increment of necromass [21]. Aboveground annual grass primary production was found to be 794 g m^{-2} yr^{-1} and 1073 g m^{-2} yr^{-1} under dense tree clump canopy and in the open, respectively.

8.4.3 Tree effect on grass shoot/root ratio

The comparison between the grass shoot/root ratio (i.e., ratio of the grass aboveground phytomass to belowground phytomass) observed in the open and shrub clump situations provides further information about the tree influence on light, soil water and nutrient availability for grasses. Indeed, according to

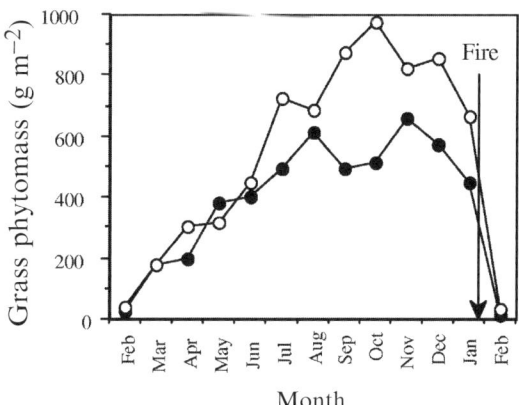

Fig. 8.13. Seasonal dynamics of the aboveground grass phytomass in the canopy and open situations ($n = 12$) (after [56], with permission of Opulus Press).

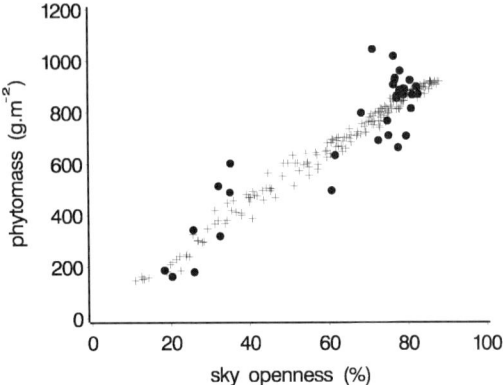

Fig. 8.14. Relationships between sky openness measured by fish-eye photographs taken skyward above the grass layer and grass aboveground phytomass either measured (dots) or simulated by the TREEGRASS model (+) for 1 m × 1 m grass plots at the end of the year (after [69], with permission of the Ecological Society of America).

the functional equilibrium theory [15, 13, 75], plants invest more in organs that will allow them to acquire the most limiting resource. For instance, if a plant is mainly limited by light, it will invest more in shoots than roots, while the reverse is true if water and/or nutrient are more limiting.

In Lamto, the shoot/root ratio is higher beneath the tree canopy during most of the year (Fig. 8.15). This result is consistent with the strong reduction of light availability, the lack of important changes in soil water availability and the slightly improved nutrient availability observed under a tree canopy when compared to open areas (Sect. 8.2). A similar change in the shoot/root ratio has been reported for a Sahelian savanna in Senegal [26].

8.5 Conclusion

In Lamto, trees and grasses are both shallow rooted and largely share the same soil resources. The occurrence of trees results in (i) a strong decrease in the light availability for grasses and thus a decrease in their photosynthetic carbon assimilation, (ii) unchanged soil water regime in the topsoil but improved grass water status due to the shading effect, and (iii) an improved soil nutrient status in the topsoil for grasses growing beneath a tree canopy. The outcome of these complex tree/grass interactions (Fig. 8.16) is a decrease in grass standing crop and production and an increase in the grass shoot/root ratio in the presence of a tree canopy. This shows that in this humid savanna, the improved soil nutrient availability and plant water status beneath a shrub clump do not counter-balance the negative effect of the strong decrease in the light resource.

8 Tree/Grass Interactions   155

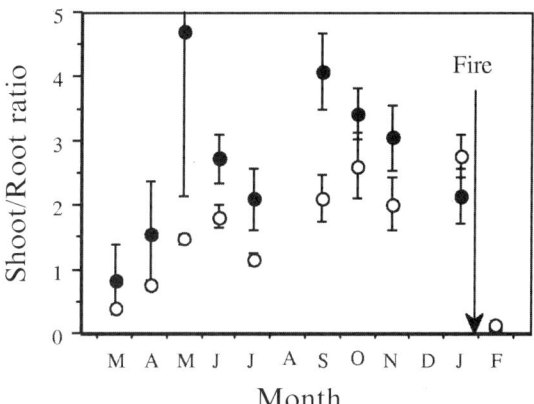

Fig. 8.15. Monthly variations in the grass shoot/root ratio in the open and canopy situations. Bars are standard errors ($n = 2$) (after [52]).

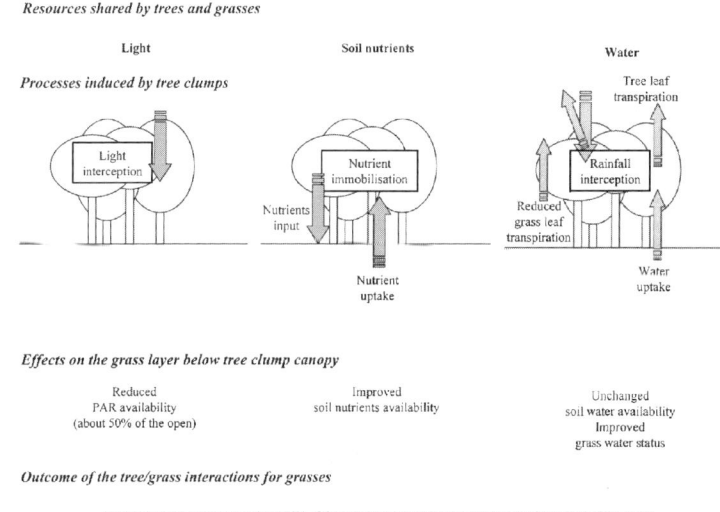

Fig. 8.16. Summary of the relative importance of the actual changes in light, water and nutrient availability for grasses observed under tree canopy as compared to open situations in Lamto, and impact on grass production and shoot/root ratio.

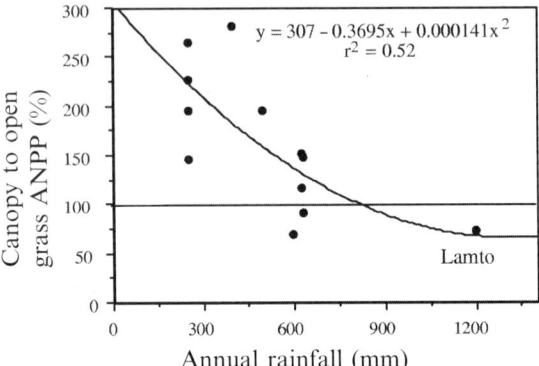

Fig. 8.17. Ratio of annual aboveground grass primary production observed under a tree canopy to the production observed in the open situation, along a rainfall gradient, calculated from various papers dealing with African savanna ecosystems [26, 6, 7, 27, 36, 56].

The outcome of the tree effect on grass production observed at Lamto is probably typical of humid, nutrient-poor savannas, and cannot be generalized to other savannas. Indeed, along a rainfall gradient in the savanna domain, woody plants alter the light, water and nutrient availability on various ways, and the outcome of the tree effect on grasses changes with changing rainfall regime. This is illustrated by a review of published results dealing with the influence of trees on the grass production, as observed in different savanna environments (Fig. 8.17). Broadly speaking, in arid environments, trees generally improve the water budget beneath their canopy (e.g., [34]). Since water is a major limiting resource in these environments, this favors a higher grass primary production beneath a tree canopy than in the open. In addition, in arid savannas, the studies are most often performed on single trees that are generally legume trees. These trees thus decrease light availability less than shrub clumps do in humid savanna and are liable to improve significantly the nutrient soil conditions. On the opposite, in humid savannas, woody plants generally do not alter markedly the available water for the grass layer. In this case, the decrease in the availability of the light resource is of major importance and results in a lower grass primary production beneath a tree canopy than in the open (an increase in nutrient availability under the tree canopy could however mitigate this effect). As proposed by Simioni et al. [71], predicting the effect of tree density and spatial distribution on the production of the grass layer in savanna ecosystems thus implies to develop a modeling tool that properly accounts for the effects of the structure of the tree layer on changes in the spatial distribution of the light, water and nutrient availability (see Chap. 9).

References

1. L. Abbadie, M. Lepage, and X. Le Roux. Soil fauna at the forest-savanna boundary: role of the termite mounds in nutrient cycling. In J. Proctor, editor, *Nature and dynamics of forest-savanna boundaries*, pages 473–484. Chapman & Hall, London, 1992.
2. T. Bariac, A. Klamecki, C. Jusserand, and R. Létolle. Evolution de la composition isotopique de l'eau (^{18}O) dans le continuum sol-plante-atmosphère. Exemple d'une parcelle cultivée en blé, Versailles, France, juin 1984. *Catena*, 14:55–72, 1987.
3. S. Barot, J. Gignoux, and J.-C. Menaut. Demography of a savanna palm tree: predictions from comprehensive spatial pattern analyses. *Ecology*, 80(6):1987–2005, 1999.
4. I.F. Beale. Tree density effects on yields of herbage and tree components in south west Queensland mulga (*Acacia aneura* F. Muell.) scrub. *Tropical Grasslands*, 7(1):135–142, 1973.
5. A. Belsky. Influence of trees on savanna productivity: test of shade, nutrients, and tree-grass competition. *Ecology*, 75:922–932, 1994.
6. A.J. Belsky, R.G. Amundson, J.M. Duxbury, S.J. Rika, A.R. Ali, and S.M. Mwonga. The effects of trees on their physical, chemical, and biological environment in a semi-arid savanna in kenya. *Journal of Applied Ecology*, 26:1005–1024, 1989.
7. A.J. Belsky, S.M. Mwonga, R.G. Amundson, J.M. Duxbury, and A.R. Ali. Comparative effects of isolated trees on their undercanopey environments in a high- and low rainfall savannas. *Journal of Applied Ecology*, 30:143–155, 1993.
8. R. Ben-Shahar. Abundance of trees and grasses in a woodland savanna in relation to environmental factors. *Journal of Vegetation Science*, 2:345–350, 1991.
9. F. Bernhard-Reversat. Les cycles des éléments minéraux dans un peuplement à *Acacia seyal* et leur modification en plantation d'Eucalyptus au Sénégal. *Acta Oecologica, Oecologia Generalis*, 8:3–16, 1987.
10. S.P. Bratton. Resource division in an understorey herb community: responses to temporal and microtopographic gradients. *The American Naturalist*, 110:679–692, 1976.
11. D.D. Breshears, J.W. Nyhan, C.E. Heil, and B.P. Wilcox. Effect of woody plants on microclimate in a semiarid woodland: soil temperature and evaporation in canopy and intercanopy patches. *International Journal of Plant Science*, 159(6):1010–1017, 1998.
12. D.D. Breshears, P.M. Rich, F.J. Barnes, and K. Campbell. Overstory-imposed heterogeneity in solar radiation and soil moisture in a semiarid woodland. *Ecological Applications*, 7(4):1201–1215, 1997.
13. R. Brouwer. Functional equilibrium: sense or nonsense? *Netherlands Journal of Agricultural Sciences*, 31:335–348, 1983.
14. C. Dancette and J.F. Poulain. Influence of *Accacia albida* on pedoclimatic factors and crop yields. *African Soils*, 14:143–184, 1969.
15. R.L. Davidson. Effects of soil nutrients and moisture on root/shoot ratios in *Lolium perenne* L. and *Trifolium repens* L. *Annals of Botany*, 33:571–577, 1969.
16. P. Drechsel, B. Glaser, and W. Zech. Effect of four multipurpose tree species on soil amelioration during tree falow in central Togo. *Agroforestry Systems*, 16:193–202, 1991.

17. K.M. Dunham. Comparative effects of *Acacia albida* and *Kigelia africana* trees on soil characteristics in Zambezi riverine woodlands. *Journal of Tropical Ecology*, 7:215–220, 1991.
18. P. Eagleson and R. Segarra. Water-limited equilibrium of savanna vegetation systems. *Water Resource Research*, 21(10):1483–1493, 1985.
19. J. Ehleringer and T. Dawson. Water uptake by plants: perspectives from stable isotope composition. *Plant Cell and Environment*, 15:1073–1082, 1992.
20. J.R. Ehleringer and C.B. Osmond. Stable isotopes. In R.W. Pearcy, J.R. Ehleringer, H.A. Mooney, and P.W. Rundel, editors, *Plant physiological ecology. Field methods and instrumentation.*, pages 281–300. Chapman & Hall, London, 1989.
21. A. Fournier and M. Lamotte. Estimation de la production primaire des milieux herbacés tropicaux. *Annales de l'Université d'Abidjan, Série E*, 16:7–38, 1983.
22. P. Frost, E. Medina, J.-C. Menaut, O. Solbrig, M. Swift, and B. Walker. Responses of savannas to stress and disturbance. *Biology International*, 10:1–82, 1986.
23. W.E. Frost and S.B. Edinger. Effects of tree canopies on soil characteristics of annual rangeland. *Journal of Range Management*, 44:286–288, 1991.
24. G. Simioni, A. Walcroft, X. Le Roux, and J. Gignoux. Leaf gas exchange characteristics and water- and nitrogen-use efficiencies of dominant grass and tree species in a West African savanna. *Plant Ecology*, 173:233–246, 2003.
25. H. Gauthier. *Echanges radiatifs et production primaire dans une savane humide d'Afrique de l'Ouest (Lamto - Côte d'Ivoire)*. Dess de télédétection, Université de Toulouse, 1993.
26. M. Grouzis and L.E. Akpo. Influence of tree cover on herbaceous above- and below-ground phytomass in the Sahelian zone of Senegal. *Journal of Arid Environments*, 35:285–296, 1997.
27. J.O. Grunow, H. Groeneveld, and S. Du Toit. Above-ground dry matter dynamics of the grass layer of a South African tree savanna. *Journal of Ecology*, 68:877–889, 1980.
28. B.I. Hesla, H.L. Tieszen, and T.W. Boutton. Seasonal water relations of savanna shrubs and grasses in Kenya, East Africa. *Journal of Arid Environments*, 8:15–31, 1985.
29. J.I. House, S. Archer, D.D. Breshears, R.J. Scholes, M.B. Coughenour, M.B. Dodd, J. Gignoux, D.O. Hall, N.P. Hanan, R. Joffre, X. Le Roux, J.A. Ludwig, J.-C. Menaut, R. Montes, W.J. Parton, J.J. San José, J.C. Scanlan, J.M.O. Scurlock, G. Simioni, and B. Thorrold. Conundrums in mixed woody-herbaceous plant systems. *Journal of Biogeography*, 30:1763–1777, 2003.
30. A.O. Isichei and J.I. Muoghalu. The effects of tree canopy cover on soil fertility in a Nigerian savanna. *Journal of Tropical Ecology*, 8:329–338, 1992.
31. L.E. Jackson, R.B. Strauss, M.K. Fireston, and J.W. Bartolome. Influence of tree canopies on grasslands productivity and nitrogen dynamics in deciduous oak savanna. *Agriculture, Ecosysems and Environment*, 32:89–105, 1990.
32. P.C. Jackson, F.C. Meinzer, M. Bustamante, G. Goldstein, A. Franco, P.W. Rundel, L. Caldas, E. Igler, and F. Causin. Partitioning of soil water among tree species in a Brazilian cerrado ecosystem. *Tree Physiology*, 19:717–724, 1999.
33. R. Joffre and S. Rambal. Soil water improvement by trees in the rangelands of southern Spain. *Acta Oecologica, Oecologia Plantarum*, 9(4):405–422, 1988.
34. R. Joffre and S. Rambal. How tree cover influences the water balance of mediterranean rangelands. *Ecology*, 74(2):570–582, 1993.

35. M. Kellman. Soil enrichment by neotropical savanna trees. *Journal of Ecology*, 67:565–577, 1979.
36. D.G. Kennard and B.H. Walker. Relationships between tree canopy cover and *Panicum maximum* in the vinicity of Fort Victoria. *Rhodesian Journal of Agricultural Research*, 11:145–153, 1973.
37. J.J. Kessler and H. Breman. The potential of agroforestry to increase primary production in the Sahelian and Sudanian zones of West Africa. *Agroforestry Systems*, 13:41–62, 1991.
38. J.I. Kinyamario, M.J. Trlica, and T.J. Njoka. Influence of tree shade on plant water status, gas exchange, and water use efficiency of *Panicum maximum* Jacq. and *Themeda triandra* Forsk. in Kenya savanna. *African Journal of Ecology*, 33:114–123, 1995.
39. W.T. Knoop and B.H. Walker. Interactions of woody and herbaceous vegetation in a southern african savanna. *Journal of Ecology*, 73:235–253, 1985.
40. B. Koechlin, S. Rambal, and M. Debussche. Rôle des arbres pionniers sur la teneur en eau du sol en surface de friches de la région méditéranéenne. *Acta Oecologica, Oecologia Plantarum*, 7:177–190, 1986.
41. S. Konaté, X. Le Roux, D. Tessier, and M. Lepage. Influence of large termitaria on soil characteristics, soil water regime, and tree leaf shedding pattern in a West African savanna. *Plant and Soil*, 206:47–60, 1999.
42. G.W. Lawson, J. Jenik, and K.O. Armstrong-Mensah. A study of a vegetation catena in Guinea savanna at mole game reserve (Ghana). *Journal of Ecology*, 56:505–522, 1968.
43. X. Le Roux and T. Bariac. Seasonal variations in soil, grass and shrub water status in a West African humid savanna. *Oecologia*, 113:456–466, 1998.
44. X. Le Roux, T. Bariac, and A. Mariotti. Spatial partitioning of the soil water resource between grass and shrub components in a West African humid savanna. *Oecologia*, 104(2):147–155, 1995.
45. M. Lepage, L. Abbadie, and A. Mariotti. Food habits of sympatric termites species (Isoptera, Macrotermitinae) as determined by stable isotope analysis in a Guinean savanna (Lamto, Côte d'Ivoire). *Journal of Tropical Ecology*, 9:303–311, 1993.
46. M. Ludlow, J. Troughton, and R. Jones. A technique for determining the proportion of C_3 and C_4 species in plant samples using natural isotopes of carbon. *Journal of Agricultural Science*, 87:625–632, 1976.
47. F. Ludwig. *Tree-grass interactions on an East African savanna: the effects of competition, facilitation and hydraulic lift*. Ph.D. thesis, Wageningen Universiteit, 2001.
48. J.C. Menaut. *Etude de quelques peuplements ligneux d'une savane guinéenne de Côte d'Ivoire*. Ph.D. thesis, Université de Paris, 1971.
49. J.C. Menaut. The vegetation of african savannas. In F. Bourlière, editor, *Tropical savannas.*, volume 13 of *Ecosystems of the world.*, pages 109–149. Elsevier, Amsterdam, 1983.
50. J.C. Menaut, J. Gignoux, C. Prado, and J. Clobert. Tree community dynamics in a humid savanna of the Côte d'Ivoire: modelling the effects of fire and competition with grass and neighbours. *Journal of Biogeography*, 17:471–481, 1990.
51. C. Montana, B. Cavagnaro, and O. Briones. Soil water use by co-existing shrubs and grasses in the Southern Chihuahuan desert, Mexico. *Journal of Arid Environments*, 31:1–13, 1995.

52. P. Mordelet. *Influence des arbres sur la strate herbacée d'une savane humide (Lamto, Côte d'Ivoire)*. Ph.D. thesis, Université Paris 6, 1993.
53. P. Mordelet. Influence of tree shading on carbon assimilation of grass leaves in Lamto savanna, Côte d'Ivoire. *Acta Oecologica*, 14(1):119–127, 1993.
54. P. Mordelet, L. Abbadie, and J.C. Menaut. Effects of tree clumps on soil characteristics in a humid savanna of West Africa (Lamto, Côte d'Ivoire). *Plant and Soil*, 153:103–111, 1993.
55. P. Mordelet, S. Barot, and L. Abbadie. Root foraging strategies and soil patchiness in a humid savanna. *Plant and Soil*, 182:171–176, 1996.
56. P. Mordelet and J.C. Menaut. Influence of trees on above-ground production dynamics of grasses in a humid savanna. *Journal of Vegetation Science*, 6:223–228, 1995.
57. P. Mordelet, J.C. Menaut, and A. Mariotti. Tree and grass rooting patterns in an African humid savanna. *Journal of Vegetation Science*, 8:65–70, 1997.
58. D.U.U. Okali, J.B. Hall, and G.W. Lawson. Root distribution under a thicket clump on the Accra plains, Ghana: its relevance to clump localization and water relations. *Journal of Ecology*, 61:439–454, 1973.
59. C. Ovale and J. Avendano. Interactions de la strate ligneuse avec la strate herbacée dans les formations d'*Acacia caven* (Mol.) Hook. et Arn. au Chili. II. Influence de l'arbre sur quelques éléments du milieu: microclimat et sol. *Acta Oecologica, Oecologia Plantarum*, 9:113–134, 1988.
60. V.T. Parker and C.H. Muller. Vegetational and environmental changes beneath isolated live oak trees (*Quercus agrifolia*) in a California annual grassland. *The American Naturalist*, 107:69–81, 1982.
61. J.A.D. Parrish and F. A. Bazzaz. Underground niche separation in successional plants. *Ecology*, 57:1281–1288, 1976.
62. A.J. Pressland. Productivity and management of mulga in south-western Queensland in relation to tree structure and density. *Australian Journal of Botany*, 23:965–976, 1975.
63. O. E. Sala, R.A. Golluscio, W.K. Lauenroth, and A. Soriano. Resource partitioning between shrubs and grasses in the Patagonian steppe. *Oecologia*, 81:501–505, 1989.
64. G. Sarmiento. *The ecology of neotropical savannas*. Harvard University Press, Cambridge, MA, 1984.
65. R. Scholes and B. Walker. *An African savanna. Synthesis of the Nylsvley study*. Cambridge University Press, Cambridge, 1993.
66. R.J. Scholes. *Response of three semi-arid savannas on contrasting soils to the removal of the woody component*. Ph.D. thesis, University of Witwatersrand, Johannesburg, 1987.
67. R.J. Scholes and S. Archer. Tree-grass interactions in savannas. *Annual Review of Ecology and Systematics*, 28:517–544, 1997.
68. G. Simioni. *Importance de la structure spatiale de la strate arborée sur les fonctionnements carboné et hydrique des écosystèmes herbes-arbres*. Ph.D. thesis, Université Pierre et Marie Curie, Paris, 2001.
69. G. Simioni, J. Gignoux, and X. Le Roux. Tree layer spatial structure can affect savanna production and water budget: Results of a 3-D model. *Ecology*, 84(7):1879–1894, 2003.
70. G. Simioni, J. Gignoux, X. Le Roux, R. Appé, and D. Benest. Spatial and temporal variations in leaf area index, specific leaf area, and leaf nitrogen of two co-occurring savanna tree species. *Tree Physiology*, 24:205–216, 2002.

71. G. Simioni, X. Le Roux, J. Gignoux, and H. Sinoquet. Treegrass: a 3D, process-based model for simulating plant intercations in tree-grass ecosystems. *Ecological Modelling*, 131:47–63, 2000.
72. B.N. Smith and S. Epstein. Two categories of $^{13}C/^{12}C$ ratios for higher plants. *Plant Physiology*, 47:380–384, 1971.
73. R.M. Strang. Soil moisture relations under grassland and under woodland in the Rhodesian highveld. *Commonwealth Forestry Review*, 48(1):26–40, 1969.
74. T.J. Svejcar and T.W. Boutton. The use of stable carbon isotope analysis in rooting studies. *Oecologia*, 67:205–208, 1985.
75. D. Tilman. Constraint and tradeoff: toward a perspective theory of competition and succession. *Oikos*, 58:3–15, 1990.
76. R. Valentini, G.E. Scarascia Mugnozza, and J.R. Ehleringer. Hydrogen and carbon isotope ratios of selected species of a mediterranean macchia ecosystem. *Functional Ecology*, 6:627–631, 1992.
77. B. Walker and I. Noy-Meir. Aspects of stability and resilience of savannas ecosystems. In B.J. Huntley and B.H. Walker, editors, *Ecology of tropical savannas*, pages 556–590. Springer-Verlag, Berlin, 1982.
78. C.D. Walker and S.B. Richarson. The use of stable isotopes of water in characterising the source of water in vegetation. *Chemical Geology (Isotope Geoscicences Section)*, 94:145–158, 1991.
79. H. Walter. *Natural savannas. Ecology of tropical and subtropical vegetation.* Oliver and Boyd, Edinburgh, 1971.
80. J.F. Weltzin and M.B. Coughenour. Savanna tree influence on understorey vegetation and soil nutrients in northwestern Kenya. *Journal of Vegetation Science*, 1:325–334, 1990.
81. J.W.C. White, E.R. Cook, J.R. Lawrence, and W.S. Broecker. The D/H ratios of sap in trees: implications for water sources and tree ring D/H ratios. *Geochimica et Cosmochimica Acta*, 49:237–246, 1985.

9

Modeling the Relationships between Vegetation Structure and Functioning, and Modeling Savanna Functioning from Plot to Region

Xavier Le Roux, Jacques Gignoux, and Guillaume Simioni

9.1 Introduction

As anticipated by Scholes and Walker [19] and House et al. [6], the herbaceous layer in a savanna can be considered as a mosaic of microenvironments, dependent on the spatial distribution of the tree canopies and rooting zones [16]. The overall functioning (and, moreover, the spatial variations in the functioning) of the grass layer depends on the relative extent and properties of the component subhabitats. Similarly, the functioning of the whole tree layer (and, moreover, the frequency distribution of the functioning of tree individuals or tree parts, e.g., [12, 26]) also depends on the tree spatial distribution. Such a complexity, in which the spatial structure of the vegetation plays a major role, has led some authors to admit that modeling was probably the best approach to predict savanna functioning [20].

However, for sake of simplicity and because of computation requirements, most current models aiming at describing savanna functioning are not spatially explicit (but see [7, 8, 4]). Furthermore, no savanna model has already tested the potential importance of the fine scale spatial structure of the vegetation (e.g., the spatial location and shape of tree individuals) on water balance and net primary production (NPP) in these ecosystems.

This context has spurred on the development of a spatially explicit (representing each individual in space) and mechanistic (reproducing explicitly the main processes driving tree/grass interactions, see Chap. 8) modeling approach to analyze structure-functioning relationships in savanna ecosystems, and more particularly to test whether the fine scale spatial structure of the vegetation can influence net primary production and water balance in Lamto savanna.

This chapter presents a short overview of previous models developed for simulating the functioning of homogeneous savanna areas in Lamto. Then, the 3-dimensional (3D) model TREEGRASS [25], devoted to study the relationships between savanna structure and functioning, is presented. Its ability to simulate the seasonal, interannual, and spatial variations of the savanna water balance and NPP are discussed. Approaches to model the savanna functioning at larger scales are also presented.

9.2 Models previously developed for predicting the functioning of Lamto savannas

To assist our understanding of the Lamto savanna functioning, several models were proposed in the 1980s and used for "homogeneous" areas.

Clément [3] simulated the soil water balance and soil thermal regime of a grass-dominated savanna site as part of a soil fauna activity model [17]. This process-based model considered 5 or 10 cm soil layers. The model adequately predicted the seasonal and vertical variations in soil temperature and soil moisture.

Prado [18] and Boulier [1] simulated the primary production and water balance of Lamto savannas, distinguishing C_3 and C_4 species. Primary production was computed as the product of a potential NPP and reduction factors accounting for the effects of water stress and light regime. Due to lack of information, these models applied empirical and sometimes inadequate treatments of some processes (e.g., radiation absorption by plants, evapotranspiration, and water uptake by roots). However, these models provided first insights into the analysis of the seasonal and interannual variability of savanna NPP and water balance.

The experiment dealing with savanna-atmosphere interactions recently carried out at Lamto (Sect. 6.2) provided new insights into essential but poorly documented processes, such as radiation transfer and energy budget. This has paved the way to the development and validation of models simulating savanna functioning that explicitly link radiation exchanges, water balance, primary production, and development of the canopy leaf area for annual or longer periods. The model PEPSEE-grass (Production Efficiency and Phenology in Savanna EcosystEms) [11, 14, 15] was developed to study the seasonal and interannual variations in the production, water balance, and leaf area development of grass-dominated areas in Lamto. This model included three submodels that simulate (1) radiation absorption by the canopy, (2) primary production and phenology, and (3) evapotranspiration and soil water balance. The model adequately simulated the seasonal courses of green and dead leaf areas, soil moisture in two layers, and evapotranspiration rate. Accounting for the dependency of grass productivity on canopy nitrogen status and/or for root/shoot allocation significantly improved the simulation of NPP at the beginning of the vegetation cycle [15]. The model was also used to assess the impact of increasing grazing intensity on primary production of Lamto grasslands ([15], see Chap. 10). However, PEPSEE-grass was not designed for simulating the complex interactions between the tree and grass layers in heterogeneous shrubby areas (Chap. 8).

9.3 TREEGRASS: A 3D model for simulating structure-functioning relationships in savanna ecosystems

9.3.1 Overview of the model TREEGRASS

TREEGRASS [25, 22] is a spatially explicit, individual-based ecosystem model that simulates water fluxes and primary production for small savanna plots (from 100 to 10000 m^2) over one or a few vegetation cycles, with a hourly to daily time step. Competition for light and water are treated mechanistically. It includes submodels (Fig. 9.1) derived from the 3D RATP model (Radiation Absorption, Transpiration and Photosynthesis [26]) that computes

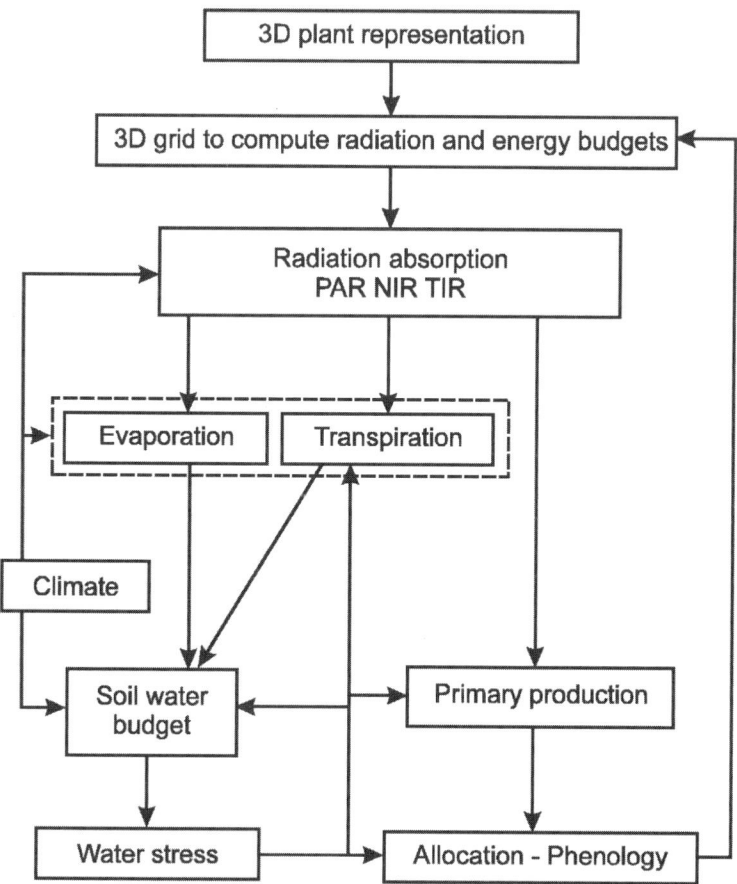

Fig. 9.1. Processes computed in the TREEGRASS model (after [25], copyright (2000), with permission of Elsevier).

radiation transfer and energy budget within vegetation canopies, and from the PEPSEE-grass model [15] that simulates primary production and soil water balance. It works within the MUSE simulation framework (MUltistrata Spatially Explicit model) [5] that provides background algorithms for spatially explicit ecosystem modeling. Two major objectives of the development of TREEGRASS were (1) to help understand the ecosystem functioning in relation to its structure, by providing a framework within which to analyze data, test hypotheses, and identify areas of poor understanding in process description and (2) to provide a predictive tool for assessing savanna functioning under different environmental conditions, and particularly future environmental scenarios. The main features of TREEGRASS are the following:

1. Space is divided into a 3D grid of cells (Fig. 9.2) with user-definable dimensions. Aboveground cells can contain grass and/or tree foliage elements. There are two layers of belowground cells: layer 1 defined so that it contains 90% of grass roots, and layer 2 underneath is defined according to the maximum plant rooting depth.
2. The continuous grass layer is divided into plots according to cell basal dimensions. Grass foliage is located in one or several above ground cells. Grass roots are located in the two soil cells underneath. The height, foliage and root crown radii, and maximal leaf area index of individual trees are linked by allometric relationships. In the current version of TREEGRASS, tree foliage and root crowns are assumed to be cylindrical, but other shapes can be prescribed. Tree crowns can overlap with grass cells and other tree crowns. Because the model is not applied over periods longer

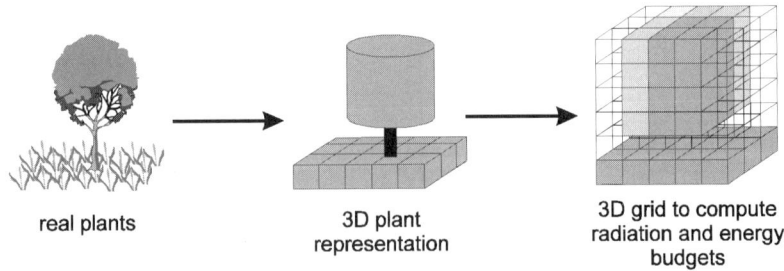

Fig. 9.2. 3D representation of the spatial distribution of the vegetation in the model TREEGRASS. The second picture shows the simple plant structural features used to represent trees (i.e., simple cylindrical crown, crown radius, and total height and bole height) and grasses (grass individuals are lumped into homogeneous plots and they form a continuous layer). Roots are represented in a similar way (not shown). The third picture represents the 3D grid used to compute the spatial distribution of plant foliage (different levels of grey correspond to different values of leaf area density (LAD)) (after [25], copyright (2000), with permission of Elsevier).

than 1 to 3 years in its present version, tree dimensions are assumed to be constant, and only the leaf area density (LAD) and root mass density change within individual foliage and root volumes.

3. A radiation absorption submodel calculates the amounts of photosynthetically active radiation, near infrared radiation, and thermal infrared radiation that are absorbed or emitted at each time step by the foliage in each vegetation cell and by the soil surface. This submodel accounts for daily and yearly solar course, for diffuse radiation, and for reflection and transmission by foliage entities and soil surface. It depends on the 3D distribution of grass and tree leaf area, leaf angle distribution, and leaf optical properties defined for each plant species.

4. The transpiration and evaporation submodel computes the energy budget for each entity in each cell, i.e., the balance between absorbed and emitted radiation, sensible heat flux, and latent heat flux (leaf transpiration or soil evaporation). Transpiration and evaporation are derived from the energy budget equation.

5. Amounts of transpired water for each entity in each vegetation cell are summed to obtain plant individual transpiration rates. The water stress experienced by each plant depends on the soil water content in the cells where its roots are present. Water transpired by a given individual plant is extracted in layers 1 and 2 according to the amount of roots present in these layers and to the intensity of the water stress experienced by the plant.

6. When rainfall occurs, runoff is computed according to the degree of soil coverage by vegetation and rainfall amount. Drainage from layer 1 to layer 2, and below layer 2, is computed by a simple bucket model.

7. In the first version of the model [25], total absorbed PAR by each plant individual is converted into dry matter using the light use efficiency (LUE) approach. The actual LUE is computing according to a species-dependent, maximum LUE, and a reducing factor accounting for water stress. In the second version of the model [22], biochemically based modules of leaf photosynthesis are used to compute carbon gain by C_4 grass secies and C_3 tree species. Net primary production of plant individual is then computed by using species-specific values of the net-to-gross production ratio [22].

8. The ratio of the amount of dry matter produced allocated to roots to the amount allocated to shoots is computed as a function of the ratio of actual to maximum LUE values and thus depends on water stress. As water stress gets stronger, the proportion of dry matter allocated to roots increases.

9. Grass or tree leaf mortality is a function of vegetation phenology and water stress, while the decomposition rate of dead grass leaves is assumed to be constant.

A full description of TREEGRASS and its parameterization for Lamto savanna is presented by Simioni et al. [25, 24, 22].

9.3.2 Ability of TREEGRASS to predict the temporal and spatial variations in savanna functioning

Temporal variations in water balance and primary production

TREEGRASS was tested against field data documenting the seasonal variations in radiation absorption by the tree and grass layers, and the seasonal variations in soil water balance and primary production in grass-dominated or shrubby areas in Lamto savannas [25, 24]. The seasonal dynamics of grass biomass, grass necromass, and soil water content were adequately simulated. An example of model testing against field data of seasonal variations in soil water content is given in Fig. 9.3.

The database acquired during field studies carried out at Lamto since the early 1960s allows accurate testing of the ability of TREEGRASS to correctly simulate the seasonal, and moreover interannual, variations in the savanna functioning. For instance, it has been possible to compare the model outputs with the seasonal variations in grass biomass, necromass, or phytomass as observed in *Andropogon/Hyparrhenia* dominated sites over 13 different vegetation cycles. TREEGRASS generally predicts the interannual variations in grass standing crop reasonably well, although it underestimates grass standing

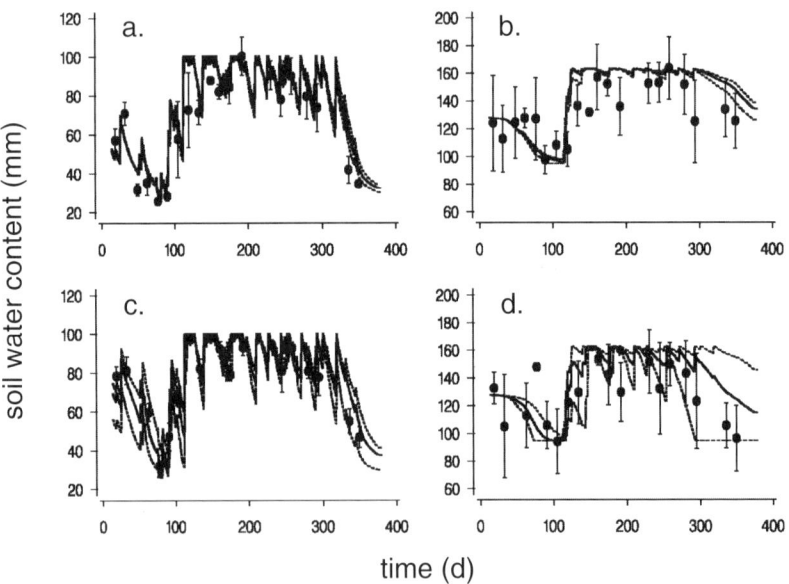

Fig. 9.3. Observed (circles) and simulated (lines) soil water contents in layers 1 (a and c) and 2 (b and d) for open (a and b) and tree clump (c and d) areas. Bars are standard deviations. Solid lines represent average simulated values and dashed lines represent maximum and minimum simulated values (after [22]).

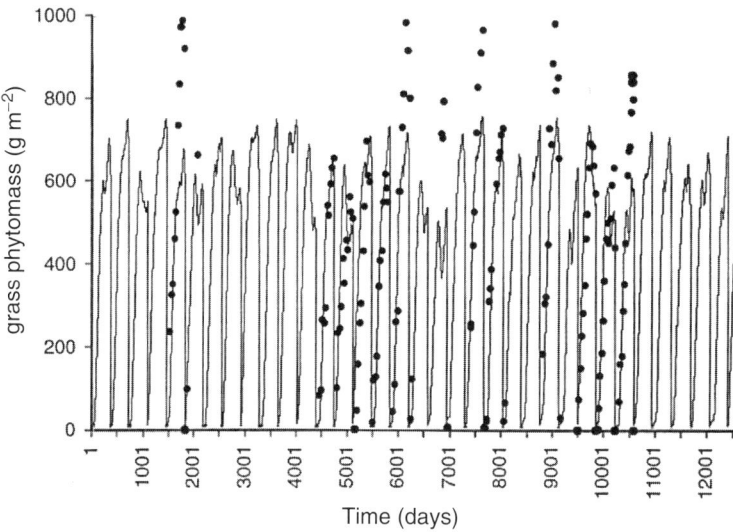

Fig. 9.4. Comparison of the observed (symbols) and simulated (lines) variations in total standing crop along 34 annual cycles in *Hyparrhenia/Andropogon* dominated areas of Lamto. Observations are available for 13 annual cycles (after Gignoux, Simioni, and Le Roux, unpublished).

crop during a few vegetation cycles (Fig. 9.4). Even if sites with similar vegetation were selected, site effects such as local soil fertility could partly explain these discrepancies. Similar tests have been conducted for soil water balance and in *Loudetia* dominated sites (unpublished).

Spatial variations in water balance and primary production and impact of tree cover

The ability of TREEGRASS to predict the spatial variations in savanna functioning has been tested by comparing model outputs and field data obtained over a same period in a pure grass area and a dense shrubby area nearby [22]. More recently, field data documenting the temporal and spatial variations in grass production according to tree cover (quantified by sky openness measured by fish-eye photographs) were used to test the model [25]. TREEGRASS simulates the spatial variations in grass biomass, necromass, or total standing crop reasonably well for all the levels of tree cover (Fig. 9.5). TREEGRASS has also been tested successfully in drier savanna environments, namely dry Australian savannas with contrasting *Eucalyptus* cover, part of a savanna model comparison exercise [6]. Thus, TREEGRASS is a reliable tool for predicting both the temporal and spatial variations of savanna functioning. In particular,

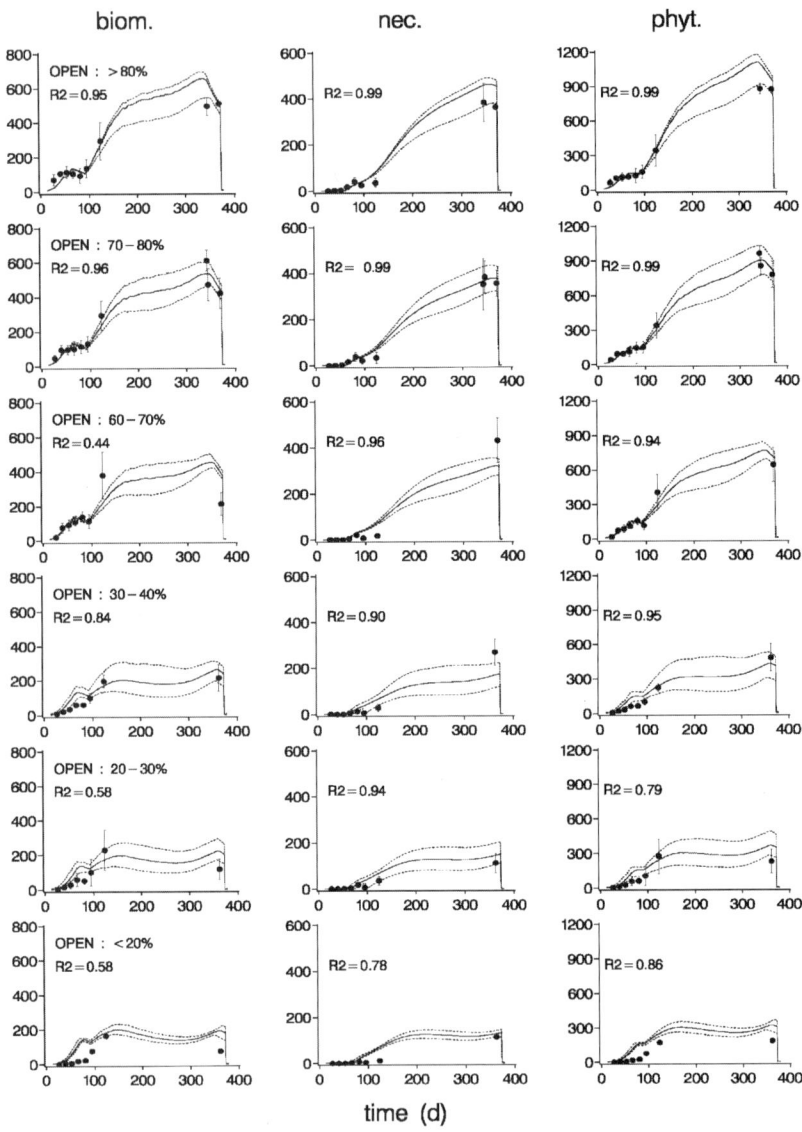

Fig. 9.5. Temporal dynamics of measured (circles) and simulated (lines) grass aboveground biomass, necromass, and phytomass (g m^{-2}) for different sky openness classes (OPEN). Solid line represents the average simulated value for an openness class and dotted lines are the extremes (after [22]).

it can be used to simulate the horizontal variations of the savanna functioning (e.g., grass NPP or transpiration, soil water balance) on a given savanna plot according to the fine scale structure of the tree layer (Fig. 9.6). Because spatial distribution of grass production is important for herbivores, TREEGRASS can provide an original support to study plant-herbivore interactions in tree/grass systems. The model is also very useful to analyze the determinants controlling the outcome of tree/grass interactions in savanna-like systems.

9.3.3 Analyzing the factors driving the outcome of tree/grass interactions

Simulations using the model TREEGRASS were used to unravel the effects of (1) grass shading by trees and (2) tree cover-induced changes in soil moisture for humid and dry West African savannas [22]. Plant nutrient statuses were prescribed from field observations for this study. For the two annual rainfall treatments tested (1200 mm for humid savannas and 600 mm for dry savannas), grass production was reduced as tree cover increased (Fig. 9.7). However, the decrease in grass phytomass with increasing tree cover (i.e., decreasing gap fraction) was much less in the dry environment than humid environment. This was due to a larger beneficial effect of trees on soil water for the low rainfall regime (Fig. 9.7). As observed in Lamto (see Sect. 8.2), the presence of a tree cover weakly influences soil moisture during most of the year (Fig. 9.8). In contrast, the length of good water periods was substantially increased under tree cover as compared to open areas under dry conditions, particularly at the end of the rainy season (Fig. 9.8). Tree/grass ecosystem processes are thus strongly structured in time and space, and this structure has to be taken into account to understand contrasting responses of grass production to changes in tree abundance and rainfall.

9.3.4 Studying structure-functioning relationships in Lamto savanna

TREEGRASS was used to assess the effects of the tree spatial structure on radiation absorption, NPP, and water balance for small savanna areas (around 1000 m^2) in Lamto [23]. The effects of tree density, tree spatial distribution, and crown size distribution on NPP and water fluxes were tested at the ecosystem scale (i.e., using total NPP and evapotranspiration as the target variables) and at the tree- or grass-component scale (i.e., distinguishing total tree NPP and transpiration vs. total grass NPP and transpiration). Among other results, the simulations showed the following: (i) grass NPP decreases non-linearly with increasing tree density, while total NPP of Lamto savanna remains nearly constant with increasing tree density (Fig. 9.9); in this case, the increase in tree NPP compensates the decrease in grass NPP; (ii) annual total evapotranspiration increases with increasing tree density; (iii) total NPP

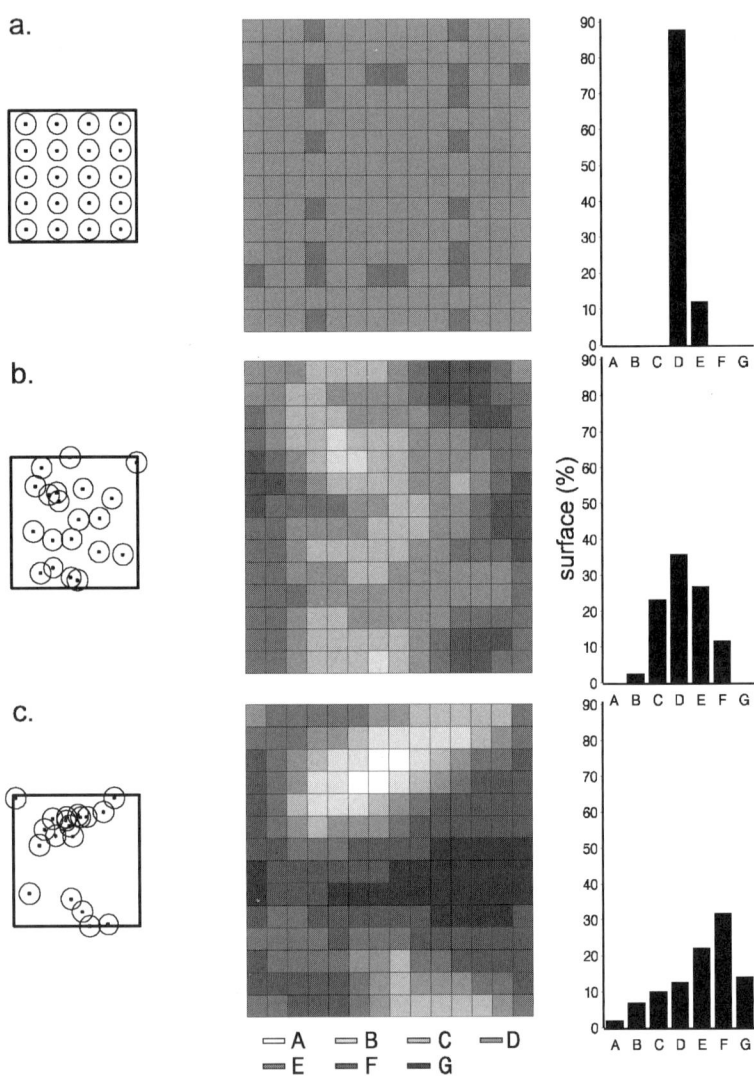

Fig. 9.6. Spatial variability of grass aboveground net primary production with (a) regular, (b) random, and (c) aggregated tree spatial distribution. Left, site maps (14 × 14 m) representing tree canopy locations. Center, distribution of grass above ground production per square meter (A, <300; B, 300-500; C, 500-700; D, 700-900; E, 900-1100; F, 1100-1300; and G, >1300 g m^{-2}). Right, fraction of total site area occupied by each grass production class (after [24]).

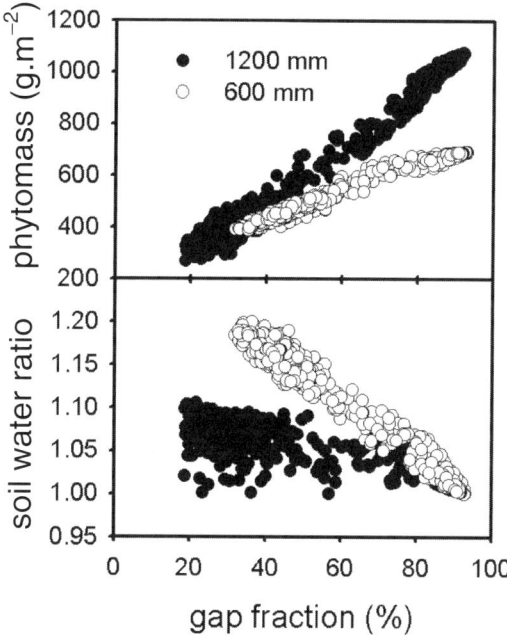

Fig. 9.7. Simulated variations in grass aboveground phytomass and soil water content in the 0–60 cm layer with gap fraction, for a Lamto savanna (annual rainfall 1200 mm) and a Sudano-Sahelian savanna (annual rainfall 600 mm). Values are for 1 × 1 m pixels. Soil water values correspond to annual means and are relative to values in open areas (after [22]).

decreases slightly with increasing tree aggregation, because the increase in grass NPP under-compensates the decrease in tree NPP (Fig. 9.10); (iv) total evapotranspiration remains constant whatever the type of tree aggregation; (v) NPP and transpiration of grasses, trees, and grass plus tree vegetation were largely insensitive to tree size class distribution. Simioni et al. [23] showed that tree LAI is not the best predictor of tree, grass, and tree plus grass functioning. In contrast, the fractional coverage of soil by tree canopy is a reliable predictor of the functioning of tree, grass, and tree plus grass layers (in terms of primary production or transpiration), whatever the fine-scale structure of the tree cover. The fractional coverage of soil by tree canopy can thus be used in savanna models operating at large scales in order to better account for the relationships between vegetation structure and functioning.

9.4 Modeling the functioning of savannas at large scales

Most of the work conducted at Lamto during the last four decades has focused on ecological processes that occur at small spatial scales (from cm^2

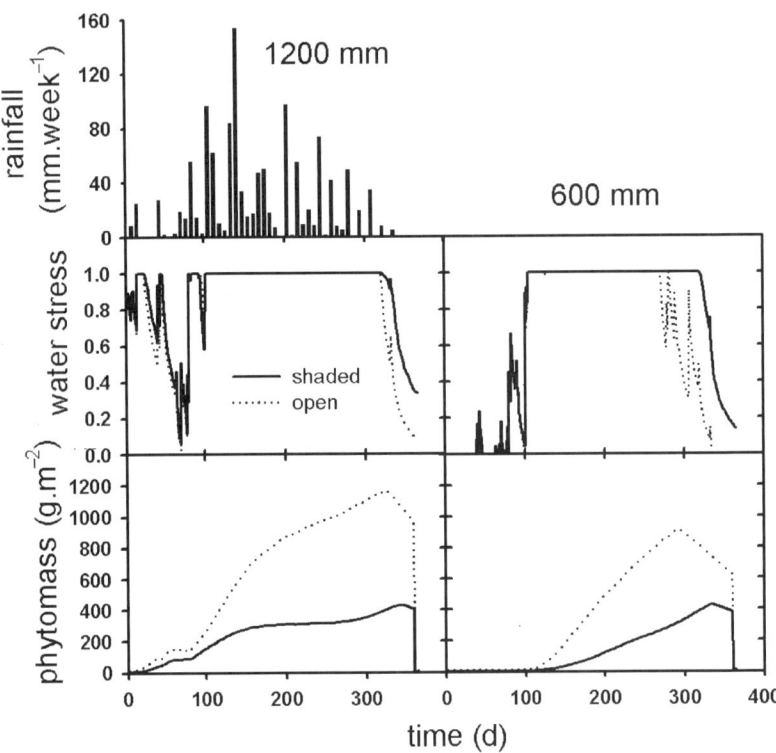

Fig. 9.8. Simulated temporal variations in grass water stress index and grass aboveground phytomass under two rainfall regimes (see caption of Fig. 9.7) and two tree cover classes: water stress and above ground phytomass are average values for grass pixels with gap fractions above 90% (after [22]).

to a few hectares) and did not aim at representing ecological processes at larger scales (landscape, region). Thus, models predicting the functioning of the savanna vegetation were developed to assess this functioning at the ecosystem scale, without particular strategy to scale up to larger scales. However, a hybrid modeling-remote sensing approach was developed by Kergoat et al. [10] in order to apply a process-based savanna function model at the regional scale. In addition, studies were conducted to couple a simple savanna model within general circulation models (GCMs). Such studies at large scales will allow interfacing phenomena that occur at different scales of space, time, and ecological organization. In the medium term, such studies should benefit a biologically sound understanding of the savanna ecosystem functioning.

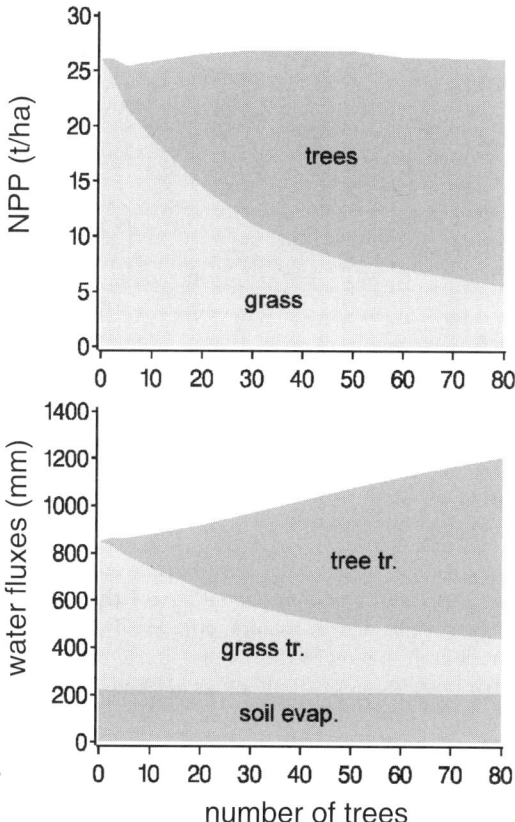

Fig. 9.9. Simulated effects of tree density on annual net primary production (NPP) and transpiration of the grass and tree components or of the whole savanna system (tr., transpiration; evap., evaporation) (after [23], with permission of the Ecological Society of America).

9.4.1 Modeling primary production and water balance in the West African savanna zone: A regional-scale approach using satellite data assimilation

Ecosystem functioning models are generally developed and validated for one or a few test sites, and mix both process-based and empirical formulations (e.g., use of an empirical relationship for simulating carbon allocation between roots and shoots in the largely mechanistic model TREEGRASS). Furthermore, ground or literature estimates of some parameters used can be missing or inaccurate. This severely hampers model application beyond the conditions encountered on test sites during the periods studied. This thus restricts the

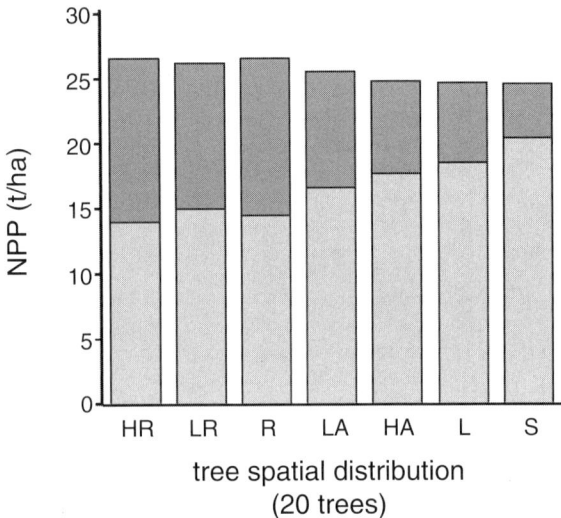

Fig. 9.10. Simulated effects of tree spatial distribution on annual net primary production (NPP) of the grass and tree components or of the whole savanna system. Tree spatial distributions are coded as follows: HR, highly regular; LR, loosely regular; R, random; LA, loosely aggregated; HA, highly aggregated; L, in line; S, in square (after [23], with permission of the Ecological Society of America).

assessment of ecosystem functioning at large scales. Additional information is needed to allow such application. In this context, remote sensing technology offers instantaneous, multi-spectral measurements with adequate spatial coverage and temporal frequency. These measurements provide surrogates of some vegetation canopy charcateristics [21] and can be used to test or constrain models.

A hybrid modeling-remote sensing approach [9] was applied to assess the interannual functioning of the Lamto savanna [10]. The philosophy of the approach, namely satellite data assimilation in a surface water and carbon balance model, is described by Kergoat et al. [9]. Remotely sensed data are used to control the temporal evolution of model simulations. The model simulates the canopy LAI development, from which a spectral signature time series is computed and compared to the measured signature time series. Model simulations are controlled through retrieval of ill-known model parameters (e.g., date of fire occurrence or root:shoot allocation) or slowly evolving state variables (e.g., LAI, absorbed PAR). This procedure, when applied on different pixels within a region, enables to estimate the spatial variations in the retrieved parameters and allows one (i) to analyze the possible cause of these variations and (ii) to extrapolate predictions from pixel to region.

Kergoat et al. [10] tested the assimilation technique at plot scale with data acquired during the 1991-1994 field campaign carried out at Lamto (see

Sect. 6.2). A model simulating carbon budget and water balance of tropical perennial grasslands was developed. The model is based on daily gas exchanges, allocation of photosynthates, and water flows. It simulates the seasonal courses of soil water content and green and dead LAI reasonably well for three contrasted years [10]. The assimilation of ground based radiometric data was used to retrieve two parameters: date of grass emergence after fire, and the relative amount of photosynthates allocated to shoots during the early growing season. Two (low and high) realistic values were chosen a priori for each parameter. The four a priori parameter sets result in four a priori simulations significantly different from observations (Fig. 9.11). The assimilation of remotely sensed data in the model leads to improved simulated LAI seasonal trends (Fig. 9.11) and primary production [10]. These results act as an incentive for further effort devoted to the satellite data assimilation technique.

This approach can clearly be used to increase the accuracy of carbon and water flux estimates at the regional scale (e.g., for humid savannas of West Africa). A major drawback is that the approach does not reinforce the mech-

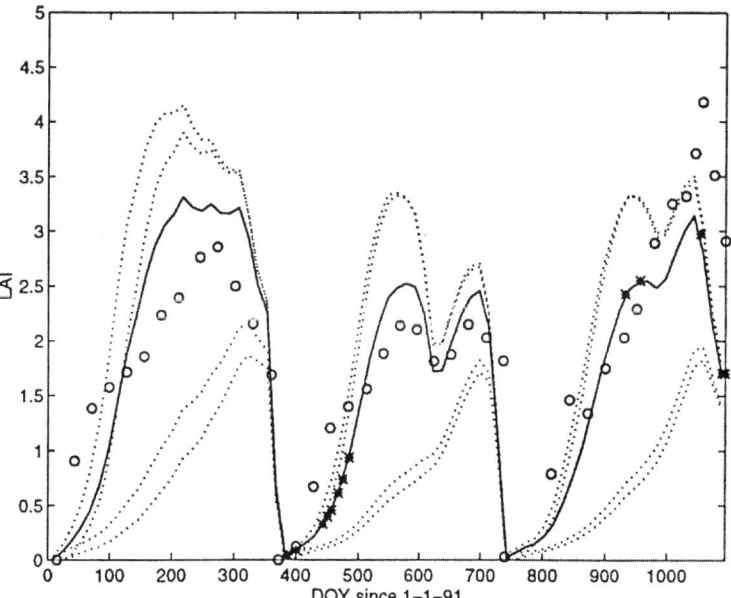

Fig. 9.11. Seasonal courses of green LAI in Lamto for years 1991 to 1993 measured in situ (circles), simulated for a priori parameter sets (dotted lines), and simulated after assimilation of radiometric data (solid line). A priori values were chosen for the day of grass emergence and ratio of root/shoot allocation of photosynthates (after [10], with permission of the Centre National d'Etudes Spatiales).

anistic properties and predictive capabilities of ecosystem functioning models, since satellite data are needed to apply the models to a wide range of environmental conditions. However, identification of a few ill-documented key processes and retrieval of spatially distributed associated parameters (that can be analyzed in the light of concurrent spatially distributed environmental variables) could help in defining key functioning hypotheses. Along with new direct measurements in the field on test sites, this approach can greatly improve our understanding of the savanna functioning and thus the mechanistic bases of savanna models in the future.

9.4.2 Modeling the primary production and phenology of the savanna biome: A global-scale approach in the context of General Circulation Models

Another approach for simulating the savanna functioning at large scales is to include a model of savanna ecosystem within a General Circulation Model (GCM). Such models simulate the climate and water balance of terrestrial surfaces (typically around $1 \times 1°$ or less), but their representation of ecosystem structure and functioning is still crude due to the study scale and the high number of biomes represented. Coupling a savanna ecosystem model to a GCM is a challenging task that could greatly enhance the representation of surface/atmosphere feedbacks in the tropics by GCMs (and perhaps improve the predictions of the effects of surface disturbance on climate) and contribute to better understand the determinants of savanna functioning world-wide.

A first, simple approach was tested by Ciret et al. [2]. A savanna module derived from PEPSEE-grass, representing production, green and dead leaf area development, and fire occurrence of Lamto savanna, was tested with the climate observed in the field. It was then tested by using the forcing climate computed by the Laboratoire de Météorologie Dynamique GCM as an input for the simple savanna model. Tests were done for GCM grid cells corresponding to the Lamto region, but also for sites located in the Northern Territory (Australia) and experiencing annual precipitation ranging from 676 to 953 mm (Fig. 9.12). It was concluded that the approach generates reasonably realistic temporal trends in canopy development and fire occurrence in these two regions, despite the existing biases in the climate variables simulated by the GCM [2]. This exercise showed that the models developed for the Lamto savannas are general enough for application in other biogeographic contexts.

9.4.3 Current limitations of savanna models operating at regional or global scale

A prerequisite for modeling savanna functioning at the regional or global scale is the availability of reliable input variables at this scale. Among climatic variables, precipitation and incoming solar radiation at the surface are

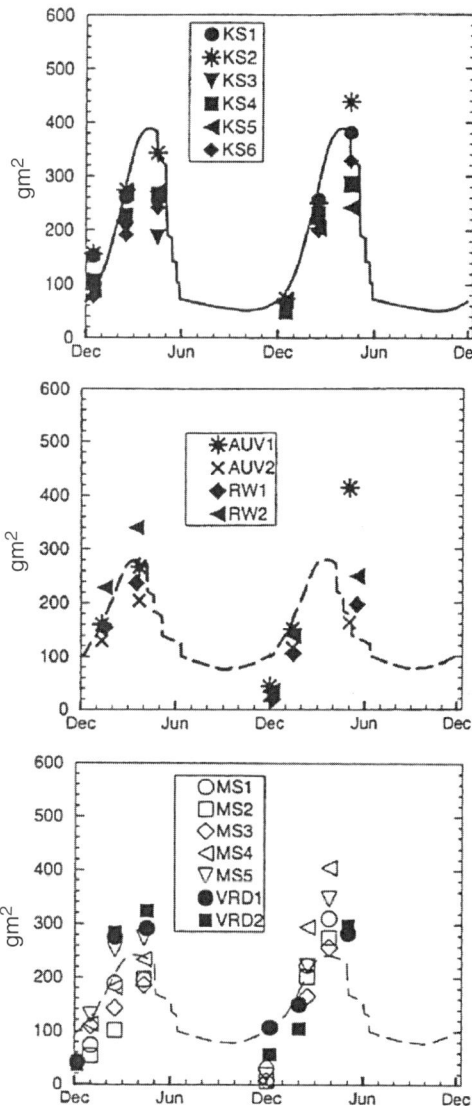

Fig. 9.12. Comparison of the total standing crop simulated for three different grid cells of a General Circulation Model (lines) and field observations in sites located in these grid cells (symbols). Simulations were obtained by using the forcing climate computed by the GCM as an input for the simple savanna ecosystem model. All the sites are located in the Northern territories, Australia, and experience annual precipitation ranging from 676 to 953 mm (after [2], copyright 1999 American Geophysical Union, modified with permission of the American Geophysical Union).

arguably the most important. At large scales, solar radiation is obtained either as GCMs outputs or as estimations from satellite data. However, Le Roux et al. [13], Le Roux [11], and Ciret et al. [2] showed that both GCMs and satellites strongly overestimate solar radiation at the surface in the Guinea region. Similar overestimations by several GCMs were reported over other regions (see references in [13]). Similarly, precipitation patterns derived from GCMs or remote sensed data are sometimes inaccurate for some regions ([13, 2] for the Guinea region). Thus, along with other requirements, simulations or estimates of incoming solar radiation and precipitation at the surface have to be improved before ecosystem models can be applied at large scales with some confidence.

Regional simulation approaches also require soil variables such as soil field capacity and wilting point and soil nutrient status, along with vegetation variables such as the regional distribution of tree and grass fractional cover. If relevant for the whole West African region, results of the TREEGRASS model show that the predicted total NPP does not depend on the relative coverage by trees and grasses, while the predicted total evapotranspiration does. Thus, regional-scale data documenting the spatial variation in tree cover should be necessary particularly for models that aim at coupling vegetation functioning with water fluxes or at distinguishing the contribution of the grass and tree components to NPP.

9.5 Conclusion

9.5.1 State of the art of the modeling of the Lamto savanna

In the past few years, a comprehensive savanna model, namely TREEGRASS, has been developed to explore the relationships between savanna structure and functioning. It has already provided, and will continue to provide, authenticated knowledge and formulations for functional aspects of the Lamto savanna.

At the regional scale, the hybrid modeling-remote sensing approach developed by Kergoat et al. [10] offers a powerful tool for extending the knowledge acquired at Lamto. In this context, the availability of satellite data in the Guinea zone, the impact of noise in available temporal series of multispectral signals, and the surface cover heterogeneity within satellite pixels are major issues that will need to be addressed in the future.

At the regional or global scale, including a savanna model within General Circulation Models should allow the exploration of the feedbacks between the savanna biome and the atmosphere. In particular, the strong link between the seasonality of soil water regime, vegetation development, and fire occurrence in the savanna domain could exacerbate vegetation-atmosphere interactions, but this remains to be tested.

9.5.2 Perspectives

In the context of savanna modeling, three major issues will be adressed in the near future.

First, the model TREEGRASS will be coupled to the soil organic matter/nitrogen dynamics model SOMKO (see Sect. 12.5). This way, competition for soil nitrogen and the effect of soil nutrient availability on the functioning of tree and grass components will be quantified. In particular, the role of nutrient-rich soil patches located on termite mounds or below tree clump canopy in the savanna functioning will be assessed.

Second, a simplified version of TREEGRASS will be developed for application at the landscape/small catchment scale. This will imply coupling the simpler model with an hydrological model. This work is currently in progress, using the model RWF [27] that operates over a few square kilometers and uses 20×20 m elementary pixels. This will allow the assessment of the role of savanna functioning on water balance at a small catchment scale and the role of water redistribution by lateral flows in savanna functioning along a catena.

At least, a simplified version of TREEGRASS will be coupled to a plant demography model (Chap. 19) in order to examine the long term changes in savanna functioning and plant species composition. Particular attention will be paid to the complex feedbacks existing between the spatial pattern of plant species and the corresponding spatial variations in functional processes and disturbances such as fire (Chap. 18) or herbivory (Chap. 10).

Acknowledgments

The authors would like to thank L. Kergoat and G. Dedieu (CESBIO, CNES, Toulouse), B.A. Monteny (Hydrologie ORSTOM, Montpellier), J. Polcher and C. Ciret (Laboratoire de Météorologie Dynamique, CNRS-ENS, Paris), and H. Sinoquet (INRA Bioclimatologie, Clermont-Ferrand) for their dedication to some aspects of the work presented in this chapter. Their input and insights are greatly appreciated. Dr. S.J. Allen and Dr. J.H.C. Gash (Institute of Hydrology, Wallingford, U.K.) critically reviewed a previous version of the manuscript.

References

1. F. Boulier. *Modélisation de bilan hydrique et de production primaire dans quelques savanes de l'Afrique de l'Ouest*. M.Sci. thesis, Université de Paris, Paris, 1985.
2. C. Ciret, J. Polcher, and X. Le Roux. An approach to simulate the phenology of savanna ecosystems in the Laboratoire de Météorologie Dynamique general circulation model. *Global Biogeochemical Cycles*, 13(2):603–621, 1999.

3. D. Clément. Modélisation des flux hydriques et thermiques dans la savane de Lamto (Côte d'Ivoire). *Publications du Laboratoire de Zoologie (E.N.S.)*, 23:1–176, 1982.
4. M.B. Coughenour. Savanna - landscape and regional ecosystem model. Documentation, Colorado State University, Fort Collins, CO, 1994.
5. J. Gignoux, J.C. Menaut, I.R. Noble, and I.D. Davies. A spatial model of savanna function and dynamics: model description and preliminary results. In D.M. Newbery, H.H.T. Prins, and N.D. Brown, editors, *Dynamics of tropical communities*, volume 37 of *Annual symposium of the BES*, pages 361–383. Blackwell Scientific Publications, Cambridge, 1998.
6. J.I. House, S. Archer, D.D. Breshears, R.J. Scholes, M.B. Coughenour, M.B. Dodd, J. Gignoux, D.O. Hall, N.P. Hanan, R. Joffre, X. Le Roux, J.A. Ludwig, J.-C. Menaut, R. Montes, W.J. Parton, J.J. San José, J.C. Scanlan, J.M.O. Scurlock, G. Simioni, and B. Thorrold. Conundrums in mixed woody-herbaceous plant systems. *Journal of Biogeography*, 30:1763–1777, 2003.
7. F. Jeltsch, S.J. Milton, W.R.J. Dean, and N. Van Rooyen. Analysing shrub encroachment in the southern Kalahari: a grid-based modelling approach. *Journal of Applied Ecology*, 34:1497–1508, 1997.
8. F. Jeltsch, S.J. Milton, W.R.J. Dean, N. Van Rooyen, and K.A. Moloney. Modelling the impact of small-scale heterogeneities on tree-grass coexistence in semi-arid savannas. *Journal of Ecology*, 86:780–793, 1998.
9. L. Kergoat, A. Fischer, S. Moulin, and G. Dedieu. Satellite measurements as a constraint on estimates of vegetation carbon budget. *Tellus*, 47B:251–263, 1995.
10. L. Kergoat, X. Le Roux, H. Gauthier, and G. Dedieu. Assimilation of time series of radiometric measurements in a vegetation model. A case study for a humid savanna site. In G. Guyot, editor, *Proceedings of the International Symposium on Photosynthesis and Remote Sensing*, pages 457–464, Montpellier, 1996. Centre National d'Etudes Spatiales.
11. X. Le Roux. *Etude et modélisation des échanges d'eau et d'énergie sol-végétation-atmosphère dans une savane humide (Lamto, Côte d'Ivoire)*. Ph.D. thesis, Université de Paris 6, Paris, 1995.
12. X. Le Roux, A. Lacointe, A. Escobar-Gutierrez, and S. Le Dizes. Carbon-based models of individual tree growth: A critical appraisal. *Annals of Forest Science*, 58(5):469–506, 2001.
13. X. Le Roux, J. Polcher, G. Dedieu, J.C. Menaut, and B. Monteny. Radiation exchanges above West African moist savannas: seasonal patterns and comparison with a GCM simulation. *Journal of Geophysical Research*, 99(D12):25,857–25,868, 1994.
14. X. Le Roux, A. Tuzet, O. Zurfluh, J. Gignoux, A. Perrier, and B. Monteny. Modélisation des interactions surface/atmosphère en zone de savane humide. In M. Hoepffner, T. Lebel, and B. Monteny, editors, *Interactions surfaces continentales/atmosphère: l'expérience HAPEX-Sahel.*, pages 303–317. ORSTOM, Paris, 1996.
15. H. Leriche, X. Le Roux, J. Gignoux, A. Tuzet, H. Fritz, L. Abbadie, and M. Loreau. Which functional processes control the short-term effect of grazing on net primary production in grasslands? Assessment by modelling. *Oecologia*, 129(1):114–124, 2001.
16. S.N. Martens, D.D. Breshears, and C.W. Meyer. Spatial distribution of understory light along the grassland/forest continuum: effects of cover, height, and spatial pattern of tree canopies. *Ecological Modelling*, 126:79–93, 2000.

17. S. Martin and P. Lavelle. A simulation model of vertical movements of an earthworm population (*Millsonia anomala* Omodeo, Megascolecidae) in an African savanna (Lamto, Ivory Coast). *Soil Biology and Biochemistry*, 24(12):1419–1424, 1992.
18. C. Prado. *Modèle de la production primaire herbacée d'une savane à Lamto (Côte d'Ivoire)*. M.Sci. thesis, Université de Paris 6, 1984.
19. R. Scholes and B. Walker. *An African savanna. Synthesis of the Nylsvley study*. Cambridge University Press, Cambridge, 1993.
20. R.J. Scholes and S. Archer. Tree-grass interactions in savannas. *Annual Review of Ecology and Systematics*, 28:517–544, 1997.
21. P.J. Sellers. Canopy reflectance, photosynthesis, and transpiration. ii. the role of biophysics in the linearity of their interdependence. *Remote Sensingand Environment*, 21:143–183, 1987.
22. G. Simioni. *Importance de la structure spatiale de la strate arborée sur les fonctionnements carboné et hydrique des écosystèmes herbes-arbres*. Ph.D. thesis, Université Pierre et Marie Curie, Paris, 2001.
23. G. Simioni, J. Gignoux, and X. Le Roux. Tree layer spatial structure can affect savanna production and water budget: Results of a 3-D model. *Ecology*, 84(7):1879–1894, 2003.
24. G. Simioni, J. Gignoux, X. Le Roux, and H. Leriche. Treegrass, a 3d process-based model for simulating the functioning of tree-grass ecosystems: potential for herbivory studies. In *Proc. IUFRO workshop on Agroforestry, San José, Costa Rica, April 2001.*, page 6, 2002.
25. G. Simioni, X. Le Roux, J. Gignoux, and H. Sinoquet. Treegrass: a 3d, process-based model for simulating plant interactions in tree-grass ecosystems. *Ecological Modelling*, 131:47–63, 2000.
26. H. Sinoquet, X. Le Roux, B. Adam, T. Ameglio, and F. A. Daudet. RATP: a model for simulating the spatial distribution of radiation absorption, transpiration and photosynthesis within canopies: application to an isolated tree crown. *Plant Cell and Environment*, 24(4):395–406, 2001.
27. B.E. Vieux and N. Gauer. Finite-element modeling of stormwater runoff using GRASS GIS. *Microcomputers in Civil Engineering*, 9:263–270, 1994.

10

Modification of the Savanna Functioning by Herbivores

Xavier Le Roux, Luc Abbadie, Hervé Fritz, and Hélène Leriche

10.1 Introduction

Beyond consumption of a given amount of the net primary production (NPP), herbivores may have major effects on ecosystem structure, functioning and dynamics (e.g., [35]). Many authors have represented plant-herbivore interactions by predation-like relationships [8] assuming that herbivory has a purely negative impact on plant growth. It is now recognized that grazing may be not detrimental, and even favorable for plants. In particular, herbivory can promote grassland soil nitrogen cycling [43, 35, 29] which strongly influences plant responses to grazing [9]. Furthermore, herbivores can largely influence the temporal changes in tree/grass balance, directly through the reduction in competition intensity or indirectly through the reduction in fire frequency and intensity (e.g., [25, 10, 38]). In Lamto, the large herbivore biomass has often been regarded as low as for other West African savannas [2, 11]. The scarcity of large herbivores in Lamto was assumed to be the result of man's actions that have, for instance, virtually exterminated elephants and hippopotami. This historical pattern is consistent with the results of several comparative studies showing that apparent low herbivore biomass in West African savannas were due to the loss of large mammal species, i.e., truncated herbivore communities, primarily in relation with human demography and development in the course of the last century [15, 7]. Studies on herbivory in Lamto savannas have mainly addressed two issues to date: (1) quantifying large and small herbivore densities, biomasses and green grass consumption rate and (2) testing the effect of grazing on the functioning and primary production of the grass layer.

10.2 Herbivore densities, biomasses, and green grass consumption rate in Lamto savannas

10.2.1 Invertebrate herbivores

The density and biomass of invertebrate herbivores have been accurately determined in Lamto savannas, mainly during the 1970s. Acridids (mainly

Anablepia granulata, Catantopsilus taeniolatus, Orthochtha brachycnemis and *Rhabdoplea munda*) represent around 16,000 individuals ha^{-1}, i.e., 554 g ha^{-1} dry mass [18]. Acridid biomass as high as 1,060 g ha^{-1} has been reported in December. Grass consumption rate in the field were estimated from daily consumption rate of different species in the laboratory and field estimates of biomass. Annual consumption was around 69.6 kg ha^{-1} [18], as compared to the total primary production values ranging from 30 to 60 t ha^{-1} (see Sect. 7.4). Extrapolation of these results to other herbivore arthropods living in the aboveground grass layer (which biomass was estimated at Lamto) leads to annual grass consumption estimates around 240 kg ha^{-1} (excluding ants; Fig. 10.1) [27].

In addition, consumption by termites ranges from 30 to 50 kg ha^{-1} [26]. However, termites consume dead rather than green material. As a whole, the biomass of animal consumers excluding large mammals (around 80 kg ha^{-1}) is quite low at Lamto [27] (Fig. 10.1). In contrast to detritivorous organisms such as termites and earthworms that have a huge impact on the savanna functioning (Fig. 10.1) (see Chaps. 13 and 16), such animal herbivores thus do not play an important role for the savanna functioning.

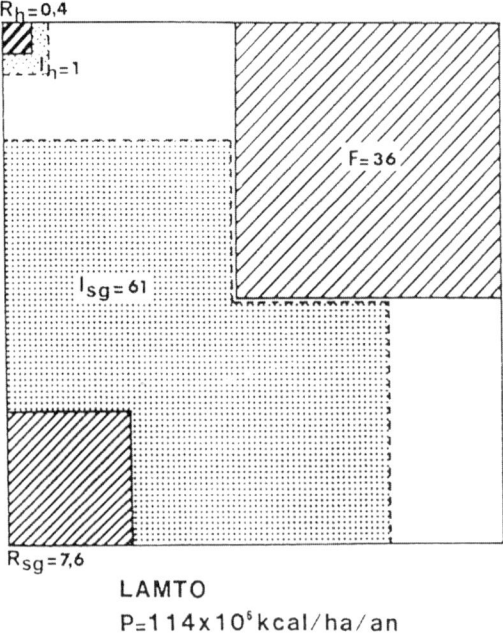

Fig. 10.1. Simple representation of utilization of primary production (P) by herbivores (I_h) and detritivores (I_{sg}). Losses by fire (F), respiration by herbivores (R_h) and respiration by detritivores (R_{sg}) are indicated (after [28]).

10.2.2 Large mammal herbivores

The number of large grazers has increased recently due to hunting prohibition on the reserve area: antelope (e.g., 0.025 *Kobus kob kob* ha^{-1}) and buffalo (0.0204 *Syncerus caffer nanus* ha^{-1}; Fig. 10.2) densities both fall within the range of values found in protected areas in Western Africa (Table 10.1).

The moist savanna grasslands of Lamto however correspond to selected habitat for the kob, which can reach densities as high as 0.7 individuals ha^{-1} in preferred habitats (e.g., [42], in the Comoe National Park). The other ungulates encountered at Lamto are the bushbuck (*Tragelaphus scriptus*) and

Fig. 10.2. The Lamto buffaloes (photography by L. Abbadie).

Table 10.1. Densities of kob antelope (*Kobus kob*, average population body mass 50 kg) and buffalo (*Syncerus caffer nanus*, average population body mass 450 kg) in Lamto savannas and other protected savanna areas of West Africa (after [19]).

Protected area	Kob (km^{-2})	Buffalo (km^{-2})
Lamto (Côte d'Ivoire)	2.49	2.40
Arly (Burkina faso)	0.65	10.90
Bénoué (Côte d'Ivoire)	1.14	1.55
Comoé (Côte d'Ivoire)	2.15	4.35
Deux-Bale (Burkina Faso)	0.07	0.18
Kainji (Nigeria)	0.03	1.18
Manovo-Gouda (Rép. Centrafricaine)	1.80	2.60
Niokolokoba (Sénégal)	0.33	0.55
Penjari (Bénin)	1.27	14.70

Maxwell's duiker (*Cephalophus maxwellii*). The red-flanked duiker (*Cephalophus rufilatus*), black-fronted duiker (*Cephalophus nigrifrons*) and bushpig (*Potamochoerus porcus*) may still exist in very low numbers. Recently, spoors of hartebeest (*Alcephalus busephalus major*) were observed in the reserve. The density values of large grazers obtained for the Lamto savanna are low compared to high densities of large herbivores reported in some East and Southern African savannas. However, the roles of soil nutrient and associated plant nutrient concentration are crucial in patterns of ungulate community abundance [2, 16]. Thus, herbivore biomass in West African savannas, i.e., nutrient poor ecosystems (see Sect. 4.3), should be compared with herbivore biomass in other savanna ecosystems with similar low soil nutrient availability. Fritz [15] compared herbivore biomass in different savanna ecosystems accounting for their soil nutrient richness and found that medium-size ungulates are more abundant in West Africa than their counterparts in nutrient-poor East and Southern African savannas for a given level of primary production. Although large herbivore biomass is correlated to aboveground net primary production in both West African and East and Southern African savannas (Fig. 10.3), the major differences in ungulate biomass between subregions are not strictly related to differences in primary production. Missing large predators (such as lion and hyena) and the quasi-absence of very large herbivores (megaherbivores, such as elephants [39]), i.e., of key competitors [17], probably explain this pattern of higher medium-size ungulate biomass in nutrient-poor West

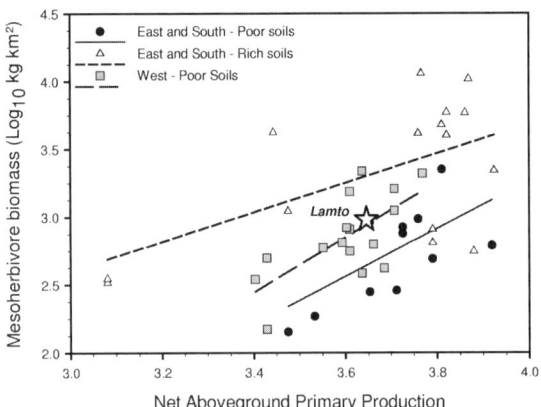

Fig. 10.3. Relationship between aboveground net primary production (ANPP) and large herbivore biomass (LHB) of truncated herbivore communities (i.e., without megaherbivores and buffalo, for (squares) West African savannas, east and southern African savannas with low soil nutrient availability and east and southern African savannas with high soil nutrient availability (after [15], with permission of Blackwell Publishing).

African savannas compared to that of nutrient-poor East and Southern savannas. Thus, in contrast to views prevailing during the 1970s and 1980s, the potential for high large herbivore biomass is now recognized for such nutrient-poor ecosystems. Furthermore, megaherbivores probably represented a large fraction of the primary consumer trophic level in the past at Lamto, as in most ecosystems with high rainfall and low soil nutrient status [2, 17]. Their density could still increase if more drastic protection rules are applied. In the meantime, an increase in protection would certainly benefit the medium-size ungulates. This has spurred on development of field experiments and modeling approaches to better understand the potential role of mammal herbivores (i.e., mainly grazers) on the functioning of the grass layer in the Lamto savanna.

10.3 Field studies of grazing effect on the savanna functioning

10.3.1 Response of grass production to grazing

Grazing, as a removal of living tissue, has first been considered as detrimental to plants. However, experimental results (e.g., [32, 33, 22, 6]) showed that net primary production, NPP, can be maintained (compensatory growth) or stimulated (overcompensatory growth) in response to grazing (i.e., the herbivory optimization hypothesis HOH). Some authors [32, 23, 22] suggested that an optimal plant removal level should occur beyond which production is reduced. The ecological significance and generality of these findings were jeopardized by critical appraisals of published data [3, 4, 5].

The HOH was tested in the Lamto savanna by analyzing the growth of grasses in response to a clipping (3 levels) × fertilization (2 levels) trial in the field [30]. During 3 months, the effects of clipping and fertilization on the dry matter and nitrogen yields to producers (i.e., mass or nitrogen amount of residual phytomass at the end of the experiment) and to grazers (i.e., mass or nitrogen amount of clipped-off tissues during the experiment) were surveyed. Total phytomass yield and yield to grazers were maintained under moderate clipping frequency and fertilization as compared to control conditions (Fig. 10.4). Both clipping frequency decreased total phytomass yield to producers as compared to control plots. Clipping frequency significantly increased nitrogen concentrations in the total yield, in the yield to producers and in the yield to grazers [30]. Total nitrogen yield and nitrogen yield to grazers were 65% and 91%, respectively, higher on the plots experiencing moderate clipping frequency with fertilization as compared to control plots (Fig. 10.5). Root phytomass was not influenced by clipping frequency and fertilization [30]. These results provide evidence that the HOH is realistic for the Lamto grass layer in the short term. The observed changes in the pattern of mass- and nitrogen-yield distribution concurrently to the improved grass

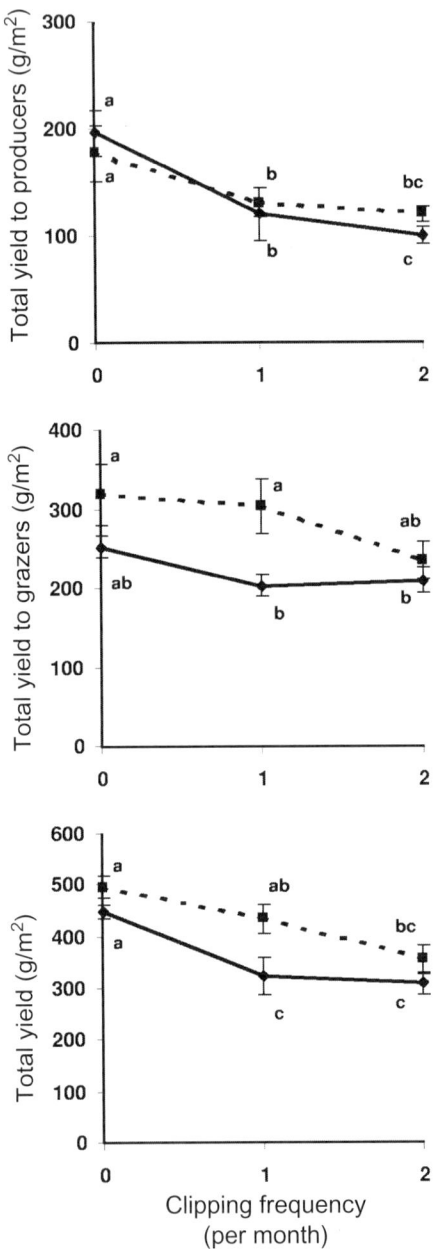

Fig. 10.4. Dry matter yield to producers (below 0-10 cm height), yield to grazers (over 10 cm height) and total yield as a function of clipping frequency (per month) with (dotted line) and without (solid line) fertilization (after [30], with permission of the Ecological Society of America).

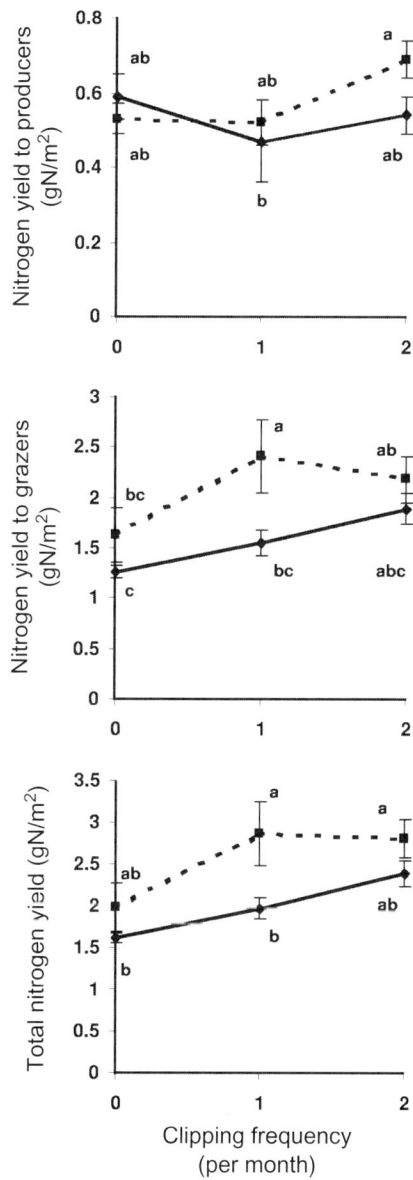

Fig. 10.5. Nitrogen yield to producers (below 0-10 cm height), yield to grazers (over 10 cm height) and total yield as a function of clipping frequency (per month) with (dotted line) and without (solid line) fertilization (after [30], with permission of the Ecological Society of America).

quality observed (i.e., higher nitrogen concentration) show that grazers can modify ecosystem processes in such a way as to alleviate nutritional deficiencies in the Lamto savanna. Longer-term experiments testing the sustainability of such a response of the grass layer to grazing are under progress at Lamto.

10.3.2 Response of soil microbial activities following a grazing event

Herbivory has important effects on soil nitrogen cycling [43, 35], which strongly influences plant responses to grazing. Grazer-induced decreases in microbial immobilization and increases in N mineralization have generally been observed [44, 24, 45, 46, 12, 47, 36, 29]. Such changes generally result in improved N availability to plants in intensively grazed sites [24, 34, 21]. However, few studies have quantified the effects of grazing on two key processes involved in soil N cycling, i.e., nitrification and denitrification [13, 14, 40, 20]. These two microbially mediated processes largely control the balance between the forms of soil mineral nitrogen (NO_3^- versus NH_4^+) available to plants and nitrogen conservation at the ecosystem level. The short-term response of nitrification and denitrification to a clipping event was tested in a *Hyparrhenia* grassland at Lamto [1]. The grass layer was clipped once on 0.5 m^2 plots during the early rainy season in April, and potential nitrification and denitrification were surveyed before and after clipping. Soil was sampled in the 0-10 cm layer below clipped *Hyparrhenia* tufts at the center of clipped plots and below control, unclipped *Hyparrhenia* tufts. The main results of this study were (1) clipping did not influence the water regime in the upper soil layer (Fig. 10.6) and (2) denitrification was significantly higher below clipped *Hyparrhenia* tufts than below unclipped tufts 6 days after the clipping event (Fig. 10.6). This study showed the potential of one effect of grazing, i.e., defoliation, on soil processes driving N cycling. Enhancement of root exudation by clipped plants (e.g., [41]) could partly explain the observed increased denitrification activity because denitrifiers are heterotrophs for C. Experiments for quantifying the long-term effect of grazing on key soil microbial activities and the diversity of associated microbial functional groups are under progress at Lamto. This will greatly improve our understanding of the effect of grazing on savanna functioning.

10.4 Modeling approaches for understanding grazing effect on the savanna functioning

Many functional processes controlling NPP are affected by grazing, such as modification of light availability, reduction of water stress, accelerated nutrient recycling, changed allocation of assimilates within the plant and enhanced photosynthetic rates [33, 37]. Modeling is thus a useful tool to unravel the roles

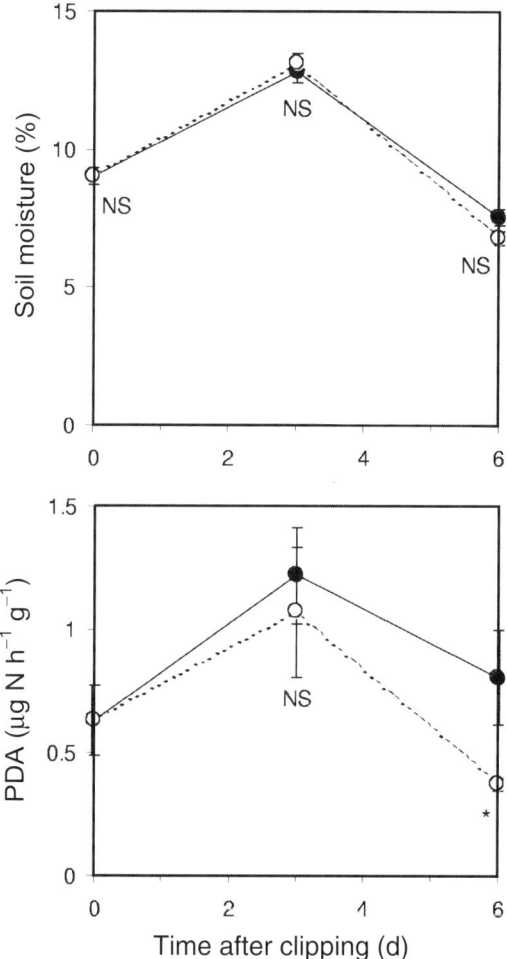

Fig. 10.6. Changes in soil moisture and potential denitrifying activity (PDA) following a clipping event. $50 \times 50\,cm^2$ areas centered on a *Hyparrhenia* individual tuft were clipped at the beginning of the experiment. Soil moisture and microbial activities were surveyed for the 0-10 cm soil layer below the *Hyparrhenia* individuals (7 replicates per treatments) (M. Bardy and X. Le Roux, unpublished).

of all these processes in grass response to grazing. Leriche et al. [31] used a modeling approach to better understand the impact of grazing on grass production in Lamto by simulating the response of grass NPP to plant biomass removal. The process-based PEPSEE-grass model (see Sect. 9.2) parameterized for Lamto grasslands was used (1) to quantify the relative importance of key functional processes (i.e., changes in light absorption efficiency, reduction of water stress, improved canopy nitrogen status and ensuing productivity rate, changes in the pattern of root/shoot allocation) in the response of NPP to grazing and particularly those that can lead to compensatory growth and (2) to test the grazing optimization hypothesis under different functional hypotheses at the canopy and annual scales for the Lamto savanna. Simulations were performed using a constant or resource-driven root/shoot allocation coefficient and assuming a dependence or independence of conversion efficiency of absorbed light into dry matter on nitrogen availability [31]. Main results were as follows: (i) the response of NPP to grazing intensity emerged as a complex result of both positive and negative, direct and indirect, effects of biomass removal on light absorption efficiency, soil water availability, grass nitrogen status and productivity, and root/shoot allocation pattern; (ii) overcompensation was observed for aboveground NPP when assuming a nitrogen-dependent conversion efficiency and a resource-driven root/shoot allocation (Fig. 10.7); (iii) the response of NPP to grazing was mainly controlled by the effect of plant nitrogen status on conversion efficiency and by the root/shoot allocation pattern (Fig. 10.7), while the effects of improved water status and reduced light absorption were secondary. The originality of this work was to provide a comprehensive representation of the functional response of grasslands to grazing. Given the assumptions made in the model, this study does not provide evidence for or against the grazing optimization hypothesis in West African humid grasslands. However, the changes in plant nutrient status and productivity and the response of the root/shoot allocation pattern were identified as the two key interacting processes controlling the response of Lamto grassland NPP to increasing grazing intensity. Thus, predicting the response of Lamto grass NPP to increasing grazing intensity requires one to couple a model simulating the savanna functioning (as TREEGRASS, see Sect. 9.3) to a model simulating the nitrogen dynamics in the soil-plant system (as SOMCO, see Sect. 12.5). Such an approach should accurately represent the interactions between plant functional processes (N uptake, litter or exsudate inputs to the soil, dependence of grass productivity on nutrient availability and root/shoot allocation pattern), soil microbial activities (soil organic matter dynamics and soil nutrient availability, mineralization from urine and feces, nitrification and denitrification) and soil water balance (which controls both soil and plant functioning).

10.5 Conclusion

The density, biomass and consumption rate of animal herbivores are presently low in the Lamto savanna, but the density of mammal herbivores is higher than that reported in other nutrient-poor savannas. According to Lamotte [27], "earthworms, termites and decomposing microorganisms occupy the role played elsewhere [in eastern and southern African savannas] by large herbivorous mammals" because they "discreetly insure, together with fire, the mineralization of organic matter produced every year by green plants."

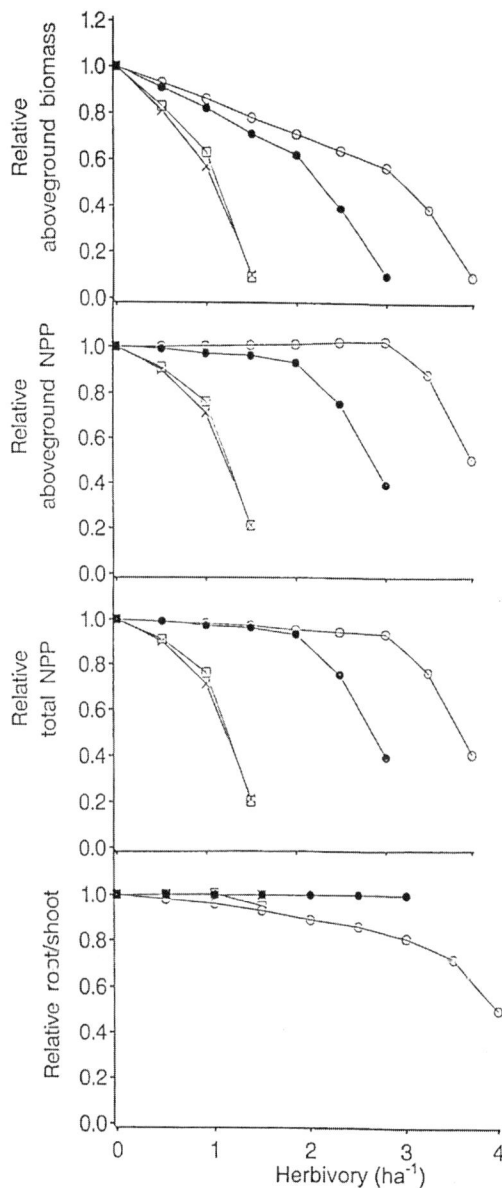

Fig. 10.7. Simulated response of grass biomass, aboveground net primary production (ANPP), total net primary production (TNPP), and grass root/shoot ratio to grazing intensity. Four versions of the PEPSEE model were used (crosses: constant conversion efficiency and constant root/shoot ratio; squares: constant conversion efficiency and resource-driven root/shoot ratio; solid circles: nitrogen-dependent conversion efficiency and constant root/shoot ratio; empty circles: nitrogen-dependent conversion efficiency and resource-driven root/shoot ratio). All values are normalized to values simulated without herbivory (after [31], with kind permission of Springer Science and Business media).

However, the density of mammal herbivores could still increase if more drastic protection rules are applied, which could have major influences on the savanna functioning and dynamics, because the way large herbivores consume and process plant matter is very different from that of decomposers. Studies of the impact of large grazers on vegetation functioning and dynamics will thus develop at Lamto. In particular, a better understanding of processes driving the response of savanna functioning to grazing on the long term (years to decades) is needed. Studies on the role of grazers on the tree/grass equilibrium should also be initiated. This will help a better understanding of the role of large grazers in the past in West African humid savannas. This will also help better management of the Lamto savanna if conservation priorities allow substantial increase in the density/diversity of large herbivores at Lamto in the future.

References

1. M. Bardy. *Impact d'un pâturage simulé sur la nitrification et la dénitrification en savane humide*. M.Sci. thesis, Ecole Normale Supérieure de Lyon, 2002.
2. R.H.V. Bell. The effect of soil nutrient availability on the community structure in African ecosystems. In B. J. Huntley and B. H. Walker, editors, *Ecology of tropical savannas*, pages 193–216. Springer-Verlag, Berlin, 1982.
3. A.J. Belsky. Does herbivory benefit plants? A review of the evidence. *The American Naturalist*, 127(6):870–892, 1986.
4. A.J. Belsky. The effects of grazing: confounding of ecosystem, community, and organism scales. *The American Naturalist*, 129(5):777–783, 1987.
5. A.J. Belsky, W.P. Carson, C.L. Jensen, and G.A. Fox. Overcompensation by plants: herbivore optimization or red herring? *Evolutionary Ecology*, 7:109–121, 1993.
6. M.E. Biodini, B.D. Patton, and P.E. Nyren. Grazing intensity and ecosystem processes in a northern mixed-grass prairie, USA. *Ecological Applications*, 8(2):469–479, 1998.
7. J.S. Brashares, P. Arcese, and K.S. Sam. Human demography and reserve size predict wildlife extinction in West Africa. *Proceedings of the Royal Society of London, Series B*, 286:2473–2478, 2001.
8. M.J. Crawley. *Herbivory: The dynamics of animal-plant interactions*. University of California Press, Berkeley, 1983.
9. C. de Mazancourt, M. Loreau, and L. Abbadie. Grazing optimisation and nutrient cycling: When do herbivores enhance plant production? *Ecology*, 79(7):2242–2252, 1998.
10. H.T. Dublin. Vegetation dynamics in the Serengeti-Mara ecosystem: The role of elephants, fire, and other factors. In A.R.E. Sinclair and P. Arcese, editors, *Serengeti II*, pages 71–90. University of Chicago Press, Chicago, 1995.
11. R. East. Rainfall, soil nutrient status and biomass of large African savanna mammals. *African Journal of Ecology*, 22:245–270, 1984.
12. D.A. Frank and P.M. Groffman. Denitrification in a semi-arid grazing ecosystem. *Oecologia*, 117(4):564–569, 1998.

13. D.A. Frank and P.M. Groffman. Ungulate vs. landscape control of soil C and N processes in grasslands of Yellowstone national Park. *Ecology*, 79(7):2229–2241, 1998.
14. D.A. Frank, P.M. Groffman, R.D. Evans, and B.F. Tracy. Ungulate stimulation of nitrogen cycling and retention in Yellowstone Park grasslands. *Oecologia*, 123(1):116–121, 2000.
15. H. Fritz. Low ungulate biomass in West African savannas: Primary production or missing megaherbivores or large predator species. *Ecography*, 20:417–421, 1997.
16. H. Fritz and P. Duncan. On the carrying capacity for ungulates of African savanna ecosystems. *Proceedings of the Royal Society of London, Series B*, 256:77–82, 1994.
17. H. Fritz, P. Duncan, I.J. Gordon, and A.W. Illius. Megaherbivores influence trophic guilds structure in African ungulate communities. *Oecologia*, 131:620–625, 2002.
18. Y. Gillon and D. Gillon. Recherches écologiques dans la savane de Lamto (Côte d'Ivoire): Cycle annuel des effectifs et des biomasses d'Arthropodes de la strate herbacée. *La Terre et la Vie*, 21:262–277, 1967.
19. S. Glémin. *Mise au point d'une méthode de recensement de grands herbivores dans une savane de type mosaïque*. M.Sci. thesis, Université de Paris 6, 1997.
20. P.M. Groffman, C.W. Rice, and J.M. Tiedje. Denitrification in a tallgrass prairie landscape. *Ecology*, 74(3):855–862, 1993.
21. E.W. Hamilton and D.A. Frank. Can plants stimulate soil microbes and their own nutrient supply? Evidence from a grazing tolerant grass. *Ecology*, 82(9):2397–2402, 2001.
22. D.S. Hik and R.L. Jefferies. Increases in the net above-ground primary production of a salt-marsh forage grass: A test of the predictions of the herbivore-optimization model. *Journal of Ecology*, 51:180–195, 1990.
23. D.W. Hilbert, D.M. Swift, J.K. Detling, and M.I. Dyer. Relative growth rates and the grazing optimization hypothesis. *Oecologia*, 51:14–18, 1981.
24. E.A. Holland and J.K. Detling. Plant response to herbivory and belowground nitrogen cycling. *Ecology*, 71(3):1040–1049, 1990.
25. N.J. Huntly. Herbivores and the dynamics of communities and ecosystems. *Annual Review of Ecology and Systematics*, 22:477–504, 1991.
26. G. Josens. *Etudes biologique et écologique des termites (Isoptera) de la savane de Lamto-Pakobo (Côte d'Ivoire)*, volume 42 of *Mémoires de la classe des Sciences, series 2*. Académie royale de Belgique, Bruxelles, 1977.
27. M. Lamotte. The structure and function of a tropical savannah ecosystem. In E. Medina and F.B. Golley, editors, *Tropical ecological systems*, pages 179–222. Springer-Verlag, New York, 1975.
28. M. Lamotte. Présentation des travaux des chercheurs de Lamto (Côte d'Ivoire), 1962-1989. *ENS, Publications du laboratoire de Zoologie*, 36:1–158, 1990.
29. X. Le Roux, M. Bardy, P. Loiseau, and F. Louault. Stimulation of nitrification and denitrification by grazing in grasslands: Do changes in plant species composition matter? *Oecologia*, 137(3):417–425, 2003.
30. H. Leriche, X. Le Roux, F. Desnoyers, D. Benest, and L. Abbadie. Response of grass dry matter- and nitrogen-yields to clipping in an African savanna: An experimental test of the herbivory optimisation hypothesis. *Ecological Applications*, 13:1346–1354, 2003.

31. H. Leriche, X. Le Roux, J. Gignoux, A. Tuzet, H. Fritz, L. Abbadie, and M. Loreau. Which functional processes control the short-term effect of grazing on net primary production in grasslands? Assessment by modelling. *Oecologia*, 129(1):114–124, 2001.
32. S.J. McNaughton. Grazing as an optimization process: grass-ungulate relationships in the Serengeti. *The American Naturalist*, 113(5):691–703, 1979.
33. S.J. McNaughton. Compensatory plant growth as a response to herbivory. *Oikos*, 40:329–336, 1983.
34. S.J. McNaughton, F.F. Banyikwa, and M.M. McNaughton. Promotion of the cycling of diet-enhancing nutrients by African grazers. *Science*, 278:1798–1800, 1997.
35. S.J. McNaughton, R.W. Ruess, and S.W. Seagle. Large mammals and process dynamics in African ecosystems. *Biosciences*, 38(11):794–800, 1988.
36. E. M. Molvar, R. T. Bowyer, and V. Vanballenberghe. Moose herbivory, browse quality, and nutrient cycling in an Alaskan treeline community. *Oecologia*, 94:472–479, 1993.
37. I. Noy-Meir. Compensating growth of grazed plants and its relevance to the use of rangelands. *Ecological Applications*, 3(1):32–34, 1993.
38. H. Olff and M.E. Ritchie. Effects of herbivores on grassland plant diversity. *Trends in Ecology and Evolution*, 13:261–265, 1998.
39. N. Owen-Smith. *Megaherbivores. The influence of very large body size on ecology.* Cambridge University Press, Cambridge, 1988.
40. A.J. Parsons, R.J. Orr, P.D. Penning, D.R. Lockyer, and J.C. Ryden. Uptake, cycling and fate of nitrogen in grass clover swards continuously grazed by sheep. *Journal of Agricultural Science*, 116:47–61, 1991.
41. E. Paterson and A. Sim. Rhizodeposition and C-partitioning of *Lolium perenne* in axenic culture affected by nitrogen supply and defoliation. *Plant and Soil*, 216:155–164, 1999.
42. P. Poilecot. Un écosystème de savane soudanienne: le Parc National de la Comoé. Côte d'Ivoire. Technical report, UNESCO/PNUD, 1991.
43. R.W. Ruess. The role of large herbivores in nutrient cycling of tropical savannas. In B.H. Walker, editor, *Stress and disturbance in tropical savannas*, pages 67–91. IUBS, Oxford, 1986.
44. R.W. Ruess and S.J. McNaughton. Grazing and the dynamics of nutrient and energy regulated microbial processes in the Serengeti grasslands. *Oikos*, 49(1):101–110, 1987.
45. R.W. Ruess and S.W. Seagle. Landscape patterns in soil microbial processes in the Serengeti national-park, Tanzania. *Ecology*, 75(4):892–904, 1994.
46. A.R. Shariff, M.E. Biondini, and C.E. Grygiel. Grazing intensity effects on litter decomposition and soil- nitrogen mineralization. *Journal of Range Management*, 47(6):444–449, 1994.
47. B.F. Tracy and D.A. Frank. Herbivore influence on soil microbial biomass and nitrogen mineralization in a northern grassland ecosystem: Yellowstone National Park. *Oecologia*, 114(4):556–562, 1998.

Part III

Carbon Cycle and Soil Organic Matter Dynamics

11

Origin, Distribution, and Composition of Soil Organic Matter

Luc Abbadie and Hassan Bismarck Nacro

11.1 Introduction

The soil organic matter (SOM) dynamics results from two opposite processes: the accumulation of organic compounds, mainly depending on the rate of primary production, and their degradation, mainly depending on environmental conditions. A very important agent of SOM dynamics is the heterotrophic micro-flora. The rate of microbial mineralization determines both the quantity and quality of SOM and the concentration of the soil solution in mineral compounds such as nitrogen, phosphorus, potassium, etc. In the long term, the SOM is a pool of chemical nutritive elements, potentially usable by plants, which partly constrains the structure and dynamics of plant cover. In the short term, it is a major determinant of soil fertility because it is almost the only source of nutrients for plant growth and microbial activity.

A conceptual view of the biochemical transformation of organic matter in soil can be that of quanta of organic matter going through different stages of degradation, from coarse dead plant material to evolved humified organic matter tens or hundreds of years old. Soil organic matter can thus be considered as a continuum of compounds more or less accessible to micro-organisms, more or less abundant, each contributing to mineral nutrients production. All micro-organisms are obviously not able to degrade all types of organic matter. From a theoretical point of view, it should be possible to predict the potential of microbial activity in a soil from a good knowledge of the chemical composition of organic matter. Indeed, the real activity is only a modulation of this potential by the soil micro-climate and nature and organization of mineral particles. The fine particles act as colloidal material and their large surface area allows them to adsorb large quantities of organic compounds in the interlayer of clays. Organic compounds also bind the mineral particles to each other, which results in the additional sequestration of carbon compounds in completely closed aggregates. The newly incorporated plants residues are generally associated to coarse particles, while the old organic matter is associated to fine particles [12, 6, 34].

Many steps of the carbon cycle and SOM dynamics processes have been investigated at Lamto. The distributions of organic compounds, chemical quality of SOM and heterotrophic microbial activity have been studied at both the landscape and organomineral particles scale. The objective was to highlight the mechanisms which control the accumulation and mineralization of carbon in the Lamto savanna ecosystem in order to better assess the interactions between soil processes and plant cover dynamics.

11.2 The inputs of organic matter to soil

The variation of the quantity of SOM with time can be described by the simple equation proposed by Jenny (1941) in [26]: $\frac{dC}{dt} = A - kC$, where C is the organic matter pool at time t, A is the mass of organic matter added and k is the coefficient of decomposition of the organic matter (first order rate constant). A valid estimate of the SOM pool therefore requires a good measurement of the primary production and, above all, of the input of plant debris to soil (A). The size of the SOM pool, as the rate of primary production, is generally positively correlated to rainfall and temperature [23, 11, 21]. But it is also locally related to the species composition of plant cover and biological type of the dominant species that control the chemical composition (e.g., lignin and nitrogen concentrations) of the dead organic matter, i.e., its nutritive value for micro-organisms. Moreover, the spatial structure of the vegetation at different scales has a strong impact on the horizontal distribution of SOM and nutrients and, consequently, on the ecosystem functioning and dynamics.

The grass compartment dominates the carbon cycle in the savannas and, more generally, in the grassland systems. Even if the tree biomass is higher than that of grass in nearly all savanna types of the Lamto reserve, most of the yearly primary production is made up by grasses particularly belowground [27] (see Sect. 7.4). The grass layer in Lamto is dominated almost exclusively by perennial species with a C_4 photosynthetic pathway (Table 11.1). Their

Table 11.1. Natural abundance of ^{13}C (as $\delta^{13}C$ in ‰) in the most widespread trees and grasses from the Lamto reserve (after [24], with permission of Cambridge University Press).

Group	Species	Coarse roots	Fine roots	Leaves
Trees	Crossopteryx febrifuga	−26.6	−27.9	−27.9
	Piliostigma thonningii	−28.0	−29.7	−29.2
	Bridelia ferruginea	−27.9	−28.5	−28.5
	Cussonia arborea	−27.6	−27.1	−28.0
Grasses	Hyparrhenia diplandra		−12.4	−13.1
	Hyparrhenia smithiana		−12.4	−12.6
	Loudetia simplex		−12.3	−12.6
	Andropogon schirensis		−13.4	−12.5

natural abundance in ^{13}C show very weak variations, with a δ^{13}C only ranging between –12.1‰ and –13.4‰. The difference between the δ^{13}C of their aboveground parts compared to that of roots is never greater than 1‰. The δ^{13}C of the trees is much lower and typical of plants with a C_3 photosynthetic pathway. It ranges between –29.7‰ and –26.6‰. The difference of natural abundance of ^{13}C between the coarse and fine roots can reach 1.7‰. Tree leaves are more depleted in ^{13}C than roots.

The two components of the plant cover in Lamto, the trees and grasses, can be therefore accurately distinguished by their natural abundance in ^{13}C that differs by ca. 16‰ (see Sect. 8.3). Moreover, within each of the components, the δ^{13}C of the different species is very close to each other. It is consequently easy to estimate the contribution of the grasses and trees to the SOM pool by a simple measurement of the natural abundance of ^{13}C in soil since many works have shown the lack of strong isotopic enrichment of carbon during the decomposition process (e.g., [25]). In the areas where tree density is maximum, such as in the center of the old mounds of Macrotermitinae, the δ^{13}C of the SOM can reach –22‰, showing that 50-60% of the organic matter originates from trees. But, in most of the savanna soils, the δ^{13}C is between –14‰ and –12‰, clearly demonstrating that ca. 90% of the SOM originates from C_4 plants, i.e., from grass roots [1].

Because of the burning of the aboveground part of grasses and tree leaf litter, the contribution of the aboveground production to SOM is negligible. Consequently, it can be hypothesized that almost all the organic compounds stored in the soil come from belowground primary production. The belowground primary production is very difficult to assess due to rapid root biomass variations, sampling problems and lack of reliable data about root life span (see Sect. 7.3). However, many studies have been performed in Lamto on this point. For grasses, the estimates range from ca. 900 to 3000 $g\,m^{-2}$, and belowground primary production is higher than aboveground production in all the savanna types (see Sect. 7.4). Minimum estimates of tree root production in Lamto range from 5-10 $g\,m^{-2}\,yr^{-1}$ in the open shrub savanna to 40 $g\,m^{-2}\,yr^{-1}$ in the savanna woodland [27]. If we hypothesize that the Lamto savannas are steady state ecosystems, we can therefore estimate that the input of organic carbon from roots to soil is between 360 and 1200 $g\,m^{-2}\,yr^{-1}$ (with an average concentration of carbon in roots at 40% dry matter).

11.3 Soil organic matter distribution

11.3.1 Variations of soil organic matter distribution at landscape scale

The plant cover is highly variable with the topography in the region of Lamto and it is common to see a rapid change of vegetation over only 200-300 m long transects. The plant cover ranges from semi-deciduous dry forest on plateau to evergreen wet gallery forest in thalwegs through various more or less densely wooded savannas on slopes (see Sect. 5.2). Although macroclimate and soil physical characteristics are comparable everywhere, a great variability of SOM content is observed along transects, obviously due to plant species composition and primary production variability. In the first 20 cm, the concentrations and pools of organic carbon and nitrogen are always significantly higher under forests than under savannas, up to threefold between 0 and 5 cm depth (Table 11.2). This relative poverty of the surface savanna soil in carbon and nitrogen is undoubtfully related to the absence of aboveground litter due to annual fire. Indeed, while the contribution of the aerial parts of plants to SOM is negligible in savannas, the tree leaf and twig falls amount to 500-700 $g\,m^{-2}\,yr^{-1}$ in the semi-deciduous plateau forest and 600-800 $g\,m^{-2}\,yr^{-1}$ in the gallery forest [8]. This dead plant matter accumulates on soil surface in a litter layer. Bacteria, fungi and micro-arthropods quickly decompose a part of this litter, but another part is incorporated to soil and contributes to increase the SOM pool.

The difference of SOM content between forests and savannas vanishes rapidly with depth. In fact, forest and savanna soils differ as much by their total quantity of organic matter as by the vertical distribution of carbon pools. This is particularly obvious at the forest-savanna boundary. A study of soil carbon content has been performed in the north of the Lamto reserve ([35]; Mareuil, unpublished data), where a semi-deciduous plateau forest moves into the grass savanna at the rate of ca. 0.6 $m\,yr^{-1}$ [16, 17]. Under the ancient savanna, the organic matter is nearly regularly distributed on the whole soil

Table 11.2. Concentration in particles smaller than 2 mm (% of total dry soil weight), apparent density (a.d.), concentrations in carbon and nitrogen (% C and N dry fine earth dry weight, average and standard error of 3 samples), carbon and nitrogen pools and C:N ratios in the superficial layers of forest and savanna soils. PF: plateau forest; AS: Andropogoneae savanna; LS: *Loudetia* savanna; GF: gallery forest (after [30]; Abbadie unpublished data).

Soil	<2 mm		C		N		C:N
	%	a.d.	%	$g\,m^{-2}$	%	$g\,m^{-2}$	
PF 0–5 cm	98	1.2	2.17 (0.23)	1302	0.23 (0.02)	138	9.4
PF 5–10 cm	92	1.4	1.33 (0.05)	931	0.13 (0.01)	91	10.2
PF 10–20 cm	84	1.5	0.83 (0.08)	1245	0.07 (0.01)	105	11.9
AS 0–5 cm	97	1.4	1.22 (0.06)	854	0.08 (0.01)	56	15.2
AS 5–10 cm	89	1.5	1.06 (0.07)	795	0.07 (0.01)	53	15.1
AS 10–20 cm	85	1.6	0.98 (0.02)	1568	0.06 (0.01)	96	16.3
LS 0–5 cm	99	1.4	0.70 (0.04)	490	0.04 (0.01)	28	17.5
LS 5–10 cm	99	1.5	0.61 (0.03)	458	0.03 (0.01)	23	20.3
LS 10–20 cm	99	1.6	0.53 (0.02)	848	0.03 (0.01)	48	17.7
GF 0–5 cm	99	1.3	2.96 (0.27)	1924	0.29 (0.03)	189	10.2
GF 5–10 cm	99	1.5	1.40 (0.22)	1050	0.14 (0.03)	105	10.0
GF 10–20 cm	99	1.6	0.50 (0.04)	800	0.04 (0.02)	64	12.5

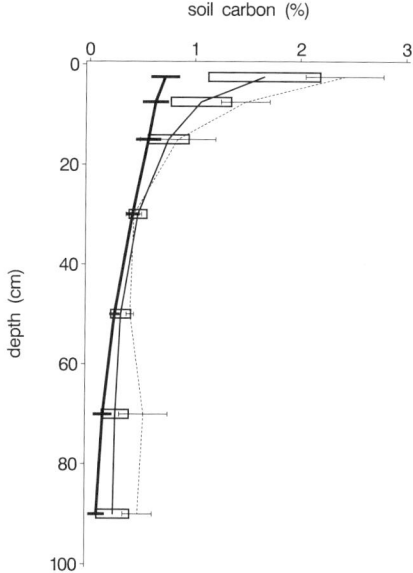

Fig. 11.1. Vertical distribution of soil carbon concentration along a transect located at a savanna-forest boundary. Thick solid line, savanna ($n = 12$); thin solid line, transition zone ($n = 22$); dotted line, forest ($n = 3$). Bars represent one standard error (after [35]).

profile height, whereas under ancient forest, it is more concentrated in surface layers (Fig. 11.1). In the intermediate zone, where forest has recently pushed back savanna, an intermediate distribution of carbon is observed until 40 cm approximately; below this depth, the concentrations in organic carbon are close to those measured under ancient forest and savannas, showing that the impact of vegetation cover on SOM content in this area is limited to the upper soil. This difference in carbon concentration variation with depth between forests and savannas results in a higher accumulation of carbon in forest soils than in savanna soils. In the first 20 cm, 3480 g C m^{-2} are accumulated in the plateau forest and 3770 g C m^{-2} in the gallery forests, while these amounts are only 1800 g C m^{-2} in the *Loudetia simplex* grass savanna and 3220 g C m^{-2} in the Andropogoneae shrub tree savanna (Table 11.2).

11.3.2 Variations of soil organic matter distribution at organomineral particle scale

The fractionation of soil into size classes of organomineral particles [18, 5] is a useful tool to study SOM dynamics at the micro-organism scale because it integrates both chemical and spatial determinants of microbial distribution and activity. Soil samples from the open shrub tree savanna dominated by

Hyparrhenia diplandra were studied by this method [31]. Their total organic carbon and nitrogen concentrations strongly differ among size classes, with more C and N in fine than in coarse size classes. The C:N ratiodecreases with particle size: the highest is measured in the 100-250 μm particles and the lowest in clays (0-2 μm) (Table 11.3). One-third (37.8%) of the organic carbon is linked to the clay fraction and more than one-third (42.5%) to the fine silt fraction (2-20 μm) (Fig. 11.2). Plant debris larger than 50 μm account for 7% of the total soil carbon pool. Half (50.1%) of the organic nitrogen pool is associated with clay particles and more than one third (35.9%) with fine silt fraction. Plant debris only account for 4.8% of the total soil nitrogen pool (Fig. 10.3). The coarse fractions are generally a more important pool of organic matter in savanna sandy soils, even in cultivated soils [13]. The particular situation in Lamto is likely due to the length of the wet season (see

Table 11.3. Organic carbon and nitrogen concentrations and C:N ratios of whole soil and size fractions from the 0-10 cm layer of an *Hyparrhenia diplandra* shrub savanna (μg g^{-1} fraction weight) (after [31], copyright (1997), with permission of Elsevier).

Fraction (μm)	% of whole soil	Carbon	Nitrogen	C:N ratio
0.05–2	6.59	34,850	3,430	10.2
2–20	9.43	11,750	2,460	16.1
20–50	4.28	3,800	635	18.5
50–100	8.57	3,800	220	17.3
100–250	24.07	1,550	60	25.8
250–2000	48.27	1,150	55	20.9
Whole soil	100	8,400	530	15.9

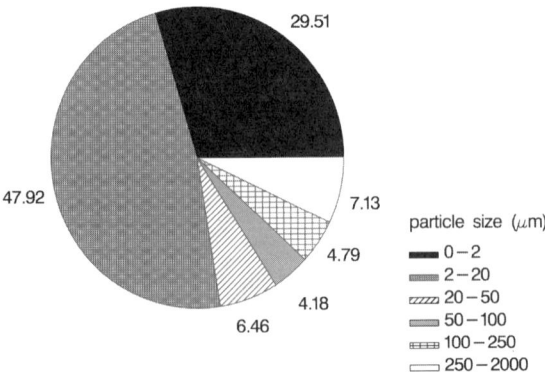

Fig. 11.2. Distribution of organic carbon between different size classes of organo-mineral particles of a shrub tree savanna dominated by *Hyparrhenia diplandra* (data from [31]).

Chap. 3). A long wet season is favorable to micro-organisms and induces a high rate of litter decomposition (above and belowground). This results in a low quantity of plant matter entering the soil and consequently induces a low contribution of the > 50 μm fraction to the total soil organic pool. Conversely, most of the organic matter is linked to fine mineral particles. Therefore, most of SOM in Lamto is physically protected against biodegradation.

11.4 Chemical composition of soil organic matter

The decomposition rate of organic compounds is a key parameter determining soil organic matter content. It is obviously very dependent on soil physical conditions at the scale of micro-organisms: temperature, water and oxygen availability, clay mineralogy, pH [37] and texture, the latter more or less controlling the others [3, 14, 20]. But, these factors only control the expression of a potential of decomposition, which is first determined by the chemical nature of organic compounds [36]. The rate of microbial mineralization is therefore frequently correlated to the lignin, fiber and polysaccharides concentrations of SOM.

The chemical quality of SOM can be assessed by different methods. One of the simplest is the measurement of the carbon mineralization ratio (ratio of the carbon evolved as CO_2 to total initial organic carbon) that, in laboratory conditions, depends almost exclusively on the nature of organic compounds metabolized if environmental conditions (temperature and soil humidity) are optimal. Another method is the measurement of the natural abundance of ^{15}N in soil as an indicator of the past changes of organic matter. It is also possible to identify and measure some of the molecules that are known to influence the rate of microbial activity, such as humic acids or polysaccharides. In Lamto, all these studies have been performed at two scales in the shrub tree savanna: at the landscape scale in order to understand the relationships between SOM and vegetation structure, and at the organomineral particle scale in order to understand the mechanisms of the transformation of the organic matter and regulation of microbial activity. Indeed, the physical separation of the organomineral particles results in the separation of organic compounds classes with different chemical properties and bearing different microbial activities. The physical fractionation of soil into size classes of organomineral particles is therefore frequently used for the study of the variability of organic matter quality and associated microbial activities.

11.4.1 Variations of the chemical composition of soil organic matter at landscape scale

Potential soil respiration

Comparative studies of the carbon mineralization ratios (ratio of carbon evolved as CO_2 to initial organic carbon) [9] enable one to assess the relative physiological value of organic matter from different soils if laboratory incubations are performed in identical and optimal conditions for bacteria and fungi activity (constant temperature, high water content, high oxygen partial pressure). A study of this type [31] has been conducted on four Lamto soils: a semi-deciduous forest on a plateau from the north of the reserve, an evergreen gallery forest, a *Loudetia* savanna and an Andropogoneae savanna. Soil cores have been collected between 0 and 5 cm depth, 5 and 10 cm, and 10 and 20 cm, then air-dried. Then, 20 g aliquots were incubated in the dark at 28°C and 80% of the water holding capacity. The production of CO_2 was measured by gas chromatography after one week of incubation. The results (Table 11.4) were quite surprising since the carbon mineralization ratios were very close to each other, clearly indicating that the availability of organic matter to microorganisms does not differ between forest and savannas. They also show that the origin of the organic matter, C_3 (in forest) or C_4 plants (in savanna) has no impact on the biodegradability of SOM.

The natural abundance of ^{15}N in SOM

The organic matter in almost all the soils in the world is enriched in ^{15}N compared to fresh plant debris. As decomposition proceeds, ^{15}N accumulates in SOM: the measurement of the natural abundance of ^{15}N in the soil is therefore a good estimate of the stage of SOM humification. In Lamto, the $\delta^{15}N$ of total SOM ranges from 2.9‰ to 6.8‰ (Fig. 11.3).

The plateau forest and Andropogoneae savanna soils have comparable and high $\delta^{15}N$ values while that of *Loudetia* savanna are quite low, the gallery forest soils having medium values. In all the soil, the $\delta^{15}N$ increases with depth. To understand the meaning of these results, it is necessary to know the $\delta^{15}N$ of the plants from which SOM originates. As expected, all the $\delta^{15}N$ values of the soils are higher than that of the fresh plant debris. But the savanna grasses show negative values while that of forest trees are positive (Table 11.5). The difference between the soil and plant $\delta^{15}N$ is 7.4‰ to 7.8‰ in the Andropogoneae savanna, 4.9‰ to 6.4‰ in the *Loudetia* savanna, but only 3.5‰ to 5.2‰ in the forest, strongly suggesting that plant debris have been more transformed, i.e., more humified, in the savanna than in the forest and, consequently, that SOM is less accessible to micro-organisms in the savanna than in the forest.

Table 11.4. Carbon mineralization ratios (ratio of carbon evolved as CO_2 to initial organic carbon in %) after 7 days incubation of soil samples from a plateau forest (PF), a gallery forest (GF), a *Loudetia* grass savanna (GS) and an Andropogoneae shrub savanna (SS) (after [31], copyright (1997), with permission of Elsevier).

Depth (cm)	PF	GF	GS	SS
0–5	1.95	1.76	2.01	1.92
5–10	2.45	2.13	2.48	2.04
10–20	2.56	3.36	2.68	2.42

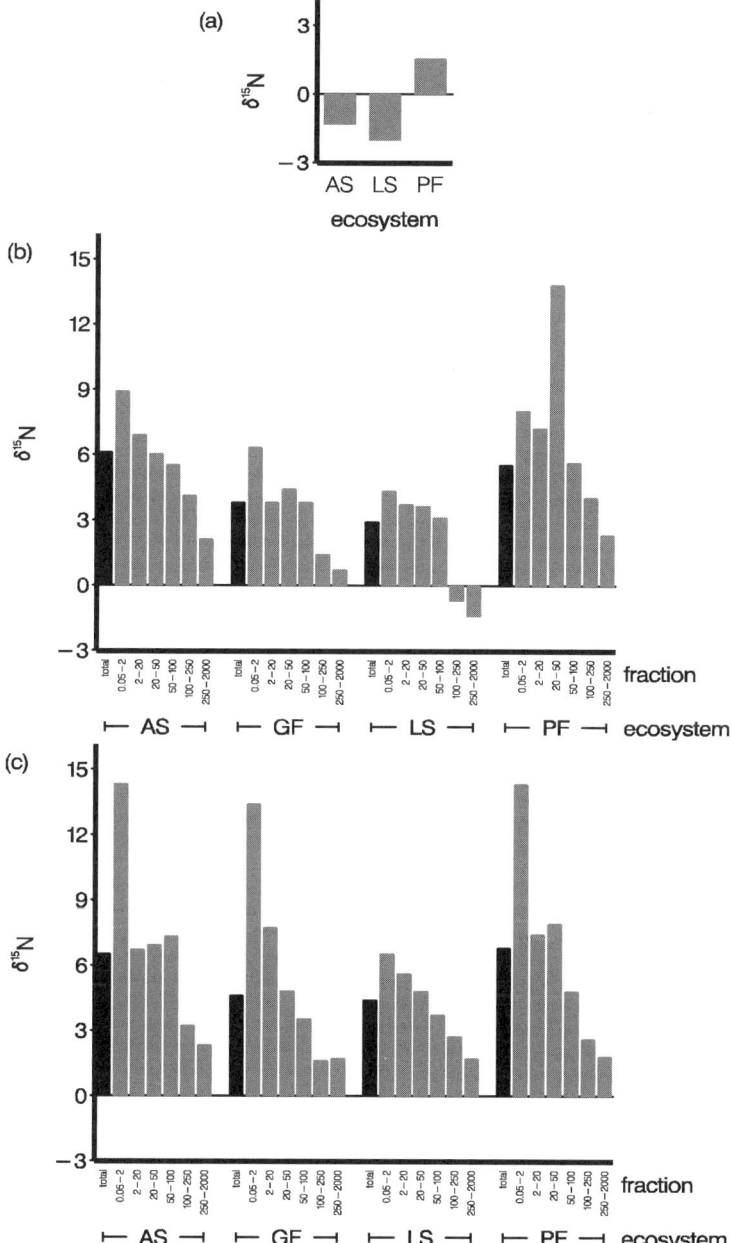

Fig. 11.3. Natural abundance of ^{15}N (as δ^{15}N, in ‰) in the (a) fresh plant debris, (b) organic matter at 0–10 cm depth, and (c) 10–20 cm depth in the Lamto soils. Granulometric fractions in µm. LS, *Loudetia* savanna; AS, Andropogoneae shrub savanna; PF, plateau forest; GF, gallery forest (data from [33]).

Table 11.5. Natural abundance of ^{15}N in Lamto savanna grass and forest tree leaves (after [2]; Abbadie and Mariotti, unpublished data).

Group	Species	δ^{15}N(‰)
Savanna grasses	*Loudetia simplex*	−2.0
	Andropogon schirensis	−1.9
	Hyparrhenia diplandra	−1.3
	Hyparrhenia smithiana	−1.3
Forest trees	*Erythroxylum emarginatum*	1.1
	Olax subscorpioides	0.9
	Celtis prantlii	2.8

Table 11.6. Chemical composition of soil organic carbon after alkali extraction in Lamto savanna soils. ppmC as μg C g^{-1} soil. Data from 4 different Andropogoneae shrub savanna sites (J. Delmas, unpublished data).

Sample	Total C (ppmC)	Extractable C (%)	Fulvic acids (ppmC)	Brown humic acids (ppmC)	Gray humic acids (ppmC)
Plateau 1	8333	24	528	600	812
Plateau 2	6200	49	984	901	1162
Slope 1	6980	27	566	520	1054
Slope 2	9900	35	1056	953	1482

The composition of SOM after alkali extraction

The high degree of organic matter humification in the Lamto soils is confirmed by the low chemical extractability of organic compounds [10]. Less than one-third of the total organic can be extracted by $Na_4P_2O_7$ under the form of fulvic acids and brown and grey humic acids (Table 11.6). Humic acids are dominant, as in many tropical soils: the high temperatures and length of the wet season induce a rapid polymerization of the grey humic acids on one hand and quick leaching and mineralization of the low molecular weight compounds (fulvic acids and brown humic acid) on the other hand. Consequently, most of the SOM in Lamto is composed of highly polymerized compounds, that are probably rich in aromatic rings and costly to degrade by micro-organisms.

SOM extractable sugars

The origin of soil carbohydrates is very variable. Some of them originate from dead plant matter and enter the soil matrix through the microbial degradation of plant residues. Others are basically extracellular compounds produced in great quantities by soil bacteria and fungi. They form a complex mixture of polysaccharides, oligosaccharides and monosaccharides that generally contribute to 5-25% of the total SOM pool. They are suspected of playing an important role on soil fertility because they act as a source of carbon and energy for soil micro-organisms and as stabilizing agents for soil aggregates.

They are therefore an important factor of regulation of the heterotrophic microbial activity, including that associated with the humified organic matter stored on the fine particles. They have been studied many times in agricultural soils, but only few works are available on the relationships between the natural vegetation and soil concentration in sugars [19, 15]. The quantity and nature of extractable sugars of Lamto soils have been determined by a two-step acidic hydrolysis and analysis by capillary gas chromatography after sylilation [22]. Twelve of the 14 sugars usually present in soils have been detected in Lamto (Table 11.7). The hexoses (glucose, galactose, mannose; fructose has not been detected), pentoses (arabinose, xylose and ribose) and deoxyhexoses (rhamnose) contribute respectively to 52-65%, 33-43% and 2-4% of the total soil extractable monosaccharides. The fucose, acidic sugars (galacturonic and glucoronic acids) and basic sugars (galactosamine and glucosamine) are only present in very low quantities [31]. The concentration in total extractable sugars varies along the toposequence: in the first 5 cm, the highest concentration is in the gallery forest (2.3 mg C g^{-1} soil) and the lowest in the *Loudetia simplex* savanna (0.8 mg C g^{-1} soil). There is a positive order 2 polynomial relationship ($R^2 = 0.51$) between soil organic carbon concentration and soil extractable monosaccharides concentration. However, the proportion of total carbon as extractable sugars is lower under the forest that under the savanna: under the forest, only 3.0% to 4.0% of the total carbon is made of extractable monosaccharides vs. 4.8% to 7.5% under the savanna. In forest soils, the total monosaccharide concentration decreases with depth but not in savanna soils, likely because most of the dead plant matter added to soil is deposited on the soil surface in the forest whereas most of plant debris come from the root system in the savanna.

Table 11.7. Pentose, hexose and deoxyhexose concentrations (μg C g^{-1} soil) and contribution to total soil organic carbon between 0 and 20 cm depth in forest and savanna soils (PF: plateau forest; AS: Andropogoneae savanna; GS: *Loudetia* grass savanna; GF: gallery forest) (after [32], reproduced with permission of the Australian Journal of Soil Research. Published by CSIRO PUBLISHING, Melbourne Australia (http://www.publish.csiro.au/journals/ajsr)).

	Depth	Hexoses	Pentoses	Deoxyhexoses	Other	Total	% soil C
GF	0–5 cm	1367.2	825.0	82.5	73.1	2347.8	3.2
	5–10	845.3	461.6	41.4	24.6	1372.9	4.0
	10–20	297.1	151.9	14.0	0.0	463.1	3.7
PF	0–5 cm	948.1	604.3	52.3	43.8	1648.5	3.0
	5–10	610.2	499.9	46.0	12.5	1168.6	3.5
	10–20	385.4	275.5	20.3	0.0	681.2	3.3
AS	0–5 cm	1361.2	845.4	54.3	12.0	2272.9	7.5
	5–10	902.0	536.6	33.5	0.0	1472.1	5.6
	10–20	1068.9	586.8	38.0	8.0	1701.7	7.0
GS	0–5 cm	536.7	286.1	15.3	6.4	844.5	4.8
	5–10	684.7	382.0	19.9	6.1	1092.7	7.1
	10–20	510.1	262.3	17.4	0.0	789.8	6.0

Table 11.8. Monosaccharide composition of forest and savanna soils ($\mu g\, g^{-1}$ soil) between 0 and 10 cm depth. Glu: glucose; Gal: galactose; Man: mannose; Ara: arabinose; xyl: xylose; Rib: ribulose; Rha: rhamnose; Fuc: fucose; Gal-ac: galacturonic acid; Glu-ac: glucuronic acid; Gal-N: galactosamine; Glc-N: glucosamine (PF: plateau forest; AS: Andropogoneae shrub savanna; GS: *Loudetia* grass savanna; GF: gallery forest) (data from [31] and [30]).

	Depth	Glu	Gal	Man	Ara	Xyl	Rib	Rha	Fuc	Gal-ac	Glu-ac	Gal-N	Glc-N
GF	0–5 cm	857.0	156.3	353.9	107.4	147.0	570.6	82.5	58.3	0	6.4	0	8.4
	5–10	508.7	97.8	238.8	60.6	86.9	314.1	41.4	0	12.4	4.5	0	7.7
	10–20	179.1	32.7	85.3	22.3	32.0	97.6	14.0	0	0	0	0	0
PF	0–5 cm	628.1	111.7	208.3	84.1	88.4	431.8	52.3	15.7	0	0	0	28.1
	5–10	356.9	82.9	170.4	66.1	69.9	363.9	46.0	0	12.5	0	0	0
	10–20	218.8	50.9	115.7	39.2	44.5	191.8	20.3	0	0	0	0	0
AS	0–5 cm	845.3	157.8	358.1	87.7	220.6	537.1	54.3	0	0	0	6.1	0
	5–10	546.4	99.6	256.0	61.8	127.4	347.4	33.5	0	0	0	0	0
	10–20	635.2	116.4	317.3	67.6	134.4	384.8	38.0	0	8.0	0	0	0
GS	0–5 cm	344.8	61.7	130.2	34.9	83.3	167.9	15.3	6.4	0	0	0	0
	5–10	442.7	75.4	166.6	46.1	112.7	223.2	19.9	0	6.1	0	0	0
	10–20	323.2	52.6	134.3	35.0	77.5	149.8	17.4	0	0	0	0	0

The most abundant sugars are the same in all the soils studied: 31% to 41% of the total of extractable sugars is glucose, 21% to 31% is ribose, 13% to 19% is mannose, whatever the depth (Table 11.8). This gives information about the origin of soil carbohydrates. Glucose is abundant in plant tissues, but is also produced in large amounts by micro-organisms. Galactose, mannose, rhamnose, and fucose are mainly synthesised by micro-flora, but xylose and arabinose are quite exclusively of plant origin [29, 28]. In order to identify the origin of soil carbohydrates, several authors have proposed calculating various ratios using the concentration of soil in mannose, galactose, arabinose and xylose [38, 4, 29]. In Lamto, all of them indicate that soil extractable sugars are mainly of microbial origin.

11.4.2 Variations of the chemical composition of soil organic matter at organomineral particles scale

Potential soil respiration and carbon mineralization ratio

The carbon mineralization ratios (ratio of the carbon evolved as CO_2 to total initial organic carbon) were not assessed through the incubations of separated size fractions as in many previous works. The contribution of each size class of organomineral particles was estimated in Lamto by comparing the mineralization in incomplete soil lacking one size class to that of recombined soil containing all classes [31]. This approach was chosen in order to make the conditions of incubation of each size class as close as possible to those of whole soil, keeping most of the interactions between fractions. The sum of calculated productions of CO_2 in each fraction was almost exactly equal to that of CO_2 in the completely reconstituted soil (Table 11.9). This result suggests that the

Table 11.9. CO_2 production in 21 days incubation ($\mu g\,C\,CO_2$ per 20 g dry soil), carbon mineralization ratios and calculated net contributions to total CO_2 production (% of the total CO_2 accumulated) of separate granulometric size classes from a shrub tree Andropogoneae savanna (after [30]).

Fraction	C-CO_2	C-CO_2:C	%CO_2
0.05–2 μm	1375	3.2	31.1
2–20 μm	1012	1.5	22.9
20–50 μm	606	6.1	13.7
50–100 μm	265	3.9	6.0
100–250 μm	359	4.4	8.1
150–2000 μm	803	8.6	18.2
Total	4420		100.0
Recombined soil	4423	2.8	
Control soil	5643	3.4	

process of carbon mineralization in a single fraction is independent of what occurs in other fractions: the mineralization of carbon is therefore a strictly additive process. The lowest carbon mineralization ratios were recorded in clays and fine silts, and the highest in 250-2000 μm particles. The accessibility of organic matter to micro-organisms was 3 to 4 times lower in fine that in coarse particles, suggesting important differences of chemical quality between fractions and a physical protection of organic compounds against microbial degradation by clays and fine silts. However, particles less than 20 μm produce half of the soil CO_2 because they contain more than 75% of the total organic carbon pool.

The natural abundance of ^{15}N in the organomineral fractions from an open shrub tree savanna dominated by *Hyparrhenia diplandra* was measured as an estimate of the stage of SOM humification, i.e., of the chemical accessibility of organic compounds to soil micro-organisms. The δ^{15}N increases with decreasing particle size, ranging from 2.1‰ in the coarse sands (250-2000 μm) to 8.9‰ in clays (0.05-2.0 μm) (Fig. 11.3). The δ^{15}N of the larger particles is close to that of living grasses, i.e., -1.3‰ [2], clearly showing that the organic matter associated to coarse sands has undergone only few microbial transformations. On the contrary, there are ca. 10‰ between the most humified organic matter, associated with clays, and fresh pant matter, indicating a high degree of past biochemical transformations, probably resulting in a low biodegradability for present soil organisms.

Extractable sugars in soil organomineral fractions

The distribution of monosaccharides in organomineral fractions has also been studied on soil samples from the open shrub tree savanna [31]. The monosaccharide concentrations are higher in fine than in coarse fractions and higher

Table 11.10. Monosaccharide concentration ($\mu g\,g^{-1}$ soil) of size separates and whole soil (0-10 cm depth) from an Andropogoneae shrub savanna, contribution of monosaccharides in each fraction to total soil organic carbon (% of the pool of carbon in the whole soil) (after [33], with kind permission of Springer Science and Business media).

Fraction	Pentoses	Hexoses	Deoxy-hexoses	Other	Total	% C fraction	% C soil
Whole soil	699.0	1137.3	40.9	0	1877.2		7.7
0.05–2 μm	1112.8	2445.6	168.4	57.4	3784.2	4.4	0.7
2–20 μm	1710.0	3988.4	188.5	156.4	6043.3	6.1	0.7
20–50 μm	1571.4	2476.1	77.3	48.6	4173.4	14.3	3.0
50–100 μm	793.6	1056.6	20.0	0	1870.1	19.8	2.3
100–250 μm	479.9	615.5	11.9	0	1107.3	29.8	1.2
150–2000 μm	138.9	245.2	11.7	0	395.8	14.4	0.3

in fine silt than in clays (Table 11.10). Monosaccharides contribute to 7.7% of total soil carbon. This contribution is among the lowest reported worlwide. It clearly indicates a low chemical availability of SOM to micro-organisms (see data in [7]). The highest contribution values are reported for the 20-50 μm (3.0%) and 50-100 μm (2.3%) fractions. The carbohydrate composition differs between size separates (Fig. 11.4). The ratios (galactose + mannose): (arabinose + xylose), mannose : xylose and mannose : (arabinose + xylose) all decrease with increasing particle size, indicating that micro-organisms derived carbohydrates accumulate more in fine than in coarse fractions. This is in good agreement with the high microbial activity of carbon degradation measured on these fractions, and likely indicates that most micro-organisms are located on fine fractions. Even if the degradation of the organic matter linked to fine fractions is intrinsically costly for micro-organisms, the nitrogen richness of these fractions allows the maintenance and reproduction of the microbial community that is, consequently, preferentially located on fine silts and clays.

11.5 Conclusion

The savanna soils from Lamto are poor in organic matter, probably due to the burning of most of the aboveground primary production, rapid decomposition of plant debris and soil texture. Moreover, this organic matter is quite regularly distributed in a soil profile: at 1 m depth, the concentration in organic matter still amounts to 0.2-0.3% vs. 0.8-0.9% in surface layers [35]. This is a strong difference with tropical forest ecosystems where organic matter is very concentrated in the first centimeters of the soil profile (2.2-2.8% at 5 cm depth in the plateau forest soil from Lamto). In other words, a part of organic matter in the savanna is out of the root strata and, consequently, a part of the nutrients. This clearly suggests that SOM can only contribute in a

11 Origin, Distribution, and Composition of Soil Organic Matter   215

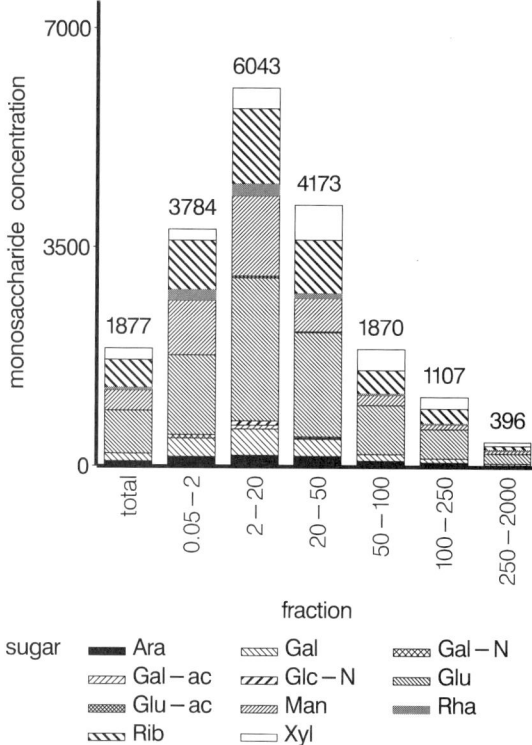

Fig. 11.4. Monosaccharide composition (μg g^{-1} soil) of separate size classes (in μm) and whole soil (0-10 cm depth) from an Andropogoneae shrub savanna. Glu: glucose; Gal: galactose; Man: mannose; Ara: arabinose; xyl:xylose; Rib: ribulose; Rha:rhamnose; Fuc:fucose; Gal-ac:galacturonic acid; Glu-ac: glucuronic acid; Gal-N:galactosamine; Glc-N: glucosamine (data from [33]).

minor way to the grass requirements in nutrients, whatever the rate of SOM mineralization. Moreover, the adsorption of most of SOM on silt and clays and the low chemical degradability of SOM do not allow a high microbial activity of carbon and nitrogen mineralization in Lamto soils. One can therefore expect a strong impact of plant litter mineralization, especially grass root mineralization, on primary production.

References

1. L. Abbadie, M. Lepage, and X. Le Roux. Soil fauna at the forest-savanna boundary: Role of termite mounds in nutrient cycling. In P.A. Furley, J. Proctor, and J.A. Ratter, editors, *Nature and dynamics of forest-savanna boundaries*, pages 473–484. Chapman & Hall, London, 1992.
2. L. Abbadie, A. Mariotti, and J.-C. Menaut. Independence of savanna grasses from soil organic matter for their nitrogen supply. *Ecology*, 73:608–613, 1992.

3. D.W. Anderson, S. Saggar, J.R. Bettany, and J.W.B. Stewart. Particle-size fractions and their use in studies of soil organic matter. I. The nature and distribution of forms of carbon, nitrogen and sulfur. *Soil Science Society of America Journal*, 45:767–772, 1981.
4. J.A. Baldock, B.D. Kay, and M. Schnitzer. Influence of cropping treatment on the monosaccharide content of the hydrolysates of a soil and its aggregate fractions. *Canadian Journal of Soil Science*, 67:489–499, 1987.
5. J. Balesdent, J.-P. Pétraud, and C. Feller. Effets des ultrasons sur la distribution granulométrique des matières organiques des sols. *Science du sol*, 29(2):95–106, 1991.
6. S. Brucker. Analyse des complexes organo-minéraux des sols. In M. Souchier and B. Bonneau, editors, *Pédologie, volume 2: Propriétés et constituants du sol*. Masson, Paris, 1979.
7. M.V. Cheshire. *Nature and origin of carbohydrates in soil*. Academic Press, New York, 1979.
8. J.L. Devineau. Données préliminaires sur la litière et la chute des feuilles dans quelques formations forestières semi-décidus de moyenne Côte-d'Ivoire. *Oecologia Plantarum*, 11:375–395, 1976.
9. Y. Dommergues. La notion de coefficient de minéralisation du carbone dans les sols. *L'agronomie tropicale*, 15(1):54–60, 1960.
10. Ph. Duchaufour. *Précis de pédologie. Détail des méthodes d'analyse*. Masson, Paris, 1965.
11. C. Feller. *La matière organique dans les sols tropicaux à argile 1:1. Recherche de compartiments organiques fonctionnels. Une approche granulométrique*. Doctorat ès sciences naturelles, Université Louis Pasteur, Strasbourg, 1994.
12. C. Feller, F. Bernhard-Reversat, J.L. Garcia, J.J. Pantier, S. Roussos, and B. Van Vliet-Lanoe. Etude de la matière organique de différentes fractions granulométriques d'un sol sableux tropical. effet d'un amendement organique (compost). *Cahiers de l'ORSTOM, série Pédologie*, 20:223–238, 1983.
13. C. Feller, C. François, G. Villemin, J.M. Portal, F. Toutain, and J. L. Morel. Nature des matières organiques associées aux fractions argileuses d'un sol ferrallitique. *Comptes rendus de l'Académie des Sciences*, 312(II):1491–1497, 1991.
14. C. Feller, E. Fritsch, R. Poss, and C. Valentin. Effet de la texture sur le stockage et la dynamique des matières organiques dans quelques sols ferrugineux et ferrallitiques (Afrique de l'Ouest, en particulier). *Cahiers de l'ORSTOM, série Pédologie*, XXVI(1):25–36, 1991.
15. B.L. Folsom, G.H. Wagner, and C.L. Scrivner. Comparison of soil carbohydrate in several prairie and forest soils by gaz-liquid chromatography. *Soil Science Society of America Proceedings*, 38:305–309, 1974.
16. L. Gautier. Contact forêt- savane en Côte d'Ivoire centrale : Evolution de la surface forestière de la réserve de Lamto. *Bulletin de la Société Botanique de France*, 136(3/4):85–92, 1989.
17. L. Gautier. Contact forêt-savane en Côte d'Ivoire centrale: Evolution du recouvrement ligneux des savanes de la réserve de Lamto (sud du V baoulé). *Candollea*, 45:627–641, 1990.
18. E.G. Gregorich, R.G. Kachanoski, and R.P. Voroney. Carbon mineralization in soil size fractions after various amounts of aggregate disruption. *Journal of Soil Science*, 40:649–659, 1989.

19. U.C. Gupta, F.J. Sowden, and P.C. Stobbe. The characterization of carbohydrate constituents from different soil profiles. *Soil Science Society of America Proceedings*, 27:380–382, 1963.
20. J. Hassink. Effects of soil texture and structure on carbon and nitrogen mineralization in grassland soils. *Biology and Fertility of Soils*, 14:126–134, 1992.
21. M.J. Jones. The organic matter content of the savanna soils of West Africa. *Journal of Soil Science*, 24(1):42–53, 1973.
22. M.C. Larré-Larrouy and C. Feller. Determination of carbohydrates in two ferrallitic soils : analysis by capillary gas chromatography after derivatization by silylation. *Soil Biology and Biochemistry*, 29:1585–1589, 1997.
23. H. Laudelout, J. Meyer, and A. Peeters. Les relations quantitatives entre la teneur en matières organiques du sol et le climat. *Agricultura*, 8:103–140, 1960.
24. M. Lepage, L. Abbadie, and A. Mariotti. Food habits of sympatric termite species (Isoptera, Macrotermitineae) as determined by stable carbon isotope analysis in a Guinean savanna (Lamto, Côte d'Ivoire). *Journal of Tropical Ecology*, 9:301–311, 1993.
25. A. Mariotti and J. Balesdent. ^{13}C natural abundance as a tracer of soil organic matter turnover and paleoenvironment dynamics. *Chemical Geology*, 84:217–219, 1990.
26. W.B. McGill. Review and classification of ten soil organic matter (SOM) models. In D.S. Powlson, P. Smith, and J.U. Smith, editors, *Evaluation of soil organic matter models*, volume 38, pages 111–132. Springer-Verlag, Berlin, 1996.
27. J.C. Menaut and J. César. Structure and primary productivity of Lamto savannas, Ivory Coast. *Ecology*, 60(6):1197–1210, 1979.
28. M.E.C. Moers, M. Baas, J.W. De Leeuw, J.J. Boon, and P.A. Schenk. Occurence and origin of carbohydrates in peat samples from a red mangrove environment as reflected by abundances of neutral monosacchardies. *Geochimica et Cosmochimica Acta*, 54:2463–2472, 1990.
29. S. Murayama. Microbial synthesis of saccharides in soils incubated with ^{13}C labelled glucose. *Soil Biology and Biochemistry*, 20:193–199, 1988.
30. H.B. Nacro. *Hétérogénéité de la matière organique dans un sol de savane humide (Lamto, Côte d'Ivoire): Caractérisation chimique et étude in vitro des activités microbiennes de minéralisation du carbone et de l'azote*. Ph.D. thesis, Université de Paris 6, Paris, 1997.
31. H.B. Nacro, D. Benest, and L. Abbadie. Distribution of microbial activities and organic matter according to particle size in a humid savanna soil (Lamto, Côte d'Ivoire). *Soil Biology and Biochemistry*, 28:1687–1697, 1996.
32. H.B. Nacro, M.C. Larré-Larrouy, C. Feller, and L. Abbadie. Hydrolysable carbohydrate in tropical soils under adjacent forest and savanna vegetation in Lamto, Côte d'Ivoire. *Australian Journal of Soil Research*, 43:1–7, 2005.
33. H.B. Nacro, M.C. Larré-Larrouy, A. Mariotti, C. feller, and L. Abbadie. Natural nitrogen-15 abundance and carbohydrate content and composition of organic matter particle-size fractions from a sandy soil. *Biology and fertility of soils*, 40:171–177, 2004.
34. J.M. Oades and J.N. Ladd. Biochemical properties. In J.S. Greacen and F.L. Russel, editors, *Soil factors in crop production in a semi-arid environment*, pages 126–160. University of Queensland Press, Brisbane, 1977.
35. C. Pirovano. Studio di dinamiche del carbonio e dell'azoto in suoli tropicali sotto savana e sotto foresta in Costa d'Avorio. Laurea in science ambientali, Università degli Studi di Milano, 1996.

36. M. J. Swift, O. W. Heal, and J.M. Anderson. *Decomposition in terrestrial ecosystems*, volume 5 of *Studies in Ecology*. Blackwell Scientific Plublications, Oxford, 1979.
37. B.K.G. Theng, K.R. Tate, and P. Sollins. Constituents of organic matter in temperate and tropical soils. In D.C. Coleman, J.M. Oades, and G. Uehara, editors, *Dynamics of soil organic matter in tropical ecosystems*, pages 5–32. University of Hawaii Press, Honolulu, 1989.
38. L.W. Turchenek and J.M. Oades. Fractionation of organo-mineral complexes by sedimentation and density techniques. *Geoderma*, 21(4):311–344, 1979.

12

Soil Carbon and Organic Matter Dynamics

Luc Abbadie, Hassan Bismarck Nacro, and Jacques Gignoux

12.1 Soil micro-organisms

In Lamto, the biodiversity of soil micro-organisms is still poorly documented. The only available data are those obtained with classical methods focusing on culturable micro-organisms by Pochon and Bacvarov [18] about the bacterial community from *Loudetia simplex* and *Hyparrhenia diplandra* savannas and by Rambelli et al. [20, 21] about microfungi. Pochon and Bacvarov [18] collected samples in the first 10 cm of the soil, between tussocks, in the shrub tree savanna. The dilution and counting of organisms was performed directly on fresh samples or after 3 weeks incubation. They observed that total microflora in the non rhizospheric soil is quite high, ranging from 7×10^6 to 90×10^6 organisms per g soil in the Andropogoneae savanna and from 25×10^6 to 80×10^6 organisms per g soil in the *Loudetia* savanna. Rambelli et al. [21] and Schaefer [23] noticed that most of these organisms are actinomycetes. Among bacteria, all the functional groups are present, but only at very low densities (Table 12.1). In the unburned savanna dominated by *Hyparrhenia* spp., 59 species of fungi have been identified [20]; this number is slightly lower in the burned savanna.

In the rhizosphere of *Loudetia simplex*, the average number of microorganisms (total microflora) is 66×10^6 organisms per g soil in the yearly burned savanna, but reaches up to 927×10^6 organisms per g soil in the unburned shrub-tree savanna, with an average value of 94×10^6 organisms per g soil [13]. The number of micro-organisms seems highly variable with season: the above densities, measured in January, i.e., during the dry season, fall to 3×10^6 organisms per g soil in March, i.e., in the beginning of the wet season [21]. The species richness of fungi in the *Loudetia* rhizosphere is higher in unburned than in burned savanna. In both savannas, the most common species of fungi are *Penicillium cyclopum*, *Penicillium lilacinum* and *Spicaria griseola* [13]. All the fungi associated to *Loudetia simplex* are vesicular arbuscular mycorrhizae [20].

The microbial biomass has never been measured in the field in Lamto. But, Nacro [16] assessed it on air dried soil samples in laboratory conditions. He

Table 12.1. Bacterial densities (10^3 organisms g^{-1} dry soil) in non-rhizospheric 0–10 cm soils (reprinted from [18], copyright (1973), with permission of Elsevier).

Facies	Date	N fixers	Cellulolytic	Nitrifiers
Andropogoneae	Oct. 1969	7.0	6.5	1.5
tree/shrub	May 1970	7.0	13.2	2.3
savanna	Feb. 1971	4.0	2.2	4.0
	Mar. 1971	1.3	3.5	1.9
	May 1971	0.9	0.4	4.1
Loudetia simplex	Oct. 1969	23.0	11.5	3.2
savanna	May 1970	4.0	18.4	4.7

used the fumigation and nihydrin reactive nitrogen method [2], after 14 days incubation at 28°C and 80% of the water holding capacity, i.e., in roughly steady state, close to the conditions occurring in the field during the wet season. In these conditions, the carbon and nitrogen microbial biomass are respectively 129.8 $\mu g\,C\,g^{-1}$ soil and 19.2 $\mu g\,N\,g^{-1}$ soil in the Andropogoneae shrub tree savanna, i.e., 1.5% and 3.6% of the pools of organic carbon and nitrogen in soils, respectively. Theses values are in the range of what is frequently measured in organic matter-poor soils. They confirm that soil microorganisms in Lamto have medium densities even if they generally exhibit low activities.

12.2 The limitation of soil microbial activity by the supply of organic and mineral compounds

A characteristic feature of savanna soils in Lamto is their low microbial respiration and nitrogen mineralization rates despite favorable climatic conditions (soil temperature and relative water content) during most of the year for micro-organisms heterotrophic to carbon. Several studies have been performed in Lamto in order to identify the exact mechanisms controlling the heterotrophic activity rate in savanna soil [1, 7, 30, 16]. The basic principle of these studies is the supply of organic (carbon alone, proteins or a mixing of both) or mineral compounds to soil samples collected between 0 and 10 cm depth in a tree shrub savanna, incubated in optimal conditions (28°C and 80% of the water holding capacity) for determination of carbon and nitrogen mineralization rates.

In all cases, the supply of carbon compounds results in the significant increase of soil respiration. For example, Nacro [16] supplied 500 $\mu g\,C\,g^{-1}$ soil and 50 $\mu g\,N\,g^{-1}$ soil of various carbon and nitrogen substrates to air dried soil samples (79, 97 and 1081 $\mu g\,N\,g^{-1}$ soil for gelatine, leucine and urea). He observed a production of CO_2 twice higher in the soil samples amended with glucose and leucine than in control soil after 2 weeks incubation, and between 1.2 and 1.5 times higher in those amended with tannin, lignin, gelatine, urea or cellulose (Fig. 12.1). Under the hypothesis of an insensitive effect of these

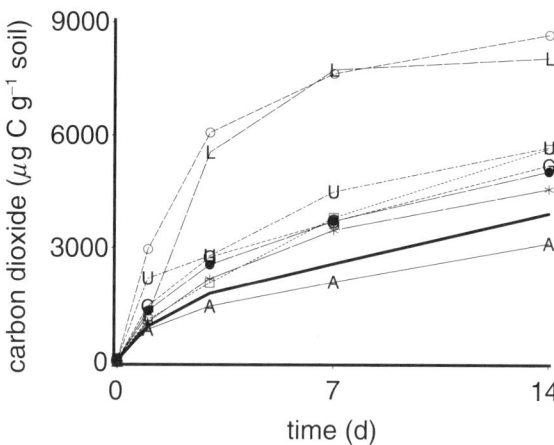

Fig. 12.1. Carbon dioxyde accumulation in control and amended soils from the 0–10 cm layer from the *Hyparrhenia diplandra* shrub tree savanna during aerobic incubation. Substrate added: solid line, control; A, ammonium nitrate; squares, cellulose; G, gelatine; open circles, glucose; L, leucine; solid circles, lignin; *, tannins; U, urea (after [16]).

compounds on the mineralization of humified SOM, the mineralization ratio of the supplied carbon, i.e., the ratio of carbon evolved as CO_2 to initial organic carbon supplied, is 54% and 53% for leucine and glucose, 22% for urea, 15% for cellulose, 12% for gelatine, 10% lignin and 6% for tannin. In the control, the carbon mineralization ratio is only 2.3%. On the contrary, the supply of mineral nitrogen under the form of ammonium [1] and ammonium-nitrate induced a light decrease of respiration. This could be due to the intrinsic toxicity of ammonium or its negative effect on the synthesis of some enzymes. These results suggest that the soils in Lamto have a functional and diverse microbial community. In other words, the total heterotrophic activity is always low due to the lack of degradable substrates and not to the lack of microorganisms.

In the same experiment, the quantity of mineral nitrogen accumulated in the soils amended with urea, leucine and gelatine is higher than that in control soil, but is lower in soils amended with glucose, cellulose, lignin and tannin (Fig. 12.2). Assuming that the effect of these compounds on the mineralization of humified soil organic nitrogen is negligible, it is possible to calculate the mineral nitrogen accumulation rate, i.e., the ratio of mineral nitrogen accumulated to the initial pool of organic nitrogen. This ratio is 99% for urea, 60% for leucine and 32% for gelatine after 7 days (a slow decrease of the pool of mineral nitrogen is observed during the following 7 days, likely as a result of microbial immobilization). The mineral nitrogen concentration of the soil amended with ammonium-nitrate strongly varies with time: 60% is

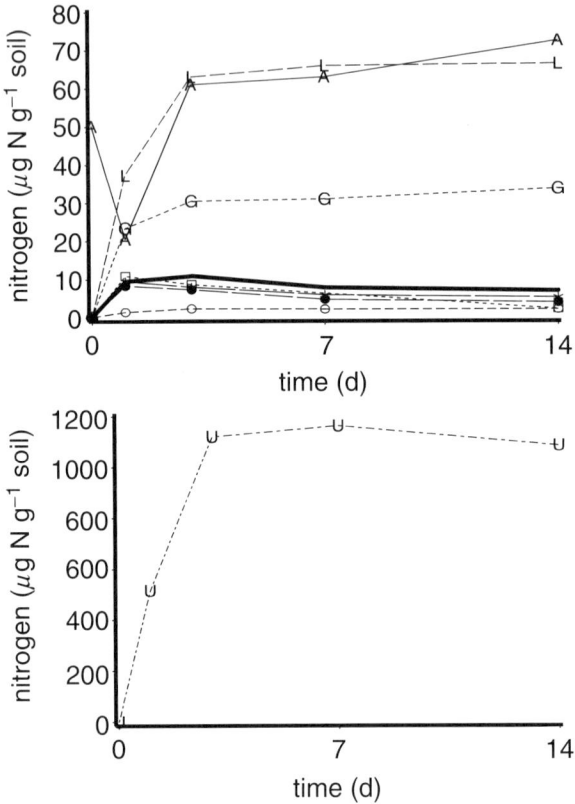

Fig. 12.2. Mineral nitrogen accumulation in control and amended soils from the 0–10 cm layer from the Andropogoneae shrub tree savanna during aerobic incubation. Legend of substrates as in Fig. 12.1 (after [16]).

immobilized after 24 h incubation, but, after 3 days incubation, it exceeds the initial quantity and reach ca. 140% of the initial supply at day 0 [16]. Clearly, the supply of ammonium nitrate induced an additional mineralization of the soil organic nitrogen, i.e., a priming effect of the mineral nitrogen on the humified organic matter degradation.

Basically, all these experiments confirm the hypothesis of [7] and [23] about the limitation of the production of mineral nitrogen in savanna soils by the lack of metabolizable organic matter. Indeed, when easily metabolizable compounds, such as urea, leucine and gelatine, containing both carbon and nitrogen, are supplied to microflora, they are quickly decomposed. A microbial potential of organic compounds decomposition therefore exists in the soils, but is usually not expressed, or expressed at a low rate, because the organic matter available is too costly to degrade. Finally, the soil in the Lamto savanna can be viewed as an "energetic desert", where micro-organisms are in a state of starvation, unable to produce substantial amounts of mineral nitrogen.

12.3 Plant litter decomposition

The decomposition of the aboveground plant material is classically studied in situ by the regular measurement of mass loss and chemical changes of leaves enclosed in net bags deposited on soil surface. It is also possible to carry a demographic approach by following in situ cohorts of dead leaves. This second method was used in Lamto and gave an average decomposition rate of the standing dead matter at 0.015 $g\,g^{-1}$ dead matter d^{-1} [22]. For roots, the burying of root litter bags in soil is also possible, but is quite difficult due to the mixing of plant debris with soil aggregates. But all these methods, if they enable one to measure the real decomposition rate, do not allow one to separate climatic from chemical effects. If the objective is to understand the chemical determinants of plant litter decomposition, it is necessary to control the physical conditions to get generalizable results and knowledge. In order to build up a generic model of soil organic matter (SOM) dynamics (see Sect. 12.5), such an experimental approach was chosen in Lamto where different types of plant material were incorporated to soil and the induced activities of heterotrophic micro-organisms measured.

Dead leaves and roots collected in November, i.e., at the end of the growing period, were incubated at 28°C and 80% of the water holding capacity in tree-shrub savanna soil collected between 0 and 10 cm depth. The study was conducted on leaves of the most common tree species in Lamto: *Piliostigma thonningii*, *Bridelia ferruginea*, *Cussonia arborea*, *Crossopteryx febrifuga*, the herbaceous legume *Indigofera polysphaera* and leaves and roots of the two grass species *Andropogon schirensis* and *Loudetia simplex*. The chemical composition of plant materials strongly differs among these species. The lowest C:N ratio is measured in the legume species (22) while the highest are observed in grass leaves and roots with values only varying from 89 to 97. Tree leaves show intermediate C:N values (52-76). The lignin and lignocellulose concentrations are higher in tree than in grass leaves and the cellulose concentrations are lower in tree than in grass leaves. Generally, the leaves contain less cellulose and lignin than roots (Table 12.2).

In this experiment, Nacro [16] crumbled the plant materials to 80 μm and supplied them to soil samples (50 $\mu g\,N\,g^{-1}$ soil added). The accumulations of CO_2 and mineral nitrogen and microbial biomass (fumigation and nihydrin reactive nitrogen method) were regularly measured [2]. The flow of CO_2 evolved from the plant debris has been estimated as the difference between the flow of CO_2 evolved from amended soil and that evolved from the non-amended soil. This estimate is valid only if there is no priming effect of plant material on soil humus mineralization, i.e., if there is no stimulating

Table 12.2. Chemical characteristics (% of plant dry matter weight) and quantity of carbon supplied (mg C 20 g^{-1} dry soil) to soil samples in the plant materials decomposition experiment. A: dead leaves of *Andropogon schirensis*; L: dead leaves of *Loudetia simplex*; RA: roots of *Andropogon schirensis*; RL: roots of *Loudetia simplex*; In: leaves of *Indigofera polysphaera*; B: leaves of *Bridelia ferruginea*; Cr: leaves of *Crossopteryx febrifuga*; Cu: leaves of *Cussonia arboreai*; P: leaves of *Piliostigma thonningii* (after [16]).

Substrate	C (%)	N (%)	C:N	Cellulose (%)	Lignocellulose (%)	Lignin (%)	C supplied (mg)
A	37.6	0.39	96.5	35.2	43.4	10.4	97.2
L	42.0	0.47	89.4	32.9	41.2	12.2	88.7
RA	41.6	0.46	90.4	36.2	48.5	51.4	19.5
RL	48.9	0.55	88.8	48.5	51.4	19.5	88.4
In	42.7	1.97	21.7	14.6	42.4	26.7	21.8
B	45.1	0.85	53.1	23.4	51.0	37.5	53.2
Cr	47.6	0.92	51.8	21.9	49.1	41.6	51.4
Cu	45.8	0.60	76.4	16.2	28.6	21.0	76.6
P	47.6	0.72	66.2	30.2	54.2	35.4	66.2

effect of the carbon supplied on humified soil carbon mineralization. The data from Fig. 12.3 have therefore to be considered carefully because they could be overestimated. The most decomposable materials are obviously the *Indigofera polysphaera* leaves, whose 17.1% of the carbon is mineralized in 14 days (vs. 2.5% in control soil). Grass roots and tree leaves are also rapidly decomposed (6.9% to 12.8% of their carbon mineralized in 14 days). The most resistant to biodegradation are the grass leaves (5.9% to 6.5%). As expected, a strong microbial immobilization of mineral nitrogen occurred in all the amended soils (ammonia volatilization and denitrification are negligible in Lamto soils; see Sect. 14.6), including those amended with *Indigofera polysphaera* leaves. The microbial biomass grows during the first 24 h and then remains almost constant all along the incubation period. It is significantly different from that of non-amended control soil only when the plant material supplied is grass root. For the other substrates, the biomass is not significantly different from that of control soil. Around 2% of the grass root carbon are transformed in microbial biomass carbon. Compared to the data from literature [28, 29], this value is quite low and could mean that grass roots have an intermediate decomposability.

Finally, neither the origin of plant litter (tree or grass) nor the measured chemical characteristics (lignin, lignocellulose and cellulose contents) allow reliable prediction of the mineralization rates of the various plant debris. The concentration in low mass polysaccharides could be more meaningful and further research is needed on this point. Anyway, grass litter (particularly roots) is highly decomposable. This result is important because in the savanna, fire destroys most of the aboveground grass biomass. Consequently,

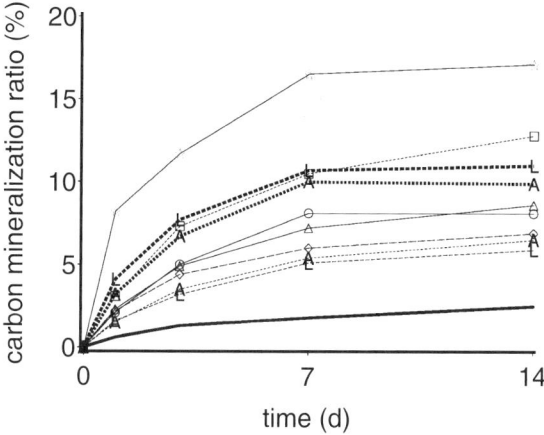

Fig. 12.3. Carbon mineralization ratios of soil organic carbon (ratio of C-CO_2 evolved to initial soil organic carbon in % of soil initial carbon dry weight). Substrates: thick solid line, soil organic carbon; A: *Andropogon schirensis* (thin dotted line, dead leaves; thick dotted line, roots); L: *Loudetia simplex* (thin dotted line, dead leaves; thick dotted line, roots); stars, leaves of *Indigofera polysphaera*; circles, leaves of *Bridelia ferruginea*; triangles, leaves of *Crossopteryx febrifuga*; squares, leaves of *Cussonia arborea*; diamonds, leaves of *Piliostigma thonningii* (after [16]).

the only important source of SOM is the root compartment that is potentially quickly decomposed i.e., which is potentially able to quickly produce a part of the nutrients required by the grass cover.

12.4 Soil organic matter mineralization and turnover

The field production of CO_2 by soils basically depends on the chemical composition of SOM (see Sect. 11.4) and secondarily on climatic conditions and activity of roots and soil fauna (see Chap. 13). Incubation of soil samples in optimal conditions for heterotrophic micro-organisms, i.e., 28°C and 80% of soil water holding capacity, allows theoretical assessment of the maximum organic matter decomposition rates. In soils sampled in May, in the heart of the wet season, it ranges between 1.8% and 2.0% of the initial soil organic carbon mineralized in 7 days in the 0-5 cm layer, between 2.0% and 2.5% in the 5-10 cm layer and between 2.4% and 3.4% in the 10-20 cm layer. This increase with depth is not always observed: on soils sampled during the dry season, Abbadie (unpublished data) observed the opposite trend, with carbon maximum mineralization ratios higher in surface than in deep layers. Moreover, he also observed strong variations with season in 0-10 cm depth samples, from 0.8% to 3% in 30 days incubations. This suggests that the maximum respiration rate probably depends on the physical conditions that occurred in the soil during the previous days or weeks. Therefore, it cannot be considered as a simple baseline, which can be used directly for modeling activities (see Sect. 12.5). It has only a comparative value for soils sampled at the same date and only allows one to sort the soils according to their carbon mineralization abilities. Last, the potential carbon mineralization ratio only gives indications on soil functioning pathways, but not on the actual field heterotrophic activity.

In the Lamto savanna soils, most of the SOM is linked to silts and clays that physically prevent microbial attacks (see Sect. 11.3); moreover, its chemical composition make it costly to degrade for micro-organisms (see Sect. 11.4). Consequently, the average SOM mineralization is quite low, except where energy is supplied such as in the rhizosphere zone or sites where soil is re-handled by invertebrates (see Chap. 13). Villecourt [25] estimated the yearly emission of carbon by soil in the grass savanna at 800 $g\,m^{-2}\,y^{-1}$. Bauzon in [23] measured the field CO_2 production by the *Loudetia* and Andropogoneae savanna soils in small chambers set up on the soil surface. The average daily respiration rates were respectively 1.500 and 1.579 $g\,C\,m^{-2}\,d^{-1}$, i.e., 548 and 576 $g\,C\,m^{-2}\,y^{-1}$. Le Roux and Mordelet [12] measured the diurnal carbon evolution from the soil surface on 0.49 m^2 areas (including standing dead matter and tussock base) using a chamber technique. The soil respiration rates range from 6.6 to 9.6 $\mu mol\,CO_2\,m^{-2}\,s^{-1}$ (6.8 to 9.9 $g\,C\,m^{-2}\,d^{-1}$) during the wet season, and were around 4.0 $\mu mol\,CO_2\,m^{-2}\,s^{-1}$ (4.1 $g\,C\,m^{-2}\,d^{-1}$) at the beginning of the dry season. All these values are in the range of those published previously for productive tropical grasslands [24]. Soil respiration rate in Lamto exhibits hourly variations, strongly correlated with those of air temperature, and monthly variations, values recorded in November (2.100 $g\,C\,m^{-2}\,d^{-1}$) being twice higher than those recorded in February (1.110 $g\,C\,m^{-2}\,d^{-1}$), when the soil is dry.

The soil respiration rate also shows variations in space. It is ca. 60% higher under than between grass tussocks, suggesting a strong contribution of root respiration to total CO_2 emission. But this high heterotrophic activity under tussocks can also result from a high dead root decomposition rate and a stimulation of SOM mineralization by root exudates. Field measurements of CO_2 emission by grassland soils have shown that the contribution of living roots to total soil respiration ranges from ca. 20% to 40% [24, 11, 6, 27]. In Lamto, this means that SOM mineralization and root decomposition should range between 346 and 461 $g\,C\,CO_2\,m^{-2}\,y^{-1}$. Under the hypothesis of a steady state of the plant cover in Lamto at the scale of the year, root decomposition should be equal to root production. The minimum estimate of belowground production in the open Andropogoneae savanna is 986 $g\,DM\,m^{-2}\,y^{-1}$, i.e., about 400 $g\,C\,m^{-2}\,y^{-1}$ (see Sect. 7.4). The mineralization of SOM alone should therefore produce only a low quantity of $C-CO_2$, not more than 60 $g\,C\,m^{-2}\,y^{-1}$. For the 0-20 cm layer, that contains ca. 3217 g C m^{-2} (Table 11.2), the average yearly mineralization rate of total soil organic carbon should therefore be 1.9 % per year.

The mineralization rate of the total SOM has often been assessed through the measurement of the natural abundance of ^{13}C in soils when plant cover is naturally or experimentally changed. Such a change occurs in Lamto in two cases: in the unburned savanna near the buildings of the station and in the north of the reserve, where forest regularly advances over grass savanna. The plant cover dynamics at this forest savanna boundary has been studied by Gautier [8, 9]. He estimated the forest surface increase rate at 0.6% a year, i.e., a forest advance rate of 60 cm y^{-1}. On two transects ([17] and Mareuil, unpublished data), we studied SOM dynamics through δ^{13}C measurements. Indeed, forest plants all belong to the C$_3$ photosynthetic pathway while grasses of the neighbouring grass savanna all belong to the C$_4$ type (Table 11.1).

A strong change of the natural abundance of ^{13}C occurs from the savanna-forest boundary toward the center of the forest. In the first 20 cm, the δ^{13}C shifts from –13.7‰ to –26.4‰ on only a 10 m distance. If we consider – 13.5‰ and –27.0‰ as reference values for respectively the savanna and forest soils and we hypothesize that the 10 m sampled correspond to a duration of 12.5 years, we can estimate the proportion and rate of C$_4$ organic matter loss (Table 12.3). This percent loss decreases with depth, clearly showing that the turnover rate of organic matter is higher in surface than in deep layers. In the 0-5 cm layer, 100% of the C$_4$ organic matter disappeared in 12.5 years. Within the same period, 85.2% and 77.8% of the C$_4$ organic matter is lost in the 5-10 and 10-20 cm layers, respectively. In this experiment, the yearly mineralization rate of SOM can therefore be estimated at 6.25%, 5.3% and 4.9% at 0-5, 5-10 and 10-20 cm depth, respectively.

These results are surprising because they indicate a high carbon mineralization rate, which is quite contradictory with other indicators of organic matter degradation rate such as the chemical composition of organic compounds, potential soil respiration or even actual soil respiration in the field. It could be due (i) to the particularly low clay concentration of the savanna soil

Table 12.3. Natural abundance of ^{13}C (δ^{13}C in ‰) and percent of C$_4$ organic matter loss (parentheses) in soil samples at three depths from a transect in a transition zone between grass savanna (0 m) and spreading forest (10 m) (unpublished data by N. Mareuil).

Depth (cm)	Location on transect (m)					
	0	2.5	5	7.5	8.5	10
0–5	–14 (3.3%)	–18 (33.3%)	–22.9 (69.6%)	–25.8 (91.1%)	–27.2 (101.5%)	–26.9 (99.3%)
5–10	–13.5 (0%)	NA	–20.1 (48.9%)	NA	NA	–25.0 (85.2%)
10–20	–13.3 (0%)	NA	–16.5 (22.2%)	NA	NA	–24.0 (77.8%)

on which forest spreads, which induces a low physical protection of organic compounds, (ii) to a change in soil microclimatic conditions under the forest which could induce an acceleration of micro-organisms activity, or (iii) to an overestimate of the forest advance rate. However, the study by Martin et al. [14] also demonstrates a high organic matter turnover rate in the Lamto region. Their study was conducted in a plot unburned for 25 years. The original savanna vegetation with low tree density was replaced by a dense woodland, with forest trees dominating savanna trees [26]. A sudden shift of C_4 to C_3 vegetation occurred 9 years after fire protection. After 16 years, the upper (10 cm) soil layer lost 53-71% of the C_4 plant originated organic matter, i.e., 800 g C m^{-2} and the 10-25 cm layer lost 34-44%, i.e., 630 g C m^{-2}. The corresponding annual SOM mineralization rates are 3.9% and 2.4%, respectively, an estimate closer to that calculated with field respiration data (1.9%). The reason why this carbon mineralization rate is lower than that measured by Pirovano [17] and Mareuil (unpublished data) is still unclear. Soil climate conditions could account for these differences since soil temperature and relative water content are more constant and, probably, closer to optimal conditions for heterotrophic activity in forest than in savanna. Fine particles (clay + silt) concentration could also be important since it is 11-12% in the study by Pirovano [17] and Mareuil (unpublished data) vs. 19% in that by Martin et al. [14]. However, the actual residence time of SOM in Lamto seems short in the upper layers and should not exceed 30 years. All the information acquired on soil organic matter status and cycling at Lamto was also used as a basis to develop a comprehensive model of soil organic matter dynamics.

12.5 Modeling organic matter dynamic in Lamto soils

A review by Powlson et al. [19] showed the main limitations of current models aiming at understanding the coupling between physical and biological processes affecting SOM dynamics. Most of the process-based models with analytical solutions like QSOIL [32, 3, 4, 31] are usually very difficult to parameterize and, very often, the decomposition process is based on a distribution of SOM in quality pools, without indication of what a measurement of it could actually be. In compartment-based models, pools are often not experimentally testable [5]. Moreover, most models do not consider the microbial community as the driving factor of decomposition, even if microbial biomass is considered as an organic matter pool. McGill [15] suggested that future developments of SOM models could benefit from a more mechanistic treatment of the role of soil organisms. He proposed to make a distinction between microbial biomass and active forms of SOM and to simulate changes in SOM dynamics as a result of changes in activity or characteristics of soil organisms.

In order to benefit from the many soil processes studies performed in Lamto, a model based on a cohort approach like QSOIL was built, but without imposing any type of distribution in cohorts or quality types, therefore leaving the SOM pools free to evolve according to the litter input regime. The objective was to make this model adapted to deal with transient vegetation. SOMKO was developed on four basic assumptions [10]:

- At each time step, a new cohort of fresh organic matter is incorporated into the soil and its fate is followed individually until it reaches a very low contribution to the total SOM content. The metabolic quality of this fresh matter and of any cohort is defined by its chemical composition.
- Decomposition is performed by microbes that have energy and nutrient requirements that they must satisfy with the available material. Microbial biomass growth only occurs if the pool of decomposable carbon in the cohort is larger than that necessary to ensure microbial maintenance respiration. Microbial growth is limited by available nitrogen, i.e., nitrogen contained in organic matter decomposed or in soil solution.
- Micro-organisms synthesize biomass with a given C:N ratio from material with a different C:N ratio. The result of this constraint (the exact amount of SOM decomposed) is managed with an offer/demand formalism. When organic matter contains more nitrogen than required, mineral nitrogen is released in soil (net mineralization); when organic matter contains less nitrogen than required, mineral nitrogen is uptaken from the soil solution (nitrogen microbial immobilization).
- The chemical quality of organic matter in a cohort and characteristics of the microbial biomass associated define a potential decomposition rate of a cohort. The actual decomposition rate is a function of potential decomposition, soil temperature, relative water content and texture.

The overall dynamics of the model is described by the following system of equations:

$$\frac{dC_{i,j}}{dt} = -f_c k'_j (A_i, C_i) \alpha_i \left(A_i, D_i, C_{i,j}, N_{i,j}, [\text{N}_{\min}]\right) C_{i,j} + \delta p_j (A_i + D_i)$$

$$\frac{dN_{i,j}}{dt} = -f_c k'_j (A_i, C_i) \alpha_i \left(A_i, D_i, C_{i,j}, N_{i,j}, [\text{N}_{\min}]^*\right) N_{i,j} + \frac{\delta p_j}{\beta} (A_i + D_i)$$

$$\frac{dA_i}{dt} = -\delta A_i - \Delta_i (A_i, D_i, C_{i,j}) + \Gamma_i \left(A_i, D_i, C_{i,j}, N_{i,j}, [\text{N}_{\min}]^*\right)$$

$$\frac{dD_i}{dt} = -\delta D_i + \Delta_i (A_i, D_i, C_{i,j})$$

$$\frac{d\,[\text{CO}_2]}{dt} = \sum_{i=1}^{c} \rho \left((1-\delta) A_i - \Delta_i (A_i, D_i, C_{i,j})\right)$$

$$\frac{d\,[\text{N}_{\min}]}{dt} = \sum_{i=1}^{c} M_i \left(A_i, D_i, C_{i,j}, N_{i,j}, [\text{N}_{\min}]^*\right) - \lambda_P [\text{N}_{\min}]^*$$

$$-\nu \max \left(0, [\text{N}_{\min}] - \sum_{i=1}^{c} [\text{N}_{\min}]_i^D - [\text{N}_{\min}]_P^D \right)$$

with the following notations:

$C_{i,j}$ SOM carbon in cohort i and pool j (g C) $i = 1,...,c$, $j = 1,...,q$
$N_{i,j}$ SOM nitrogen in cohort i and pool j (g N)
A_i active microbial biomass (g C)
D_i dormant microbial biomass (g C)
$[CO_2]$ CO_2 pool (g C)
$[N_{min}]$ mineral nitrogen pool (g N)
k'_j decomposition rate of each pool $(g^{-1} d^{-1})$
f_c soil climate reduction of decomposition factor $[0,...,1]$
α_i nitrogen limitation reduction factor $[0,...,1]$
δ microbial mortality rate (d^{-1})
p_j proportion of microbial biomass returning to quality pool j after death
β microbial C:N ratio
$\Delta_i()$ active microbial biomass becoming dormant ("dormancy")
$\Gamma_i()$ new active microbial biomass ("growth")
ρ microbial respiration rate
$M_i()$ flux to/from mineral nitrogen pool ("mineralization/immobilization")
λ_P plant competitivity relative to microbes for mineral nitrogen
ν fraction of remaining mineral nitrogen leached (d^{-1})
$[N_{min}]^*$ minimum of the total available mineral nitrogen and the total demand in mineral nitrogen
$[N_{min}]^D_i$ demand in mineral nitrogen of cohort i (if $i=P$, plant demand).

Because SOMKO aims at being a generic model, most of basic processes were parameterised with data from literature. SOMKO was tested for the Lamto savanna soils (Table 12.4). It was run until equilibrium was reached (15,000 days) with yearly cohorts and average values of soil climatic factor and litter inputs derived from climate data (see Sect. 3.5) and savanna productivity measurements (see Sect. 7.4). Since the ecosystem modeled was a humid savanna with yearly fires, there was no aboveground litter input in the soil. Measured

Table 12.4. Comparison of SOMKO simulation results at equilibrium between litter inputs and SOM decomposition vs. measured SOM variables for Lamto soils. Measured values from [16]. The relative error (ratio of (measured-simulated):measured value) averaged for all variables was 77% (after [10], with permission of Blackwell Publishing).

Variable	Measured value	Simulated value
Total SOM C	1200 g m^{-2}	486
SOM C:N ratio	12	10.9
Mineral:organic nitrogen ratio	$2\,10^{-3}$	$6\,10^{-4}$
% extractable sugar in SOM C	5 (max)	4.1
% microbes in SOM C	0.5–3%	2.9
CO_2 flux (proportion of SOM C)	0.001 d^{-1}	$2\,10^{-3}$
Litter:(litter+SOM) ratio (%)	23	35

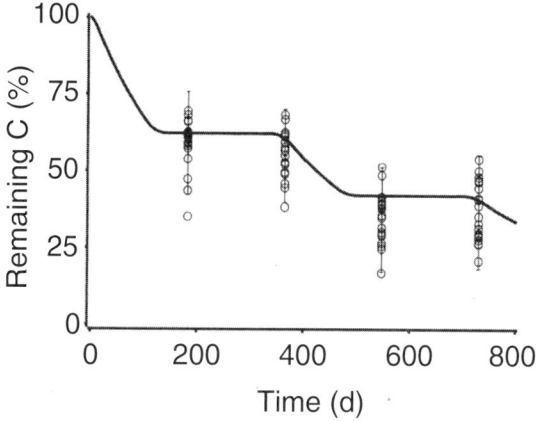

Fig. 12.4. Comparison of SOMKO simulations (line) with litter bag experiments (circles) (after [10], with permission of Blackwell Publishing).

and simulated main SOM variables are given in Table 12.4. The fit was optimized on all seven variables altogether, not on a particular variable. The fit is relatively good, particularly for total accumulated soil organic carbon, SOM C:N ratio (litter excluded), mineral to organic nitrogen ratio, proportion of microbial biomass in SOM, proportion of fresh litter, proportion of sugars and ratio of CO_2 flux to total soil organic carbon. SOMKO was also able to properly reproduce the dynamics of decomposition observed in litter bag experiments for Sahelian soils (Fig. 12.4).

This good fit on seven independent variables integrative of the behavior of the whole system constitutes a good test of the model's ability to generate realistic soils. The next step will require the coupling of the SOMKO model with a vegetation model such as TREEGRASS (see Sect. 9.3) and simulation of plant cover change like at the savanna-forest boundary in the north of the Lamto reserve. SOMKO will also be used to test the effects of climate change (rainfall amount and distribution, see Sect. 3.5), altered litter input regime and cultivation after savanna clearing on SOM dynamics.

12.6 Conclusion

In Lamto, like in most savanna ecosystems, the accumulation of organic matter in soil is low compared to the high primary productivity. The burning of aboveground biomass strongly decreases the input of plant debris to soil. Twenty to 30% of the total annual production (above and below ground) is instantaneously destroyed by fire and cannot be incorporated to soil where it could be stabilized. But the main factor leading to a low concentration of soil in organic matter is probably the climate. Indeed, the length of the wet season and constant high air temperature (see Chap. 3) result in optimal conditions for heterotrophic micro-organisms most of time. Moreover, the intrinsic chemical characteristics of fresh plant debris are favorable to micro-organisms, especially grass roots that have a high potential biodegradability (ca. 10% in 2 weeks). The decomposition of aboveground and belowground litters in the field is consequently rapid and the ratio particulate-to-non-particulate SOM is therefore low. In other words, most of SOM in Lamto is humified and adsorbed on fine soil particles (silt and clay). This is the reason why the maximum mineralization rate of the whole soil organic carbon is quite low, ca. 2% a week in the first 20 cm (this rate is 2 to 3 times higher in the adjacent forest soils). This is also the reason why the whole soil appears as unfavorable environment for micro-organisms that show low density and biomass. However, the climate conditions compensate the low biodegradability of SOM, allowing a mineralization rate of 2% to 6% a year.

Due to the burning of aboveground biomass, grass roots are the quite exclusive source of SOM in Lamto. But, as a result of the tufted architecture of grass roots, the input of carbon to soil is very heterogeneous in space. It is very intense just under the tufts (10-20% of soil volume) and very slow between tufts (80-90% of soil volume). Consequently, the Lamto soils show large horizontal variations of organic matter quantity and quality, particulate:nonparticulate organic matter ratio, microbial biomass and mineralization activities, which are all higher under tufts than between tufts. In other words, soil nitrogen and carbon cycles are concentrated in 10-20% of soil volume while 80-90% of soil volume is rather empty of biogeochemical activities. Organic carbon accumulation and mineralization processes therefore show a patchy structure, resulting from and inducing a patchy structure of grass cover.

References

1. L. Abbadie and R. Lensi. Carbon and nitrogen mineralization and denitrification in a humid savanna of West Africa. *Acta Oecologica*, 11(5):717–728, 1990.
2. M. Amato and J.N. Ladd. Assay for microbial biomass based on milydrinrective nitrogen in extracts of fumigated soils. *Soil Biology and Biochemistry*, 20:107–114, 1988.
3. E. Bosatta and G.I. Ågren. Dynamics of carbon and nitrogen in the organic matter of the soil: a generic theory. *American Naturalist*, 138(1):227–245, 1991.
4. E. Bosatta and G.I. Ågren. Theoretical analysis of microbial biomass dynamics in soils. *Soil Biology and Biochemistry*, 26(1):143–148, 1994.
5. B.T. Christensen. Matching measurable soil organic matter fractions with conceptual pools in simulation models of carbon turnover: revision of model structure. In D.S. Powlson, P. Smith, and J.U. Smith, editors, *Evaluation of soil organic matter models*, volume 38, pages 143–160. Springer-Verlag, Berlin, 1996.
6. D.C. Coleman. Compartement analysis of "total soil respiration": an explanatory study. *Oikos*, 24:361–366, 1973.

7. C. Darici. *Effet du type d'argile sur quelques activités microbiennes dans divers sols tropicaux. Comparaison d'un sol à allophane, d'un vertisol à montmorillonite et d'un sol ferrugineux tropical à illite et kaolinite.* Ph.D. thesis, Université de Paris-Sud, 1978.
8. L. Gautier. Contact forêt- savane en Côte d'Ivoire centrale : Evolution de la surface forestière de la réserve de Lamto. *Bulletin de la Société Botanique de France*, 136(3/4):85–92, 1989.
9. L. Gautier. Contact forêt-savane en Côte d'Ivoire centrale: Evolution du recouvrement ligneux des savanes de la réserve de Lamto (sud du V baoulé). *Candollea*, 45:627–641, 1990.
10. J. Gignoux, D. Hall, J. House, D. Masse, H.N. Nacro, and L. Abbadie. Design and test of a generic cohort model of soil organic matter decomposition: The SOMKO model. *Global Ecology and Biogeography*, 10:639–660, 2001.
11. S.R. Gupta and J.S. Singh. The effect of plant species, weather variables and chemical composition of plant material on decompsition in a tropical grassland. *plant and Soil*, 59:99–117, 1981.
12. X. Le Roux and P. Mordelet. Leaf and canopy CO_2 assimilation in a West African humid savanna during the early growing season. *Journal of Tropical Ecology*, 11:529–545, 1995.
13. O. Maggi, A. Bartoli, G. Puppi, S.G. Albonetti, and A. Rambelli. Première contribution à la reconnaissance de la microflore rhizosphérique de *Loudetia simplex*, graminée typique de la savane de Lamto. *Revue d'Ecologie et de Biologie du Sol*, 14:403–419, 1977.
14. A. Martin, A. Mariotti, J. Balesdent, P. Lavelle, and R. Vuattoux. Estimate of organic matter turnover rate in a savanna soil by ^{13}C natural abundance measurements. *Soil Biology and Biochemistry*, 22(4):517–523, 1990.
15. W.B. McGill. Review and classification of ten soil organic matter (SOM) models. In D.S. Powlson, P. Smith, and J.U. Smith, editors, *Evaluation of soil organic matter models*, volume 38, pages 111–132. Springer-Verlag, Berlin, 1996.
16. H.B. Nacro. *Hétérogénéité de la matière organique dans un sol de savane humide (Lamto, Côte d'Ivoire): Caractérisation chimique et étude in vitro des activités microbiennes de minéralisation du carbone et de l'azote.* Ph.D. thesis, Université de Paris 6, Paris, 1997.
17. C. Pirovano. *Studio di dinamiche del carbonio e dell'azoto in suoli tropicali sotto savana e sotto foresta in Costa d'Avorio.* Laurea in science ambientali, Università degli Studi di Milano, 1996.
18. J. Pochon and I. Bacvarov. Données préliminaires sur l'activité microbiologiques des sols de la savane de Lamto. *Revue d'Ecologie et de Biologie du Sol*, 10(1):35–43, 1973.
19. D.S. Powlson, P. Smith, and J.U. Smith. *Evaluation of soil organic matter models*, volume 38 of *Global environmental change*. Springer-Verlag, Berlin, 1996.
20. A. Rambelli. recherches mycologiques préliminaires dans les sols de forêt et de savane en Côte d'Ivoire. *Revue d'Ecologie et de Biologie du Sol*, 8(2):219–226, 1971.
21. A. Rambelli, G. Puppi, A. Bartoli, and S.G. Albonetti. Deuxième contribution à la connaissance de la microflore fongique dans les sols de Lamto. *Revue d'Ecologie et de Biologie du Sol*, 10(1):13–18, 1973.

22. X. Le Roux. *Etude et modelisation des échanges d'eau et d'énergie sol-végétation-atmosphère dans une savane humide (Lamto, Côte d'Ivoire)*. Ph.D. thesis, Université de Paris 6, Paris, 1995.
23. R. Schaefer. Activité métabolique du sol. Fonctions microbiennes et bilans biogéochimiques dans la savane de Lamto. *Bulletin de Liaison des Chercheurs de Lamto*, 5:167–184, 1974.
24. J.S. Singh and S.R. Gupta. Plant decomposition and soil respiration in terrestrial ecosystems. *Botanical Review*, 43:449–528, 1977.
25. P. Villecourt. Contribution à l'étude du bilan du carbone dans un sol de la savane de Lamto en Côte d'Ivoire. *Revue d'Ecologie et de Biologie du Sol*, 10(1):19–23, 1973.
26. R. Vuattoux. Contribution à l'étude de l'évolution des strates arborée et arbustive dans la savane de Lamto. *Annales de l'Université d'Abidjan, Série C*, 12:35–63, 1976.
27. F.R. Warembourg and B.A. Paul. Seasonal transfers of assimilated ^{14}C in grassland: Plant production and turnover, soil and plant respiration. *Soil Biology and Biochemistry*, 9:295–301, 1977.
28. J. Wu, P.C. Brookes, and D.S. Jenkinson. Formation and destruction of microbial biomass during the decomposition of glucose and ryegrass in soil. *Soil Biology and Biochemistry*, 25(10):1435–1441, 1993.
29. J. Wu, A.G. O'Donnell, and J.K. Syers. Microbial growth and sulfur immobilization following the incorporation of plant residues into soil. *Soil Biology and Biochemistry*, 25(11):1567–1573, 1993.
30. Z. Zaidi. *Recherches sur les modalités de l'interdépendance nutritionnelle entre vers de terre et microflore dans la savane guinéenne de Lamto (Côte d'Ivoire). Esquisse d'un système interactif*. Ph.D. thesis, Université de Paris Sud, Paris, 1985.
31. G.I. Ågren, M.U.F. Kirschbaum, D.W. Jonson, and E. Bosatta. Ecosystem physiology - soil organic matter. In A.I. Breymeyer, D.O. Hall, J.M. Melillo, and G.I. Ågren, editors, *Global change: Effects on coniferous forests and grasslands*, volume 56, pages 207–228. John Wiley & Sons, Chichester, 1996.
32. G.I. Ågren, R.E. Macmurtrie, W.J. Parton, J. Pastor, and H.H. Shugart. State of the art of models of production-decomposition linkages in conifer and grassland ecosystems. *Ecological Applications*, 1(2):118–138, 1991.

13

Perturbations of Soil Carbon Dynamics by Soil Fauna

Michel Lepage, Luc Abbadie, Guy Josens, and Souleymane Konaté, and Patrick Lavelle

13.1 Introduction

In this chapter, we focus on the soil macrofauna compartment and its role in soil carbon dynamics. A large number of studies have been conducted on other fauna groups in Lamto. Detailed information on those groups can be found in the academic works and papers made by the research scientists who worked in Lamto. Synthetic data are in [35, 33, 37, 34] and in the volume edited in 1983 by F. Bourlière about mammals [13], snakes [8], amphibians [36], invertebrates [18], role of fire on invertebrates [17], earthworms [43], ants [61], termites [27], energy flow and nutrient cycling [39].

With regard to soil fauna, data are available on microarthropods [4, 6, 5], cockroaches [3], ants [58, 57, 60], Coleoptera Scarabeidae [16, 15], Coleoptera Carabidae [52, 53], Coleoptera with endogeic larvae [20, 19]. Athias et al. [7, 5] gave a synthetic overview of the soil fauna endogeic populations. Lavelle [42] presented the community structure of the soil fauna in tropical savannas, with several references to Lamto. More recently, Lamotte [38] described the position of the detritivore animals and microbial decomposers in the African savanna food chains and energy fluxes.

This chapter is devoted to the soil animals known as "ecosystem engineers" [22], mainly earthworms and termites. This choice was first determined by the importance of those two groups in terms of abundance and biomasses. In a synthetic table Lavelle [42], estimated the annual average biomass of microfauna (Protozoa, Nematoda) at $0.56\,\mathrm{g\,m^{-2}}$ fresh weight (FW) and that of mesofauna (Enchytreidae, Acari, Collembola and other micro-arthropods) at $1\,\mathrm{g\,m^{-2}}$ (FW). Within the macrofauna, earthworms represented $49\,\mathrm{g\,m^{-2}}$ and termites $1.9\,\mathrm{g\,m^{-2}}$ (but, as we shall see below, this figure was largely underestimated), while the other groups (Chilopoda, Diplopoda, Arachnida, Coleoptera, Diptera and others) represented a fresh biomass of $0.80\,\mathrm{g\,m^{-2}}$. The other group of soil fauna that could be considered an ecosystem engineer is that of ants (fresh biomass $2\,\mathrm{g\,m^{-2}}$), but very few data are available to precisely assess their role in soil processes.

It has been shown that the impact of macrofauna on the decomposition process is very strong in tropical ecosystems compared to temperate ones [73]. The diversity of ecosystem engineers deeply influences the soil function through nested interactions with groups of smaller sizes (microflora, micro-predators and litter transformers) [47]. Earthworms and termites are major regulators of microbial activities and determine the abundance and activities of smaller groups of soil fauna [47].

13.2 Earthworms and termites: Abundances and spheres of influence

13.2.1 Densities and biomasses

All data are given in fresh weights. The fresh weight/dry weight ratio ranges from 4 to 5 for termites and from 7 to 20 for earthworms. Earthworm biomass in Lamto amounts to 39.0 - 57.0 $g\,m^{-2}$ (FW) (Table 13.1). The two main trophic groups are the epigeics earthworms (from 2.03 $g\,m^{-2}$ in the grass savanna to 12.9 $g\,m^{-2}$ in the Andropogonae savanna) and the endogeic species (from 42.4 to 26.3 $g\,m^{-2}$). Termite biomass amounts to 3.4 - 19.6 $g\,m^{-2}$ (FW) (Table 13.2). The main trophic groups are the fungus-growing termites (from 1.15 $g\,m^{-2}$ in grass savanna to 5.93 $g\,m^{-2}$ in woody savanna), the humivorous species (from 2.14 to 11.65 $g\,m^{-2}$) and the xylophageous and foraging species (from 0.14 to 1.98 $g\,m^{-2}$).

13.2.2 Biogenic structures

Soil macrofauna has two major impacts on the carbon cycle and soil organic matter (SOM) dynamics: a direct effect on soil micro-organisms, related to

Table 13.1. Densities and biomasses (fresh weight) of earthworm species in different savanna types at Lamto, during the year 1972 (after [40]).

Species	Grass savanna Density (m^{-2})	Grass savanna Biomass ($g\,m^{-2}$)	Andropogoneae sav. Density (m^{-2})	Andropogoneae sav. Biomass ($g\,m^{-2}$)	Unburned savanna Density (m^{-2})	Unburned savanna Biomass ($g\,m^{-2}$)
Millsonia lamtoiana	0.067	0.43	0.64	4.99	1.56	9.8
Dichogaster agilis	9.5	1.60	8.1	0.8	1.0	3.1
Millsonia anomala	21.5	25.00	23.5	21.7	18.0	14.7
Dichogaster terrae-nigrae	0.24	1.3	2.43	18.0	0.5	2.6
Chuniodrilus zielae	32.0	0.80	148.0	3.8	245.0	5.46
Stuhlmania porifera	112.0	3.0	109.0	2.8	57.0	1.28
Agastrodrilus spp.	2.2	2.6	1.96	1.7	1.38	0.8
Millsonia ghanensis	1.5	9.7	0.72	3.2	0.49	1.1
Total	180	44	295	57	340	39

Table 13.2. Densities and biomasses (fresh weight) of the different termite species in three savanna types of Lamto (after [24, 25, 74, 31] and unpublished data).

Species	Grass savanna		Andropogoneae sav.		Savanna woodland	
	Density (m^{-2})	Biomass ($g\,m^{-2}$)	Density (m^{-2})	Biomass ($g\,m^{-2}$)	Density (m^{-2})	Biomass ($g\,m^{-2}$)
Ancistrotermes cavithorax	540	0.58	1160	1.25	2730	2.95
Microtermes toumodiensis	15	0.01	5430	3.06	2605	1.47
Odontotermes nr. *pauperans*	225	0.56	350	0.87	255	0.63
Pseudacanthotermes militaris	0	0.00	55	0.15	330	0.88
Basidentitermes potens	0	0.00	720	1.90	425	1.12
Promirotermes holmgreni	0	0.00	1350	2.03	510	0.77
Adaiphrotermes sp.	730	1.10	3745	5.62	4555	6.84
Aderitotermes sp.	385	1.04	215	0.58	800	2.17
Astratotermes nr. *prosenus*	0	0.00	0	0.00	70	0.75
Microcerotermes spp.	0	0.00	300	0.23	300	0.23
Amitermes evuncifer	16	0.02	6	0.01	3	0.005
Trinervitermes spp.	30	0.14	160	0.77	365	1.75
Total	1941	3.45	13491	16.5	12948	19.6

their feeding activities, and indirect interactions associated with their burrowing, building activities and bioturbation, which deeply affect the physical and chemical characteristics of the soil environment. In this respect, soil macrofauna—mainly ants, termites and earthworms—may be viewed as true ecosystem engineers [22]. Their activities create particular pathways of energy and matter and increase the patchiness of resource distribution. They create specific environments in their sphere of influence called functional domains [45] that strongly impact the micro-organisms' activities. The soil influenced by the activities of the two main groups of macrofauna is commonly referred to as drilosphere (earthworms) or termitosphere (termites).

Earthworms in Lamto produce large amounts of casts, whose size and shape vary according to the size of the species, the soil texture and the feeding habits of the species [47, 10]. Annual cast production on the soil surface has been estimated at 2.5 to 3.0 $kg\,m^{-2}$ in the Lamto savanna [40, 48]. But a much larger proportion of these casts are produced within the soil profile and never appear on the soil surface, since most of the species are endogeic [40, 11]. Surface casts only represent 1.7% to 3.5% of the 800 to 1250 Mg of dry soil ingested annually by those endogeic species (Fig. 13.1).

The physical structure of earthworm casts deeply influences the further dynamics of SOM over time. In that respect, two different types of cast may be distinguished: the globular shape, surrounded by a hard cortex, belonging to the "compacting" species, and the "granular" shape, belonging to the "decompacting" species [10, 46]. In the Lamto savanna, it has been assumed that the opposite effects of both "compacting" and "decompacting" species could regulate the soil aggregation [11].

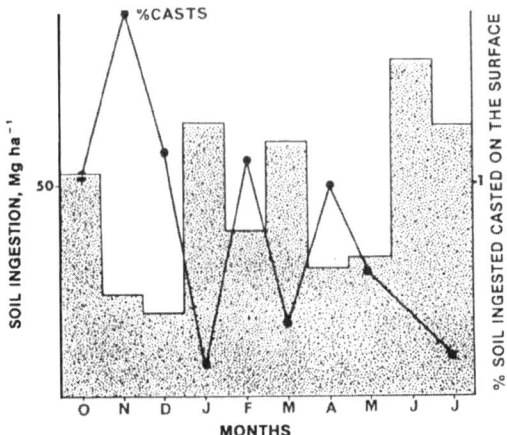

Fig. 13.1. Seasonal variations of soil ingestion (blocks) and surface cast (solid line) production by earthworms at Lamto (from [40]).

Table 13.3. Densities of termite mounds in three savanna types in Lamto. GS, *Loudetia* grass savanna; AS, Andropogoneae savanna; SW, savanna woodland. The lenticular termite mounds are mostly occupied by *Odontotermes* nr *pauperans*. After [24, 25, 31, 29].

Biogenic structure	Density (ha^{-1})			Average height (m)	Average diameter (m)
	GS	AS	SW		
Small termitaria of *Trinervitermes* spp.	51	69	72	0.3	0.3
Lenticular termite mounds	8	12	12	0.6	3.5
Large termitaria of *Macrotermes subhyalinus*	0	0.03	0.03	2	2

The termite structures are represented aboveground in Lamto by the small mounds belonging to the foraging species (genus *Trinervitermes*) (Table 13.3) or by the lenticular mounds, mostly built by the fungus-growing termite *Odontotermes* nr. *pauperans* (Table 13.3 and Fig. 13.2). The large termitaria (so-called "cathedral" type), typical of many savanna landscapes in West Africa, are virtually absent from Lamto savannas (Table 13.3). Another negative feature of these savannas, not yet fully explained, is the absence of the "mushroom" mound shape of the humivorous genus *Cubitermes*. Most of the termite population in Lamto savanna have belowground nests. In the fungus-growing species, each belowground nest system comprises a variable number (147 to 560 according to Josens [26]) of chambers containing fungus-comb, scattered around the queen chamber and connected by a diffuse gallery network. The total density of such chambers varies from 0.95 m^{-2} in the grass

Fig. 13.2. Cross section in belowground fungus-comb chambers within a lenticular nest of *Odontotermes* nr. *pauperans* (photography S. Konaté).

Table 13.4. Densities and dry masses of fungus combs in three savanna types in Lamto (after [24, 25, 29, 31, 32] and unpublished data).

Species	Grass savanna		Andropogoneae sav.		Savanna woodland	
	Density (m^{-2})	Dry mass ($g\,m^{-2}$)	Density (m^{-2})	Dry mass ($g\,m^{-2}$)	Density (m^{-2})	Dry mass ($g\,m^{-2}$)
Ancistrotermes cavithorax	0.5	1.9	2.2	8.2	4.5	17.0
Microtermes toumodiensis	0.04	0.02	8.1	4.4	5.5	3.0
Odontotermes nr *pauperans*	0.4	2.2	0.7	4.1	0.4	2.29
Pseudacanthotermes militaris	0.01	0.06	1.2	7.3	1.1	6.8
Total	0.95	4.2	12.3	24.0	11.6	29.0

savanna to 12.3 m^{-2} in the Andropogonae savanna, representing a dry weight of 4.2 to 29 $g\,m^{-2}$ (Table 13.4). The nests of most belowground humivorous species are located in the top 10 or 20 cm of soil.

13.3 Carbon distribution and storage

Soil macrofauna has been classified into three main functional groups according to their habitat and main food source: epigeic species live and feed on the litter; anecic species live in burrows they build but feed at night on litter they collect at the soil surface; endogeic or geophageous species live in the soil and feed on SOM [12, 41]. The main species in Lamto belong to the anecic (earthworms and termites) and geophageous (earthworms) groups.

13.3.1 Anecic species

As previously outlined (Chap. 7), roots comprise the major part of the primary production that enters the soil system and directly contributes to the accumulation of organic matter in soil. Most of the aboveground production is burned every year and lost for the SOM, except the part consumed by detritivores and herbivores. Lamotte [34] estimated that ca. 5% of the annual primary production is consumed and degraded through respiration by invertebrates, mainly under the form of decaying litter. The quantity of plant matter saved from fire is therefore important. Anecic earthworms annually incorporate 18-51 $g\,m^{-2}$ of dry litter to the soil in the regularly burned savanna [51, 75]. Josens [24, 25], Konaté [29], Kouassi [32] and Konaté et al. [31] estimated the amount of plant material stored as fungus combs in Macrotermitinae between 4.2 and 29.0 $g\,m^{-2}$ dry weight (Table 13.4), which requires an input of 36 to 216 $g\,m^{-2}$ of dry litter to the soil in the burned grass savanna and savanna woodland respectively [23, 26]. A large proportion of this material comes from the aboveground production, tree and grass litter. With an average 40% carbon content in fungus combs [68], the Macrotermitinae subfamily could incorporate from 14.4 to 86.4 $g\,C\,m^{-2}\,yr^{-1}$ into the soil compartment (to be compared with the 3200 $g\,C\,m^{-2}$ in the 0-20 cm soil layer in a typical Andropogoneae savanna (Chap. 11). Food availability dramatically changes with seasons: when the grass layer has been almost completely destroyed by the annual fire, tree leaves fall down to the ground and become the main food source, but during the wet season, the grass material generally dominates in the diet of litter-feeding termites [56].

In spite of these figures, the main effect of termites on the carbon cycle is probably a consequence of the accumulation of clay minerals in their constructions [2]. The impact of termites on clay physics remains unclear, but some laboratory experiments suggest that fungus-growing termites (*Odontotermes* nr. *pauperans*) can modify in a more or less irreversible way the mineralogical properties of silicate clays [29], as hypothesized by Boyer [14] after in situ observations. In this respect, termites could be considered as weathering agents of clays by creating expandable clay minerals [28]. This effect over decades and centuries likely has tremendous consequences on SOM dynamics in the Lamto soil environment that is dominated by low activity clays such as kaolinite.

In the long term, this modification of clay physics in addition to the accumulation of clay particles that induce an increase in the cation exchange capacity, and to the input of plant material to nest induce particular chemical characteristics in termite soil. For example, Lopez-Hernandez and Febres [59] found that *Trinervitermes geminatus* mounds contained 30 to 60 times more potassium than the average savanna soil (i.e., 270 to 600 $mg\,kg^{-1}$, compared to 10 to 45 mg kg^{-1} in control soil). In Lamto, the largest lenticular termite mounds contain 1.75-2.24% total organic C while the surrounding soil contains only 0.96-1.17% [2]. Tree density is generally higher on these large mounds (Chap. 18). The mound soil actually exhibits an isotopic composition close to that of C$_3$ plants (δ^{13}C between –21.8‰ and –20.6‰) while the isotopic composition of carbon in standard savanna soil is characteristic of C$_4$ plants (δ^{13}C between –13.2‰ and –13.4‰) [2, 71]. *Borassus aethiopum* palm trees can extend their root system more than 20 m away and proliferate in the nutrient-rich termite mounds [70]. In this environment, large lenticular termite mounds deeply modify soil characteristics and soil water regime, influencing the composition and structure of plant communities [29, 30].

13.3.2 Geophageous species

The major effect of macrofauna on the carbon cycle in Lamto savannas is due to the geophageous trophic group. Humivorous termite populations contribute to more than half of the whole termite community (Table 13.2). No estimates of their consumption rates of organic matter are available and their feeding habits remain unexplored. However, the role of soil-feeding termites is certainly very limited compared to that of geophageous earthworms. At Lamto, 100 to 150 kg m^{-2} of soil annually passes through the guts of the earthworms. Since this soil contains 1500 g m^{-2} of organic matter [51], that is one-third of the soil reserves. The ingestion rate in *Millsonia anomala* is influenced by the quality and quantity of organic matter [51, 75]. Earthworms ingest large amounts of soil (up to 24 g g^{-1} dry soil) when fed with soil poor in assimilable organic matter. With an addition of 0.25% water soluble grass extracts, the soil ingestion rate drops down to half this value. This behavior has a tremendous effect on the microbial activity in worm casts and the dynamics of SOM, as outlined below.

13.4 Carbon mineralization

13.4.1 Biological systems of regulation

The organic matter dynamics and carbon release as a consequence of soil fauna activities are related to the adaptive strategies evolved by the ecosystem engineers (earthworms and termites). As stressed by Lavelle [44], the biological systems of regulation associate soil invertebrates and micro-organisms. In the Lamto savanna, as well as in most seasonal ecosystems, the microbial communities are unable to move into compact soil and to have access to new substrates when resources have been exhausted in their close environment. They are therefore mostly dependent upon the mechanical activities of soil macrofauna and roots, the only living agents able to transport them. On the other hand, this macrofauna has a limited ability to degrade complex organic substrates and then depends upon micro-organisms to digest the complex compounds that make up the greatest part of the SOM. This mutualistic relationship can develop inside the guts of earthworms and termites (internal rumen strategy) or outside in their biogenic structures (fecal pellets, nest material, fungus-combs and other structures). As a result of ecosystem engineers' activities, free living soil micro-organisms or specific gut dwelling microbes are directly harnessed by the intestinal machinery as mutualists. They degrade organic material, at a rate that is consistent with the metabolic demands of the host and which can be optimized by adjustment of the physiological conditions under which the growth and metabolism of the microbes take place [9].

The main effect of soil macrofauna is the mixing of soil mineral particles with organic compounds, with the addition of water and assimilable organic matter, either by salivary secretions (termites) that bind particles in the mound and gallery constructions, or by mucus secretion (in the earthworm gut [9]). The undigested materials are voided as excrements in physical structures with specific microenvironmental conditions that may dramatically influence the biological and physico-chemical processes [65]. In Lamto, it has been demonstrated that the lack of assimilable carbon was the main reason for the low level of potential microbial activity in soils [54]. However, higher rates of carbon mineralization have been measured in the structures built by macrofauna. They probably result from a priming effect of low molecular weight compounds, the intestinal mucus produced in the anterior part of the gut, on stable SOM. Soil micro-organisms are stimulated by small amounts of assimilable organic matter supplied by termites and earthworms and their production of extra-cellular enzymes increased, which induces an accelerated depolymerization and mineralization of soil organic carbon [49]. This process typically occurs in the gut of endogeic earthworms and in the walls of the fungus-comb chambers built by *Ancistrotermes cavithorax* in which Abbadie and Lepage [1] measured a twice higher potential carbon mineralization rate than in the control soil (Fig. 13.3).

As far as the carbon cycle is concerned, two main strategies have been developed by the ecosystem engineers in Lamto savannas. Geophageous earthworms produce intestinal mucus, rich in energetic and easily metabolizable compounds [63], able to induce a priming effect that stimulates the SOM decomposition. In the fungus-growing termites that utilize dead plants as food resources with a high C:N ratio, an external symbiosis with specialized fungi, *Termitomyces* sp., reduces the C:N of their food to that of their own body tissue [21]. This strategy eliminates C as CO_2 in the fungal respiration.

Mutualistic relationships between endogeic earthworms and soil microflora have been described in *Millsonia anomala* and *Dichogaster terrae-nigrae*, two major species in the Lamto savannas [9, 63]. Suitable conditions for microbial growth are first created in the anterior part of the earthworm gut [9]: a neutral pH, high water content (100-150% of the soil dry weight), and large quantities of mucus (ca. 5-7% of the dry weight of ingested soil) that can be readily used as carbon source by soil micro-organisms. This mucus is a mixture of

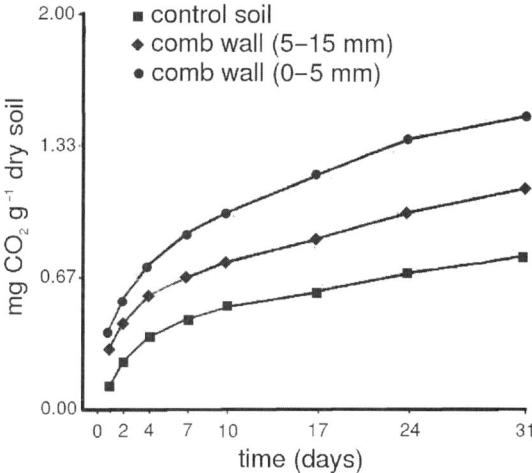

Fig. 13.3. Cumulative curves of soil respiration, in laboratory incubations, between layers around *Ancistrotermes cavithorax* fungus-comb chamber and the control soil, in the second part of the dry season in Lamto (after [1], copyright (1989), with permission of Elsevier).

low molecular weight aminoacids and sugars, and a glycoprotein [63]. Barois and Lavelle [9] showed that microbial activity at 28°C was 6-10 times higher in soil from the anterior part of the gut of *Millsonia anomala* than in the control soil and ca. 30 times higher than in an undisturbed soil (i.e., field conditions) [47]. In the second half of the gut, (1) mucus is no longer found in the gut and we hypothesize that the worm must have reabsorbed it and (2) micro-organisms having recovered their metabolic capacities would digest SOM and release assimilable compounds that are absorbed in the last part of the gut. In other words, 3-19% of the SOM could be assimilated during the gut transit, i.e., in 0.5 to 4 h, depending on species and soil feeding value. Assimilation was estimated at 2-6% in the species *Millsonia anomala* [66, 64]. Although all particle size fractions seem to be equally digested, the coarse fraction ($> 250\,\mu$m) is decreased by 33% as a result of fragmentation into smaller particles and digestion.

13.4.2 Termites and CO_2 release

In termites, an external symbiosis exists between Macrotermitinae species and fungi. Fungus-growing termites house fungi either in the central part of their epigeous mounds or in subterranean chambers. The fungi are grown on partially digested plant material which has transited once through the gut of worker termites [26]. The material is extensively colonized and degraded by the fungus [47, 55]. The net effect is a very efficient degradation of organic carbon into CO_2. Field measurements done with a closed container system exhibited significantly higher CO_2 fluxes from locations with *Odontotermes* fungus-comb chambers underneath (Fig. 13.4). At the landscape scale, the CO_2 emitted by termites was around 1.58 mol m^{-2} yr^{-1} (or 18.9 g C m^{-2} yr^{-1}) for *Odontotermes* nr. *pauperans* and 0.69 mol m^{-2} yr^{-1} (or 8.3 g C m^{-2} yr^{-1}) for *Ancistrotermes cavithorax* [31]. Josens [23, 24, 26] estimated the fungus-growing termite consumption in this ecosystem from the turnover of the fungus-comb mass. However, this first assessment was based on an extrapolation from mass to production that did not take into account the portion continuously lost by respiration. From Konaté et al. [31], it can be inferred that the fungus-comb respiration represents roughly 42% of the carbon consumed by the Macrotermitinae populations. Thus, new figures of consumption incorporating the portion subsequently lost in respiration have been computed: in a shrubby savanna dominated by Andropogonae, it amounts to 180 g m^{-2} yr^{-1} (dry mass of litter). From the data given in Lepage et al. [56], this amount included 41% of grass litter and 59% of tree litter and is equivalent to 77.4 g C m^{-2} yr^{-1}. Furthermore, a noticeable proportion of the carbon ingested is assimilated into termite biomass and dispersed, mostly by predation and swarms, or incorporated into termite structures [28], as galleries and soil sheeting, which was not taken into account.

At the whole community level, termite consumption represents 13.3% of the total aboveground primary production (grass and woody species), i.e., 558 g C m^{-2} yr^{-1} (from the figures given in Menaut and César [69], balanced with the proportions of the savanna types in Lamto and an average plant carbon content of 43%). If we take into account the proportion of the aboveground production burned by annual fires (57% after [69]), termite consumption amounted to 31% of the carbon in the aboveground primary production

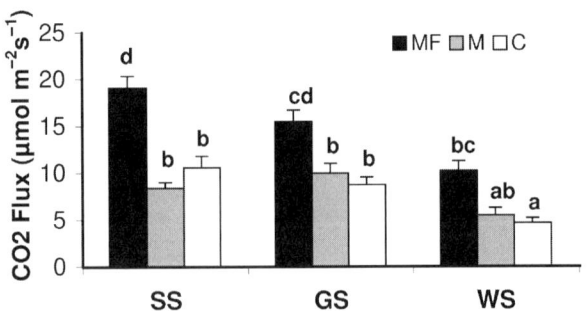

Fig. 13.4. In situ soil respiration rates measured in *Loudetia* grass savanna (GS), Andropogoneae savanna (SS) and savanna woodland (WS) savannas, for control savanna soil (C), for locations without termite chambers on termite mounds (M) and for locations with *Odontotermes* fungus-comb chambers on termite mounds (MF) (after [31], with permission of Blackwell Publishing).

that was not mineralized by annual fires, a figure similar to the one given by Sugimoto et al. [72] for other savanna ecosystems [31]. The results also stress the importance of the aerobic respiration from the termite food reserves, as the fungus-comb mass was responsible for 51% and 82% of the CO_2 flux emitted by chambers for *Odontotermes* nr *pauperans* and *Ancistrotermes cavithorax*, respectively.

13.4.3 Scaling

These effects of ecosystem engineer biogenic structures on SOM and carbon mineralization should be considered within a time-scale framework. In earthworms casts, the short-term effect is characterized by a nutrient release in the drilosphere. The medium-term effect results in a protection of the organic matter in casts [50]. In freshly deposited casts, a significant decrease in SOM in the early days of the measurements is followed by a rapid stabilization and a further arrest of mineralization in the compact structure of casts, where organic matter becomes physically protected from microbial decomposition. In the long term, natural ^{13}C labeling shows a rather fast turnover of SOM in soils colonized by these earthworms: in 16 years, 65% of the SOM in the upper 10 cm of soil turned over (as compared with a 40% turnover in the 10-25 cm layer) [67].

13.5 Conclusion

Soil fauna has contrasted effects on soil carbon dynamics that must be considered on different scales in both time and spatial scales. There is a clear stimulation of soil micro-organisms at short scales, with high carbon mineralization rates in small volumes of soil. This effect occurs during the earthworm gut transit and in the soil rehandled by termites in nest walls. This stimulation probably results from a strong priming effect induced by additional water and intestinal mucus to soil in the case of earthworms [50] or by mixing mineral particles with proteins and sugars in the case of termites [1]. Consequently, the activities of the soil fauna increase the heterogeneity of the soil CO_2 emission to atmosphere in Lamto savannas.

On a medium scale, a protection of organic matter in earthworm casts has been observed [62, 50]. If a rapid mineralization of the cast organic matter occurs in the first days following ingestion, a stabilization of carbon is observed during the following months, probably due to soil compaction in the casts. On a larger scale (years), the decomposition process is dramatically slowed down in the structures resulting from soil fauna activities. In aging casts, the soil organic matter is physically protected from microbial degradation (Fig. 13.5).

A similar conclusion can be drawn about termite effect. At the landscape scale, lenticular termite mounds can be viewed as spots of organic matter accumulation with an increased nutrient availability by a mass effect (a large

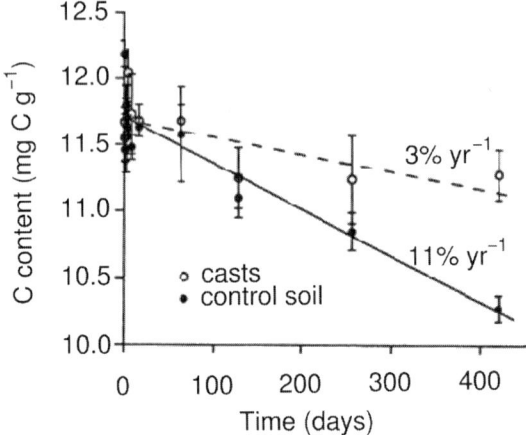

Fig. 13.5. Variation in total C content of casts of the endogeic earthworm *Millsonia anomala* and a 2 mm-sieved soil in long-term laboratory incubations (after [48], reprinted from [62], with kind permission of Springer Science and Business media).

quantity of low degradable organic matter leads to the production of a high quantity of nutrients), but also as a result of the colonization by earthworms and underground fungus-growing termites. In other words, lenticular termite mounds accumulate carbon and nutrients in a first step, when they are inhabited by epigeous Macrotermitinae, and they release carbon and nutrients in a second one, when they are colonized by hypogeic invertebrates.

It can be concluded that earthworms and termites play a key role in the savannas of Lamto, in structuring the soil, enhancing the patch scale heterogeneity, with a large influence on microbial and other soil processes [47]. The consequences of their activities on soil physical properties (mainly soil texture and aggregation) increase the water holding capacity and nutrient storage in the long term, while on a micro/meso-scale, their intimate mutualistic and symbiotic relationships with the micro-organisms locally enhance CO_2 emission and organic matter mineralization.

References

1. L. Abbadie and M. Lepage. The role of subterranean fungus comb chambers (Isoptera, Macrotermitinae) in soil nitrogen cycling in a preforest savanna (Côte d'Ivoire). *Soil Biology and Biochemistry*, 21(8):1067–1071, 1989.
2. L. Abbadie, M. Lepage, and X. Le Roux. Soil fauna at the forest-savanna boundary: Role of termite mounds in nutrient cycling. In P.A. Furley, J. Proctor, and J.A. Ratter, editors, *Nature and dynamics of forest-savanna boundaries*, pages 473–484. Chapman & Hall, London, 1992.

13 Perturbations of Soil Carbon Dynamics by Soil Fauna   247

3. J. Arbeille. *Recherches biologiques et écologiques sur les blattes de la région de Lamto (Côte d'Ivoire)*. Ph.D. thesis, Université de Paris 6, Paris, 1986.
4. F. Athias. Recherches écologiques dans la savane de Lamto (Côte d'Ivoire): Etude quantitative préliminaire des microarthropodes du sol. *La Terre et la Vie*, 3:395–409, 1971.
5. F. Athias. Influence du feu de brousse annuel sur le peuplement endogé de la savane de Lamto (Côte d'Ivoire). In *5th International colloquium on soil zoology*, pages 389–397, Prague, 1975.
6. F. Athias. Recherche sur les microarthropodes du sol de la savane de Lamto (Côte d'Ivoire). *Annales de l'Université d'Abidjan, Série E*, 9:193–271, 1976.
7. F. Athias, G. Josens, P. Lavelle, and R. Schaefer. V. Les organismes endogés. In *Analyse d'un écosystème tropical humide: La savane de Lamto (Côte d'Ivoire)*, pages 1–187. Laboratoire de Zoologie, ENS, Paris, 1974.
8. R. Barbault. Reptiles in savanna ecosystems. In F. Bourlière, editor, *Tropical savannas*, pages 325–336. Elsevier Scientific Publishing Company, Amsterdam, 1983.
9. I. Barois and P. Lavelle. Changes in respiration rate and some physicochemical properties of tropical soil during transit through *Pontoscolex corethrurus* (Glossoscolecidae, Oligochaeta). *Soil Biology and Biochemistry*, 18:539–541, 1986.
10. E. Blanchart, A. Bruand, and P. Lavelle. The physical structure of casts of *Millsonia anomala* (Oligochaeta : Megascolicidae) in shrub savanna soils (Côte d'Ivoire). *Geoderma*, 56:119–132, 1993.
11. E. Blanchart, P. Lavelle, E. Braudeau, Y. Le Bissonais, and C. Valentin. Regulation of soil structure by geophagous earthworm activities in humid savannahs of Côte d'Ivoire. *Soil Biology and Biochemistry*, 29:431–439, 1997.
12. M.B. Bouché. Stratégies lombriciennes. *Ecological Bulletin*, 25:122–132, 1977.
13. F. Bourlière. Mammals as secondary consumers in savanna ecosystems. In F. Bourlière, editor, *Tropical savannas*, volume 13 of *Ecosystems of the world*, pages 463–475. Elsevier, Amsterdam, 1983.
14. H. Boyer. Quelques aspects de l'action des termites du sol sur les argiles. *Clays and Minerals*, 17:453–462, 1982.
15. Y. Cambefort. *Etude écologique des Coléoptères Scarabaeidae de Côte d'Ivoire*. Doctorat d'état, Université de Paris 6, 1984.
16. Y. Cambefort. Rôle des Coléoptères Scarabaeidae dans l'enfouissement des excréments en savane guinéenne de Côte d'Ivoire. *Acta Oecologica, Oecologia Generalis*, 7(1):17–25, 1986.
17. D. Gillon. The fire problem in tropical savannas. In F. Bourlière, editor, *Tropical savannas*, volume 13 of *Ecosystems of the world*, pages 617–641. Elsevier, Amsterdam, 1983.
18. Y. Gillon. The invertebrates of the grass layer. In F. Bourlière, editor, *Tropical savannas*, volume 13 of *Ecosystems of the world*, pages 289–311. Elsevier, Amsterdam, 1983.
19. C. Girard. Etude des peuplements de Coléoptères Ténébrionides de la savane de Lamto (Côte d'Ivoire). *Annales de la Société d'Entomologie française*, 11(2):335–381, 1975.
20. C. Girard. *Etude écologique de Coléoptères à larves endogées dans une savane préforestière de Côte d'Ivoire*. Thèse d'université, Université de Paris 6, Paris, 1983.

21. M. Higashi, T. Abe, and T.P. Burns. Carbon-nitrogen balance and termite ecology. *Proceedings of the Royal Society of London, Series B*, 249:303–308, 1992.
22. C.G. Jones, J.H. Lawton, and M. Shachak. Organisms as ecosystem engineers. *Oikos*, 69:373–386, 1994.
23. G. Josens. Le renouvellement des meules à champignons construites par quatre Macrotermitinae (Isoptères) des savanes de Lamto-Pacobo. *Comptes rendus de l'Académie des Sciences, Série D*, 272:3329–3332, 1971.
24. G. Josens. *Etudes biologique et écologique des termites (Isoptera) de la savane de Lamto-Pakobo (Côte d'Ivoire)*. Ph.D. thesis, Université Libre de Bruxelles, Bruxelles, 1972.
25. G. Josens. Etude fonctionnelle de quelques groupes animaux: Les termites. *Bulletin de liaison des chercheurs de Lamto*, S5:91–131, 1974.
26. G. Josens. *Etudes biologique et écologique des termites (Isoptera) de la savane de Lamto-Pakobo (Côte d'Ivoire)*, volume 42 of *Mémoires de la classe des Sciences, Series 2*. Académie royale de Belgique, Bruxelles, 1977.
27. G. Josens. The soil fauna of tropical savannas. III. The termites. In F. Bourlière, editor, *Tropical savannas*, volume 13 of *Ecosystems of the world*, pages 505–524. Elsevier, Amsterdam, 1983.
28. P. Jouquet, L. Mamou, M. Lepage, and B. Velde. Effect of termites on clay minerals in tropical soils: fungus-growing termites as weathering agents. *European Journal of Soil Science*, 53:1–7, 2002.
29. S. Konaté. *Structure, dynamique et rôle des buttes termitiques dans le fonctionnement d'une savane préforestière (Lamto, Côte d'Ivoire): Le termite Odontotermes comme ingénieur de l'écosystème*. Ph.D. thesis, Université de Paris 6, Paris, 1998.
30. S. Konaté, X. Le Roux, D. Tessier, and M. Lepage. Influence of large termitaria on soil characteristics, soil water regime and tree leaf shedding pattern in a West African savanna. *Plant and Soil*, 206:47–60, 1999.
31. S. Konaté, X. Le Roux, B. Verdier, and M. Lepage. Effect of underground fungus-growing termites on carbon dioxide emission at the chamber- and landscape-scales in an African savanna. *Functional Ecology*, 17:305–314, 2003.
32. P. Kouassi. *Structure et dynamique des groupes trophiques de la macrofaune du sol d'écosystèmes naturels et transformés de Côte d'Ivoire*. Ph.D. thesis, Université de Cocody, Abidjan, 1999.
33. M. Lamotte. *The structure and function of a tropical savannah ecosystem*. Springer-Verlag, New York, 1975.
34. M. Lamotte. La savane préforestière de Lamto, Côte d'Ivoire. In M. Lamotte and F. Bourlière, editors, *Problèmes d'écologie: Structure et fonctionnement d'écosystèmes terrestres*, pages 231–311. Masson, Paris, 1978.
35. M. Lamotte. Structure et fonctionnement des écosystèmes de la savane de Lamto (Côte d'Ivoire). In Unesco, PNUE, and FAO, editors, *Ecosystèmes pâturés tropicaux*, pages 529–580. UNESCO, Paris, 1981.
36. M. Lamotte. Amphibians in savanna ecosystems. In F. Bourlière, editor, *Tropical savannas*, volume 13 of *Ecosystems of the world*, pages 313–323. Elsevier, Amsterdam, 1983.
37. M. Lamotte. Research on the characteristics of energy flows within natural and man-altered ecosystems. In H.A. Golley and M. Godron, editors, *Disturbance in ecosystems*, volume 44 of *Ecological studies*. Springer-Verlag, Berlin, 1983.

38. M. Lamotte. Place des animaux détritivores et des microorganismes décomposeurs dans les flux d'énergie des savanes africaines. *Pedobiologia*, 33:17–35, 1989.
39. M. Lamotte and F. Bourlière. Energy flow and nutrient cycling in tropical savannas. In F. Bourlière, editor, *Tropical savannas*, volume 13 of *Ecosystems of the world*, pages 583–603. Elsevier, Amsterdam, 1983.
40. P. Lavelle. *Les vers de terre de la savane de Lamto (Côte d'Ivoire): Peuplements, populations et fonctions dans l'écosystème*. Ph.D. thesis, Université de Paris 6, Paris, 1978.
41. P. Lavelle. Stratégies de reproduction chez les vers de terre. *Acta Oecologica, Oecoogia Generalis*, 2:117–133, 1981.
42. P. Lavelle. The soil fauna of tropical savannas. I. The community structure. In F. Bourlière, editor, *Tropical savannas*, volume 13 of *Ecosystems of the world*, pages 477–484. Elsevier, Amsterdam, 1983.
43. P. Lavelle. The soil fauna of tropical savannas. II. The earthworms. In F. Bourlière, editor, *Tropical savannas*, volume 13 of *Ecosystems of the world*, pages 485–504. Elsevier, Amsterdam, 1983.
44. P. Lavelle. Faunal activities and soil processes: adaptative strategies that determine ecosystem function. *Advances in Ecological Research*, 27:93–132, 1997.
45. P. Lavelle. Functional domains in soils. *Ecological Research*, 17:441–450, 2002.
46. P. Lavelle, I. Barois, E. Blanchart, G. Brown, L. Brussaard, T. Decaëns, C. Fragoso, J.J. Jimenez, K.K. Kajondo, M.D.L.A. Martinez, A. Moreno, B. Pashanasi, B. Senepati, and C.Villenave. Earthworms as a resource in tropical ecosystems. *Nature and Resources*, 34(1):26–41, 1998.
47. P. Lavelle, D. Bignell, M. Lepage, V. Wolters, P. Roger, P. Ineson, O.W. Heal, and S. Dhillion. Soil function in a changing world: the role of invertebrate ecosystem engineers. *European Journal of Soil Biology*, 33(4):159–193, 1997.
48. P. Lavelle, E. Blanchart, A. Martin, A.V. Spain, and S. Martin. Impact of soil fauna on the properties of soils in the humid tropics. In R. Lal and P.A. Sanchez, editors, *Myths and science of soils of the tropics*, volume 29 of *Special Publications*, pages 157–185. Soil Science Society of America, Madison, WI, 1992.
49. P. Lavelle and C. Gilot. Priming effects of macroorganisms on microflora : a key process of soil function? In K. Ritz, J. Dighton, and K.E. Giller, editors, *Beyond the biomass*, pages 173–180. John Wiley and Sons, Chichester, 1994.
50. P. Lavelle and A. Martin. Small-scale and large-scale effects of endogeic earthworms on soil organic matter dynamics in soils of the humid tropics. *Soil Biology and Biochemistry*, 24(12):1491–1498, 1992.
51. P. Lavelle, R. Schaefer, and Z. Zaïdi. Soil ingestion and growth in *Millsonia anomala*, a tropical earthworm, as influenced by the quality of the organic matter ingested. *Pedobiologia*, 33:379–388, 1989.
52. C. Lecordier. *Les peuplements de Carabiques (Coléoptères) dans la savane de Lamto (Côte d'Ivoire)*. Doctorat d'université, Université de Paris 6, Paris, 1975.
53. C. Lecordier and C. Girard. Peuplements hypogés de Carabiques (Col.) dans la savane de Lamto (Côte d'Ivoire). *Bulletin de l'I.F.A.N., Série A*, 35(2):361–392, 1973.
54. B. Legay and R. Schaefer. Modalities of the energy-flow in different tropical soils, as related to their mineralization capacity of organic carbon and to the type of clay. II. the degradation of various substrates. *Zentralblatt für Mikrobiologie*, 139:389–400, 1984.

55. M. Lepage. The role of fungus-growing termites in savanna organic matter cycling. In D. Bellan, G. Bonin, and C. Emig, editors, *Functioning and dynamics of natural and perturbed ecosystems*, pages 88–98. Lavoisier, Paris, 1995.
56. M. Lepage, L. Abbadie, and A. Mariotti. Food habits of sympatric termite species (Isoptera, Macrotermitinae) as determined by stable carbon isotope analysis in a Guinean savanna (Lamto, Côte d'Ivoire). *Journal of Tropical Ecology*, 9:303–311, 1993.
57. J.M. Leroux. Densité des colonies et observations sur les nids de Dorylines *Anomma nigricans* Illiger (Hym. Formicidae) dans la région de Lamto (Côte d'Ivoire). *Bulletin de la Société de Zoologie Française*, 102:51–62, 1977.
58. J.M. Leroux. *Ecologie des populations de Dorylines* Anomma nigricans *Illiger (Hym. Formicidae) dans la région de Lamto (Côte d'Ivoire)*. Doctorat d'université, Université de Paris 6, Paris, 1982.
59. D. Lopez-Hernandez and A. Febres. Changements chmiques et granulométriques produits dans des sols de Côte d'Ivoire par la présence de trois espèces de termites. *Revue d'Ecologie et de Biologie du Sol*, 21(4):477–489, 1984.
60. J. Lévieux. Etude du peuplement en fourmis terricoles d'une savane préforestière de Côte d'Ivoire. *Revue d'Ecologie et de Biologie du Sol*, 10(3):379–428, 1972.
61. J. Lévieux. The soil fauna of tropical savannas. IV. The ants. In F. Bourlière, editor, *Tropical savannas*, volume 13 of *Ecosystems of the world*, pages 525–540. Elsevier, Amsterdam, 1983.
62. A. Martin. Short- and long-term effects of the endogeic earthworm *Millsonia anomala* (Omodeo) (Megascolecidae, Oligochaeta) of tropical savannas on soil organic matter. *Biology and Fertility of Soils*, 11:234–238, 1991.
63. A. Martin, J. Cortez, I. Barois, and P. Lavelle. Les mucus intestinaux de ver de terre moteur de leurs interactions avec la microflore. *Revue d'Ecologie et de Biologie du Sol*, 24(4):549–558, 1987.
64. A. Martin and P. Lavelle. Effect of soil organic matter quality on its assimilation by *Millsonia anomala*, a tropical geophagous earthworm. *Soil Biology and Biochemistry*, 24(12):1535–1538, 1992.
65. A. Martin and J.C.Y. Marinissen. Biological and physico-chemical processes in excrements of soil animals. *Geoderma*, 56:331–347, 1993.
66. A. Martin, A. Mariotti, and J. Balesdent. Mise en évidence de l'impact des vers de terre sur la dynamique de la matière organique du sol par la biogéochimie isotopique du carbone. *Cahiers de l'ORSTOM, Série Pédologie*, XXVI(4):349–356, 1991.
67. A. Martin, A. Mariotti, J. Balesdent, P. Lavelle, and R. Vuattoux. Estimate of organic matter turnover rate in a savanna soil by ^{13}C natural abundance measurements. *Soil Biology and Biochemistry*, 22(4):517–523, 1990.
68. T. Matsumoto. The role of termites in an equatorial rain forest ecosystem of West Malaysia. I. Population density, biomass, carbon, nitrogen and calorific content and respiration rate. *Oecologia*, 22:153–178, 1976.
69. J.C. Menaut and J. César. Structure and primary productivity of Lamto savannas, Ivory Coast. *Ecology*, 60(6):1197–1210, 1979.
70. P. Mordelet, S. Barot, and L. Abbadie. Root foraging strategies and soil patchiness in a humid savanna. *Plant and Soil*, 182:171–176, 1996.
71. A.V. Spain and P. Reddell. δ^{13}C values of selected termites (Isoptera) and termite-modified materials. *Soil Biology and Biochemistry*, 28:1585–1593, 1996.
72. A. Sugimoto, D.E. Bignell, and J.A. McDonald. Global impact of termites on the carbon cycle and atmospheric trace gases. In T. Abe, D.E. Bignell, and

M. Higashi, editors, *Termites: Evolution, sociality, symbioses, ecology*, pages 409–436. Kluwer, Dordrecht, 2000.

73. M. J. Swift, O. W. Heal, and J.M. Anderson. *Decomposition in terrestrial ecosystems*, volume 5 of *Studies in Ecology*. Blackwell Scientific Plublications, Oxford, 1979.

74. A. Yapi. *Biologie, écologie et métabolisme digestif de quelques espèces de termites humivores de savane*. Thèse de 3e cycle, Université d'Abidjan, Abidjan, 1991.

75. Z. Zaïdi. *Recherches sur les modalités de l'interdépendance nutritionnelle entre vers de terre et microflore dans la savane guinéenne de Lamto (Côte d'Ivoire). Esquisse d'un système interactif.* Ph.D. thesis, Université de Paris-Sud Orsay, Paris, 1985.

Part IV

The Nitrogen Cycle

14

Nitrogen Inputs to and Outputs from the Soil-Plant System

Luc Abbadie

14.1 Introduction

Soil nitrogen is not derived from the bedrock as other nutrients. It only comes from the atmosphere. It enters the ecosystem by rainfall or dust deposition, or by biological fixation performed by specialized bacteria. The development rate of an ecosystem, as during plant succession, is often limited by the rate of nitrogen accumulation. More generally, nitrogen is the first factor limiting primary production in most of mature terrestrial and aquatic ecosystems [54], particularly in those periodically submitted to perturbations leading to a significant output of nitrogen (crops in the fields, fire in the grasslands). In savannas, primary production is also considered as strongly limited by nitrogen. This assumption is supported by many experiments showing positive yield responses of grasses to nitrogen fertilizer application [10] and by the strong link between photosynthetic capacity and leaf nitrogen content [23, 47]. In Lamto, the supply of ammonium nitrate in March induced a 15% increase of the biomass measured at the end of June [27], suggesting that annual production depends on mineral nitrogen availability in soil. This latter partly results from the rates of antagonist processes, such as dry and wet depositions, biological fixation of atmospheric dinitrogen, soil leaching and denitrification.

14.2 Dry and wet depositions

14.2.1 Nitrate and ammonium concentrations

The quantity of nitrogen brought by precipitation depends both on the rain and fog frequencies, the intensity of rains and the concentration of nitrogen in rainwater. It has been hypothesized that organic or mineral nitrogen in precipitation would mainly come from the leaching of solid nitrogen in suspension in the atmosphere, the dissolution of various compounds in droplets and the adsorption of ammonium on clay particles smaller than 2 μm and produced by wind erosion. Mineral nitrogen originates from marine surfs, transformation of atmospheric N_2 by electrical, chemical and photochemical processes, aquatic or terrestrial denitrification, industrial and agricultural activities and combustion of plant matter, noticeably in savanna regions. Organic nitrogen is present in rainwater under the form of microorganisms and humic compounds associated to clay particles.

In areas with low precipitation, the water rainfall is generally more loaded in nitrogen compounds than in areas where precipitation is high [50]. In Lamto, as in many other tropical regions [20], rains with highest N concentrations are observed in February-March at the beginning of the wet season [51], when the monthly amount of water is still low (Fig. 14.1). Rains with lowest N concentrations occur in the heart of the wet season, when the monthly rainfall amounts to around 200 mm of water. The absolute minimums are recorded when rainfall exceeds 300 mm per month, at 0.10 $g\,N\,m^{-3}$ of ammonium and 0.01 $g\,N\,m^{-3}$ of nitrate.

Atmospheric dusts bring different forms of nitrogen to the ecosystem: organic nitrogen on particles of various sizes, and ammonium and nitrate on clay particles of around 2 μm in size. The concentrations of NH_4^+ and NO_3^- in the air have been studied in Lamto during one year by Servant et al. [44]. They are very variable with time, 15-fold in the case of ammonium (from 0.19 $\mu g\,m^{-3}$ air in August to 3.01 $\mu g\,m^{-3}$ air in December) and 24-fold in the case of nitrate (from 0.14 $\mu g\,m^{-3}$ in September to 3.36 $\mu g\,m^{-3}$ air in December) (Fig. 14.2). Average monthly values are clearly higher during the dry season than during the rainy season. Sixty percent of the dry deposition of ammonium is due to particles smaller than 0.4 μm in diameter; on the

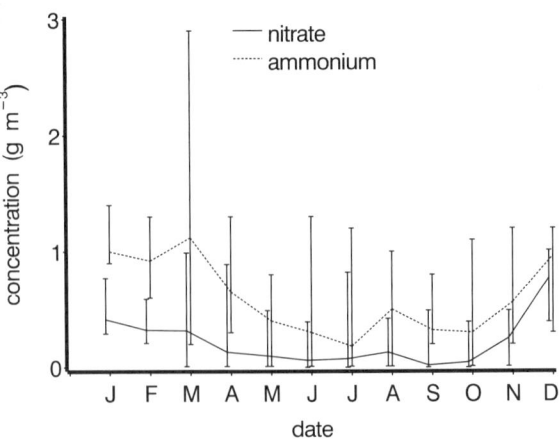

Fig. 14.1. Mean monthly concentrations of ammonium and nitrate in rainfall at Lamto in 1972 and 1973. Error bars indicate minimum and maximum measured values (data from [51]).

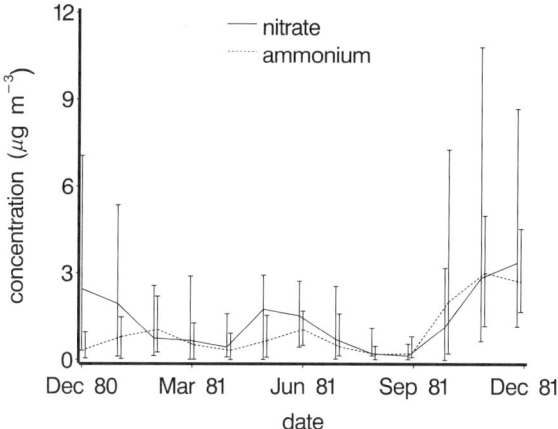

Fig. 14.2. Mean monthly concentrations of ammonium and nitrate in air at Lamto. Error bars indicate minimum and maximum measured values (data from [44]).

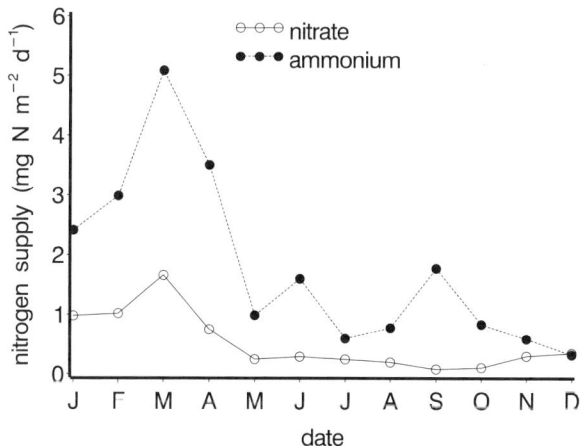

Fig. 14.3. Mean monthly supply of ammonium and nitrate by rainfall at Lamto (data from [51]).

contrary, nitrate is brought to the ecosystem indifferently by particles ranging from 0.1 μm to 10 μm in diameter.

14.2.2 Inputs of nitrogen to the ecosystem

The flow of nitrogen entering the ecosystem is very variable with season (Fig. 14.3). It is maximum in the beginning of the long rainy season at 3.51 to 5.08 mg N m^{-2} d^{-1} of ammonium and 0.75 to 1.67 mg N m^{-2} d^{-1}, and minimum in the end of the short rainy season at 0.33 to 0.59 mg N m^{-2} d^{-1} of ammonium and 0.07 to 0.10 mg N m^{-2} d^{-1} of nitrate. On a yearly basis, the total input of nitrogen in the ecosystem through rainfall is 0.65 to 0.66 g N m^{-2} of ammonium and 0.14 to 0.24 g N m^{-2} of nitrate [51], i.e., 0.79 to 0.90 g N m^{-2}. It is a high value, but close to the quantities recorded in non-polluted regions, where the annual amounts of nitrogen brought by rainwater are in the range of 0.1 to 0.5 g N m^{-2} [9, 55].

A typical feature of the Lamto region is the high flow of organic nitrogen brought by precipitation. This flow is approximately three times higher than that of mineral nitrogen and accounts for a large additional input of nitrogen to the ecosystem at ca. 1.4 g N m^{-2} y^{-1} [52]. The quantity of total nitrogen, mineral and organic brought to the Lamto savannas is therefore between 2.2 and 2.3 g N m^{-2} y^{-1}. This very high proportion of organic nitrogen as compared to the mineral nitrogen has been observed only in a small number of situations throughout the world, noticeably in Côte d'Ivoire in the north of Abidjan (Roose 1972 in [52]; [11]). This particularity is still badly explained. It could be related to the erosion of soil by wind, which is particularly active in this region due to deforestation and farming of large areas.

The dry deposition rate being constant throughout the year, the flow of nitrogen transported by dusts to soil only depends on nitrogen concentration in the air. The monthly flow is very variable with season. It is maximum during the wet season (June, July and November) and during the beginning of the dry season (December and January).

On an annual basis, the dry and wet depositions were estimated to 1 g N m^{-2} under the form of ammonium and to 0.4 g N m^{-2} under the form of nitrate by Servant et al. [44]. Since precipitations bring ca. 0.65 g N m^{-2} of ammonium and ca. 0.19 g N m^{-2} of nitrate [51], the dry deposition alone would be 0.35 g N m^{-2} under the form of ammonium and 0.21 g N m^{-2} under the form of nitrate.

14.3 Biological fixation of atmospheric dinitrogen

The air molecular nitrogen is the major source of nitrogen entering the soil-plant system in most of the ecosystems in the world. Different types of microorganisms are involved in this process: Cyanobacteria that make crusts on the surface of soil, symbiotic bacteria that are generally located in the roots of higher plants and free N_2-fixing bacteria, more or less scattered in the soil profile.

14.3.1 The dinitrogen fixation by Cyanobacteria

Many algal crusts can grow between grass tussocks in all the savanna types from Lamto during the dry season. The highest densities are observed in the open shrub savannas dominated by Andropogoneae, and in the *Loudetia* savanna downslope. These crusts are made of filamentous Chlorophyceae [43] and Cyanobacteria with heterocysts. The most common Cyanobacteria belong to the genera *Nostoc*, *Tolyopothrix* and *Calothrix*, all well known for their ability to fix molecular nitrogen; some non-fixing Cyanobacteria such as *Coccomyxa* and *Lyngbia* are also found. Raud [39] hypothesized that these non-fixing Cyanobacteria and Chlorophyceae depend on the nitrogen supplied by fixing species. Balandreau [8, 6] measured an important nitrogen fixation by Cyanobacteria crusts by the acetylene reduction method. The rate of acetylene reduction was between 0.21 and 0.38 μmol m^{-2} h^{-1}, i.e., 1.18 - 2.12 mg N m^{-2} d^{-1}. Considering a 3-month period of continuous activity, the annual flux of nitrogen entering the savanna ecosystem can be estimated as 0.2 g N m^{-2} y^{-1}.

14.3.2 The symbiotic fixation of molecular nitrogen

The herbaceous legumes community in Lamto is very diverse. Around 60 species can be found in the savanna [14, 33], which account for ca. 16% of the total number of herbaceous species [42]. The legumes weakly contribute to the aboveground herbaceous biomass, between 0.02% and 1.05% on average [14], with strong seasonal variations [1], but they can contribute up to 13.5% of the pool of nitrogen in the aboveground herbaceous layer due to their high concentration in proteins [1] (Fig. 14.4). Balandreau [5] showed that young *Vigna* sp. and *Cassia mimosoides* were able to respectively reduce 1.9 and 0.9 μmol [C$_2$H$_4$] h^{-1} per individual, but that adult plants in the field did not show any measurable activity of the nitrogenase enzyme because of the lack of fixing nodules. This strongly suggests that a non-negligible part of the nitrogen assimilated by the legumes comes from the soil solution and not from the fixation. In Lamto, the input of nitrogen to the soil-plant system through the symbiotic fixation can therefore be considered as negligible, probably below 0.1 g N m^{-2} y^{-1}, contrary to what is observed in many other terrestrial ecosystems, noticeably in other savannas [9, 19].

The reason why the biomass of legumes is weak in Lamto is not clear, but fire seems to play a major role as shown by an experiment conducted by Abbadie, Menaut and Vuattoux in 1986 (unpublished data). A quarter of a hectare was protected against fire in a grass savanna dominated by *Loudetia simplex*, downslope on a sandy soil. The most spectacular consequence of the protection against fire was the huge increase of the legume biomass (Table 14.1). At the end of the rainy season, after almost 2 years without fire (in November), the total aerial biomass of legumes was 23 times higher in the unburned plot than in the yearly burned control plot. The aboveground and underground biomass of *Indigofera polysphaera* were respectively 35 and 70 times higher in the unburned plot than in burned plot; most of the increase in root mass was due to coarse roots larger than 2 mm in diameter. In the burned plot, the number of nodules was very low while it could be counted several tens of nodules per individual in the protected plot. The density of the legume community was decreased in the burned plot (5.3 m^{-2}) compared

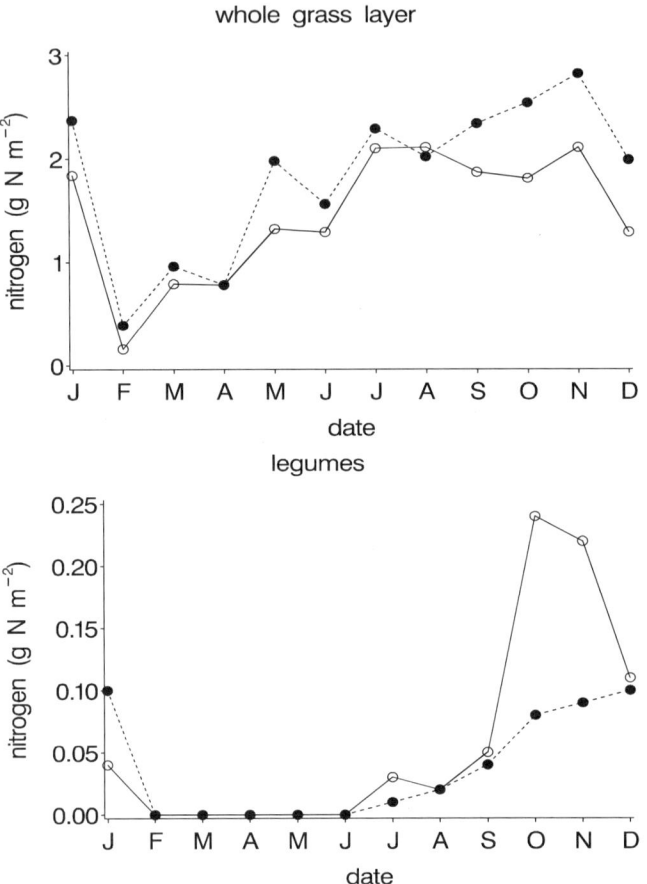

Fig. 14.4. Mean monthly nitrogen contained in the whole grass layer and in legumes for a *Loudetia* grass savanna (open circles) and an Andropogoneae shrub savanna (solid circles) (data from [2]).

Table 14.1. Legume aerial and root biomass 666 days after the last fire (November) in unburned grass savanna, in $g\,m^{-2}$ (Abbadie, Menaut and Vuattoux, unpublished data).

Biomass	Fire treatment	*Indigofera polysphaera*	Other legumes	Total
Aboveground	Burned	3.0	2.8	5.8
	Unburned	108.0	28.6	136.6
Belowground	Burned	6.5	Na	Na
	Unburned	47.3	Na	Na

Table 14.2. Variation of the legume density with time in individuals m^{-2} (Abbadie, Menaut and Vuattoux, unpublished data).

Time since last fire (d)	Fire treatment	*Indigofera polysphaera*	Other legumes
666	Burned	2.8	5.3
765		0.0	0.0
939		0.6	0.6
666	Unburned	0.7	1.7
765		0.1	0.1
939		0.0	0.5

to that in the unburned plot (1.7 m^{-2}) (Table 14.2). *Indigofera polysphaera* showed the same trend, switching from 2.8 m^{-2} when protected to 0.7 m^{-2} when burned.

Portères [37] and César [14] had already pointed out that *Indigofera polysphaera* could remain over 1 year in the absence of fire or when fire had a low intensity. Monfort [33] also reported that the date of fire strongly influences the growth rate and vegetative development of legumes, the highest proportion of *Indigofera polysphaera* showing secondary branches being observed in early burned plots, i.e., on 16-17 month-old plants. In the burned areas, the population of *Indigofera polysphaera* is renewed each year and is exclusively established from seeds. It seems that yearly fire imposes an annual biological cycle to *Indigofera polysphaera* through the burning of the aerial organs. But, when the plant cover is protected against fire, *Indigofera polysphaera* is able to carry out its complete cycle: leaflets appear on secondary branches, sometimes on tertiary branches, and a new generation of flowers is produced in November. Some plants (16% of the population) are even able to begin a third year of growth, but none exceed 1000 days of age. At this stage, after 1000 days without fire, only a very low number of plants of the year can survive because most of the seedlings do not succeed in going through the thick layer of accumulated litter due to the absence of fire.

The weakness of the legume biomass in Lamto is obviously the result of the yearly burning of the savanna that prevents the biennial herbaceous legumes from achieving a complete development cycle. On the contrary, the suppression of the regular fire allows a considerable increase of the aerial parts and, especially, roots. The lack of annual fire can therefore be interpreted as a favorable perturbation to the legume community. It is probable that a biennial fire is the most favorable regime for the legumes since it allows both a maximal vegetative development and a fair regeneration of the population from seeds. In these conditions, the germination and recruitment of the herbaceous legumes are not hampered by a too high accumulation of litter to soil since it is burned every 2 years. In other words, the ability of the legume community to symbiotically fix atmospheric dinitrogen is clearly limited by the yearly occurrence of fire through a low development of legume roots and nodules. Nevertheless, during humid years, the fire has a low intensity: many zones are badly burned and sometimes completely escape to burning. In this case, a significant input of nitrogen to the ecosytem through symbiotic fixation probably occurs; its quantitative and qualitative impact on the nitrogen cycle and plant cover dynamics has not been assessed to date.

14.3.3 The non-symbiotic fixation of molecular nitrogen

The activity of non-symbiotic nitrogen fixers is generally limited by the lack of assimilable carbon and by repression of the nitrogenase enzyme by the partial pressure in oxygen, generally too high in sandy soils. In Lamto, the number of free-fixing bacteria is ca. 10,000 cells per gram of dry soil [36] and their supply of nitrogen to soil is quite negligible. On the contrary, the soil very close to the roots is a very favorable environment for N_2-fixing bacteria. These micro-organisms find there an abundant source of energy under the form of root exudates and a low partial pressure in atmospheric oxygen as a result of root respiration. Consequently, the fixation of dinitrogen in the grass and tree rhizospheres is often an important contribution to the supply of nitrogen to savanna ecosystems [17, 29, 41].

The rhizospheric fixation of nitrogen in Lamto has been studied in the field by the acetylene reduction technique. In the grass savanna dominated by *Loudetia simplex*, the nitrogenase activity varied between 2.36 and 39.39 μmol $[C_2H_4]$ m^{-2} h^{-1}. The highest values were recorded during the wettest months and the lowest during the dry season. The same trend was observed in the shrub savanna dominated by *Hyparrhenia diplandra*, with reduction rates varying from 3.26 to 17.23 μmol $[C_2H_4]$ m^{-2} h^{-1} [5]. The water status in plant and soil humidity seem to be the main factors controlling these seasonal variations [5]. The rate of dinitrogen fixation was also highly variable during the course of the day. Two maximums were generally observed (Fig. 14.5): at 13:00 as a result of the exudation of soluble sugar by roots due to the high photosynthetic activity that occurred during the morning, and at 1:00 as a result of the night hydrolysis of the starch, a physiological mechanism typical of C_4 plants [7, 5]. The rhizospheric fixation was also studied at the scale of individual plants. Maxima at 54, 382 and 136 nmol $[C_2H_4]$ h^{-1} g $[dry\ root]^{-1}$ were reported for *Loudetia simplex*, *Andropogon* sp. and *Brachiaria brachylopha*, respectively. A high nitrogenasic activity was also measured for other monocotyledons: 46.14 μmol $[C_2H_4]$ m^{-2} h^{-1} for *Eleusine indica*, 106.57 for *Bulbostylis* sp., 34.86 for *Fimbristylis* sp., 115.14 for *Cyperus zollingeri* and 46.86 for *Paspalum* sp.

The non-symbiotic fixation of molecular nitrogen by the bacteria living in the herbaceous plants from Lamto is therefore a widespread phenomenon and it noticeably contributes to the input of nitrogen to savanna ecosystem. By hypothesizing that soil is humid during 5 months in the *Loudetia simplex* savanna and during 6 months in the Andropogoneae savanna, and taking into

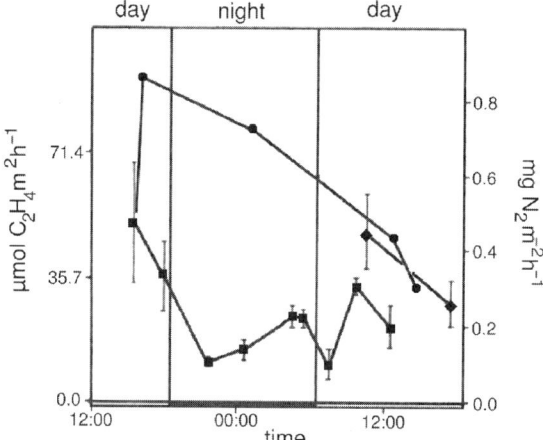

Fig. 14.5. Daily variations of [C_2H_2] fixation of nitrogen as produced C_2H_4 and fixed N_2 in a *Loudetia* grass savanna. Squares, whole plant, 21-22/07/1972; circles, after removing aboveground parts, 21-22/07/1972; diamonds, whole plant, 18/07/1972 (after [5], copyright (1976), with permission of Elsevier).

account the minimum and maximum rates of acetylene reduction, Balandreau [5] estimated the annual nitrogen flow entering the ecosystem through the rhizospheric micro-organisms at respectively 1.3 and 0.9 $g\,N\,m^{-2}\,y^{-1}$.

14.4 Grass cover leaching

The interception of rainwater by plant cover induces a transport of both organic and mineral compounds from grass and tree leaves to soil. These compounds originate from dry and wet deposition, leaf exudation and activity of the phyllosphere micro-organisms. Most of them can be therefore considered as an input of nitrogen to the soil-plant system, even if they partly overlap the flow of compounds coming from the atmosphere. Under the grass covers dominated by *Loudetia simplex* and *Hyparrhenia diplandra*, the throughfall is 1.2, 1.6 and 1.3 times richer in total nitrogen, ammonium and nitrate, respectively, than the incident rainwater. Hypothesizing an average monthly concentration of rainwater in organic nitrogen at 1.46 $g\,N\,m^{-3}$, ammonium at 0.515 $g\,N\,m^{-3}$ and nitrate at 0.148 $g\,N\,m^{-3}$, Villecourt and Roose [52] estimated the average monthly concentration of throughfall at 1.679 $g\,N\,m^{-3}$ of organic nitrogen, 0.834 $g\,N\,m^{-3}$ of ammonium and 0.194 $g\,N\,m^{-3}$ of nitrate. For an annual rainfall of around 1200 mm (Chap. 3), the input of nitrogen to soil by grass foliage leaching is therefore ca. 1.68 $g\,N\,m^{-2}\,y^{-1}$ of organic nitrogen, 0.83 $g\,N\,m^{-2}\,y^{-1}$ of ammonium and 0.19 $g\,N\,m^{-2}\,y^{-1}$ of nitrate. By comparing these data with those of the rainfall content in nitrogen, the additional input of nitrogen from the aboveground parts of grasses to soil can be estimated at 0.2-0.3 $g\,N\,m^{-2}\,y^{-1}$ under the organic form and 0.2 $g\,N\,m^{-2}\,y^{-1}$ under the form of ammonium (the additional supply of nitrate is negligible).

14.5 Soil leaching

The infiltration of rainwater in soil induces the leaching of many organic and mineral compounds, which get out from the root zone, becoming unavailable for plant growth. In Lamto, the average annual height of water drained was measured at 55 and 212 mm for respectively 1089 and 1324 mm rainfalls [52](see Chap. 6). The drainage water is very poor in mineral nutrients, except silica. Its concentrations in calcium, magnesium, potassium and phosphorus range from 1 to 3 $g\,m^{-3}$, but total nitrogen exceeds 4 $g\,m^{-3}$. This nitrogen is almost completely of the organic form, likely coming from micro-organisms or recently dead plant matter. The average concentration of drainage water in ammonium is 0.15 $g\,m^{-3}$ and that in nitrate is 0.05 $g\,m^{-3}$. The corresponding flows are 0.2 to 0.4 $g\,N\,m^{-2}\,y^{-1}$ for calcium, magnesium, potassium and phosphorus, and 0.5 and 0.02 $g\,N\,m^{-2}\,y^{-1}$ for organic nitrogen and ammonium respectively; the flow of nitrate is negligible.

Two major points must be emphasized. First, the Lamto savannas only lose very small quantities of nutrients through soil leaching. In fact, there is a positive balance between the input of nutrients to the soil-plant system by rainfall and the output by soil leaching, particularly for calcium, phosphorus and nitrogen. The soil plant system should therefore accumulate nutrients in the absence of fire. Second, the lack of nitrate in drainage water, the most labile form of soil mineral nitrogen that is often lost by the ecosystem through leaching, could result from the microbial immobilization or weakness of the nitrification process. In Lamto, the lack of nitrification is clearly the main process involved (see Chap. 15).

14.6 Denitrification

Denitrification is an anaerobic respiration. When oxygen is lacking, nitrate and derived oxides are used as final electron acceptors during the oxidation of organic compounds. The intensity of field denitrification depends on three factors: the number of denitrifying micro-organisms, the rate of synthesis of the enzymes involved in the reduction of nitrate and other nitrogen oxides and the physical and chemical characteristics of soil such as the oxygen partial pressure, temperature, soil water content, nitrate availability and assimilable carbon availability [48, 28]. Denitrification can be performed by a great number of anaerobic bacteria, heterotrophic to carbon for most of them. These micro-organisms are widespread in all types of soils, including those oxygenated because of the presence of anaerobic microsites, such as clay aggregates, which are sufficient to maintain significant populations of denitrifyers. Denitrification is very often a major route of nitrogen loss by ecosystem, but few studies have been devoted to this process in savannas to date.

Since denitrification is performed by bacteria heterotrophic to carbon, its intensity is closely dependent on the quantity and quality of soil organic matter [13]. It is consequently highly variable in space, between soils and, within one soil, between locations. This is why it was studied in Lamto at two scales: that of the landscape by a study of the potential of denitrification along a topographic transect, and that of the soil type by a study of the potential denitrification in the microenvironments created by roots and soil fauna. Soil cores were randomly collected between 0 and 10 cm and between 40 and 50 cm deep during dry and wet seasons in the riparian forest along the Bandama river, a gallery forest and a dry plateau forest in the north of the reserve, a grass savanna dominated by *Loudetia simplex* and *Andropogon schirensis* and a shrub savanna dominated by *Hyparrhenia diplandra* and *Hyparrhenia smithiana*. The measurements of the potential denitrification were performed according to the protocol of Lensi and Chalamet [25].

The results show that the most denitrifying soils are the forest soils, at least in the surface layer, whatever the season [3]. In the savanna soils, the potential denitrifying activity is very low at all depths (Fig. 14.6). When all the forest and savanna soils from Lamto are considered, a positive correlation is found between the intensity of denitrification and the initial quantity of organic carbon ($R^2 = 0.83$; $P < 0.01$). It was shown in another experiment

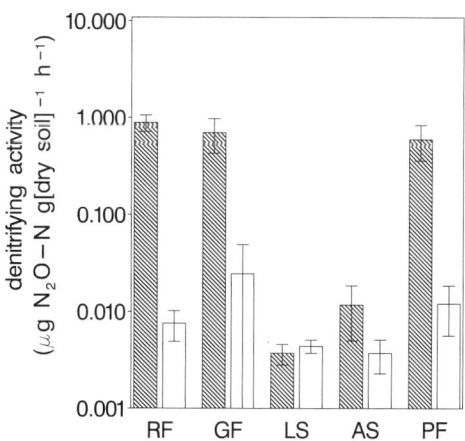

Fig. 14.6. Denitrifying activity along a topographic gradient after 5 h of incubation. GF, gallery forest; RF, riparian forest; PF, plateau forest; AS, Andropogoneae shrub savanna; LS, *Loudetia* grass savanna. hatched bars: measurement at 10 cm deep; empty bars: measurements at 50 cm deep. Note the logarithmic scale (data from [26]).

that the potential mineralization rate of carbon in those soils is quite constant, i.e., that carbon availability for heterotrophic micro-organisms is the same in forest as in savanna soils (see Chap. 12). In other words, in Lamto, the rate of denitrification mainly depends on the quantity of mineralizable carbon, supporting previous results [13, 12, 49]. In some of the samples, glucose was added but did not induce any change in the rate of potential activity. This suggests that the level of carbon availability does not control the expression of the pre-existing potential activity but the level of denitrifying enzymes.

A general feature of the Lamto soils is the strong limitation of microbial activity by the low carbon availability. But, in many microsites, this availability is significantly increased, such as in the rhizosphere, earthworm casts or soil worked by termites (see Chap. 13). A study of denitrification in these microsites showed a non-negligible activity in fresh soil samples collected during the wet season, where the availability of carbon assessed as the rate of CO_2 production in controlled conditions was 2 to 5 times higher than in the average soil, i.e., soil devoid of recent root or soil fauna activity [26]. The same study conducted on air dried samples [3] did not show any stimulation of denitrification in these microsites, suggesting that the control of denitrification by carbon is complex and dependent on climate seasonal variations.

It is very difficult to estimate actual field microbial activities by using laboratory incubations since they are always much lower than potentials. However, it could be hypothesized that the ratio of potential to actual activities is the same for denitrification and carbon mineralization because the two processes are performed by heterotrophic micro-organisms that use the same carbon substrates. But, in Lamto, the first limiting factor of denitrification is the availability of nitrate that is always at the detection limit [40]. Moreover, the potential of nitrification itself is also extremely low due to a direct control of nitrifying bacteria by the dominant grass species. In these conditions, it can be concluded that the actual flow of nitrogen from the soil to the atmosphere through denitrification is negligible. However, it has to be noticed that a significant potential of denitrification exists. This could lead to local nitrogen losses in particular soil patches in relation to soil fauna activity (see Chap. 16) or in case of NPK mineralization for agricultural purposes.

14.7 Nitrogen monoxide emission

The emission of nitrogen monoxide (NO) often occurs during the nitrification and denitrification processes [16]. It is generally considered that tropical soils produce larger amounts of NO than temperate soils [18]. Several authors have reported emission rate values from tropical forests ecosystems, but very few studies have been conducted in tropical savannas. The emission rates from the Lamto soils are among the first published about the humid savannas worldwide. Le Roux et al. [22] showed that in the Lamto region, the highest NO fluxes occur in the nutrient rich environments downslope (gallery forest), that

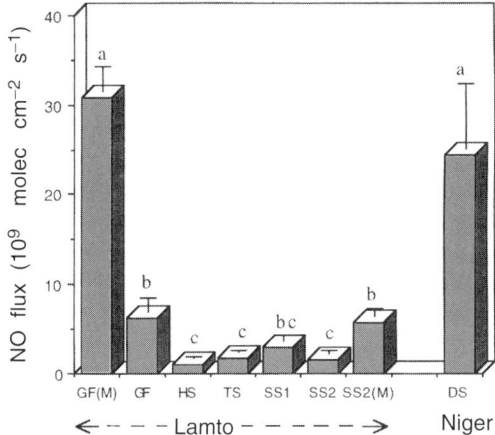

Fig. 14.7. Mean NO emission rates from the soils of the catena in Lamto and of a dry savanna in Niger. Bars are standard errors. Values with the same letter not significantly different at the 0.05 level. GF, gallery forest; GF(M), termite mound in gallery forest; HS, *Loudetia* savanna; TS, Andropogoneae-*Loudetia* transition savanna; SS1 and SS2, Andropogoneae shrub savanna; SS2(M), termite mound in Andropogoneae savanna; DS, dry savanna from Niger (after [22], copyright 1995 American Geophysical Union, modified with permission of American Geophysical Union).

also exhibit high denitrification potentials and fairly high nitrate accumulation potentials. The lowest NO fluxes were measured in the Andropogoneae and *Loudetia* savannas (Fig. 14.7). A study at the scale of the African continent [45] confirmed the low potential of NO emission of the Lamto savannas, as in all the humid savannas dominated by *Hyparrhenia* in the Guinean domain.

The great interest of the study by Le Roux et al. [22] was the assessment of the spatial variability of the NO emission rates. In very small areas, dominated by a particular ecotype of *Hyparrhenia diplandra* (see Sect. 17.4), or covered by old epigeous termitaries colonized by subterranean Macrotermitinae, high NO emissions rates were measured (Fig. 14.7). In the termite mounds, the potential of nitrification was high, while that of denitrification was low, suggesting that the NO emitted by the mounds was produced during nitrification rather than denitrification. On the contrary, in the average soil from the forest or savanna, the NO emission seems to occur during the denitrification process [22] but, in the field, is generally strongly limited by the low availability of NO_3^-.

14.8 Impact of fire on the nitrogen cycle

Fire has many impacts on nutrient cycles: (i) directly through the burning of living and dead plant matters that do not enter the soil and, consequently, decrease its organic matter content, and through the stimulation of the functioning of micro-organisms by the supply of ashes to soil (see Chap. 4) and (ii) indirectly through the perturbation of the tree-grass balance and, consequently, the primary production (see Sect. 7.4). The major short and long term effects of biomass burning on savanna functioning is the loss of nutrients which, in the absence of fire, could contribute to the fertility of the soil-plant system. Quantitatively, this loss of mineral resources depends on the extent and intensity of the fire, themselves related to the date of burning and status of vegetation [32, 31]. In Lamto, the fire generally occurs in January, i.e., in the heart of the dry season and around 80% of the savanna soil surface is burned yearly (see Sect. 4.5).

The impact of fire on the tree nitrogen pool largely depends on the date of burning. Early fires occur before the leaf fall and induce a negligible volatilization of nitrogen, while late fires induce an important loss of nitrogen from the leaf litter. In Lamto, Menaut [30] showed the great interannual variability of the leaf fall since 30% to 85% of the total leaf fall can occur after the fire. With an average concentration of nitrogen at 0.7% before leaf shedding [46] and a leaf fall ranging from 90 to 180 $g\,m^{-2}\,y^{-1}$, the fire can therefore induce a loss of 0.63-1.26 $g\,N\,m^{-2}\,y^{-1}$ from the tree leaf compartment in Lamto.

The impact of fire on the grass nitrogen pools is also variable. In the heart of the dry season, all the biomass from the herbaceous layer is destroyed, except some grass stems in the downslope humid environments. At this time, the proportion of burned biomass is close to 100% in some areas and can be estimated at ca. 85% at the scale of the whole Lamto reserve [32]. But, with early fire in December, this proportion can fall to 12%. Of course, many local variations have been observed relative to the force and direction of wind, topography and, mainly, plant water content which directly controls the burning efficiency of fuel [21]. When fire occurs, most of the plants from the herbaceous layer are at the end of their growing period: their nitrogen concentration is low (see Sect. 15.3) and varies weakly between months. Nevertheless, the average input of nitrogen to the atmosphere shows huge interannual variations, ranging from 0.9 to 2.4 $g\,N\,m^{-2}\,y^{-1}$ (Table 14.3). It is always lower in the savannas dominated by *Loudetia simplex* than in that dominated by

Table 14.3. Flux of nitrogen from the grass biomass burning to the atmosphere in the Lamto savannas.

Savanna type	Year	N flux ($g\,N\,m^{-2}$)	% of available N	Source
Loudetia sav.	1972	1.0	90	[53]
	1982	1.7	95	[1]
	1987	0.9	N.A.	[38]
	1995	2.0	N.A.	[15]
Andropogoneae sav.	1982	2.3	93	[1]
	1987	1.4	N.A.	[38]
	1995	2.4	76	[15]

Table 14.4. Total aboveground fuel load (dry matter) before burning and burned fuel, pools of N in the grass layer before burning and volatilized to atmosphere and nitrogen compounds emitted by fire from the grass layer (after [15], with kind permission of Springer Science and Business media).

	Loudetia savanna		Andropogoneae savanna	
	$g\,m^{-2}$	$g\,kg[DM]^{-1}$	$g\,m^{-2}$	$g\,kg[DM]^{-1}$
Fuel load	863	—	955	—
Burned fuel	733	—	620	—
Total N before fire	2.62	—	2.57	—
Total N volatilized	1.98	—	2.38	—
N-NO$_x$	0.39	0.42	0.44	0.33
N-NO$_2$	0.14	0.15	0.16	0.12
N-NH$_3$	0.012	0.01	0.014	0.01
N-N$_2$O	0.07	0.07	0.08	0.06

Hyparrhenia diplandra. Most of this nitrogen is emitted as N_2 (unknown proportion, but probably higher than 50% [15]); 19% is emitted as NO_x, 6.9% as NO_2, 0.6% as NH_3 and 3.9% as N_2O (Table 14.4). Lacaux et al. [21] also measured an emission of 6 mg N g [dry plant matter]$^{-1}$ as NO_3^-.

A part of the nutrients contained in the burned biomass comes back to the soil-plant system through ash deposition. The ashes from living and dead parts of *Loudetia simplex* contain respectively 0.60 and 0.62 mg N g^{-1} ashes and those of *Hyparrhenia diplandra* contain 0.70 and 0.70 mg N g^{-1} [15]. This supply of mineral nutrients to soil, i.e., the supply of nutrients immediately available for micro-organisms, is undoubtedly the most important indirect effect of biomass burning on the soil functioning through a priming effect on soil organic matter mineralization. It is very probable that, in the field, the supply of ashes to soil has a significant impact on both soil respiration and mineral nitrogen accumulation, but for a short time. An unpublished experiment by H.B. Nacro in optimal laboratory conditions for heterotrophic micro-organisms activity showed a small increase in soil respiration by 8-11% 14 days after the supply of *Loudetia simplex* or *Hyparrhenia diplandra* ashes, but a decrease in the concentration of ammonium in soil by 20-55% and in nitrate concentration by 6-13%. The supply of ashes clearly induced an immobilization of nitrogen by the heterotrophs (Fig. 14.8). The supply of ashes to the soil after burning is thus able to increase the yearly mineralization rate of carbon in the savanna and to strengthen during a few days the competitivity of soil micro-organisms for nitrogen at the expense of newly growing plants at the very beginning of the rainy season.

Fire hardly affects soil temperature in the very short term, even in the superficial horizons (see Sect. 4.5). On the contrary, the ash mat remaining several days or weeks after the occurrence of fire induces a long-lasting change in the soil albedo and upper soil temperature. In Lamto, the albedo is

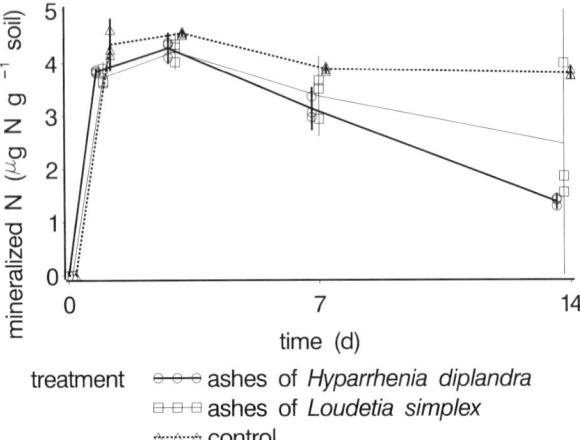

Fig. 14.8. Effect of ashes on the mineralization of nitrogen. Bars represnet one standard error (H.B. Nacro and L. Abbadie, unpublished data).

decreased by 50% in the visible 3 days after the fire occurrence [35, 24]. This blackening of the soil surface induces an increase in soil temperature up to 5°C at 5 cm depth [34]. This heating of soil undoubtedly affects the functioning of micro-organisms, positively by the increase of temperature, negatively by the drying of soil. The resulting effect on carbon and nitrogen cycles is difficult to assess due to the lack of reliable field or laboratory data. But other works published in literature (see, e.g., [4]) seem to indicate a stimulation of the mineralization activities, even if it is difficult to separate the ash effect from the soil climate effect.

14.9 N fluxes associated to grass consumption by animals

The consumption of living (herbivory) and dead (detritivory) grass leaves by animals has been assessed respectively at around 1% and 7% of the primary production, i.e., 24 and 110 g DM m^{-2} (see Chap. 10). For average concentrations of nitrogen of 0.5% in the living and 0.4% in dead grass leaves, the corresponding nitrogen flows are 0.12 and 0.44 g m^{-2} y^{-1}. These values have to be seen as minimum because the living grass consumption by invertebrates is not fully documented and because that of megaherbivores, such as Buffon's kob and buffaloes, has not been fully evaluated (Chap. 10). A part of the ingested plant nitrogen obviously comes back to the soil-plant system, but its impact on primary production is difficult to assess because the spatial pattern of urine and dung deposition is not necessary that of vegetation. Moreover, most of the nitrogen contained in dung is likely lost through leaching, ammonia volatilization or denitrification.

14.10 Conclusion: The input-output balance of nitrogen

The major inputs of nitrogen to the soil-plant system are the dry and wet depositions, symbiotic and non-symbiotic fixation of atmospheric dinitrogen and supply of organic and mineral compounds from leaf leachates. The outputs are the leaching of soil nitrate, denitrification and biogenic emission of NO and emission of various compounds during savanna burning. In Lamto, all these processes have been studied, some of them during several years in several places. It is consequently possible to establish a tentative nitrogen budget of the most common type of savanna in the region of Lamto, i.e., the open shrub savanna dominated by the grass *Hyparrhenia diplandra*. This budget is presented in Table 14.5.

The most important contribution to the accumulation of nitrogen in the soil-plant system is rainfall with $2.25\,\mathrm{g\,N\,m^{-2}\,y^{-1}}$. Two-thirds of this nitrogen is in the organic form (likely linked to silts and clays, spores, pollens), i.e., not immediately available for plant and micro-organisms. The supply of mineral nitrogen to the soil plant system by dry deposition is ca. $0.56\,\mathrm{g\,N\,m^{-2}\,y^{-1}}$ and that of biological fixation is $1.2\,\mathrm{g\,N\,m^{-2}\,y^{-1}}$. In both cases, the supplied nitrogen is immediately available for micro-organisms and plants. Undoubtedly,

Table 14.5. Tentative nitrogen buget in the open shrub Andropogoneae savanna (all figures in $\mathrm{g\,m^{-2}\,y^{-1}}$).

	Inputs	
Dry deposition	ammonium	0.35
	nitrate	0.21
Wet deposition	organic	1.40
	ammonium	0.65
	nitrate	0.20
Dinitrogen fixation	cyanobacteria	0.20
	herbaceous legumes	0.10
	grasses	0.90
Grass cover leaching	organic	0.25
	ammonium	0.20
	nitrate	0.00
Total		**4.46**
	Outputs	
Soil leaching	organic	0.50
	ammonium	0.02
	nitrate	0.00
Denitrification, NO and NO_2 emissions		0.01
Fire	grass leaves	0.90–2.40
	tree leaves	0.73–1.26
	herbaceous legumes	0.10
Animal consumption		0.53
Total		**2.79–4.82**

the dry and wet depositions of mineral nitrogen and the biological fixation of N_2 have to be considered as major potential sources of nitrogen for the soil-plant system in Lamto since they alltogether contribute ca. 47% of the annual grass cover nitrogen requirement (8.45 $g\,N\,m^{-2}\,y^{-1}$) and 0.45% of the soil organic nitrogen pool. The total flow of nitrogen entering the soil plant-system in Lamto is therefore about 4.5 $g\,N\,m^{-2}\,y^{-1}$, i.e., 2% of the pool of organic nitrogen stored in the first 50 cm of the soil (250 $g\,N\,m^{-2}$).

The total output of nitrogen from the Lamto shrub savanna is between 2.79 and 4.82 $g\,N\,m^{-2}\,y^{-1}$. The major flow is biomass burning, as in many African grasslands [41, 32]; it amounts to ca. 60% of the whole nitrogen lost each year. The nitrogen leached from the soil, around 0.5 $g\,N\,m^{-2}\,y^{-1}$, is quite exclusively in the form of organic compounds and likely comes from microorganisms or recently dead plant matter. The denitrification, NO and N_2O emissions are negligible. The emission of ammonia is quite impossible due to soil pH. On the contrary, the flow of nitrogen through animal consumption is important and probably underestimated since our present data do not take into account the grazing by megaherbivores whose densities increase in Lamto for several years. The estimate of the part of this nitrogen recycled within the ecosystem through urine and dung depositions needs further research.

The nitrogen budget in the shrub tree savanna in Lamto is not far from equilibrium. Even if inputs and outputs show important interannual variations and if some data are not accurate, there is not a large discrepancy between the total of inputs and outputs. At the scale of the decade, it can be concluded that the nitrogen cycle is likely balanced in Lamto.

References

1. L. Abbadie. *Contribution à la biogéochimie dans des savanes de Lamto*. Ph.D. thesis, Université de Paris 6, Paris, 1983.
2. L. Abbadie. Evolution saisonnière du stock d'azote dans la strate herbacée d'une savane de Côte d'Ivoire protégée des feux de brousse. *Acta Oecologica, Oecologia Plantarum*, 4(6(20)):323–335, 1985.
3. L. Abbadie and R. Lensi. Carbon and nitrogen mineralization and denitrification in a humid savanna of West Africa. *Acta Oecologica*, 11(5):717–728, 1990.
4. F.O. Adedeji. Effect of fire on soil microbial activity in nigerian southerm guinea savanna. *Revue d'Ecologie et de Biologie du Sol*, 20(4):483–492, 1983.
5. J. Balandreau. Fixation rhizosphérique de l'azote (C_2H_2) en savane de Lamto. *Revue d'Ecologie et de Biologie du Sol*, 13(4):529–544, 1976.
6. J. Balandreau and A. Balandreau. Premières données sur la fixation de l'azote dans le sol de la savane de Lamto. In *Bulletin de Liaison des Chercheurs de Lamto*, pages 15–18. Ecole Normale Supérieure, Paris, 1971.
7. J. Balandreau and G. Villemin. Fixation biologique de l'azote moléoculaire en savane de Lamto (basse Côte d'Ivoire), résultats préliminaires. *Revue d'Ecologie et de Biologie du Sol*, 10(1):25–33, 1973.

8. J. Balandreau, P. Weinhard, G. Rinando, and Y. Dommergues. Influence de l'intensité de l'éclairement de la plante sur la fixation non symbiotique de l'azote dans sa rhizosphère. *Oecologia Plantarum*, 6:341–352, 1971.
9. G.C. Bate. Nitrogen cycling in savanna ecosystem. In F.E. Clark and T. Rosswall, editors, *Terrestrial nitrogen cycles*, volume 33 of *Ecological Bulletins*, pages 463–476. Royal Swedish Academy of Sciences, Stockholm, 1981.
10. G.C. Bate and C. Gunton. Nitrogen cycling in *Burkea* savanna. In B.J. Huntley and B.H. Walker, editors, *Ecology of Tropical Savannas*, volume 42, pages 498–513. Springer-Verlag, Berlin, 1982.
11. F. Bernhardt-Reversat, C. Huttel, and G. Lemée. La forêt sempervirente de basse Côte d'Ivoire. In M. Lamotte and F. Bourlière, editors, *Problèmes d'écologie: Ecosystèmes terrestres*, pages 313–345. Masson et Cie, Paris, 1978.
12. Bijay-Singh., J.C. Ryden, and D.C. Whitehead. Some relationships between denitrification potential and fractions of organic carbon in air-dried and field-moist soils. *Soil Biology and Biochemistry*, 20(5):737–741, 1988.
13. J.R. Burford and J.M. Bremner. Relationship between denitrification capacities of soils and total water soluble and readily decomposable soil organic matter. *Soil Biology and Biochemistry*, 7:389–394, 1975.
14. J. César. *Etude quantitative de la strate herbacée de la savane de Lamto*. Doctorat de 3e cycle, Faculté des Sciences de Paris, Paris, 1971.
15. R. Delmas, J.P. Lacaux, J.C. Menaut, L. Abbadie, X. Le, Roux, G. Helas, and J. Lobert. Nitrogen compound emission from biomass burning in tropical african savanna, FOS/DECAFE 1991 experiment (Lamto, Ivory Coast). *Journal of Atmospheric Chemistry*, 22:175–193, 1995.
16. M.K. Firestone and E.A. Davidson. Microbiological basis of NO and N_2O production and consumption in soil. In M.O. Schimel and D.S. Andreae, editors, *Exchange of trace gases between terrestrial ecosystems and the atmosphere*, pages 7–21. John Wiley & Sons, New York, 1989.
17. P. Hogberg. Soil nutrient availability, root symbioses and tree species composition in tropical Africa: A review. *Journal of Tropical Ecology*, 2(4):359–372, 1986.
18. C. Johansson. Fluxes of NOx above soil and vegetation. In D.S. Schimel and M.O. Andreae, editors, *Exchanges of trace gases between terrestrial ecosystems and the atmosphere*, pages 229–246. John Wiley & Sons, New York, 1989.
19. P. Kaiser. The role of soil micro-organisms in savanna ecosystems. In F. Bourlière, editor, *Tropical savannas*, pages 541–557. Elsevier, Amsterdam, 1983.
20. M. Kellman, J. Hudson, and K. Sanmugadas. Temporal variability in atmospheric nutrient influx to a tropical ecosytem. *Biotropica*, 14(1):1–9, 1982.
21. J.P. Lacaux, J.M. Brustet, R. Delmas, J.C. Menaut, L. Abbadie, B. Bonsang, H. Cachier, J. Baudet, M.O. Andreae, and G. Helas. Biomass burning in the tropical savannas of Ivory Coast: An overview of the field experiment fire of savannas (FOS/DECAFE 91). *Journal of Atmospheric Chemistry*, 22:195–216, 1995.
22. X. Le Roux, L. Abbadie, R. Lensi, and D. Serça. Emission of nitrogen monoxide from African tropical ecosystems: control of emission by soil characteristics in humid and dry savannas of West Africa. *Journal of Geophysical Research*, 100(D11):23,133–23,142, 1995.
23. X. Le Roux and P. Mordelet. Leaf and canopy CO_2 assimilation in a West African humid savanna during the early growing season. *Journal of Tropical Ecology*, 11:529–545, 1995.

24. X. Le Roux, J. Polcher, G. Dedieu, J.C. Menaut, and B.A. Monteny. Radiation exchanges above West African moist savannas: Seasonal patterns and comparison with a GCM simulation. *Journal of Geophysical Research*, 99(12):25,857–25,868, 1994.
25. R. Lensi and A. Chalamet. Denitrification in waterlogged soils: In situ temperature dependent variations. *Soil Biology and Biochemistry*, 14(1):51–55, 1982.
26. R. Lensi, A.M. Domenach, and L. Abbadie. Field study of nitrification and denitrification in a wet savanna of West Africa (Lamto, Côte d'Ivoire). *Plant and Soil*, 147:107–113, 1992.
27. H. Leriche, X. Le Roux, F. Desnoyers, D. Benest, G. Simioni, and L. Abbadie. Grass response to clipping in an African savanna: Testing the grazing optimization hypothesis. *Ecological Applications*, 13:1346–1354, 2003.
28. C. Lescure, P. Gamard, A. Pidello, and R. Lensi. Effect of oxygen on soil denitrifying expression and potential. *Mitteilungen der Deutschen Bodenkundl. Gesellschaft*, 60:313–318, 1990.
29. B.V. Maasdorp. Contribution of associative N_2 fixation (acetylene reduction) in some grassland ecosystems in Zimbabwe. *Soil Biology and Biochemistry*, 19(1):7–12, 1987.
30. J.C. Menaut. Chute de feuilles et apport du sol de litière par les ligneux dans une savane préforestière de Côte d'Ivoire. *Bulletin d'Ecologie*, 5:27–39, 1974.
31. J.C. Menaut. Effets des feux de savane sur le stockage et l'émission du carbone et des éléments tracés. *Sécheresse*, 4:251–263, 1993.
32. J.C. Menaut, L. Abbadie, F. Lavenu, P. Loudjani, and A. Podaire. Biomass burning in West African savannas. In J.S. Levine, editor, *Global biomass buring. Atmospheric, climatic and biospheric implications*, pages 133–142. The MIT Press, Cambridge, MA, 1991.
33. B. Monfort. *Dynamique du renouvellement des populations de deux Papilionoideae herbacées d'une savane brûlée de basse Côte d'Ivoire*. Ph.D. thesis, Université des Sciences et Techniques du Languedoc, Montpellier, 1985.
34. Y. Monnier. *La poussière et la cendre: Paysages, dynamique des formations végétales et stratégies des sociétés en Afrique de l'Ouest*. Agence de coopération culturelle et technique, Paris, 1981.
35. Y. Monnier and A. Cerf. Influence du feu de brousse sur le coefficient d'albedo de la savane de Lamto. *Annales de l'Université d'Abidjan, Série E*, 10:43–57, 1977.
36. J. Pochon and I. Bacvarov. Données préliminaires sur l'activité microbiologiques des sols de la savane de Lamto. *Revue d'Ecologie et de Biologie du Sol*, 10(1):35–43, 1973.
37. R. Portères. *Florule de la savane de Lamto*. Muséum National d'Histoire Naturelle, Paris, 1966.
38. J.P. Puyravaud. *Processus de la production primaire en savanes de Côte d'Ivoire. Mesures de terrain et approche satellitaire*. Ph.D. thesis, Université de Paris 6, Paris, 1990.
39. G. Raud. Etude des algues du sol fixatrices d'azote dans la savane de Lamto. elaboration d'une nouvelle technique. *Annales de l'Université d'Abidjan, Série E*, 9:307–335, 1976.
40. P. De Rham. Recherches sur la minéralisation de l'azote dans les sols des savanes de Lamto. *Revue d'Ecologie et de Biologie du Sol*, 10(2):169–196, 1973.
41. G.P. Robertson and T. Rosswall. Nitrogen in West Africa: The regional cycle. *Ecological Monographs*, 56(1):45–72, 1986.

42. J.C. Roland. Recherches écologiques dans la savane de Lamto: Données préliminaires sur le cycle annuel de la végétation herbacée. *La Terre et la Vie*, 3:228–249, 1967.
43. R. Schaefer. Traits généraux du peuplement endogé des savanes de Lamto : Le peuplement microbien. *Bulletin de Liaison des Chercheurs de Lamto*, S5:39–44, 1974.
44. J. Servant, R. Delmas, J. Racher, and M. Rodriguez. Aspects of the cycle of inorganic nitrogen compounds in the tropical rain forest of the Ivory Coast. *Journal of Atmospheric Chemistry*, 1(4):391–401, 1984.
45. D. Serça, R. Delmas, X. Le Roux, D.A.B. Parsons, M.C. Scholes, L. Abbadie, R. Lensi, O. Ronce, and L. Labroue. Comparison of nitrogen monoxide emissions from several African tropical ecosystems and influence of season and fire. *Global Biogeochemical Cycles*, 12(4):637–651, 1998.
46. G. Simioni, J. Gignoux, X. Le Roux, R. Appé, and Benest D. Spatial and temporal variation in leaf area index, specific leaf area, and leaf nitrogen of two co-occuring savanna tree species. *Tree Physiology*, 24(2):205–216, 2003.
47. G. Simioni, X. Le Roux, J. Gignoux, and A. S. Walcroft. Leaf gas exchange characteristics and water- and nitrogen-use efficiencies of dominant grass and tree species in a West African savanna. *Plant ecology*, 173:233–246, 2004.
48. T.J. Simpkin and W.C. Boyle. The lock of repression by oxygen of the denitrifying enzymes in activated sludge. *Water Research*, 2:201–206, 1988.
49. G. Stanford, R.A. Vander Pol, and S. Dziena. Denitrification rates in relation to total and extractable soil carbon. *Soil Science Society of America Proceedings*, 39:284–289, 1975.
50. R. Söderlund. Dry and wet deposition of nitrogen compounds. *Ecological Bulletin*, 33:123–130, 1981.
51. P. Villecourt. Apports d'azote minéral par la pluie dans la savane de Lamto. *Revue d'Ecologie et de Biologie du Sol*, 12(4):667–680, 1975.
52. P. Villecourt and E. Roose. Charge en azote et en éléments minéraux divers des eaux des pluie, de pluviolessivage et de drainage dans la savane de Lamto. *Revue d'Ecologie et de Biologie du Sol*, 15(1):1–20, 1978.
53. P. Villecourt, W. Schmidt, and J. César. Pertes d'un écosystèmes à l'occasion du feu de brousse. *Revue d'Ecologie et de Biologie du Sol*, 17(1):7–12, 1980.
54. P.M. Vitousek and R.W. Howarth. Nitrogen limitation on land and in the sea: How can it occur? *Biogeochemistry*, 13:87–115, 1991.
55. R.G. Woodmansee, I. Vallis, and J.J. Mott. Grassland nitrogen. In F.E. Clark and T. Rosswall, editors, *Terrestrial Nitrogen Cycles*, volume 33 of *Ecological Bulletins*, pages 649–669. Royal Swedish Academy of Sciences, Stockholm, 1981.

15

Nitrogen Dynamics in the Soil-Plant System

Luc Abbadie and Jean-Christophe Lata

15.1 Introduction

On an annual basis, the concept of primary production limitation by soil nutrient in natural systems is complex because of the very strong link between nutrient availability and plant cover spatial structure and species composition. In old and stable ecosystems, the plant community is probably highly adapted to the level of available nutrients. The latter is itself strongly dependent on plant community structure, and it is not obvious that annual primary production is controlled by the flux of nutrients originating from soil humus mineralization. The spatial pattern of the distribution of mineral nutrients in soil is rarely homogenous: nutrient concentrations vary rapidly at different scales, from meter (presence or absence of trees for example) to micrometer (presence or absence of bacteria). Soil fauna distribution and activity, which modify the physical and chemical environment of micro-organisms, are key factors controlling the soil organic nitrogen storage and mineral nitrogen production. The ability of plants to uptake nutrients is another key factor controlling primary productivity and, in a sense, the real soil fertility is also a function of plant distribution and plant underground architecture. This impact of soil biological characteristics on soil fertility can explain the discrepancy sometimes observed between high productivity and low mineral nutrients content, as in Lamto.

15.2 Nitrogen dynamics in the shrub-tree layer

The nitrogen cycle in the woody stratum has not been extensively studied in Lamto. Nevertheless, data about tree biomass and production [23, 25] and chemical composition of the major species [15, 28, 33] allow giving a rough estimate of the input of nitrogen in shrub and trees. The annual nitrogen requirements of shrubs and trees (Table 15.1) have been calculated by using an average nitrogen concentration at 0.7% in tree leaves at the end of the growing season, 0.2% in branches and stems and 0.3% in roots (big and small

Table 15.1. Yearly production and nitrogen requirement of the woody strata of four savanna types in Lamto. Numbers in $g\,DM\,m^{-2}\,y^{-1}$ for production and $g\,N\,m^{-2}\,y^{-1}$ for nitrogen (data from [23, 25, 15, 28]).

	Grass shrub savanna	Open shrub savanna	Dense shrub savanna	Savanna woodland
Leaf production	43	100	238	533
Leaf N requirement	0.3	0.7	1.7	3.7
Stem and branch production	12	33	42	76
Stem and branch N requirement	0.02	0.07	0.08	0.15
Root production	5	13	23	37
Root N requirement	0.02	0.04	0.07	0.11
Total N requirement	0.34	0.81	1.85	3.96

roots mixed). They range from $0.34\,g\,N\,m^{-2}$ in the grass shrub savanna to $3.96\,g\,N\,m^{-2}$ in the most densely wooded savanna. In the open shrub savanna, the annual requirement in nitrogen of shrubs and trees is $0.81\,g\,N\,m^{-2}$, i.e., 10 times lower than that of herbs and grasses (see Sect. 15.4). Most of it is allocated to leaves and may come back to soil if leaf fall occurs after fire; if leaf fall occurs before fire, most of the $0.7\,g\,N\,m^{-2}$ contained in tree leaves are volatilized and quite completely lost for the ecosystem.

15.3 Nitrogen dynamics in the grass layer

15.3.1 Nitrogen concentrations in the aboveground parts of herbaceous plants

The aerial parts of the main grass species from the *Loudetia simplex* and Andropogoneae open shrub savannas have been harvested monthly in 1981 and beginning of 1982. The living matter has been separated in five categories: *Loudetia simplex*, *Andropogon schirensis*, other grasses, other herbs and legumes on one hand, and *Hyparrhenia diplandra*, *Hyparrhenia smithiana*, other grasses, other herbs and legumes on the other hand. The dead matter has been separated into standing dead matter and litter deposited on soil. The dried plant matter was weighted and its nitrogen concentration measured by the Kjeldahl method [1].

In the *Loudetia* savanna, the concentrations in nitrogen are always slightly higher in *Loudetia simplex* than in *Andropogon schirensis*: 1.92% vs. 1.74% 33 days after fire. A very rapid decrease is then observed until July, when the nitrogen concentrations fall to 0.60% and 0.50%, respectively. The minimum concentrations are measured just before the fire, at 0.31% and 0.23%, respectively. In the Andropogoneae savanna, the trends are similar: the highest nitrogen concentrations are measured in the young leaves appearing just after the fire. A very fast decrease is observed during the beginning of the growing period, then a slow decrease is observed from July to the end of the year. The nitrogen concentrations in the aboveground parts (stems and leaves together) vary from 1.51% to 0.28% in *Hyparrhenia diplandra* and from 2.06% to 0.30% in *Hyparrhenia smithiana*. In the two savannas, the nitrogen concentrations of legumes only vary from 0.77% to 0.79% in January to 1.33% to 1.23% in September.

The variations with time of the nitrogen concentration of the living parts of the grass layer best fit the following equations ($R^2 = 0.89$):

$$\text{Log}(N) = 3.6 - 0.86 \, \text{Log}(t) \text{ for } \textit{Loudetia simplex},$$
$$\text{Log}(N) = 3.1 - 0.73 \, \text{Log}(t) \text{ for } \textit{Andropogon schirensis},$$
$$\text{Log}(N) = 2.9 - 0.74 \, \text{Log}(t) \text{ for } \textit{Hyparrhenia diplandra},$$
$$\text{Log}(N) = 2.6 - 0.63 \, \text{Log}(t) \text{ for } \textit{Hyparrhenia smithiana},$$

where N is the nitrogen content (%) and t the number of days since the last fire.

This regular decline of the nitrogen content in green grass parts is likely related to the aging of the plants: despite a continuous production of new leaves rich in nitrogen all year long [14], the proportion of old leaves, poor in nitrogen, increases with the number of days elapsed since the fire. A covariance analysis and contrast test of the variations with time of these nitrogen concentrations strongly suggest that the needs in nitrogen of individual leaves at different dates differ (Abbadie, unpublished data), even if all the species show the same dilution pathways of nitrogen in increasing mass of structural materials. This indicates that leaves appearing at different stages have a common nitrogen metabolism and sensitivity to climate fluctuations.

A major feature of the Lamto grasses is their very low concentration in nitrogen. All the perennial grasses in Lamto belong to the C_4 photosynthetic pathway. C_4 plants are known to have maximum photosynthetic rate and growth rate per unit of nitrogen and other nutrients higher than in C_3 plants [9, 10]. Le Roux and Mordelet [20] and Simioni et al. [34] showed that Lamto grasses maintain high photosynthetic capacities despite their very low nitrogen concentration. This suggests that these plants have a high efficiency of nitrogen use in the sense of Chapin [11], i.e., a high ratio of the quantity of dry matter produced to quantity of nitrogen assimilated, among the highest recorded to date world-wide [20]. This efficiency is likely the adaptive consequence of the extreme and ancient nutritive soil poverty.

In the standing dead grass matter, the nitrogen concentration varies between 0.62% and 0.25%. It is lower than that of living biomass until September or October, but is higher after this date. The nitrogen concentration in this standing dead matter is always lower than in soil litter, suggesting that nitrogen concentration variations in grass leaves after death result from a two-step process: first, a leaching and decomposition stage until the leaves reach the soil, leading to a net loss of nitrogen; second, a relative enrichment in nitrogen as soon as the leaves touch the soil, resulting of the higher decomposition rate of C than N, or the colonization of leaves by microfungi that translocate nitrogen from soil to leaves.

15.3.2 Nitrogen concentrations in the roots of herbaceous plants

Roots have been studied on the same dates by 1 mm sieving of soils samples collected between 0 and 10 cm, 10 and 30 cm and 30 and 50 cm depths. In these two yearly burned savannas, the nitrogen concentrations in roots significantly decrease with depth within each sampling date. They can reach up to 0.45% between 0 and 10 cm, but never exceed 0.37% between 30 and 50 cm. For all depths, the monthly variations are huge and rapid, but no clear trend with time can bee seen [3, 5]. Nevertheless, it has to be noted that the nitrogen concentrations in roots during the dry season are always among the highest. This is likely related to an intense rhizogenesis after several weeks of drought, as shown for many dicotyledons in temperate areas and for *Panicum maximum* [30, 29], another common grass in West Africa.

15.3.3 Nitrogen concentrations and pools in the grasses of the yearly burned savannas

In the grass savanna, the monthly average pool of nitrogen (calculated over 1 year) is 0.49 g N m^{-2} in *Loudetia simplex*, 0.21 g N m^{-2} in *Andropogon schirensis*, 0.39 g N m^{-2} in other herbaceous species and 0.04 g N m^{-2} in legumes, i.e., 1.13 g N m^{-2} for the living aboveground herbaceous biomass. For the whole aboveground herbaceous layer, the nitrogen pool is minimum just after fire and maximum in July (1.756 g N m^{-2}), and then irregularly and slightly decreases until next fire. The trend is the same in the shrub-tree savanna. The 1-year-average pool of nitrogen is 0.81 g N m^{-2} in *Hyparrhenia diplandra*, 0.14 g N m^{-2} in *Hyparrhenia smithiana*, 0.36 g N m^{-2} in other herbaceous plants and 0.03 g N m^{-2} in legumes. In the whole grass layer, the average pool is 1.34 g N m^{-2}. It rapidly increases until July, remains high until November, and then decreases at 0.96 g N m^{-2} in January, just before fire. The share of *Hyparrhenia diplandra* is always dominant. Some nitrogen also accumulates in dead leaves, mainly in the standing dead matter. It amounts up to 0.72 g N m^{-2} in the grass savanna and to 1.49 g N m^{-2} in the shrub-tree savanna.

In the two savannas, half of the plant nitrogen pool is located in the roots. Most of the root nitrogen is located in the first 10 cm because of the N concentration and mass of roots in the superficial soil layer. The average monthly pool of nitrogen in the first 50 cm is 3.72 g N m^{-2} in the *Loudetia* savanna and 1.70 g N m^{-2} in the Andropogoneae savanna. Several maxima are observed during the year at different depths; they are generally related to particular soil conditions, hard dryness or, on the contrary, flooding and are likely the result of an intense rhizogenesis.

15.3.4 Nitrogen concentrations and pools in the grasses of unburned savanna

The measurement of the nitrogen content of the herbaceous layer in a shrub-tree savanna protected from fire for 20 years (in 1981 and 1982) has been performed at the same dates as in the two above burned savannas [2]. The protected area had a high tree density, with savanna but also forest tree species coming from the Bandama riparian forest [35, 36, 24]. The grass biomass sampling was performed in particular zones where tree density had remained low in order to get data comparable to that obtained in the regularly burned *Loudetia* and shrub-tree savanna.

In all the herbaceous species, except the legumes, the nitrogen concentration in the living phytomass follows the same trend. Two maxima are observed, in May during the period of active growth (0.7% on average) and in March when plants begin a new vegetation cycle (0.9-1.3%). The lowest values are recorded in January (0.3-0.4%), i.e., when growth is almost completely stopped. In dead plant matter (all species mixed), the nitrogen concentration varies from 0.3% to 0.6%. At each date, the concentration in soil litter is slightly higher than that in standing dead matter.

The average monthly quantity of nitrogen stored in the alive aerial parts of herbaceous plants in unburned savanna is 1.11 $g\,N\,m^{-2}$ in *Loudetia simplex*, 0.43 $g\,N\,m^{-2}$ in *Andropogon canaliculatus*, 0.13 $g\,N\,m^{-2}$ in *Imperata cylindrica*, 0.19 $g\,N\,m^{-2}$ in the other herbaceous plants and 0.02 $g\,N\,m^{-2}$ in legumes. These pools vary with season, from 0.85 to 2.47 $g\,N\,m^{-2}$ in the whole living layer. The total dead matter contains 2.57 $g\,N\,m^{-2}$ on average, 1.35 $g\,N\,m^{-2}$ in the standing dead matter and 1.22 $g\,N\,m^{-2}$ in the litter accumulated on soil surface. In the whole herbaceous stratum, the pool of nitrogen is 4.45 $g\,N\,m^{-2}$ on average. It shows only weak variations between months, at a maximum of 1.7-fold. This weak variation with time of the quantity of aboveground plant nitrogen is characteristic of the unburned savanna. It is clearly the consequence of two phenomena: the important accumulation of dead matter that contributes to 30% of the total aerial phytomass and the constant presence of living phytomass that is never destroyed by fire, like in the regularly burned savannas.

In roots, the same trends are observed in the unburned as in burned savannas. The concentration of nitrogen is higher in superficial (0.4%) than in deeper roots (0.3% between 10 and 50 cm depth). The monthly variations of nitrogen concentrations in roots are rapid and wide at all depths, but the highest concentrations are always recorded during the dry season, like in unburned savannas. More than half of the root nitrogen pool is located in the first 10 cm that contain 1.7 $g\,N\,m^{-2}$ on average. The two following layers (10-30 and 30-50 cm depths) contain respectively 1.05 and 0.23 $g\,N\,m^{-2}$ on average.

15.4 Annual nitrogen requirements of grasses

The direct measurement of nitrogen flows in field conditions is difficult and, most often, the net exchange of nitrogen between two compartments is estimated by the comparison of compartment sizes between two time steps. Practically, the input of nitrogen in grass shoots and roots can be estimated as the summation of positive monthly increments of nitrogen pools in shoots and roots. For the aerial parts of grasses, this method was used for the living matter of individual species or groups of species (*Loudetia simplex*, *Andropogon schirensis*, *Hyparrhenia diplandra*, *Hyparrhenia smithiana*, other grasses, legume community), total living matter, total dead matter and total grass matter. For roots, the method was used for the three soil layers separately and for the whole soil (50 cm depth).

The different estimates are given in Table 15.2. In the grass savanna, the yearly inputs of nitrogen to the aboveground parts and roots of grasses range from 3.6 to 5.0 $g\,N\,m^{-2}$ and 3.9 to 5.1 $g\,N\,m^{-2}$, respectively, while they range from 4.9 to 5.5 $g\,N\,m^{-2}$ and from 2.7 to 3.8 $g\,N\,m^{-2}$ in the shrub-tree savanna. The total input of nitrogen entering the grass layer is therefore 7.5-10.1 $g\,N\,m^{-2}$ in the grass savanna and 7.6-9.3 $g\,N\,m^{-2}$ in the shrub tree savanna. These values are probably underestimated since the very fine roots, with high nitrogen concentrations and rapid turnovers, generally escape to underground biomass sampling. However, the grass and shrub savannas do not seem to differ in their annual nitrogen requirements in spite of their different soil characteristics, plant species composition and primary production rate.

Table 15.2. Nitrogen requirements in $g\,N\,m^{-2}\,y^{-1}$ of the grass layer in two savanna types in Lamto (computed from data in [2]).

Loudetia savanna		Andropogoneae savanna	
Loudetia simplex	2.265	*Hyparrhenia diplandra*	2.920
Andropogon schirensis	0.794	*Hyparrhenia smithiana*	0.583
Other species	1.693	Other species	1.804
Legumes	0.255	Legumes	0.104
Total	5.007	Total	5.411
Biomass	2.811	Biomass	3.243
Necromass	0.760	Necromass	1.626
Total	3.571	Total	4.869
Total phytomass	4.256	Total phytomass	5.539
0-10 cm roots	2.764	0-10 cm roots	1.855
10-30 cm roots	1.375	10-30 cm roots	1.008
30-50 cm roots	0.928	30-50 cm roots	0.890
Total	5.067	Total	3.753
Whole soil roots	3.933	Whole soil roots	2.683

15.5 Origin of grass nitrogen

The annual requirement of the grass layer in nitrogen has been evaluated in Lamto at around 8.0-9.0 $g\,N\,m^{-2}$ (see Sect. 15.4). The mineralization of soil organic matter provides approximately 0.2-0.5 $g\,N\,m^{-2}$, mainly in the form of ammonium (see Sect. 15.6). Even if the soil invertebrates, such as earthworms and termites, locally stimulate nitrogen mineralization, it is clear that soil nitrogen mineralization is too low to meet the plant cover requirements. Other possible sources of nitrogen for plants are the wet and dry deposition and the symbiotic and non-symbiotic fixation of atmospheric N_2, but they are also too weak to fulfill the difference. In order to identify the origin of the nitrogen assimilated by the grasses in Lamto savannas, integrating both its spatial and temporal variability, a study of the natural abundance of ^{15}N has been conducted in soils and plants [7]. Indeed, the isotopic fractionation of nitrogen occurring during the uptake and assimilation processes is low and the natural abundance of ^{15}N in the plant is therefore close to that in sources. A simple comparison of the abundance of ^{15}N in the plants and in the possible sources thus allows identifying the origin of the nitrogen assimilated by plants.

In Lamto, the natural abundance of ^{15}N in soil organic matter is significantly higher ($\delta^{15}N = 5.3‰$) than that of molecular nitrogen in air, whatever the depth. The $\delta^{15}N$ of the native mineral nitrogen (mineral nitrogen extracted from soil without incubations) is 7.2‰ on average, i.e., close to that of soil organic nitrogen. The natural abundance of ^{15}N in rainwater has not been measured, but many data in literature have shown that it is equal or smaller than in atmospheric N_2. The isotopic composition of nitrogen in plants is very different from that in soil. In legumes, the $\delta^{15}N$ is –2.0‰ on average, i.e., close to that of atmospheric N_2 and very typical of nitrogen fixing plants. In the grasses, the average $\delta^{15}N$ is also close to that of molecular nitrogen, i.e., 1.3‰. A particular study in the leaves of *Hyparrhenia smithiana* has shown that it is constant with time, i.e., with season and age of leaves.

If the Lamto grasses uptook and assimilated ammonium or nitrate coming from the soil, they should have a $\delta^{15}N$ close to that of soil nitrogen, i.e., 5-7‰. The similarity of $\delta^{15}N$ between grasses and fixing legumes could indicate a strong contribution of N_2 fixation to grass nitrogen nutrition. The isotopic biogeochemistry does not provide any argument to reject this interpretation. However, all the available data in literature make it difficult to accept since the contribution of non-symbiotic fixation in the grass rhizosphere never exceeds one-third of the annual requirements of the grass layer. Moreover, nitrogen fixation in the grass rhizosphere in Lamto has been estimated at 0.9-1.3 $g\,N\,m^{-2}\,y^{-1}$ in [8]: it is not large enough to meet the needs of the grass layer, i.e., to label on annual basis the isotopic composition of nitrogen in grasses as in typical fixing legumes. Another nitrogen source has to be found. Abbadie et al. [7] hypothesized that it could be the dead root litter. The dead roots have a natural abundance of nitrogen close to that of living roots and, on a yearly basis, they should provide mineral nitrogen with a $\delta^{15}N$ close to

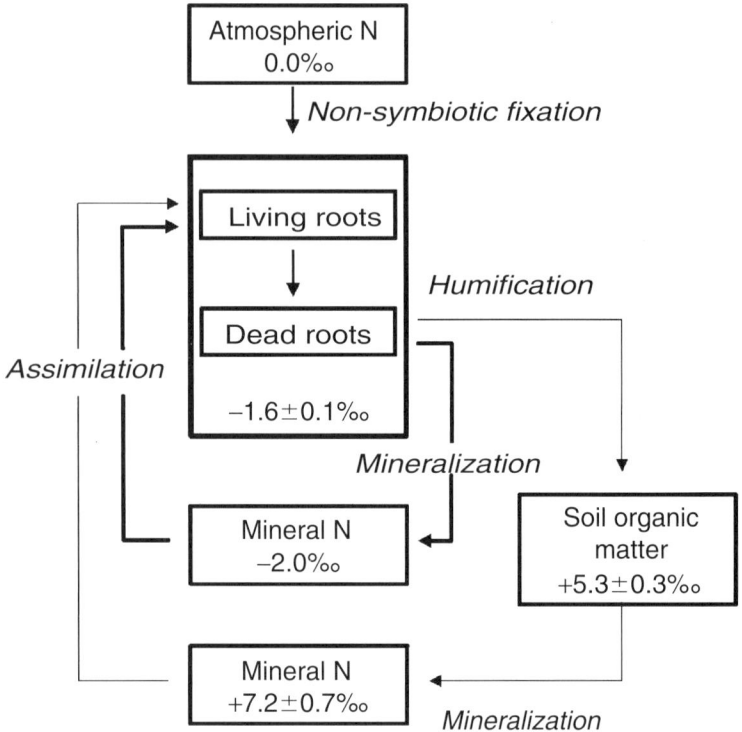

Fig. 15.1. The fast nitrogen recycling of savanna grasses as inferred from δ^{15}N measurements (data from [7]).

−1‰ or −2‰ in the steady state. In other words, there should be a short circuit of nitrogen circulating directly from dead roots to living roots, without significant contribution of the mineral nitrogen coming from the mineralization of soil organic matter (Fig. 15.1). The isotopic composition of the grass nitrogen would not reflect a strong grass rhizospheric fixation of nitrogen, but the closing up of nitrogen cycle in the soil-plant system.

The spatial structure of the grass layer is sufficient to explain this recycling of nitrogen from dead to living roots. In Lamto, 99.9% of the grass biomass is made of tufted perennial species (Chap. 5). At the scale of the square meter, the grass cover is very discontinuous since the grass bases cover only 10% of the soil, i.e., 90% of the soil surface is bare. The distribution of roots in soil is consequently very heterogeneous in space. It has been studied in 1.21 m² plots to a 50 cm depth. Soil cores were collected with a 4 cm diameter soil sampler and washed on 1000 and 250 µm sieves in order to collect roots that were dried at 80°C and weighted. Figure 15.2 shows that the root density is highly variable with depth and, above all, with the distance to tufts, the

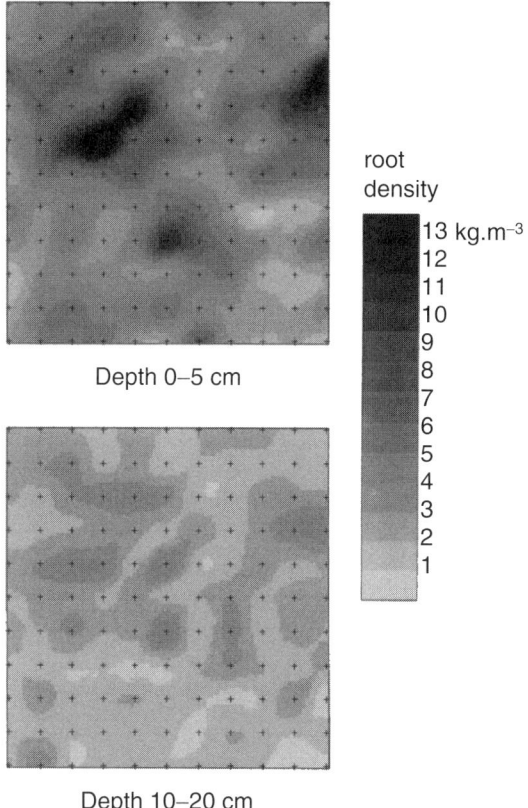

Fig. 15.2. Spatial variations of root density at the scale of the grass tuft. +'s indicate the positions of sampling points, every 10 cm on a 1.10×1.10 m plot (after [18]).

highest densities being just under the tufts whereas those between tufts are 10-fold lower.

This aggregative structure of grass roots means that dead roots, producing mineral nitrogen through decomposition, are spatially close to living roots uptaking nitrogen. One can therefore hypothesize that a molecule of mineral nitrogen produced by dead root litter has a high probability to be uptaken by a proximate living root, while the probability of a molecule of mineral nitrogen originating from humus to be absorbed by low density roots is weak. Moreover, the low microbial production of mineral nitrogen from humus could decrease, again, the contribution of soil humus to plant nutrition. This hypothesis has been tested by the measurements of the fine variations of natural abundance of nitrogen-15 in roots according to their density. Different $\delta^{15}N$ were expected between the roots sampled under the tufts and that sampled between the tufts, because the roots under tufts were expected to assimilate nitrogen coming almost exclusively from dead roots while those between tufts were supposed to partly feed on nitrogen coming from humus mineralization. Indeed, there is a significant difference of 2.2‰ between the δ^{15}N of high density and low density roots. The δ^{15}N of high density roots is 0.1‰ on average, i.e., close to the δ^{15}N of roots theoretically feeding on root litter (–1‰ to –2‰), whereas that of low density roots is 2.3‰, i.e., close to the theoretical value of humus nitrogen (5‰ to 7‰).

Two nitrogen cycles therefore exist in Lamto savannas, occurring in different places in soil: one is based on root litter mineralization and the other on humus mineralization. The relative contribution of these two cycles to the nutrition of a given root depends on the distance of this root to the tuft. The contribution of the root litter system is maximum just under the tufts while that of the humus system is maximum far from the tuft center. To the scale of the whole plant, the litter system supplies most of the nitrogen required by grasses because 80% or 90% of roots are concentrated in a tenth of the total volume of soil, just under the tufts. A real recycling of nitrogen thus occurs, directly from dead roots to alive roots, without significant contribution of soil humus to plant nitrogen nutrition. In other words, the nitrogen fertility and, more generally, the mineral nutrient fertility of the savanna soil depend less on the quantities of mineral nitrogen available than on the spatial pathway of the release of mineral nitrogen by microbial activity. In this sense, the underground grass architecture, i.e., the spatial distribution of roots, must be considered as a key factor of primary production rate.

15.6 The transformations of nitrogen in soil

15.6.1 The accumulation of organic nitrogen in soil

As in most terrestrial ecosystems, the major part of the nitrogen from the soil-plant system in Lamto is located in the soil at around 98%. This nitrogen is almost exclusively under the form of organic compounds; the concentration of the soil in both ammonium and nitrate is generally under 1 to 2 mg N g^{-1} soil. The ability of the soils from Lamto to accumulate organic matter is low because of their sandy texture. Their concentration in total organic nitrogen in the first 10 cm is generally below 0.056% in the upslope savannas dominated by *Hyparrhenia diplandra* and 0.044% in the downslope savannas dominated by *Loudetia simplex*. These concentrations rapidly decrease with depth at 0.037% in the Andropogoneae savannas and 0.026% in the *Loudetia* savannas between 30 and 50 cm (Table 15.3). In small areas where black earth has developed on amphibolitic bedrock, the soil organic nitrogen concentration amounts up to 0.081%, 0.065% and 0.058% between respectively 0 and 10, 10 and 30 and 30 and 50 cm depth [4].

The total quantity of organic nitrogen stored in these soils is low and generally does not exceed 280 g m^{-2} for the 0-50 cm layer, as shown in Table 15.3

Table 15.3. Total organic nitrogen concentrations and pools in soil, and proportion of the organic nitrogen in the light (density below 2) soil fractions (after [4], with permission of the Editions Universitaires de Côte d'Ivoire).

Savanna type	Depth (cm)	Total organic N (% N)	(g N m^{-2})	N in light fractions (% of total pool)
Andropogoneae savanna	0–10	0.056	85	8.4
	10–30	0.041	101	15.2
	30–50	0.037	64	7.7
Loudetia simplex savanna	0–10	0.044	79	15.2
	10–30	0.033	101	9.7
	30–50	0.026	92	9.3
Black earth savanna	0–10	0.081	106	7.7
	10–30	0.065	112	5.6
	30–50	0.058	61	5.4

(the calculations of the pools take into consideration the soil volume filled by the particles larger than 2 mm). Most of these pools are associated to the heavy fractions of the soil (density above 2), i.e., most of the organic nitrogen in the Lamto soils is under the form of humified compounds, physically linked to silt and clay (see Sect. 11.3). The nitrogen contained in partly decomposed plant residues (density below 2) contributes to only 4-15% of the total organic nitrogen pools and is, undoubtedly, the major source of short term metabolizable nitrogen for the micro-organisms. This share is generally higher in the soils under *Loudetia simplex* than under *Hyparrhenia diplandra* [4].

15.6.2 The production of mineral nitrogen in soil

The availability of mineral nitrogen in soil is a key determinant of the primary production. In tropical ecosystems, the nitrogen is generally considered as the first limiting factor of plant productivity. De Rham [31, 32] studied for almost 2 years the field production of ammonium and nitrate in different soils under different plant covers: savannas dominated by *Hyparrhenia diplandra* on plateau or on slope, a savanna dominated by *Loudetia simplex* on a plateau with temporary hydromorphy, a savanna unburned for 3 years. The soil concentration in NH_4^+ and NO_3^- was measured with three methods: (i) direct field measurements on fresh soil cores collected between 0 and 5 cm depth, (ii) after 6 weeks field incubations, (iii) after 6 weeks laboratory incubations. The field measurements on fresh soil cores were performed in order to assess the natural balance between mineralization and plant uptake. The actual net mineralization was assessed as the difference between the quantities of ammonium and nitrate measured on the fresh soils and those measured after the 6 weeks of field incubations. The potential net mineralization (under constant soil humidity and temperature) was assessed as the difference between the quantities of ammonium and nitrate measured on the fresh soils and those measured after the 6 weeks of laboratory incubations.

In the yearly burned savannas, the concentrations in the fresh soil samples were generally below 1 μg N g^{-1} dry soil, whatever the season, with ammonium as the dominant form. The accumulations of mineral nitrogen in the field as in the laboratory were extremely low, at the detection limit of the methods used. De Rham [32] underscored in his paper that the traces of ammonium were more constant than that of nitrate. In the unburned savannas, the results were the same, except for a peak at 2 μmg N g^{-1} dry soil of nitrate observed on one date after the laboratory incubation. In these unburned soils, the proportion of nitrate was higher than in burned soils and, sometimes, the quantity of nitrate exceeded that of ammonium. De Rham [32] suggested that this nitrification could be due to the accumulation of litter and the appearance of a forest micro-climate resulting from the growing of forest tree species allowed by the lack of fire.

The ammonification and nitrification were also measured in the soils from a small plateau forest and a riparian forest along the Bandama River [31]. The concentrations in mineral nitrogen were low as in savanna (below 2 μg N g^{-1} dry soil). But, the actual and potential accumulation after 6 weeks incubations were much higher than in savanna: for the plateau forest, the maximums were measured at 1.5 μg N g^{-1} dry soil of ammonium and 3 μg N g^{-1} dry soil of nitrate in the field and 0.5 and 3.5, respectively, in the laboratory. For the riparian forest, the maximum field accumulation was 4 μg N g^{-1} dry soil of ammonium and 4 μg N g^{-1} dry soil of nitrate, and 0.5 μg N g^{-1} dry soil and 4 μg N g^{-1} dry soil of, respectively, ammonium and nitrate in the laboratory.

A great interest of these measurements in the forests was to point out two important characteristics of the savanna soils from Lamto: (i) their very low potential of production and accumulation of mineral nitrogen and (ii) their almost absolute lack of nitrate production. A more recent study [6] performed in controlled laboratory conditions (constant temperature at 28°C and water content corresponding to 80% of field capacity in order to get the maximum activity of heterotrophic micro-organisms) confirmed the extremely low potential mineralization rate as soon as the grasses dominate over trees (Fig. 15.3). But, behind these general features, many local variations can occur due to the small scale heterogeneity of soil due to soil fauna and roots. Mordelet et al. [26] showed a higher potential accumulation of mineral nitrogen in the soils collected under tree clumps than in those collected out of the area influenced by trees. In both situations, the mineralization rate was higher than the nitrification rate, but the ratio ammonium : nitrate was lower under tree clumps than outside, underscoring, once more, the positive influence of trees on nitrification (see Sect. 8.2).

Using his field data, De Rham [32] estimated the annual production of mineral nitrogen in the soils from Lamto. The estimate is 0.2 to 0.5 g N m^{-2} y^{-1} in savannas, 3.0 g N m^{-2} y^{-1} in the plateau forest and 7.0 g N m^{-2} y^{-1} in the

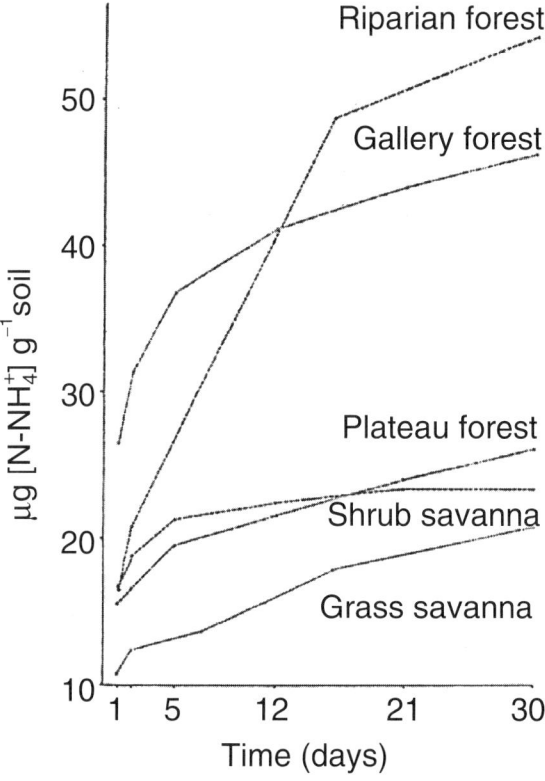

Fig. 15.3. Potential ammonium accumulation in the surface soils (0-10 cm) of the Lamto catena during 30 days of incubation (reprinted from [6], copyright (1990) with permission of Elsevier).

riparian forest. These differences cannot be explained by the physical and chemical characteristics of the soils. They are all sandy and slightly acid and the biodegradability of their organic matter is comparable, except in the Andropogoneae savanna, where it is two times lower [6]. The differences in their organic nitrogen contents are consequently not large enough to explain why the productions of mineral nitrogen in the forests are 6 to 35 times higher than in savannas. The dynamics of soil water and temperature could be important because they show daily variations lower in the forest than in the savanna soils. It is very probable that forest soils are closest to optimal conditions for micro-organisms during a longer time than savanna soils and could be consequently able to induce a highest production of mineral nutrients than in savanna soils. It is also possible that the species composition of the grass layer plays a role in the regulation of soil micro-organisms activity, especially nitrifying bacteria (see below).

It has to be noted that the production of mineral nitrogen in the topsoil does not meet the plant cover needs, especially in the savannas dominated by *Hyparrhenia diplandra* where the annual requirement in nitrogen for the alone grass layer is 8.0-9.0 $g\,N\,m^{-2}\,y^{-1}$ (see Sect. 15.4). Other sources of nitrogen are necessarily exploited by the grasses such as the nitrogen brought by the rainfall and dust deposition, the atmospheric N_2 fixed by the non-symbiotic bacteria living in the grass rhizosphere, the nitrogen coming from the dead roots decomposition or that produced by the mineralization of the deep soil organic matter. This latter has probably a weak impact on the nitrogen nutrition of plants due to (i) the weakness of the mineralization activity in the deep layers of the soil due to the lack of oxygen and the high degree of organic matter polymerisation (in optimal laboratory conditions, the accumulation of mineral nitrogen is 6-8 times lower in the soil collected between 40 and 50 cm than in that collected between 0 and 10 cm) and (ii) the low density of roots beyond 30 cm depth, which does not allow an efficient uptake of the deep nutrients. The contribution of the root litter decomposition to the annual requirements of the grass layer in nitrogen is obviously important. The study of the isotopic composition of nitrogen in the different compartments of the nitrogen cycle in Lamto [7] has shown that it is quantitatively the first (see Sect. 15.5). Under the hypothesis of a turnover of once a year, a root production at 1400 $g\,m^{-2}\,y^{-1}$ and an average concentration of nitrogen in roots at 0.35%, the input of mineral nitrogen to the soil by dead grass roots can be estimated to ca. 4.9 $g\,N\,m^{-2}\,y^{-1}$, i.e., at 50% of the annual nitrogen requirements of the grass layer.

15.6.3 Nitrification

In the soils of many ecosystems world-wide, the ammonium produced by the ammonifying bacteria, but also by fungi and animals, is used as an energy source by bacteria autotrophic for carbon: the nitrifiers. The oxidation of ammonium is a two-step process: ammonium is first transformed into nitrite and then into nitrate, but none of the micro-organisms can perform the two steps. The rate of nitrification is generally high and most of the soils are richer in nitrate than in ammonium. Many plants grow preferentially on nitrate. The humid savannas, at least those of Africa, seem to escape the rule: studies conducted in Ghana [27] and Zimbabwe [22] have revealed the lack of nitrification as soon as the grasses from the Andropogoneae group are dominant. In Lamto, De Rham [32] showed the extreme weakness of the nitrate contents of the savanna soils since 1973.

These results have been confirmed later [21] with laboratory incubations, in optimal conditions of temperature and soil water availability and with the supply of non-limiting quantities of ammonium. The work was performed on soils from two locations in savanna: a grass savanna dominated by *Loudetia simplex* and an open tree savanna dominated by *Hyparrhenia diplandra*. Samples were taken within grass tufts and outside tufts, to control for a possible

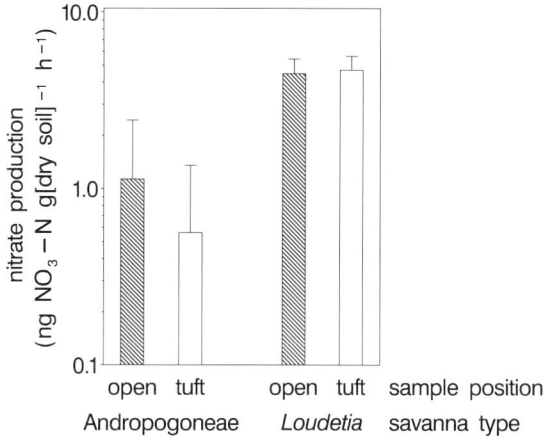

Fig. 15.4. Potential nitrification in two savanna types, measured within and outside grass tufts (after [21], with kind permission of Springer Science and Business media).

rhizosphere effect. The results showed a very low nitrifying activity in the savanna soils, with a significant 2-fold decrease in nitrification potential under *Hyparrhenia* tufts (Fig. 15.4). In forest soils, however, nitrification levels were higher than in savanna soils. These soils are close to each other by their physical and chemical characteristics. They are all sandy (clay content below 7%), except the riparian forest soil, and are consequently well ventilated. Their organic nitrogen concentration varies from onefold to threefold and they are all able to produce the ammonium necessary for ammonifiers. The physical and chemical characteristics of the soils are obviously not strongly implied in the control of the potentials of nitrification in Lamto.

This work therefore strongly supports the alternative hypothesis of a direct control of nitrification by plant cover. In the same paper, Lensi et al. [21] presented results about the potential of nitrification under a cover of *Hyparrhenia diplandra* and a cover of *Loudetia simplex*. A potential of nitrification, quite weak but significant, was observed under and between the *Loudetia* tufts, while, under and between the *Hyparrhenia* tufts, the measured potentials were to the detection threshold. Moreover, the lack of nitrification in the *Hyparrhenia* savanna was less severe between the grass tufts than below. Mordelet et al. [26] gave informations congruent with these data: in soil samples collected under tree clumps, where the density of *Hyparrhenia diplandra* is lower than in open savanna, the measured accumulation of nitrate + nitrite was higher than in the samples collected far from tree clumps.

Three mechanisms could be involved in this control of *Hyparrhenia* on the intensity of the nitrification process: (i) particular soil physical and chemical conditions making the soil unfavorable for the nitrifying organisms; (ii) a strong ability of the grasses to uptake soil ammonium, making this substrate too rare to induce a significant nitrifying activity and (iii) the production of chemical compounds by grass roots able to inhibit the activity of nitrifiers. The discovery of an area of more than 2.5 ha in the south of the reserve [19], which surprisingly exhibits a potential of nitrification 240 times higher on average than that usually measured elsewhere in the Lamto savanna [16], allowed one to both identify the mechanism of the control of *Hyparrhenia* on nitrification and assess the possible impact of the form of mineral nitrogen, either ammonium or nitrate, on the ecosystem functioning (productivity). It has to be noticed that the grass species composition of this zone is not different from that of the standard non-nitrifying savanna since the Andropogoneae group dominates, contributing from 85% to 89% of the total grass basal cover.

A cross transplantation of adult *Hyparrhenia diplandra* tussocks from both high and low nitrification soils, with individuals replanted on their own soil allowed testing the link between plant cover and nitrification potential [17]. After only 1 year, the potential of nitrification under the tufts transplanted from the high nitrifying zone to the low nitrifying zone was completely restored and was not different from that under the individuals replanted on their own nitrifying soil. Under the tussocks transplanted from the low nitrifying zone to the high nitrifying zone, the potential of nitrification was reduced threefold compared to that under the plants replanted on their own nitrifying soil. Under the tussocks replanted on their own non-nitrifying soil, the potential of nitrification was closed to that usually measured in the non-nitrifying zone, showing that the protocol itself was not responsible for the observed changes.

This experiment clearly demonstrates the control of nitrification by *Hyparrhenia diplandra* itself, independently of environment or soil conditions. But, it also shows that the depressive effect of *Hyparrhenia* on nitrification cannot be considered as a general feature. A comparative study of the potential nitrate reductase activity in the leaves of tussocks cultivated in a greenhouse under identical conditions (with non-limiting nitrate supply) from seeds sampled in nitrifying and non-nitrifying zones, confirmed that the *Hyparrhenia*-nitrifiers interactions must be considered at a plant infra-specific level [16]. During the entire growth period, the grass leaves originated from the nitrifying zone showed a significantly higher nitrate reductase activity (between 1.5 and 4.5 times) than those originated from the non-nitrifying zone, strongly suggesting that the two *Hyparrhenia diplandra* populations have long-term adaptations to different nitrogen cycles, with high and low nitrification rates in soil.

The hypothesis of a role of *Hyparrhenia* roots in the control of nitrification in Lamto was supported by a study of the variations of both grass root density and potential nitrification in small volumes of soils (cubes of 1000 cm^3) [17]. A positive significant correlation was observed in both 0-10 and 10-20 cm layers ($R^2 = 0.73$ and 0.33, respectively) in the high nitrification soil (Fig. 15.5).

This result was surprising because nitrification is generally considered not affected by roots since nitrifiers are autotrophic to carbon. It could indicate that nitrifying micro-organisms in Lamto are mixotrophic or heterotrophic to carbon [12]. It could also indicate a regulation of the nitrifier community

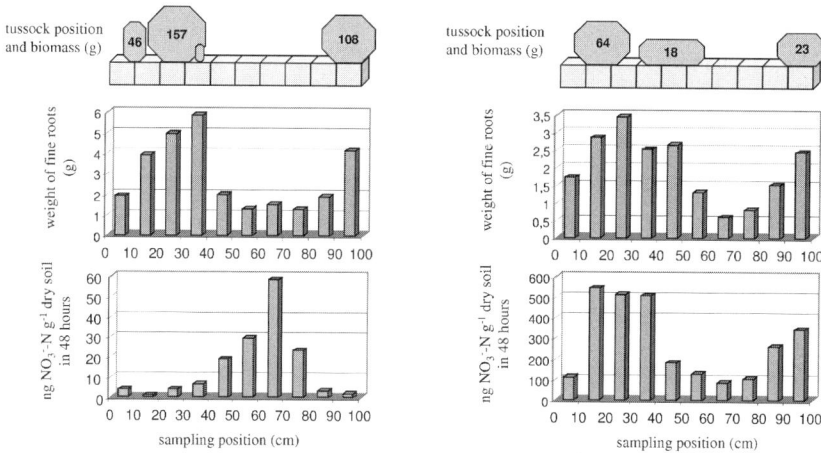

Fig. 15.5. Transects in the low nitrification site (left) and the high nitrification site (right) for the topsoil (0-10 cm). From top to bottom: tussock position and biomass in grams of dry material (each cube is $10 \times 10 \times 10$ cm); total fine roots dry weight; nitrification potential estimated from 48 hours aerobic incubation (reprinted from [17], with permission of the Royal Society).

density by the availability of the ammonium . Because ammonium mainly originates from the dead root decomposition, it should be more abundant and should induce a higher nitrification in high than in low root density zones. This positive correlation does not support the hypothesis of a strong ability of grasses to uptake soil ammonium compared to that of nitrifiers. Indeed, under this hypothesis, the correlation between the potential of nitrification and root density should be negative as a result of the low availability of ammonium in the high root density zones compared to that of low root density zones. In the common non-nitrifying soil, a negative correlation was observed in both 0-10 and 10-20 cm layers ($R^2 = 0.67$ and 0.24, respectively), showing that the intensity of the inhibition of nitrification increases with root density and supporting the Meiklejohn [22] and Munro [27] hypothesis of an allelopathic effect of organic compounds produced by the roots of Andropogoneae species on nitrifiers [17].

What could be the consequences of the lack of nitrification on the ecosystem functioning and, particularly, its productivity? It is well known that nitrate is a very labile form of mineral nitrogen: it is not fixed on clay and organic matter due to its negative charge, and it is consequently easily leached. It can also be lost under the form of gaseous compounds through the process of denitrification. On the contrary, ammonium is well adsorbed on clay and organic matter and is generally not leached. Moreover, in acid soils such as in

Lamto, the ammonia volatilization does not occur. In other words, the balance between nitrate and ammonium in the soil solution partly controls the conservation of mineral nitrogen in the soil-plant system and it can be hypothesized that primary production is less limited by nitrogen when this latter is in the ammonium form rather than in nitrate form. It could be shown in Lamto that aboveground biomass and shoot-to-root ratio are 2-fold lower and grass basal cover 5-fold lower in high than in low nitrification savanna [16] and that tussocks height, basal diameter and leaf number are also lower in high than in low nitrification sites [16]. These data do not evidence a control of nitrification rate on grass growth, but they clearly indicate that low productivity is associated with high nitrification and high productivity with low nitrification. This could result from different intrinsic structural and functional characteristics of the two *Hyparrhenia diplandra* populations and/or different availability of the mineral nitrogen. Only further investigations will allow identifying the involved mechanisms.

15.7 Conclusion: The savanna, a system that retains nitrogen and mineral nutrients

In Lamto, the alternation of dry and humid episodes, typical of the Quaternary, added to the slow uprising of the Guinean mountain chain, have given relatively young soils, sandy or very sandy, with a low capacity of hydrosoluble mineral and organic compounds retention (see Chap. 2). This scarcity of the soil mineral nutrients, inherited from the past, is reinforced by the present climate conditions. The high temperatures stimulate the activity of micro-organisms that rapidly degrade recently dead plant matter, i.e., leaves and root litter. However, the mineral nutrients produced during decomposition are seldom adsorbed on clay and organic compounds. Because Lamto soils have low contents in clay and organic matter, mineral nutrients are easily leached and carried to deep soil layers, except if they are rapidly immobilized by micro-organisms of plants. Finally, the Lamto soils have an "extreme food poverty" as said by Delmas [13] about the tropical ferruginous soils, the most common in the region.

The impact of annual fire on organic nitrogen and carbon contents of savanna soil is more difficult to assess. The loss of nitrogen resulting from the burning of aerial grass biomass and tree leaves on soil is largely balanced by rainfall and dust deposition and by the fixation of atmospheric nitrogen by grass rhizospheric bacteria. However, fire destroys more than half of the aerial primary production that, in its absence, would have been partially incorporated to soil and would have thus contributed to organic nitrogen accumulation in soil. The low content of Lamto soils in organic matter and nitrogen compounds is therefore partially the indisputable result of fire. Finally, fire, climate and soil texture induce a low availability in mineral nutrients for plants, notably in mineral nitrogen. A low primary productivity could therefore be expected because of a strong limitation of plant growing by nitrogen. Surprisingly, the opposite situation is observed and the primary production in Lamto is among the highest world-wide (Chap. 7). This performance results from two types of mechanisms: the spatial organization of the plant cover and the lack of nitrification.

On average, the first 50 cm of the tree savanna soil contain 3 kg m^{-2} of non-particulate organic matter (250 g N m^{-2}) while living and recently dead roots contain only 500-600 g m^{-2} of organic matter (1.7 g N m^{-2}). Nevertheless, for grass nutrition and, likely, tree nutrition, the nitrogen coming from the soil organic matter pool is negligible, while that coming from root meets most of the annual requirements of plants. Because of the aggregative grass root distribution and because Lamto grasses are perennials, most of the nutrients produced by decaying roots have a high probability to be uptaken by living roots, simply because living roots are very close to dead roots. Similarly, between tussocks, nitrogen originating from the slow decomposition of non-particulate humified organic matter has a low probability to be uptaken by plants because grass root density there is too low. Consequently, the relative contribution of soil organic matter to plant nitrogen nutrition is negligible, except when soil fauna supplies easily metabolizable compounds that induce a strong and localized soil organic matter degradation. The aggregative distribution of roots therefore induces a strong use of nitrogen from dead roots by living roots. Except if nitrogen is assimilated in aerial biomass, it has a low probability to get out of the system at year scale. Moreover, the form of this mineral nitrogen prevents its leaching. Indeed, *Hyparrhenia diplandra* and likely other Andropogoneae species inhibits nitrate production, i.e., makes the non-leachable ammonium dominating over nitrate, increasing again the residence time of nitrogen within the soil-plant system. It is probable that the lower nitrification is, the higher the mineral nitrogen retention is. All that results in a strong recycling of the nitrogen resource within the ecosystem: when a nitrogen atom has entered the soil-plant system, it remains in it many years and contributes to many successive primary production cycles.

In summary, the Lamto savanna shows a high ability to accumulate and retain nutrients; the losses are low, except with fire, due to the low rate of some microbial processes and spatial structure of plant cover. As in all steady state ecosystems, the inputs balance the outputs. Moreover, the uptake of nutrients is optimized by the tufted architecture of grasses and spatial aggregation of roots. The major source of nutrients for plants is not soil organic matter but litter, as in tropical humid forests. In the tropical forest, most of the primary production is allocated to the aerial biomass while mineralization of nitrogen occurs quite exclusively in aboveground leaf litter. In humid savanna, more than half of the primary production is allocated to roots and mineral nitrogen is mainly produced in underground litter.

References

1. L. Abbadie. *Contribution à la biogéochimie dans des savanes de Lamto*. Ph.D. thesis, Université de Paris 6, Paris, 1983.
2. L. Abbadie. Evolution saisonnière du stock d'azote dans la strate herbacée d'une savane soumise du feu en Côte d'Ivoire. *Acto Oecologia, Oecologia Plantarum*, 4(19):321–334, 1984.
3. L. Abbadie. Nouvelles données sur l'évolution saisonnière des phytomasses et sur la production primaire de la strate herbacée des savanes de Lamto. *Annales de l'Université d'Abidjan, Série E*, 17:83–109, 1984.
4. L. Abbadie. Variations saisonnières quantitatives de l'azote dans les sols des savanes de Lamto. *Annales de l'Université d'Abidjan*, 20:9–21, 1988.
5. L. Abbadie. *Aspects fonctionnels du cycle de l'azote dans la strate herbacée de la savane de Lamto*. Ph.D. thesis, Université de Paris 6, Paris, 1990.
6. L. Abbadie and R. Lensi. Carbon and nitrogen mineralization and denitrification in a humid savanna of West Africa. *Acta Oecologica*, 11(5):717–728, 1990.
7. L. Abbadie, A. Mariotti, and J.C. Menaut. Independence of savanna grasses from soil organic matter for their nitrogen supply. *Ecology*, 73(2):608–613, 1992.
8. J. Balandreau. Fixation rhizosphérique de l'azote (C_2H_2) en savane de Lamto. *Revue d'Ecologie et de Biologie du Sol*, 13(4):529–544, 1976.
9. R.H. Brown. A difference in N use efficiency in C_3 and C_4 plants and its implications in adaptation and evolution. *Crop Science*, 18:93–98, 1978.
10. R.H. Brown. Growth of C_3 and C_4 grass under low N levels. *Crop Science*, 25:954–957, 1985.
11. F.S. Chapin. The mineral nutrition wild plants. *Annual Review of Ecology and Systematics*, 11:233–260, 1980.
12. V. Degrange, R. Lensi, and R. Bardin. Activity, size and structure of a *Nitrobacter* community is affected by organic carbon and nitrite in a sterile soil. *FEMS Microbiology and Ecology*, 24:173–180, 1997.
13. J. Delmas. Recherches écologiques dans la savane de Lamto, premier aperçu sur les sols et leur valeur agronomique. *La Terre et la Vie*, 3:216–227, 1967.
14. A. Fournier. Dynamique foliaire chez deux espèces de graminées en savane préforestière. *Végétation*, 57:177–188, 1984.
15. L. Hédin. La valeur fourragère de la savane. *La Terre et la Vie*, 21(3):249–261, 1967.
16. J.C. Lata. *Interactions entre processus microbiens, cycle des nutriments et fonctionnement du couvert herbacé: Cas de la nitrification dans les sols d'une savane humide de Côte d'Ivoire sous couvert à Hyparrhenia diplandra*. Ph.D. thesis, Université de Paris 6, Paris, 1999.
17. J.C. Lata, K. Guillaume, V. Degrange, L. Abbadie, and R. Lensi. Relationships between root density of the African grass *Hyparrhenia diplandra* and nitrification at the decimetric scale: an inhibition-stimulation balance hypothesis. *Proceedings of the Royal Society, London, Series B*, 267:595–600, 2000.
18. E. Le Provost. *Structure et fonctionnement de la strate herbacée d'une savane humide (Lamto, Côte d'Ivoire)*. M.Sci. thesis, Universités de Paris 6 and 11 and Institut National Agronomique Paris-Grignon, Paris, 1993.
19. X. Le Roux, L. Abbadie, R. Lensi, and D. Serça. Emission of nitrogen monoxide from African tropical ecosystems: control of emission by soil characteristics in humid and dry savannas of West Africa. *Journal of Geophysical Research*, 100(D11):23,133–23,142, 1995.

20. X. Le Roux and P. Mordelet. Leaf and canopy CO_2 assimilation in a West African humid savanna during the early growing season. *Journal of Tropical Ecology*, 11:529–545, 1995.
21. R. Lensi, A.M. Domenach, and L. Abbadie. Field study of nitrification and denitrification in a wet savanna of West Africa (Lamto, Côte d'Ivoire). *Plant and Soil*, 147:107–113, 1992.
22. J. Meiklejohn. Numbers of nitrifiying bacteria in some Rhodesian soils under natural grass and improved posture. *Journal of Applied Ecology*, 5:291–300, 1968.
23. J.C. Menaut. Chute de feuilles et apport du sol de litière par les ligneux dans une savane préforestière de Côte d'Ivoire. *Bulletin d'Ecologie*, 5:27–39, 1974.
24. J.C. Menaut. Evolution of plots protected from fire since 13 years in a Guinea savanna of Ivory coast. In *Actas Del IV Symposium Internacional De Ecologia Tropical*, pages 541–558, Panama, 1977.
25. J.C. Menaut and J. César. Structure and primary productivity of Lamto savannas, Ivory Coast. *Ecology*, 60(6):1197–1210, 1979.
26. P. Mordelet, L. Abbadie, and J.C. Menaut. Effects of tree chumps on soil characteristics in a humid savanna of West Africa (Lamto, Côte d'Ivoire). *Plant and Soil*, 153:103–111, 1993.
27. P.E. Munro. Inhibition of nitrifier by grass root extract. *Journal of Applied Ecology*, 3:231–238, 1966.
28. H.B. Nacro. *Hétérogénéité de la matière organique dans un sol de savane humide (Lamto, Côte d'Ivoire): Caractérisation chimique et étude in vitro des activités microbiennes de minéralisation du carbone et de l'azote*. Ph.D. thesis, Université de Paris 6, Paris, 1997.
29. D. Picard. Dynamique racinaire de *Panicum maximum* Jacq.. 1. Emission des racines adventives primaires dans un intercoupe en liaison avec le tallage. *Cahiers de l'ORSTOM, Série Biologie*, 12(3):213–226, 1977.
30. D. Picard. Dynamique racinaire de *Panicum maximum* Jacq.. 2. Rythme annuel d'émission des racines adventives primaires et évolution de la masse racinaire pour une prairie exploitée de façon intensive. *Cahiers de l'ORSTOM, Série Biologie*, 12(3):227–245, 1977.
31. P. De Rham. L'azote dans quelques forêts, savane et terrains de culture d'Afrique tropicale humide (Côte d'Ivoire). Doctorat ès science, Faculté des Sciences, Université de Lausanne, Zurich, 1971.
32. P. De Rham. Recherches sur la minéralisation de l'azote dans les sols des savanes de Lamto. *Revue d'Ecologie et de Biologie du Sol*, 10(2):169–196, 1973.
33. G. Simioni, J. Gignoux, X. Le Roux, R. Appé, and Benest D. Spatial and temporal variation in leaf area index, specific leaf area, and leaf nitrogen of two co-occuring savanna tree species. *Tree Physiology*, 24(2):205–216, 2003.
34. G. Simioni, X. Le Roux, J. Gignoux, and A. S. Walcroft. Leaf gas exchange characteristics and water- and nitrogen-use efficiencies of dominant grass and tree species in a West African savanna. *Plant ecology*, 173:233–246, 2004.
35. Vuattoux. Observations sur l'évolution des strates et arbustive dans la savane de Lamto. *Annales de l'Université d'Abidjan, Série E*, III:285–315, 1970.
36. R. Vuattoux. Contribution à l'étude de l'évolution des strates arborée et arbustive dans la savane de Lamto. *Annales de l'Université d'Abidjan, Série C*, 12:35–63, 1976.

16

Role of Soil Fauna in Nitrogen Cycling

Michel Lepage, Luc Abbadie, Guy Josens, and Patrick Lavelle

16.1 Introduction

As stated in this volume (Chap. 15), the availability of mineral nitrogen is a key determinant of the primary production in Lamto savannas. The soil fauna distribution and activity are key factors controlling soil organic matter distribution, nitrogen storage and mineral-N release. As they strongly modify the spatial pattern of mineral nutrient distribution in soil, they reinforce the patchiness and the heterogeneity of the Lamto savanna. Their effects on the nitrogen cycle are mainly due to earthworms and termites and are related to the importance of their populations, both in terms of biomass and activity (cf. Chap. 13). Their impact on nitrogen dynamics should also be considered in relation to their mutualistic/symbiotic relationships with microorganisms and on increasing space and time scales.

16.2 Nitrogen storage and throughput in soil macrofauna and associated structures

The role of soil macrofauna species in the nitrogen cycle relates to their biomass, to the quantities of ingested soil (earthworms) or rehandled soil (termites). Background data about the macrofauna species and their biogenic structures are given in the previous part of this book (Chap. 13).

16.2.1 Nitrogen in anecic species

The average biomass of fungus-growing (feeding on dead plant matter), xylophagous and foraging termite species in the open shrubby savanna is estimated at 5.33 $g\,m^{-2}$ fresh weight (FW) and 1.00 $g\,m^{-2}$ (after [10, 11, 39, 16]). Taking a ratio fresh weight/dry weight of 4.0 for fungus-growing and 5.1 for xylophagous and foraging species [11, 10], the following values are obtained: 1.33 $g\,m^{-2}$ dry weight (DM) for the fungus-growing species and 0.2 $g\,m^{-2}$ for xylophagous and foraging termites. Nitrogen content of termite tissues was assessed by some authors [31, 7], ranging between 6.1% and 11.1% N (DM), depending on termite caste. Taking into account these values and the caste proportions in the various termite trophic groups (after [11, 10]), the nitrogen content in termite biomass is computed as follows: 0.104 $g\,m^{-2}$ for fungus-growing and 0.017 $g\,m^{-2}$ for xylophagous and foraging species. To estimate the N throughput in termite biomass, we have to know the turnover of the termite populations (their production). Unfortunately, only few estimates have been made on termites [26]. For the fungus-growing termites, the production to biomass ratio (P/B) is estimated as ca. 7 for the dominant species *Ancistrotermes cavithorax* (Josens, unpublished data), considered as being an r-strategist, giving a total nitrogen throughput in the termite biomass of 0.73 $g\,m^{-2}\,y^{-1}$. For the xylophagous and foraging species, Josens [13] calculated a P/B of 2.6 for *Trinervitermes geminatus*, considered as being a K-strategist, from which we assume a total nitrogen throughput of 0.05 $g\,m^{-2}\,y^{-1}$.

The fungus-comb biomass should also be taken into account to assess the role of termite populations in the N cycle. In the open shrub savanna, a biomass of 24 $g\,m^{-2}$ (DM) of fungus-comb was estimated [10, 11, 14, 16]. The nitrogen content of such quantity was not measured in Lamto, but several measurements are given in the literature [31, 7, 36]: average nitrogen content in the fungus-combs of 13 termite species was close to 1.3%. Thus, 24 g of the fungus-comb contain 0.31 g N. But the fungus-comb is a dynamic structure. According to Josens [9], the turnover of the fungus-comb biomass varied between 58 and 70 days. Taking into account this turnover and the matter lost by comb respiration (Chap. 13), the total amount of litter incorporated into the fungus-combs in the shrub savanna amounted to 180 $g\,m^{-2}\,y^{-1}$. From Lepage et al. [25] and Lepage and Abbadie (unpublished data), the food regime of the four main fungus-growing species includes approximately 59% of tree litter and 41% of grass litter, with an average N content of about 0.48% (calculated after Menaut et al. [32]; Chap. 15). Thus the flow of N through the combs amounts to 0.86 $g\,m^{-2}\,y^{-1}$. As in *Macrotermes*, for which 90-100% of dietary nitrogen input in the fungus-combs is used in secondary production by the termites [7], nitrogen is likely to be a limiting factor for the fungus-growing termite populations in Lamto.

Knowing that 42% of the comb carbon is released in respiration [16], simulations have shown that the symbiotic fungus cannot raise the N comb content from an initial 0.48% to an average of 1.3% (Josens, unpublished). Therefore an extra N enrichment has to occur and indeed this happens through the deposit of termite feces on the old and medium-aged parts of their combs, which can be noticed as ochre to reddish specks [12].

Some data indicated that the belowground structures (the walls of fungus-comb chambers and probably their interconnecting galleries) have higher nitrogen content than the surrounding soil: 0.58% as compared to 0.46% in

Abbadie and Lepage [1], and 0.83% as compared to 0.44% in Merdaci [33]. Here, the N enrichment can also be linked with the deposit of termite feces.

16.2.2 Nitrogen in endogeic species

In the open shrub savanna, anecic earthworms (*Millsonia lamtoiana* and *Dichogaster agilis*) incorporate between 18 and 51 $g\,m^{-2}\,y^{-1}$ (DM) of litter [19]. This quantity represents between 0.045 and 0.31 $g\,m^{-2}\,y^{-1}$ (according to a nitrogen concentration in dead grass matter between 0.62 and 0.25% (Chap. 15).

In a shrub savanna, a biomass of 48.6 $g\,m^{-2}$ (FW) earthworms was found [18]. Taking into account the ratio fresh weight:dry weight (from 7.1 to 21.0), this biomass amounted to 4.37 $g\,m^{-2}$ (DM). Nitrogen proportion in earthworm tissue is estimated at 10% on average [35, 6, 8]. That yields a nitrogen content of 0.44 $g\,m^{-2}$ in the earthworm biomass. The P/B ratios of earthworm species were estimated to vary between 1.2 and 3.6 (average: 1.9 [17, 18]). Therefore, the N throughput in earthworm populations can be estimated at about 0.84 $g\,N\,m^{-2}\,y^{-1}$.

For the soil-feeding (geophagous) termites, their living biomass is estimated to be 10.13 $g\,m^{-2}$ (FW) in a shrubby savanna (Josens [10, 11], unpublished data; [39]). Taking into account their caste proportions [10] and the N content of termite tissue, as calculated above, their fresh mass:dry mass ratio (2.7:1) and their mineral content (60% of dry mass), the N content in their biomass amounted to 0.19 $g\,N\,m^{-2}$. No estimation was made in the Lamto savanna for their production:biomass ratio, but Wood and Sands [38] gave a P/B estimation of 3:1 for non-Macrotermitinae. Using that proportion, we came to a nitrogen throughput in the humivorous termites of 0.58 $g\,m^{-2}\,y^{-1}$.

16.2.3 Importance of the soil macrofauna in the N cycle

The same computation in the *Loudetia* grass savanna and the Andropogoneae savanna woodlands of Lamto enables a comparison between savanna types (Table 16.1). Although the figures are very rough, it seems clear that both the amount of nitrogen sequestered in and the throughput of nitrogen through the soil fauna tissues are higher in the Andropogoneae savanna, whatever their tree density, than in the *Loudetia* grass savanna. This difference is higher in termites than in earthworms, the latter being less affected than the former by the long-lasting water saturation in the soil of the grass savanna during the rainy season. Moreover, the litter feeding species have more food available in Andropogoneae savannas and woodlands at a critical moment of the year, i.e. just after the annual fire. Very little grass litter is left after fire, while trees shed their leaves, providing the only aboveground litter at this time. This comparison suggests that termites have a major impact on the N cycle in Andropogoneae savannas, whereas earthworms play the major role in the grass savanna.

Table 16.1. Amounts of nitrogen sequestered in and flowing through the populations of termites (including their fungus-combs) and earthworms in different savanna types of Lamto (unpublished data by M. Lepage and P. Lavelle).

Item	Grass savanna Biomass N (g m^{-2})	Grass savanna N flow g m^{-2} y^{-1}	Andropogoneae sav. Biomass N (g m^{-2})	Andropogoneae sav. N flow g m^{-2} y^{-1}	Savanna woodland Biomass N (g m^{-2})	Savanna woodland N flow g m^{-2} y^{-1}
Macrotermitineae, termites	0.02	0.16	0.10	0.73	0.12	0.81
Macrotermitineae, fungus-combs	0.06	0.15	0.31	0.86	0.38	1.04
Macrotermitineae nest	0.08	0.16	0.41	0.80	0.50	0.93
Xylophagous and foraging termites	0.01	0.02	0.02	0.05	0.06	0.13
Geophagous termites	0.04	0.12	0.19	0.58	0.22	0.66
Earthworms	0.40	0.77	0.44	0.84	0.51	1.00
Total soil fauna	0.53	1.07	1.06	2.27	1.29	2.72

Because the digestive efficiency is far below 100%, especially in geophagous species, the amount of nitrogen flowing through their guts can be much higher than the amount flowing through their tissues. The foraging species (*Trinervitermes* spp.) consume about 5 g m^{-2} y^{-1} (DM) of grass [11], that contain 0.03 g N (Chap. 15). The xylophagous species (*Microcerotermes* spp. and *Amitermes evuncifer*) consume approximately 2 g m^{-2} y^{-1} (DM) of wood (taking an average consumption of 30 mg g^{-1} d^{-1} (FW), after [38]), that contain about 0.005 g N (Chap. 15).

According to Lavelle [17, 18], the earthworm population ingested in an average year (i.e., with 1250 mm rainfall) about 80-125 kg m^{-2} dry soil, equivalent to 18.7-29.2 kJ and containing 1.4-1.5 kg of organic matter [21] equivalent to 29.6-70 g of nitrogen (assuming a N content of 0.056% in the top 10 cm soil and 0.037% deeper; see Chap. 15).

Estimating the annual consumption of humivorous termite species is more hazardous, as we know very little about their food regime. A total biomass (FW) of 10.13 g m^{-2} would require a consumption figure of 111 g m^{-2} y^{-1} organic matter [38], equivalent to 2090 kJ. To obtain this amount (1 kg of ingested soil corresponds to 234 kJ for earthworms [17, 18]), the humivorous termite species have to process 5.4 to 7.2 kg m^{-2} y^{-1} of soil, containing roughly 1.8 to 3.6 g N m^{-2} y^{-1}.

Concerning the termite populations, an important point, very rarely treated, deserve more investigation: the nutrient pool in the alates swarming each year in the savanna and the consequences of such production in the ecosystem functioning. The reproduction in termites colonies represent 10% to 15% of the colony biomass in xylophagous species, 20% to 25% in foraging species and 45% to 55% in fungus-growing species [26]. This amount is released in the ecosystem when the grass layer has a high growth rate, well within the rainy season, since most termite species are swarming about this time (May-June), except *Pseudacanthotermes militaris* (swarming in August). According to the high nutrient content of the termite alates, their production represents a significant return to the ecosystem, since nearly all this sexual production will enter the food chain, consumed by predators or decomposed by microorganisms [24].

From these computations, it can be concluded that the earthworms and termites represent a significant quantity of nitrogen, in their living tissues as well as quantities processed with their food: 0.75 $g\,N\,m^{-2}$ in their biomass, 2.27 $g\,N\,m^{-2}\,y^{-1}$ in their production, 0.31 $g\,m^{-2}$ in the Macrotermitinae fungus-combs, 0.91 g of N consumed from tree and grass litter and 31 to 74 $g\,N\,m^{-2}\,y^{-1}$ in the soil ingested by the geophagous populations. These figures could be compared to the data computed for the average nitrogen requirements in a shrub savanna, for the woody strata and grass layer (Chap. 15): respectively 0.81 and 7.5-10.1 $g\,N\,m^{-2}\,y^{-1}$. The consumption figure by fungus-growing termite populations (0.86 $g\,N\,m^{-2}\,y^{-1}$) has to be compared to the nitrogen content in the grass standing dead matter in a shrubby savanna (around 1.4 $g\,N\,m^{-2}$; Chap. 15).

16.3 Impact of soil macrofauna on nitrogen dynamics and mineralization

The impact of soil macrofauna on nitrogen dynamics is mainly related to the importance of the biogenic structures produced in the soil environment and to the evolution of structures over time. Biogenic structures created by the soil macrofauna in Lamto differ from the control soil in their mineral composition.

16.3.1 Termites

Sequestration of organic matter in termite structures can last for several years. In some areas of the Lamto savanna, lenticular mounds represent more than 300 $m^3\,ha^{-1}$, covering about 9% of the ecosystem area [2, 15]. These large epigeous termitaria deeply modify the distribution and storage of soil organic matter locally: mounds are 3 to 4 times richer in clay (13.9-20.8% (DM), as compared to 5.6-8.4% in the surrounding soil) [2]. As these lenticular mounds are favorable sites for tree development, especially for seedlings [2], due to favorable physical and chemical changes on mound soil, such as the increase of the field capacity and water availability for plants [15], total organic carbon and nitrogen on mound top are twice as high as in the control soil. Lenticular mounds support larger densities of belowground nest chambers than the surrounding savanna soil: densities of chambers from 22.2 to 38.6 m^{-2} were found [2, 16], as compared to 0.95 to 12.3 in the surrounding savanna [16, 10]. The walls of the chambers are coated with a fine layer of reddish clay material, usually 1-3 mm thick. Under potential conditions in the laboratory, it has been found that such rehandled material could stimulate total microbial activities (CO_2 release; see Chap. 13) and, consequently, mineral-N release [1]. The more the soil is worked by termites, the more N mineralization increases. A significant mineral-N release in the walls of termite fungus-comb chambers was found (Fig. 16.1): $N-NH_4^+$ potential release was higher in such structures. After 2 days, 69 $\mu g\, g^{-1}$ were released, as compared to only 33 $\mu g\, g^{-1}$ in the control soil. Accumulation of 140 $\mu g\, g^{-1}$ of NH_4^+ in the 0-2 mm layer was also evidenced, compared to 40-60 $\mu g\, g^{-1}$ in the control soil [1]. These measurements suggest that the fungus-comb chambers are sites of strong mineralization and that termite nests could strongly impact the mineralization process at the scale of the whole ecosystem since fungus chamber walls represent around 2.5-4.3 $cm^3\, m^{-2}$ [20, 16].

From the data on nitrogen content in large lenticular mounds, together with the potential higher nitrogen mineralization in termite belowground chambers, a nitrogen cycling in two phases has been postulated: first, a nitrogen storage in the nest structures; second, a mineralization of this sequestered nitrogen, by stimulation of the soil microflora [2]. With the combination of large organic matter storage and an enhanced mineral nutrient release, mound soils have to be considered as fertile spots for plants, compared to the surrounding savanna [2].

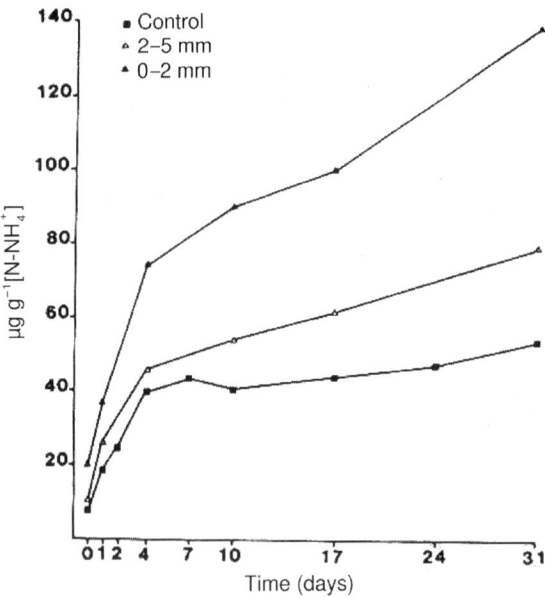

Fig. 16.1. Ammonial nitrogen production, in laboratory incubations, of layers around an *Ancistrotermes cavithorax* fungus-comb chamber and in the control soil, in the second part of the dry season in Lamto (from [1], copyright (1989), with permission of Elsevier).

16.3.2 Earthworms

In earthworms, organic matter dynamics, particularly nitrogen storage and mineralization, has been considered at three different scales of time and space [20]:

1. At the scale of the gut transit and in the short term (a few hours to a few days), freshly deposited casts of *Millsonia anomala* (the main species within the geophagous trophic group, representing 40% to 60% of the total biomass [18]) contained on average 26.4 μg N g^{-1} soil as NH_4^+ and traces of NO_3^- (as compared with the control soil with 1.5-7.5 μg N g^{-1} soil) [21]. This mineral nitrogen released in fresh casts is readily incorporated into microbial biomass in 8-16 days. Extrapolation of these data in the DRILOTROP [30] simulation model makes it possible to calculate that the populations of *M. anomala* may annually release 0.5 to 2.5 g N m^{-2} y^{-1} of assimilable nitrogen [28]. In a grass savanna, 82% of this nitrogen is deposited in the top 10 cm of the soil: 63-71% is released as ammonium in fresh casts, 11-20% as dead worms and the remaining as cutaneous mucus and urine.

   With regard to earthwoms, and probably soil-feeding termites, this short term effect is then a rapid mineralization of the organic matter in the feces and a rapid nitrogen mineralization [27, 28], probably usable by the plant roots in suitable conditions (see below and Fig. 16.2).

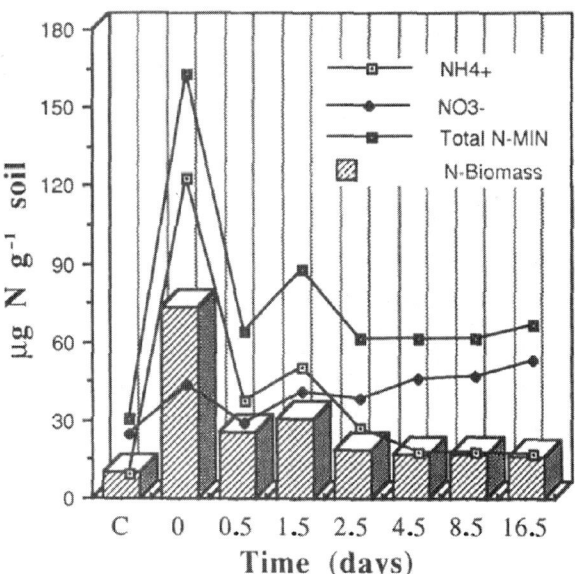

Fig. 16.2. Changes in mineral-N content and N accumulated in microbial biomass in casts of a tropical earthworm (*Pontoscolex corethrurus*) during the 16 days following deposition (from [21], copyright (1992), with permission of Elsevier).

2. At a decimetric spatial scale and in the medium term (months), the initial rapid decrease of carbon and nitrogen contents is further followed by a blockage of mineralization in older casts [21]. After 50 days, the organic content in casts was higher than in the control soil and was ca. 100% greater after 420 days [28]. Longer-term (30 months) field experiments have shown that casts were resistant structures, stable in time, in which the soil organic matter may be protected for a few years [3, 4, 5]. In Lamto, such aging casts may comprise 40% to 60% of the soil total mass in the top 10 cm [21].
3. At the scale of the whole soil profile and in the long term (tens to hundreds of years), comparisons to other locations would suggest that earthworm activities are likely to accelerate the turnover of the soil organic matter, by increasing the "active" pool from the "passive" pool. But this hypothesis has still to be tested in Lamto savannas.

16.3.3 Impact on ecosystem dynamics

In the long term, the main point to be cleared is the possible synchrony of nutrient release through the macrofauna activities with the root uptake. If such a synchrony occurs, the drilosphere and the termitosphere systems will be major sources of nutrients for plants. If not, macrofauna activities would only result in a temporary and local stimulation of microbial biomass and metabolism.

Whatever the effect of soil macrofauna in the long term, its activity profoundly modifies the heterogeneity of the savanna and therefore the distribution of nitrogen inputs and outputs from the soil-plant system (Chap. 14). The mechanisms of the nitrogen cycle are modified in sites created by soil macrofauna at different scales. In biogenic structures, anaerobic microsites of denitrification could be maintained in clay aggregates sustained by higher organic matter quality [23]. It was found that the soil samples from the Andropogoneae savanna exhibited a low to negligible potential of nitrification even after ammonium addition. After 5 h of incubation and compared with control soil, the denitrifying activities were 14.3, 5.0 and 4.3 times higher in material from termite chambers, in samples collected very close to the roots and in earthworm casts, respectively (Fig. 16.3). At the level of the large lenticular mounds, low potential of denitrification was measured while nitrification potential was high, leading to higher NO emission rates, as compared to the surrounding savanna [22].

16.4 Stimulation of plant growth by soil macrofauna

Many field studies have provided information on the impact of soil macrofauna on soil processes in Lamto savannas (e.g., Sects. 16.2 and 16.3 and

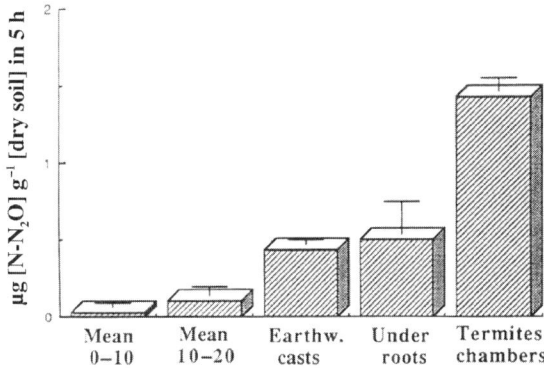

Fig. 16.3. Denitrifying activities within the Andropogoneae savanna soil, in $\mu g\, N_2O\, g^{-1}$ dry soil after 5 h incubation, between control soils and biogenic structures (roots, earthworms, termites) (after [23], with kind permission of Springer Science and Business media).

Chap. 13). But only a few quantitative data are available to evaluate the exact contribution of soil macrofauna to soil functioning and particularly to the nutrient cycles. One of the main issues to be addressed is the validity of the extrapolation on the ecosystem scale of laboratory studies conducted under optimal conditions. For instance, combining a simulation model established from background information on earthworm population dynamics and laboratory data of mineral-N release in earthworm casts, Martin [29] calculated that this fauna might contribute to nearly 15% of soil N mineralization in Lamto savannas.

Preliminary experiments have been conducted in Lamto to elucidate the possibility of this mineral N release to be used by plant (experiments on the Poaceae *Panicum maximum*), using ^{15}N-labeled earthworms [37]. The results showed a plant growth stimulation and an increased plant production related to the earthworm biomass added in the containers. Results also exhibited a change in the root:shoot ratio in the presence of earthworms: a higher aboveground production was obtained per unit mass of roots. However, the high variability of the data obtained suggested possible interactions with other mechanisms, to explain the transfer of N (and also P) from the soil microbial biomass to the plant, such as a facilitation of mycorrhization in earthworm casts.

Laboratory experiments were recently conducted to test if plant roots could take advantage of the mineral-N release in termite belowground structures (P. Jouquet et al., unpublished data). The results showed that the walls of fungus-comb chambers induce a root proliferation by herbaceous plants, and that this effect was mainly due to fine roots (<255 μm) (Fig. 16.4). Results are partly similar to those obtained by local mineral nitrogen supplies. The

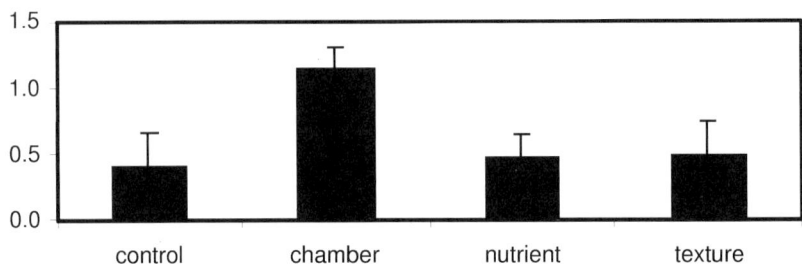

Fig. 16.4. Root biomass of grass (*Pennisetum pedicellatum*) growing in homogeneous soil (control) or in heterogeneous soil plus a patch of fungus-comb chamber wall (chamber) or plus a patch of soil supplied with an enrichment in mineral nitrogen (nutrient) or plus a patch of soil with the same texture as the fungus-comb chamber wall (texture). ($n=8$; standard error is represented as a vertical bar, Chamber-wall treatment differed significantly from the other treatments) (Jouquet, unpublished data).

walls of the fungus-comb chamber led equally to an increase of the epigeous biomass although the root:shoot ratio was higher than in the control soil.

Obviously further experiments are needed to clearly establish the link between soil macrofauna ecological strategies and plant growth. The importance of biological interactions in the soil rehandled by the soil macrofauna and the related effects on plant nutrition should be emphasized. Biogenic structures made by the soil macrofauna are sites of higher nutrient concentration but the linkage of plant nutrition to nutrient availability requires further investigations about possible association with arbuscular mycorrhiza and rhizobia. In a study conducted in another location (Senegal) about the effect of mound soil (*Cubitermes niokoloensis*) on plant growth (*Crotalaria ochroleuca*), Ndiaye et al. [34] distinguished a direct effect due to a higher nutrient availability in mound soil and an indirect microbial effect mediated on the symbionts in the soil close to the mound.

16.5 Conclusion

The rough figures obtained in this chapter clearly demonstrated the importance of soil macrofauna in the nitrogen cycle. We estimated the N-biomass sequestered in macrofauna (termites and earthworms) tissues to be $1.06 \, \mathrm{g\,m^{-2}}$, which has to be compared with a N pool in the whole grass layer of $1.4 \, \mathrm{g\,m^{-2}}$, decreasing to $0.96 \, \mathrm{g\,m^{-2}}$ just before fire, in the dry season (Chap. 15). N flow through the macrofauna compartment (as organic and mineral nitrogen) varied from $0.53 \, \mathrm{g\,m^{-2}\,y^{-1}}$ in the grass savanna to $1.29 \, \mathrm{g\,m^{-2}\,y^{-1}}$ in the woody savanna, a value that should be compared to the annual nitrogen requirements of the grass layer: from 7.5 g m^{-2} y^{-1} in the *Loudetia* savanna to 10.1 g m^{-2} y^{-1} in the Andropogoneae savanna (Chap. 15).

Of particular importance is the nitrogen flow through the belowground termite fungus-comb: 0.86 to 1.04 g m^{-2} y^{-1} in the Andropogoneae savanna. This throughput should be compared to the nitrogen flux from the grass biomass burning to the atmosphere in these savannas, from 1.4 to 2.4 g m^{-2} y^{-1} (Chap. 14). Therefore, N flow from termite belowground structures amounted 36% to 74% of the nitrogen loss due to annual fires. This is particularly important, as the food stored is mainly aboveground litter that escaped from fire.

These figures confirm that the role of the soil macrofauna in the Lamto savanna is mainly to concentrate nutrients, and particularly nitrogen, in biogenic structures. These structures could be qualified as "activation sites," since they are spots of high nutrient availability and N supply, first of all through the mutualistic relationships evolved with microorganisms. Not only spatial scale but also temporal scale should be considered to clearly assess the real importance of the fauna activity on the vegetation structure and dynamics. We do not really know, for instance, what could be the consequence for the ecosystem of the temporary storage of a non-negligible N pool in belowground nest structures (for 58 to 70 days after [9]). Nor do we know the consequence of the release in the atmosphere at the time of grass regrowth, a crucial moment for the ecosystem, of roughly 0.06 g N m^{-2} (grass savanna) to 0.39 g N m^{-2} (savanna woodland) in termite swarming alates. The possible triggering effect of such a release remains to be studied.

References

1. L. Abbadie and M. Lepage. The role of subterranean fungus comb chambers (Isoptera, Macrotermitinae) in soil nitrogen cycling in a preforest savanna (Côte d'Ivoire). *Soil Biology and Biochemistry*, 21(8):1067–1071, 1989.
2. L. Abbadie, M. Lepage, and X. Le Roux. Soil fauna at the forest-savanna boundary: Role of termite mounds in nutrient cycling. In P.A. Furley, J. Proctor, and J.A. Ratter, editors, *Nature and dynamics of forest-savanna boundaries*, pages 473–484. Chapman & Hall, London, 1992.
3. E. Blanchart. Rapport de mission. Station d'écologie tropicale de Lamto (république de Côte d'Ivoire), 4 octobre - 1 novembre 1989. Technical Report 87/007, Projet Unesco/PNUD IVC, 1989.
4. E. Blanchart, A. Bruand, and P. Lavelle. The physical structure of casts of *Millsonia anomala* (Oligochaeta : Megascolicidae) in shrub savanna soils (Côte d'Ivoire). *Geoderma*, 56:119–132, 1993.
5. E. Blanchart, P. Lavelle, E. Braudeau, Y. Le Bissonais, and C. Valentin. Regulation of soil structure by geophagous earthworm activities in humid savannahs of Côte d'Ivoire. *Soil Biology and Biochemistry*, 29:431–439, 1997.
6. M.B. Bouché. Contribution des lombriciens aux migrations d'éléments dans les sols tempérés. *Colloques Internes du CNRS*, 303:145–153, 1981.

7. N. M. Collins. Termite populations and their role in litter removal in malaysian rain forest. In *Tropical rain forest: Ecology and management*, pages 311–325. Blackwell Scientific Publications, Boston, 1983.
8. C.A. Edwards and P.J. Bohlen. *Biology and ecology of earthworms*. Chapman & Hall, London, 1996.
9. G. Josens. Le renouvellement des meules à champignons construites par quatre Macrotermitinae (Isoptères) des savanes de Lamto-Pacobo. *Comptes rendus de l'Académie des Sciences, Série D*, 272:3329–3332, 1971.
10. G. Josens. Etudes biologique et écologique des termites (Isoptera) de la savane de Lamto-Pakobo (Côte d'Ivoire). Ph.D. thesis, Université Libre de Bruxelles, Bruxelles, 1972.
11. G. Josens. Etude fonctionelle de quelques groupes animaux: les termites. *Bulletin de Liaison des Chercheurs de Lamto*, S5:91–131, 1974.
12. G. Josens. *Etudes biologique et écologique des termites (Isoptera) de la savane de Lamto-Pakobo (Côte d'Ivoire)*, volume 42 of *Mémoires de la classe des Sciences, Series 2*. Académie royale de Belgique, Bruxelles, 1977.
13. G. Josens. Le bilan énergétique de *Trinervitermes geminatus* Wassmann (Termitidae, Nasutitermitinae). I. Mesure de biomasses, d'équivalents énergétiques, de longévité et de production en laboratoire. *Insectes Sociaux*, 29(2bis):297–307, 1982.
14. S. Konaté. *Structure, dynamique et rôle des buttes termitiques dans le fonctionnement d'une savane préforestière (Lamto, Côte d'Ivoire): Le termite* Odontotermes *comme ingénieur de l'écosystème*. Ph.D. thesis, Université de Paris 6, Paris, 1998.
15. S. Konaté, X. Le Roux, D. Tessier, and M. Lepage. Influence of large termitaria on soil characteristics, soil water regime and tree leaf shedding pattern in a West African savanna. *Plant and Soil*, 206:47–60, 1999.
16. S. Konaté, X. Le Roux, B. Verdier, and M. Lepage. Effect of underground fungus-growing termites on carbon dioxide emission at the chamber- and landscape-scales in an African savanna. *Functional Ecology*, 17:305–314, 2003.
17. P. Lavelle. Etude fonctionnelle de quelques groupes animaux: Les vers de terre. *Bulletin de Liaison des Chercheurs de Lamto*, S5:133–166, 1974.
18. P. Lavelle. *Les vers de terre de la savane de Lamto (Côte d'Ivoire): Peuplements, populations et fonctions dans l'écosystème*. Ph.D. thesis, Université de Paris 6, Paris, 1978.
19. P. Lavelle. Faunal activities and soil processes: adaptative strategies that determine ecosystem function. *Advances in Ecological Research*, 27:93–132, 1997.
20. P. Lavelle, D. Bignell, M. Lepage, V. Wolters, P. Roger, P. Ineson, O.W. Heal, and S. Dhillion. Soil function in a changing world: the role of invertebrate ecosystem engineers. *European Journal of Soil Biology*, 33(4):159–193, 1997.
21. P. Lavelle and A. Martin. Small-scale and large-scale effects of endogeic earthworms on soil organic matter dynamics in soils of the humid tropics. *Soil Biology and Biochemistry*, 24(12):1491–1498, 1992.
22. X. Le Roux and P. Mordelet. Leaf and canopy CO_2 assimilation in a West African humid savanna during the early growing season. *Journal of Tropical Ecology*, 11:529–545, 1995.
23. R. Lensi, A.M. Domenach, and L. Abbadie. Field study of nitrification and denitrification in a wet savanna of West Africa. *Plant and Soil*, 147:107–113, 1992.

24. M. Lepage. Predation on termite macrotermes michaelseni reproductives and post settlement survival in the field (isoptera: Macrotermitinae). *Sociobiology*, 18(2):153–166, 1991.
25. M. Lepage, L. Abbadie, and A. Mariotti. Food habits of sympatric termite species (Isoptera, Macrotermitinae) as determined by stable carbon isotope analysis in a Guinean savanna (Lamto, Côte d'Ivoire). *Journal of Tropical Ecology*, 9:303–311, 1993.
26. M. Lepage and J.P.E.C. Darlington. Population dynamics of termites. In T. Abe, D.E. Bignell, and M. Higashi, editors, *Termites: Evolution, sociality, symbioses, ecology*, pages 333–361. Kluwer Academic Publishing, Dordrecht, 2000.
27. A. Martin. *Effets des vers de terre tropicaux géophages sur la dynamique de la matière organique du sol dans les savanes tropicales humides*. Ph.D. thesis, Université de Paris 11, Paris, 1989.
28. A. Martin. Short- and long-term effects of the endogeic earthworm *Millsonia anomala* (Omodeo) (Megascolecidae, Oligochaeta) of tropical savannas on soil organic matter. *Biology and Fertility of Soils*, 11:234–238, 1991.
29. S. Martin. *Modélisation de la dynamique et du rôle d'une population de vers de terre Millsonia anomala dans les savanes humides de Côte d'Ivoire*. Ph.D. thesis, Université de Paris 6, Paris, 1991.
30. S. Martin and P. Lavelle. A simulation model of vertical movements of an earthworm population (*Millsonia anomala* Omodeo, Megascolecidae) in an African savanna (Lamto, Ivory Coast). *Soil Biology and Biochemistry*, 24(12):1419–1424, 1992.
31. T. Matsumoto. The role of termites in an equatorial rain forest ecosystem of West Malaysia. I. Population density, biomass, carbon, nitrogen and calorific content and respiration rate. *Oecologia*, 22:153–178, 1976.
32. J.C. Menaut, L. Abbadie, and P.M. Vitousek. Nutrient and organic matter dynamics in tropical ecosystems. In P.J. Crutzen and Goldammer J.G., editors, *Fire in the environment: The ecological, atmospheric, and climatic importance of vegetation fires*, pages 215–231. John Wiley & Sons Ltd., London, 1993.
33. K. Merdaci. *Argiles et substrats incorporés par les termites et dynamique de la matière organique des sols*. M.Sci. thesis, Université de Créteil, Créteil, 1994.
34. D. Ndiaye, R. Duponnois, A. Brauman, and M. Lepage. Impact of a soil feeding termite, *Cubitermes niokoloensis* on the symbiotic microflora associated with a fallow leguminous plant *Crotalaria ochroleuca*. *Biology and Fertility of Soils*, 37:313–318, 2003.
35. R.W. Parmelee and D.A. Crossley, Jr. Earthworm production and role in the nitrogen cycle of a no-tillage agroecosystem on the georgia piemont. *Pedobiologia*, 32:351–361, 1988.
36. C. Rouland-Lefèvre. Symbiosis with fungi. In T. Abe, D.E. Bignell, and M. Higashi M, editors, *Termites: Evolution, sociality, symbioses, ecology*, pages 289–306. Kluwer Academic Publishers, Dordrecht, 2000.
37. A.V. Spain, P. Lavelle, and A. Mariotti. Stimulation of plant growth by tropical earthworms. *Soil Biology and Biochemistry*, 24(12):1629–1633, 1992.
38. T.G. Wood and W.A. Sands. The role of termites in ecosystems. In M.V. Brian, editor, *Production ecology of ants and termites*, pages 245–292. Cambridge University Press, Cambridge, 1978.
39. A. Yapi. *Biologie, écologie et métabolisme digestif de quelques espèces de termites humivores de savane*. Thèse de 3e cycle, Université d'Abidjan, Abidjan, 1991.

Part V

Plant Community Dynamics

17

Spatial Pattern, Dynamics, and Reproductive Biology of the Grass Community

Jacques Gignoux, Isabelle Dajoz, Jacques Durand, Lisa Garnier, and Michel Veuille

17.1 Introduction

The grass population dynamics in Lamto has mainly been studied during the last decade. Past studies have mainly dealt with grass primary production and ecophysiology (Chaps. 7 and 6). A few studies (Abbadie, unpublished; [37, 30]) documented the dynamics of cohorts of leaves. Recently, effort has been put on analyzing the reproductive biology and genetic structure of the dominant grass species *Hypparhenia diplandra* (Sect. 17.4). As our main goal in Lamto is to get a general understanding of the ecosystem as a whole and to predict its dynamics under changing environmental conditions, we tried as much as possible to link population dynamics with the spatial and temporal structure of populations.

17.2 Structure of the grass layer

The interest in the spatial structure of the grass layer started when the keystone role of the tufted habit of the dominant grasses in Lamto had been demonstrated (Chap. 15 and [1]).

17.2.1 Grass population structure

Population structure and spatial pattern have been studied on a 42.24 m^2 plot where all ($N = 4539$) individuals were mapped [29] (Table 17.1).

Two growth stages can be distinguished: seedlings (single stemmed individuals) and tufts. There is no way to estimate the age of grass tufts, but tuft size (number of tillers) is related to age. However, the relation is not simple: small tufts can be young or very old, resulting from the splitting of a big tuft into smaller units. In the latter case, there are often conspicuous signs of senescence (like burned culms remaining from the last fire: Fig. 17.1).

Table 17.1. Main grass species (exhaustive census of the 42.24 m² plot studied by Le Provost [29]).

Species	Density (m^{-2})	Basal cover (%)
Andropogon ascinodis	8.26	1.77
Andropogon canaliculatus	11.29	3.09
Andropogon schirensis	21.24	3.66
Brachiaria brachylopha	3.13	0.58
Hyparrhenia diplandra	10.44	2.02
Hyparrhenia smithiana	4.52	0.55
Loudetia simplex	10.35	3.08
Multi-species tufts	2.65	0.68
Undetermined	0.76	0.23
Dead tufts	0.07	0.01
Total	72.89	15.69

Fig. 17.1. Impact of fire on grass tufts. Photography 3 days after fire by J. Gignoux. Notice the dead part of the tuft in the lower left corner.

Population structure has been described by splitting the population into size classes based on the basal area of tufts, plus one class of seedlings (Fig. 17.2). For all species, the size histogram presents a classical decrease in frequency from small to large sized tufts. There is a lack of seedlings in all species except *Loudetia simplex*, with some species having less seedlings than small tufts of the first size class. This suggests either that sexual reproduction is not the major way of reproduction for these species or that a non-negligible proportion of small tufts results from the splitting of bigger tufts. A striking fact is that no tuft of size 2-8 tillers has been recorded in any species; given that data were recorded at the very beginning of the wet season, this indicates that the size of 9-10 tillers is reached in one growing season by a seedling.

17 Structure and Dynamics of the Grass Community 317

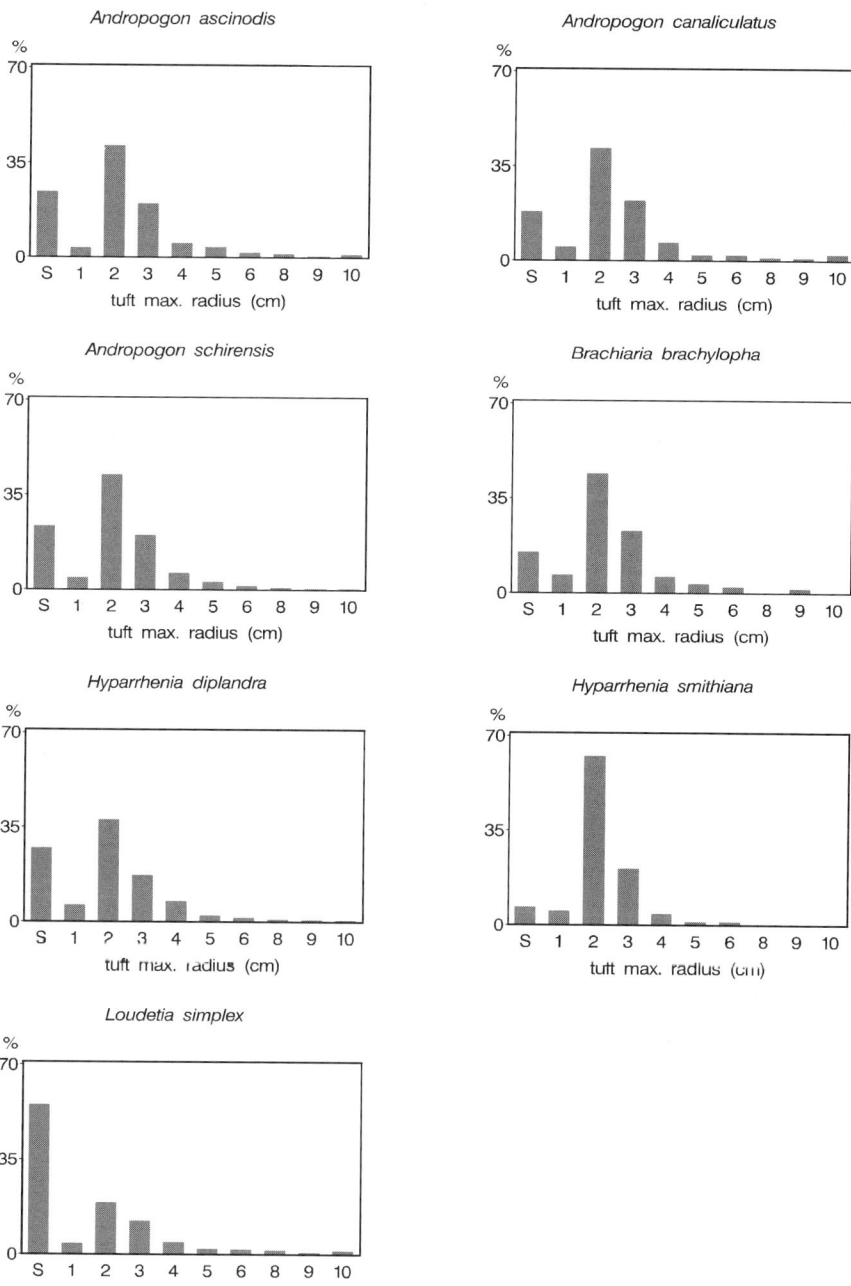

Fig. 17.2. Grass population structure per species. Frequency distribution of individuals in basal tuft radius classes. S: seedlings (redrawn from [29]).

17.2.2 Spatial distribution of grass species

We analyzed the spatial pattern of grass individuals on a 6.6 × 6.4 m plot in the "Savane du Rocher", where all individuals (seedlings and tufts) have been mapped [29] (Fig. 17.3). We used statistical methods designed for mapped point patterns, developed by Diggle [11] and Ripley [40] with a software developed by Gignoux [21]. The statistical methods rely on the distribution of distances between points, which is compared to a distribution of distances under the null hypothesis of spatial randomness or independence. This enables one to classify the spatial patterns as aggregated (clumped), random or regular. The same analysis is possible between two groups of points, leading to a classification into spatial association, independence, or repulsion. Characteristic distances related to scale of aggregation/association or repulsion distance can be estimated from the statistical tests (see Barot [5] for a discussion on the interpretation of those distances). The method enables one to detect departures from the null hypothesis with as few as 10 points, after some adaptation for small samples [22]. Since the mapped plot was large and seemed heterogeneous, we divided it into 9 more homogeneous 2.2 × 2.2 m sub-plots and performed analyses both for the whole plot and for all sub-plots.

Spatial pattern

The spatial pattern of adult tufts of all species ($N=3179$) is aggregated at the whole plot level (Table 17.2) and in a majority of sub-plots (results not shown). Patterns are extremely complex, with sometimes three scales detected in the tests.

On a per species basis, spatial pattern was aggregated for all species, for tufts as well as for seedlings (Table 17.2). Except for *Brachiaria brachylopha* and *Andropogon ascinodis*, the significant departures from randomness were observed at large distances for the K-test, which we interpret as an evidence of large scale heterogeneities on the plot probably linked to soil depth [9]. Seedlings display long distance aggregated patterns for three species only. At the sub-plot level, all significant departures from randomness indicate aggregation, either at long or short distance (note that this distinction at the sub-plot level occurs at a distance of 0.55 m only, instead of 1.6 m at the whole plot level), but the distribution of aggregation among sub-plots varies between species. No species has a clear homogeneous pattern on all sub-plots, but some facts appear:

- *Loudetia simplex* displays a tendency toward randomness (2 sub-plots out of 3 located within the *Loudetia* patch).
- *Brachiaria brachylopha* and *Hyparrhenia smithiana* can both display a random pattern (on 5 sub-plots) or aggregated patterns with short and long characteristic distances (on the remaining sub-plots).
- *Hyparrhenia diplandra* and *Andropogon ascinodis* are aggregated at short distances (0.1-0.2 m).

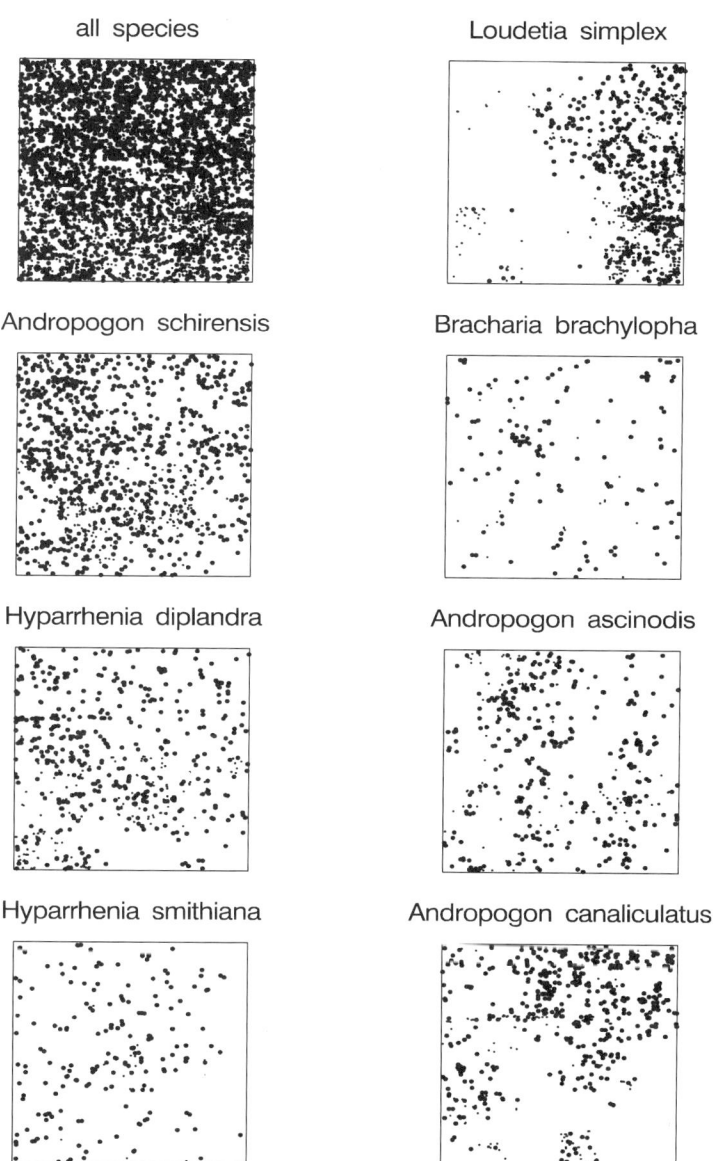

Fig. 17.3. Map of the grass spatial pattern study plot (Savane du Rocher). Large dots, adults; small dots, seedlings (after [29]).

Table 17.2. Spatial pattern analysis of the whole plot. Tufts were classified as big when their basal area was >20 cm². Results shown are: N the number of points of the sample, the test statistic (dw, dx or L), the significance level computed from 500 Monte Carlo simulations of a random pattern of the same intensity, the distance (in m) corresponding to the test statistic d_{max}, i.e., to the maximal discrepancy between observed and theoretical distributions of distances, and the conclusion of the tests (A, aggregated; R, regular; -, random pattern). Significant results are indicated in italic, and large scale patterns (i.e., when the distance is greater than half the maximal possible distance of 160 cm) in bold. * : estimated from 100 Monte Carlo simulations only because of a too large sample size. Straightforward species name abbreviations (see Table 17.1). (original results obtained from the Le Provost dataset [29]).

			G			F			K			
Species	Stage	N	dw	P	d_{max}	dx	P	d_{max}	L	P	d_{max}	Res.
All	tufts	3179	*0.028*	*0.030*	0.05	−0.045	0.070	-	*7.828*	*0.000**	**2.88**	A
	seedl.	1360	*0.401*	*0.000*	0.05	−*0.232*	*0.000*	0.16	*22.344*	*0.000**	**1.71**	A
A.a.	tufts	379	*0.327*	*0.000*	0.11	−*0.173*	*0.000*	0.27	*11.22*	*0.000*	0.43	A
	seedl.	113	*0.426*	*0.000*	0.16	−*0.278*	*0.000*	0.59	*34.06*	*0.000*	1.33	A
A.c.	tufts	495	*0.354*	*0.000*	0.11	−*0.291*	*0.000*	0.27	*46.52*	*0.000*	**2.51**	A
	seedl.	110	*0.528*	*0.000*	0.11	−*0.352*	*0.000*	0.48	*41.58*	*0.000*	0.32	A
A.s.	tufts	958	*0.075*	*0.000*	0.11	−0.068	*0.000*	0.16	*9.52*	*0.000*	**2.51**	A
	seedl.	271	*0.457*	*0.000*	0.11	−*0.347*	*0.000*	0.37	*50.68*	*0.000*	**2.08**	A
B.b.	tufts	147	*0.294*	*0.000*	0.16	−0.085	*0.000*	0.43	*13.82*	*0.002*	0.11	A
	seedl.	23	*0.398*	*0.006*	0.48	−0.142	*0.000*	0.96	*59.28*	*0.000*	0.43	A
H.d.	tufts	469	*0.248*	*0.000*	0.11	−0.104	*0.000*	0.21	*11.02*	*0.000*	**2.67**	A
	seedl.	168	*0.466*	*0.000*	0.11	−*0.449*	*0.000*	0.43	*70.91*	*0.000*	**2.08**	A
H.s.	tufts	213	*0.280*	*0.000*	0.11	−0.107	*0.000*	0.48	*19.61*	*0.000*	**2.67**	A
	seedl.	19	*0.478*	*0.000*	0.43	−0.245	*0.000*	1.17	*91.83*	*0.000*	1.01	A
L.s.	tufts	462	*0.376*	*0.000*	0.11	−*0.418*	*0.000*	0.32	*79.13*	*0.000*	**2.35**	A
	seedl.	538	*0.627*	*0.000*	0.05	−*0.509*	*0.000*	0.27	*118.60*	*0.000*	**1.85**	A

- *Andropogon canaliculatus* is also aggregated, but apparently at two different scales.

What emerges from all those analyses is an overall important heterogeneity of the grass cover at three different scales, already noticed indirectly by all people who have sampled grass biomass to estimate net primary production (discussion of this problem appears in César [9]):

- One scale of heterogeneity (ca. 2 m) is probably linked to soil variability, the plot being located on a transition between *Loudetia* and Andropogoneae savannas.
- Another (at 5-15 cm) probably has to do with the growth of different species within the same tuft and with tuft fractionation.
- A last one (ca. 1 m) is more difficult to interpret, but may be linked to dispersal: the dispersal distance of seeds is short (0.77 m for *Hyparrhenia smithiana* and 0.91 m for *Hyparrhenia diplandra* (see Sect. 17.3) and is of the order of magnitude of the long characteristic distances (0.6-1.0 m) observed for some species on the sub-plots.

Regular spatial patterns are usually supposed to result from competition. Aggregation can be due to a positive interference between individuals of a species or, more likely, to a short dispersal distance of seedlings. The important fact is that regularity is never detected at short distances, suggesting a complete absence of intraspecific and interspecific competition between adult tufts. This is compatible with the efficiency of nitrogen recycling by tufts demonstrated by Abbadie et al. [1], but also supposes an absence of competition for light or water, which is surprising given the large overlap between the leaves of neighboring tufts during the wet season.

Spatial association

We used spatial association tests based on the $G_{1,2}$, $G_{2,1}$, and $K_{1,2}$ functions of Diggle and Ripley (see Ripley [40] and Diggle [11] for a description of the methods). The limitation of these method is that one can only test the association between two groups, i.e., all pairs of species have to be analyzed (21 tests for 7 species in our case). The tests based on nearest neighbor distances ($G_{1,2}$ and $G_{2,1}$) are not symmetric, which enables one to detect asymmetric asssociation, e.g., to determine whether a species is attracted by the other or the reverse. When departure from spatial independence is detected, the characteristic distance of the test is an indication of association/inhibition distance between points of the two groups.

The study of spatial association between tufts of different species at the whole plot level lead to the following conclusions (Table 17.3):

- All tests leading to a significant repulsion involve *Loudetia simplex*. The other species concerned are *Andropogon schirensis*, *Hyparrhenia diplandra* and *H. smithiana*, i.e., the characteristic dominant species of the Andropogoneae savanna.
- The three species that show this negative association with *Loudetia* are all associated together, at long distance for some of them, but also, more significantly, at short distance (meaning that this association is not only due to the exclusion of those species from the *Loudetia*-dominated area).
- *Brachiaria brachylopha* is significantly associated with *Andropogon schirensis* and *Hyparrhenia smithiana*, and *Andropogon ascinodis* is significantly associated with *Hyparrhenia smithiana*.

What is more remarkable is that out of the 160 species pairs analyzed for spatial association on the 9 sub-plots, only 9 tests are significant (detailed results not shown here). Furthermore, 5 of these significant tests are obtained for sub-plot 2, suggesting that spatial association is more a property of the sub-plot than of the species. The significant tests are spread over all species. This indicates a surprising spatial independence at a small scale between grass tufts, apparently contradictory with the strong patterns observed for each individual species.

Table 17.3. Spatial association between species at the whole plot level. For each test (based on the $G_{1,2}$, $G_{2,1}$, or $K_{1,2}$ functions), test statistics (dw or L), P-value (P) and characteristic distances in meters are shown. Conclusions coded as R for repulsion, A for association, and - for independence. The threshold distance between long and short distances is 1.6 m. Straightforward species name abbreviations (see Table 17.1) (original results obtained from the Le Provost dataset [29]).

	$G_{1,2}$			$G_{2,1}$			$K_{1,2}$			Conclusion	
Species pair	dw	P	d_{max}	dx	P	d_{max}	L	P	d_{max}	Short d.	Long d.
L.s.×A.s.	−0.122	0.050	0.11	−0.162	0.000	0.59	−20.744	0.044	**2.40**	R	R
L.s.×B.b.	0.078	0.184	-	−0.059	0.710	-	−5.571	0.914	-	-	-
L.s.×H.d.	−0.043	0.664	-	−0.161	0.002	0.43	−22.080	0.080	-	R	-
L.s.×A.a.	0.027	0.934	-	−0.027	0.970	-	−7.734	0.908	-	-	-
L.s.×H.s.	−0.159	0.016	0.53	−0.089	0.406	-	−13.036	0.704	-	R	-
L.s.×A.c.	−0.052	0.782	-	0.134	0.192	-	25.044	0.160	-	-	-
A.s.×B.b.	0.058	0.018	0.37	0.039	0.702	-	6.247	0.408	-	A	-
A.s.×H.d.	0.076	0.000	0.16	0.093	0.004	0.16	10.599	0.048	**2.56**	A	A
A.s.×A.a.	−0.028	0.490	-	0.042	0.374	-	4.331	0.610	-	-	-
A.s.×H.s.	0.070	0.028	0.21	0.113	0.010	0.11	9.476	0.154	-	A	-
A.s.×A.c.	0.036	0.460	-	−0.023	0.796	-	7.109	0.486	-	-	-
B.b.×H.d.	0.091	0.124	-	0.045	0.348	-	8.151	0.242	-	-	-
B.b.×A.a.	0.065	0.428	-	0.035	0.704	-	6.598	0.422	-	-	-
B.b.×H.s.	−0.045	0.806	-	0.026	0.968	-	13.050	0.040	**3.17**	-	A
B.b.×A.c.	0.113	0.156	-	0.078	0.262	-	17.357	0.216	-	-	-
H.d.×A.a.	−0.034	0.538	-	−0.052	0.286	-	6.594	0.288	-	-	-
H.d.×H.s.	0.048	0.282	-	−0.049	0.486	-	12.291	0.014	**3.15**	-	A
H.d.×A.c.	0.045	0.542	-	0.030	0.826	-	8.557	0.504	-	-	-
A.a.×H.s.	0.038	0.690	-	0.050	0.520	-	13.651	0.010	**3.09**	-	A
A.a.×A.c.	0.051	0.574	-	0.041	0.488	-	8.656	0.560	-	-	-
H.s.×A.c.	0.069	0.272	-	0.117	0.052	-	10.823	0.500	-	-	-

Repulsion between plants is supposedly due to interspecific competition. Association between species can be due to a positive interference (when observed at short distance) or to soil heterogeneity at longer distances. The striking result here, compared to the previous section, is the very small number of significant tests apart from the classical separation between the *Loudetia* and Andropogoneae savanna types. Within those facies, almost no interaction between species is detected, suggesting an absence of interspecific competition.

In addition, the spatial association between adults and seedlings was studied. There is a significant repulsion between adults and seedlings at the plot level, with an average inhibition distance of 5-11 cm. This is shorter than the average distance between tufts of 21 cm and supports the hypothesis of a negative effects of adult tufts on seedlings. This is consistent with the lower survival rate of seedlings growing close to adult tufts reported by Garnier [19].

17.2.3 Discussion

Spatial pattern studies of grass-dominated communities are rarer than in tree-dominated systems, probably because grass individuals are more difficult to identify, restricting studies to systems like Lamto where tufted grasses dominate.

Spatial pattern analyses of grass-dominated communities usually lead to the conclusion that interference between adult plants is significant [32, 2], but more detailed analyses show that interspecific competition is weak compared to intraspecific competition [8, 39]. Spatial aggregation might be an explanation for this weakness [39]. We found an absence of spatial regularity at the whole community level that suggests an absence of interference at the adult stage, somewhat contradictory with the literature, but for which we have a possible explanation. On the contrary, the strong spatial structure at the species level observed in Lamto seems to be a general feature of perennial tufted grasslands.

In most perennial grassland systems, seedling recruitment is heavily impaired by adult plants (negative effect of adults on recruitment at distances < 20 cm [3] and < 15 cm [42], leading to apparently stochastic recruitment patterns at the community level [41]. Our results are consistent with those conclusions.

In conclusion, the spatial patterning of the grass layer in Lamto is comparable with those observed in other systems. Its major originality seems to be the weak intraspecific interference deduced from the random or aggregated pattern of adult plants.

17.3 Dynamics of the grass layer

17.3.1 Tuft dynamics

César [10] mapped the detailed structure of tufts on six $1\,\mathrm{m}^2$ plots and surveyed them during three consecutive years (e.g., Fig. 17.4). Tufts changed in size and exact location from year to year, due to variation in tilling. Some tufts get fractioned in smaller ones, and coalescence between tufts were observed. On the plot with the steepest slope, César observed a migration of all tufts in the same direction upslope, which reminds one at a smaller scale of the dynamics of tiger bush ecosystems [33, 43].

17.3.2 Toward demographic studies of the grass layer

Very little information is available on the population dynamics of the main grass species. A few studies enable one to estimate basic demographic parameters, either for the whole grass cover or for some particular species.

On the plot used for the spatial pattern analysis, 3 tufts out of 3179 were recorded as dead, which provides a rough estimate of the mortality rate ($9.4\times 10^{-4}\,\mathrm{y}^{-1}$). From the plots followed during 3 years by César [10] (Fig. 17.4), the fate of old large tufts is more often fractionation into smaller tufts rather than death of the whole tuft.

In a recent demographic study of *Hyparrhenia diplandra*, Garnier and Dajoz [18] found a very low adult mortality rate: during the 3 years of the study,

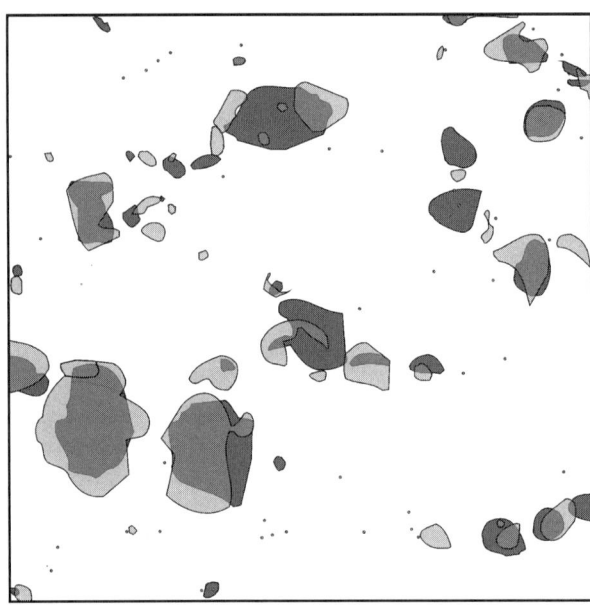

Fig. 17.4. Evolution of a $1\,\mathrm{m}^2$ plot between 1969 and 1970. Dark grey, 1969 locations; light grey, 1970 locations. Overlaps between positions appear as an intermediate color (redrawn from [10], with permission of the author and the Centre International de Recherche en Agronomie pour le Développement).

no mortality was observed among 60 plants located in a burned and unburned plot ($N = 30$ in each plot). In *Hyparrhenia diplandra*, the germination rate is 0.30 y^{-1} in burned plots and 0.20 y^{-1} in unburned plots. One year after germination, seedling survival rate was in average of 0.1 y^{-1} in burned plot and 0.34 y^{-1} in unburned plot. When seedlings had survive the first fire, their survival rate was high and was 0.73 y^{-1} in burned plot and 0.80 y^{-1} in unburned plots [18].

These demographic parameters are similar to those found by Silva and Castro [42]. They are typical of long-lived species: in a first approximation, the demographic strategy of the dominant grass species in Lamto look quite similar to that of the trees.

17.4 Reproduction system of *Hyparrhenia diplandra* and its population genetic structure as revealed by microsatellites

Spatial patterns and demography are tightly linked to the genetic polymorphism of natural populations, a fundamental property of ecosystems which is of direct importance for population dynamics and evolutionary ecology [6].

Functional ecology is also concerned in intraspecific diversity because of environmental heterogeneity. Subdivided and diverse populations are able to deal with environmental patchiness and variability [15]. The extent and structure of genetic diversity of natural populations is hence essential for assessing their evolutionary potential and understanding their biology [31]. Spatial genetic structure has also a practical interest in designing sampling strategies [15]. The genetic diversity is managed and partitioned by a system composed of three different elements: the mode of reproduction, which may be sexual or asexual or a mixture of both, the mating system of the sexual reproduction, which is characterized by the degree of inbreeding it allows for [24], and the dispersal strategies which govern migration rates and the level of sub-population differentiation, counter-balancing the genetic drift. To address these questions, *Hyparrhenia diplandra* at Lamto offers a favorable material with its large, continuous, and stable populations.

When we began this work, a few cytological data only were available on *H. diplandra* [26] that suggested tetraploidy. The observation of irregular male meiosis gave one hint of apomixis [12]. An irregular meiosis leads to poor efficiency of sexual reproduction. The persistence in the field of an organism which displays irregular meiosis suggests that this organism escapes the low efficiency of its sexual reproduction through vegetative reproduction.

During the last decade, microsatellites were characterized for *H. diplandra* and used to check if *H. diplandra* is actually apomictic in Lamto. The population genetic structure of *H. diplandra* was also studied in the field.

17.4.1 Isolation and characterization of microsatellites

Two microsatellite markers very specific to *Hyparrhenia diplandra* were designed (Table 17.4). In addition, a maternal, chloroplastic marker was determined to trace seed movements [34].

Mode of reproduction

In the conditions of Lamto, two consequences of apomixis can be verified: (1) progenies from open pollinated plants should be genetically identical to

Table 17.4. Two nuclear (Hd1 and Hd2) and one chloroplastic (ChHd1) microsatellite of *Hyparrhenia diplandra*, including repeat motif, forward and reverse primer sequences, annealing temperature (Ta), and number of alleles and gene diversity (He) observed in random samples.

Locus	Repeat motif	Primer sequence (5'-3')	Ta (°C)	Size range (bp)	# alleles	He
Hd1	$(CA)_{11}$	F: TACGTGGACGGACTGACCTC R: GATCCGTGTCGTGTCGTGTC	56	105-129	6	0.65
Hd2	$(CA)_{10}$	F: GTCATCAAGAGGCAGCTGGC R: GCAGAGGCTGAACACATGCG	60	148-224	6	0.64
ChHd1	A_{12}	F: CAGGCTCTATCCATTTATTC R: CGTCCAGACTCTTGGTAGAG	55	150-152	3	0.41

each other, and to their mother-plant [23]; (2) some siblings with 50% more chromosomes should be detected at a rather high frequency [36]. These two features of apomixis were tested in *Hyparrhenia diplandra* by comparing the genotype of several plants in the field and the genotype of their progeny produced in free pollination, and by looking for chromosome variants in a set of plantlets.

Three hundred thirty-eight siblings from 31 different families sampled in sites A, B, C, and D (Fig. 17.5) were genotyped with the 2 nuclear microsatellite markers. All but four displayed the same genotype as their mother-plant, with no other allele and no loss of one maternal allele. Two families (38 siblings) were homozygous at both loci, but the other were more informative: 22 displayed 2 alleles at 1 locus and 7 at both loci. In the case of sexual reproduction, the probability of observing only maternal genotypes is very low. When the female gamete is produced through the process of meiotic reduction, it has lost one of the two maternal alleles in every case if the plant behaves as a diploid, or in 1/3 if it is an autotetraploid of genotype A2a2, the most unfavorable case for our demonstration. The maternal genotype is recovered throught fusion with a complementary pollen grain at the maximum frequency of 0.5 in case of a diploid and an allotetraploid or 34/36 in case of a mother-plant of genotype A2a2. The probability of observing 334 progenies identical to their mother-plant under the hypothesis of sexual reproduction is at most 2.8×10^{-10} in the hypothesis of diploidy (or allotetraploidy) or 5.1×10^{-9}

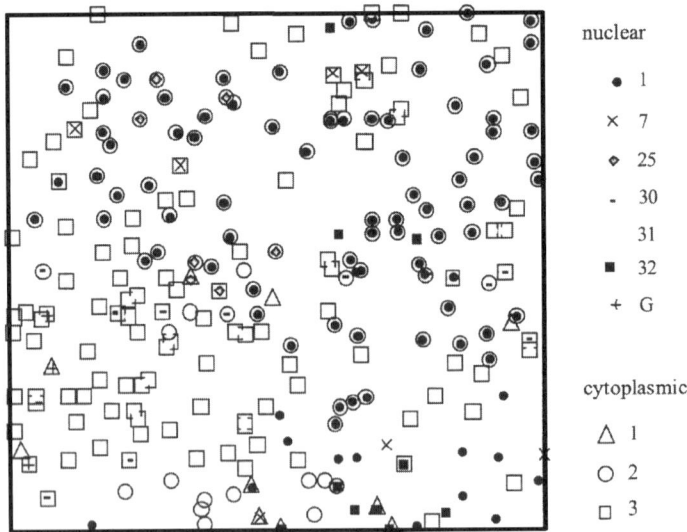

Fig. 17.5. Cytoplasmic (open symbol) and nuclear (solid symbols) genotype spatial pattern in a 10×10 m quadrat. G, nuclear genotypes with a frequency < 2% (after [20], with permission of the Botanical Society of America).

in the case of an autotetraploid. The actual probability is even lower, because this calculation does not take into account the families heterozygous at both loci. The most probable is apomixis. Four genetic variants only were detected, a single one being possibly due to a true sexual event. This allowed us to calculate a 95% confidence interval for residual sexuality: 0.0017-0.015. An analysis of DNA amount by flow cytometry [7] of 127 siblings from this sample led to the detection of 4 ploidy variants, with 50% more DNA than standard progenies, a specific mark of apomixis.

Diploid/tetraploid

The expected somatic number of chromosomes of 40 and the tetraploidy of *H. diplandra* from Lamto were confirmed. The pollen viability was 43%, as revealed by the coloration of Alexander [13], suggesting that male meiosis is indeed irregular [12]. As we observed mostly individuals displaying only one or two alleles at loci Hd1 and Hd2, we processed the data with the Genepop software [38] as if it were an allotetraploid, that is to say, functionally a diploid.

Population structure

A first sketch was obtained from studies on the first sample (Fig. 17.5). Plants were sampled in two populations: one in Site A, where nitrification is inhibited, and one 4 km away in site B, where nitrification occurs [27]. We made a hierarchical sampling: in each population, we sampled 6 groups of 5 plants separated from each other by 15 m; in each group, a central individual together with its 4 neighbors distant by 1 m (Fig. 17.5); 30 individuals were sampled in each site. As the sampling was done in December, when the seeds are mature, seeds were collected together with vegetative tissue from the central plant of each group of five.

The whole genetic diversity is rather high ($H_e = 0.646$). The populations from site A and site B differ with a genetic difference between populations (F_{ST}) of 0.05 ± 0.03, significantly ($P<0.001$) with both loci. There is a high excess of homozygotes in each population ($P<0.0001$), due to a Wahlund effect because each population is structured ($P<0.00001$ in site A and $P<0.001$ in site B). This is confirmed by the high F_{ST} value for each population: 0.257 ± 0.056 for p_{stN} ($P < 10^{-5}$) and 0.212 ± 0.011 ($P = 10^{-3}$) for p_{stS}. Indeed, when the 5 plant groups, distant from each other by 15-30 m, are compared 2 by 2, they are found significantly different ($P < 5\,10^{-2}$) in one-third of the comparisons (11/30). The structuration is even stronger at the genotype level: nearby growing plants (1 m distant) share often (30%) identical genotypes inside of the same group, but when compared to plants from other groups (15-33 m distant) in the same site, the figure drops to 4.7%, of the same order of magnitude as the comparison between the two sites (6.8%).

The two loci combined together allow for the definition of 23 different genotypes among the 59 plants genotyped (Fig. 17.5). Most (15/23) of the genotypes occur only once or twice; most are restricted to a single group or population. Only the two most frequent genotypes are encountered both in site A and site B. The parameters describing genotype richness ($G/N = 0.39$), genotype diversity ($D = 0.91$), and evenness ($E = 0.88$) all reflect that despite uniform progenies, H. diplandra manages to maintain a high genotypic diversity, without predominance of a single genotype. This diversity is encountered at the sub-population level, as despite the small sample number, groups of 5 plants are composed of at least 2 different genotypes, up to 4.

Mapping of individual plants and spatial analysis of correlation between their microsatellites revealed a strong short scale structuring of the population, with correlation below 1-2 m and no correlation above 6 m (Fig. 17.6).

The expected population structure in a facultative apomict is explained by gene dispersal through two different ways:

1. whole genotypes reproduced by apomixis are dispersed by seed in a narrow range; and
2. nuclear markers are dispersed by pollen as single genes in a larger range.

As a consequence, there are three levels of probability of allele identity at increasing distances from a plant:

1. At short distance, one expects a very high level of allele and genotype identity because the genotypes reproduced by apomixis are dispersed by seeds; the structure revealed by maternal markers should fit genotype structure.
2. At average distance, lower, but still rather high levels of allele identity, but not of genotype identity, should reflect the influence of gene dispersal by the pollen through residual sexuality.

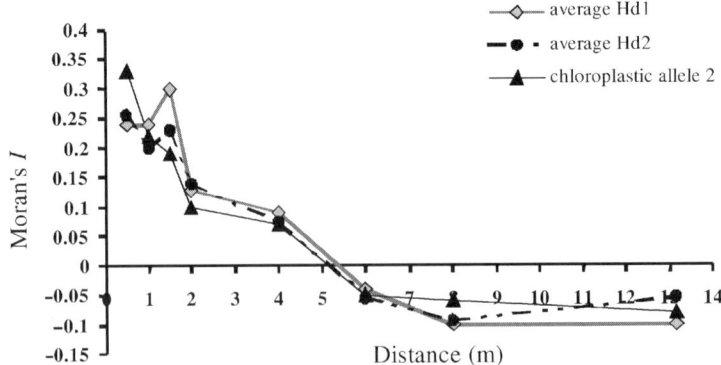

Fig. 17.6. Correlograms of Moran's I mean values at nuclear loci Hd1 and Hd2 and chloroplastic locus HdcpA in quadrat 2.

3. At long distance, this structuration due to relative kinship is erased by distance.

Our observations allow us to identify these three levels of structuration: in the 1-2 m range, predominance of apomictic reproduction and seed dispersal; in the 15-30 m range, pollen dispersal and residual sexuality; at 4 km, isolation by distance brings probability of allele identity to a low level.

Seed and pollen dispersal ability

These conclusions are corroborated by studies on morphological and ecological factors concerning the dispersal abilities of *H. diplandra* [18]. Seeds fall at a mean distance of 86 ± 40 cm from the mother-plant, in correlation to plant height, and they are buried in the soil by the large, hygroscopic awn, which is not a dispersal device. On the other hand, pollen disperses 10-12 m away from the plant, as indicated by pollen-trap experiments in the field. Floral morphology and flowering phenology enable sexual reproduction, as shown by observations in the field: anthers are exerted early in the morning; dehisce within 30 min to 1 h and are emptied by noon of their pollen. Flowering is slightly protandrous, since in most cases, stigmas are exerted between a few hours and 1 day after the anthers have dehisced. Exerted stigmas remain receptive for up to 3 days. This gives a basis for residual sexuality and shows the correlation of dispersal abilities with population genetic structure.

Characterization of residual sexuality

To verify the type of residual sexuality within the range of influence of pollen dispersal, a second sample was studied [19]: 120 plants separated by 20-30 m; 30 in site A, 30 in site B, 30 in each of two sites C and D, close to each other (separated by a road) and distant from site A by 500 m (Fig. 17.5).

Slight structuring was confirmed above 40 m ($F_{ST} = 0.055 \pm 0.015$), whatever the distance between populations is, 40 m, 500 m, or 4 km. Intrapopulation heterozygote deficiency (F_{is}) was 0.425 ± 0.029, indicating that the residual sexuality occurs mainly between related individuals. Together with the analysis of genotype diversity in the first sample, we could estimate the dispersal rate of clones through seeds: 0.040-0.123 m. Concerning genotype diversity, we encountered the same features as in the structured samples: many different two-loci genotypes (49 for 120 plants), most of them occurring only once, most limited to a single population. A single genotype occurs in the four populations.

Short scale structuring

In order to have a better estimation of the short scale structuration resulting from seed dispersal, an exhaustive sample was analyzed by Garnier [19]: every tuft of basal diameter equal or above 5 cm were mapped within two quadrats of 10×10 m, separated by 40 m in site B. There were 169 individuals in 1 quadrat, 245 in the other. They were analyzed with the two nuclear microsatellites, which revealed dispersal of whole genotypes by seeds and dispersal of genes by pollen and also with the chloroplastic microsatellite to follow the dispersal of maternal lines by seeds.

The overall genic diversity was of the same order as in the other samples ($H_e = 0.68$). The F_{ST} value calculated between the two plots was significantly lower for nuclear loci ($F_{ST} = 0.089 \pm 0.037$, $P<0.0001$) than for the chloroplastic marker ($F_{ST} = 0.417$, $P<0.0001$). Each quadrat was highly structured ($F_{ST} = 0.20 \pm 0.07$, $P < 10^{-5}$) for both type of markers. When the quadrats were subdivided into 16 squares of 2.5 × 2.5 m, the F_{ST} calculated for pairs of such squares was not different from zero when the squares were adjacent, and they took the value of 0.2 when they were distant by 5-7 m and above. The analysis by F_{ST} was confirmed by spatial autocorrelation analysis, another way of uncovering structuration among plant populations [25]. Values of Moran's autocorrelation index were high at a 1-2 m distance from each plant, and fall to 0 at a distance of 6 m, for the nuclear markers as well as the maternal marker. In this sample, the genotypic diversity was also high for such a sampling of adjacent tufts, with 58 different genotypes for 414 plants. The clones never formed homogeneous patches, but were mixed together, as the elementary squares of 2.5 × 2.5 m carried on average 5.4 different genotypes for 12 plants.

The analysis of this exhaustive sample gave precisions about the fine scale structure of *H. diplandra* populations: alleles, nuclear as well as maternal, and genotypes are highly similar within a circle of 6 m diameter and some similarity of nuclear alleles only is observed at a 40 m distance. That the first, high level of similarity is due to seed dispersal of whole genotypes is evidenced by the parallelism of the figures for the nuclear and the maternal marker. Maternal lineages are fixed in place, resulting in a patchy clonal structure [20].

Preliminary genetic characterization of the *Hyparrhenia* individuals from the low and high nitrification savanna areas

The plants used by Lata et al. [28] (see Chap. 15) in their transplantation experiment have been genotyped. These plants were sampled at random, distant from each other by 10-20 m, in the two sites A and B, and characterized for their ability to inhibit nitrification in the soil. In site A, 38 plants were inhibitors of nitrification and 7 were non-inhibitors; in site B, 7 were inhibitors and 41 plants were non-inhibitors, both types being mixed together. Amplification with the three microsatellite loci was performed successfully on both ecotypes, confirming that they belong to the same botanical species, *H. diplandra*. Gene (0.53) and genotype (36 genotypes for 86 plants) diversity was somewhat smaller than in the other samples. The same genetic differentiation was found as in the random samples ($F_{st} = 0.09 \pm 0.02$, $P = 10^{-4}$)

between geographical sites as well as between ecophysiological types. The ecophysiological criterion did not bring additional genetic differentiation to the geographical one. The ability to inhibit nitrification was not correlated to the chloroplastic marker. A linkage to the nuclear locus Hd2 was detected, associated to the predominance of one genotype in the sample from site A (19/45). This genotype was also one of the most frequent in the other samples. It is unclear if it happened by chance to be predominant in this sample from site A or if it reflects a linkage between this locus and the ability to inhibit nitrification [28].

17.4.2 Discussion

Three main consequences of apomixis have been checked on the progeny of *Hyparrhenia diplandra*:

1. Progenies are homogeneous and genetically identical to the mother-plant in almost every case.
2. Genetic variants due to possible sexual events are encountered at a frequency of about 1%.
3. Chromosomic variants, resulting from the addition by the pollen of a reduced complement to the unreduced ovule, occur also at a frequency in the 1% range.

This allowed us to infer that the reproduction system of *Hyparrhenia diplandra* is facultative apomixis.

We uncovered the main features of the genetic structure of this facultative apomict: in the 0-6 m range, the influence of dispersal by seeds of the whole genotypes produced by apomixis is predominant; between 10 and 40 m, pollen dispersal of single genes recombined by residual sexuality erases the influence of clone dispersal and suppresses the correlation of whole genotypes but allows for some level of allele identity; and above 40 m, no more structure is detected with microsatellites, indicating that dispersal by pollen is no more significant beyond that limit.

Due to conform reproduction of genotypes, low genetic diversity is expected within populations of asexually reproducing plants, with few genotypes in each population. In the reverse, we observed a high within-population allelic and genotypic diversity in *H. diplandra*, a paradoxical situation which has been reported in many asexually reproducing species [14] and which may result from the residual sexual reproduction [17]. For the residual sexuality, we found a high inbreeding rate, 0.6 in the hypothesis of an allotetraploid, at odds with classical views of apomixis as a way of propagating heterozygotes [4] and recent observations on dandelion with microsatellites: every genotype is totally heterozygote [16]. This discrepancy could be explained by considering that dandelion is representative of interspecific hybrid apomicts [35] and *H. diplandra* is representative of apomicts with residual sexuality.

17.5 Conclusion

The results obtained on the grass layer dynamics and grass demography in Lamto indicate that the former homogenous green slime view of the grass layer, used in primary production and ecophysiological studies, is not applicable to demographic and genetic processes. From these points of view, the grass layer appears at least as structured as the tree layer: very complex spatial patterns at the species level and very strong genetic differenciation in spite of a dominance of clonal reproduction. First estimates of demographic parameters suggest that the main grass species are long lived, probably as long as trees. All these results will form the basis of a more systematic study of the demographic aspects of the grass layer in the future.

References

1. L. Abbadie, A. Mariotti, and J.C. Menaut. Independence of savanna grasses from soil organic matter for their nitrogen supply. *Ecology*, 73(2):608–613, 1992.
2. M.O. Aguilera and W.K. Lauenroth. Neighborhood interactions in a natural population of the perennial bunchgrass *Bouteloua gracilis*. *Oecologia*, 94:595–602, 1993.
3. M.O. Aguilera and W.K. Lauenroth. Seedling establishment in adult neighbourhoods - intraspecific constraints in the regeneration of the bunchgrass bouteloua gracilis. *Journal of Ecology*, 81(2):253–261, 1993.
4. S.E. Asker and L. Jerling. *Apomixis in plants*. CRC Press, Boca Raton, FL, 1992.
5. S. Barot, J. Gignoux, and J.C. Menaut. Demography of a savanna palm tree: Predictions from comprehensive spatial pattern analyses. *Ecology*, 80(6):1987–2005, 1999.
6. N. Barton and A. Clark. Population structure and processes. In K. Wöhrmann and S. Jain, editors, *Evolution. Population biology; ecological and evolutionary viewpoints*, pages 115–173. Springer-Verlag, Berlin, 1990.
7. S.C. Brown, C. Bergounioux, S. Tallet, and D. Marie. Flow cytometry of nuclei for ploidy and cell cycle analysis. In I. Negrutiu and G. Gharti-Chhetri, editors, *A laboratory guide for cellular and molecular plant biology*, pages 326–345. Birkhaüser-Verlag, Basel, 1991.
8. B.S. Collins and J.E. Pinder. Spatial distribution of forbs and grasses in a South Carolina oldfield. *Journal of Ecology*, 78:66–76, 1990.
9. J. César. *Etude quantitative de la strate herbacée de la savane de Lamto*. Ph.D. thesis, Université de Paris, Paris, 1971.
10. J. César. *La production biologique des savanes de Côte d'Ivoire et son utilisation par l'homme*. CIRAD - Institut d'élevage et de médecine vétérinaire des pays tropicaux, Maisons-Alfort, 1992.
11. P.J. Diggle. Preliminary testing for mapped patterns. In P.J. Diggle, editor, *Statistical analysis of spatial point patterns*, pages 10–23. Academic Press, New York, 1983.
12. M. Dujardin. Additional chromosome numbers and meiotic behaviour in tropical African grasses from western Zaïre. *Canadian Journal of Botany*, 57:864–876, 1979.

13. J. Durand, L. Garnier, I. Dajoz, S. Mousset, and M. Veuille. Gene flow in a facultative apomictic *Poacea*, the savanna grass *Hyparrhenia diplandra*. *Genetics*, 156:823–831, 2000.
14. N. C. Ellstrand and M. L. Roose. Patterns of genotypic diversity in clonal plant species. *American Journal of Botany*, 74:123–131, 1987.
15. B. Epperson. Spatial patterns of genetic variation within plant populations. In A. Brown, M. Clegg, A. Kahler, and B. Weir., editors, *Plant population, genetics, breeding, and genetic resources*, pages 229–253. Sinauer Associates Inc., Sunderland, MA, 1990.
16. M. Falque, J. Keurentjes, J. Bakx-Schotman, and P.J van Dijk. Development and characterization of microsatellite markers in the sexual-apomictic complex *Taraxacum officinale* (dandelion). *Theoretical and Applied Genetics*, 97:283–292, 1998.
17. T.M. Gabrielsen and C. Brochmann. Sex after all: High levels of diversity detected in the artic clonal plant *Saxifraga cernua* using RAPD markers. *Molecular Ecology*, 7:1701–1708, 1998.
18. L.K. Garnier and I. Dajoz. The influence of fire on the demography of a dominant grass species of West African savannas, *Hyparrhenia diplandra*. *Journal of Ecology*, 89:200–208, 2001.
19. L.K.M. Garnier. *Stratégies démographiques et structuration génétique chez une espèce apomictique tropicale,* Hyparrhenia diplandra. Ph.D. thesis, Université de Paris 6, Paris, 2000.
20. L.K.M. Garnier, J. Durand, and I. Dajoz. Limited seed dispersal and microspatial population structure of an agamospermous grass of West African savannahs, *Hyparrhenia diplandra* (Poaceae). *American Journal of Botany*, 89:1785–1791, 2002.
21. J. Gignoux. *SPASTAT: Un logiciel pour l'analyse de répartitions spatiales par les méthodes de Diggle et Ripley*. Ecole Normale Supérieure, Paris, 1994.
22. J. Gignoux, C. Duby, and S. Barot. Comparing the performances of Diggle's tests of spatial randomness for small samples with and without edge-effect correction: Application to ecological data. *Biometrics*, 55(1):156–164, 1999.
23. W. Hanna and E. Bashaw. Apomixis: Its identification and use in plant breeding. *Crop Science*, 27:1136–1139, 1987.
24. P. Hedrick. Mating systems and evolutionary genetics In K. Wöhrmann and S. Jain, editors, *Evolution. Population biology; ecological and evolutionary viewpoints*, pages 115–173. Springer-Verlag, Berlin, 1990.
25. J. Heywood. Spatial analysis of genetic variation in plant populations. *Annual Review of Ecology and Systematics*, 22:335–355, 1991.
26. P. Kammacher, G. Anoma, E. Adjanohoun, and L. Ake-Assi. Nombres chromosomiques de Graminées de Côte d'Ivoire. *Candollea*, 28:191–217, 1973.
27. J.C. Lata. *Interactions entre processus microbiens, cycle des nutriments et fonctionnement du couvert herbacé: Cas de la nitrification dans les sols d'une savane humide de Côte d'Ivoire sous couvert à* Hyparrhenia diplandra. Ph.D. thesis, Université de Paris 6, Paris, 1999.
28. J.C. Lata, J. Durand, R. Lensi, and L. Abbadie. Stable coexistence of contrasted nitrification statuses in a wet tropical savanna ecosystem. *Functional Ecology*, 13:762–768, 1999.
29. E. Le Provost. *Structure et fonctionnement de la strate herbacée d'une savane humide (Lamto, Côte d'Ivoire)*. M.Sci. thesis, Universités de Paris 6 & 11, Institut National Agronomique Paris-Grignon, Paris, 1993.

30. X. Le Roux. *Etude et modélisation des échanges d'eau et d'énergie sol-végétation-atmosphère dans une savane humide (Lamto, Côte d'Ivoire)*. Ph.D. thesis, Université de Paris 6, Paris, 1995.
31. Y. Linhart and M. Grant. Evolutionary significance of local genetic differentiation in plants. *Annual Review of Ecology and Systematics*, 27:237–277, 1996.
32. R.N. Mack and J.L. Harper. Interference in dune annuals: spatial pattern and neighbourhood effects. *Journal of Ecology*, 65:345–363, 1977.
33. A. Mauchamp, S. Rambal, and J. Lepart. Simulating the dynamics of a vegetation mosaic: a spatialized functional model. *Ecological Modelling*, 71:107–130, 1994.
34. D. E. McCauley. The use of chloroplast DNA polymorphism in studies of gene flow in plants. *Trends in Ecology and Evolution*, 10:198–202, 1995.
35. S. Menken, E. Smit, and H.C.M. Den Nijs. Genetical population structure in plants: Gene flow between diploid sexual and triploid asexual dandelions (*Taraxacum* section *Ruderalia*). *Evolution*, 49:1108–1118, 1995.
36. G. Nogler. Gametophytic apomixis. In J. BM, editor, *Embryology of Angiosperms*, pages 475–518. Springer-Verlag, Berlin, 1984.
37. J.P. Puyravaud. *Processus de la production primaire en savanes de Côte d'Ivoire. Mesures de terrain et approche satellitaire*. Ph.D. thesis, Université de Paris 6, Paris, 1990.
38. M. Raymond and F. Rousset. GENEPOP (version 1.2): A population genetics software for exact tests and ecumenicism. *The Journal of Heredity*, 86:248–249, 1995.
39. M. Rees, P.J. Grubb, and D. Kelly. Quantifying the impact of competition and spatial heterogeneity on the structure and dynamics of a four-species guild of winter annuals. *American Naturalist*, 147(1):1–32, 1996.
40. B.D. Ripley. *Spatial statistics*. John Wiley & Sons, New York, 1981.
41. G. Rusch. Spatial pattern of seedling recruitment at 2 different scales in a limestone grassland. *Oikos*, 65(3):433–442, 1992.
42. J.F. Silva and F. Castro. Fire, growth and survivorship in a neotropical savanna grass *Andropogon semiberbis* in Venezuela. *Journal of Tropical Ecology*, 5:387–400, 1989.
43. J.M. Thiéry, J.M. d'Herbès, M. Ehrmann, S. Galle, C. Peugeot, J. Seghieri, and C. Valentin. Modèles de paysage et modèles de ruissellement pour une brousse tigrée nigérienne. In *Journées PIREVS - Les Temps de l'Environnement*, pages 557–563, Toulouse, 1997. ORSTOM.

18

Structure, Long-Term Dynamics, and Demography of the Tree Community

Jacques Gignoux, Sébastien Barot, Jean-Claude Menaut, and Roger Vuattoux

18.1 Introduction

In this chapter, we study the population and community dynamics of the major tree species of the most common savanna type in Lamto, the Andropogoneae shrub savanna (see Sect. 5.2). The most frequent tree species are: *Borassus aethiopum*, *Bridelia ferruginea*, *Crossopteryx febrifuga*, *Cussonia arborea* and *Piliostigma thonningii*. Of these, only the palm tree *Borassus* can be considered as a true tree species since it can reach 20 m in height; the other species are smaller and never grow over 10-12 m. These five species will be the main subject of our study since they comprise 90% of tree individuals. Another species, *Annona senegalensis*, is very common in Andropogoneae savannas, and although it can develop a real (small) tree morphology, it never grows over 3 m in burned savannas. Other less frequent, true tree (up to 15-20 m in height) species will sometimes be examined in this chapter: *Terminalia shimperiana*, *Zanthoxylum zanthoxyloides*, *Lannea barteri*, *Vitex doniana*, *Pterocarpus crinaceus* and *Ficus sur*. These species are usually rare in the savanna, but can be locally abundant in fire-protected sites like rocks and dense tree clumps: *Terminalia shimperiana* often dominates savanna woodlands.

Tree dynamics is driven by competition for resources and fire. As fire intensity varies considerably in space (Sect. 4.5), as nutrient availability is low in the savanna except in some rich patches (Sect. 4.3), and as the seed dispersal distances of trees are small ([12]; Gignoux, unpublished data), the population dynamics of trees is linked to the spatial structure of the ecosystem. For these reasons, we examine in this chapter tree population dynamics in relation to population spatial structure.

18.2 Factors influencing tree population dynamics

18.2.1 Competition for resources

Despite the high rainfall and the leafless state of trees during the dry season in Lamto [50], water stress can occur for tree species with a shallow rooting

Fig. 18.1. Typical *Loudetia simplex* grass savanna (top) and Andropogoneae tree savanna (bottom). Notice the presence of trees only on small mounds and on the rocky outcrop in the *Loudetia* savanna, and the presence of trees scattered everywhere in the Andropogoneae savanna (photographs by S. Barot and J. Gignoux).

depth, e.g., *Crossopteryx febrifuga*, during the dry season [48]. Konaté et al. [47] measured a slower leaf fall for *Crossopteryx febrifuga* trees growing on termite mounds, which are spots of higher shallow water availability, than for trees growing in the open; this difference was absent for a deep rooted species like *Cussonia arborea*. On the contrary, water affects community dynamics when it is too abundant, e.g., in the hydromorphic soils of flat areas dominated by the grass *Loudetia simplex* (Sect. 5.2). In these areas, tree survival seems impossible except on mounds located above the water table (Fig. 18.1). However, there is evidence that water stress in these areas is even stronger than in the Andropogoneae savannas.

The soils of Lamto are characterized by a very low average nutrient availability ([26] and Sect. 4.3), but are locally nutrient rich at different scales: termite comb chambers and grass rhizospheres concentrate nitrogen ([2, 1] and Sect. 15.6). Termite activity is concentrated in mounds of various size, visible in the landscape. Termite mounds usually have a random or regular spatial pattern [11]. Trees are spatially associated to termite mounds [11] and are more numerous on mounds than in the open [1]. Apparently, the soil of termite mounds, more nutrient-rich and with a higher water storage capacity, is able to support higher tree density than the surrounding savanna. Tree clumps, even without mounds, are also more nutrient-rich than the surrounding savanna ([55] and Sect. 8.2).

Finally, competition for light is probably important, at least in dense tree clumps. As trees of the same species may be clumped or individually scattered (Fig. 18.1), trees of the same developmental stage can experience very different light conditions.

The major characteristic of resources in Lamto is the heterogeneity of their spatial distribution: whereas no resource seems to be the main driver of the whole ecosystem, the three resources described here influence competition between trees, at least at some places or during some periods of the year. As a consequence, we expect population and community dynamics to be intimately linked to the spatial structure of the ecosystem. Tree demographic parameters (survival rates, growth rates, fecundity, etc.) should be different for trees growing on a mound, in a tree clump, or in the open.

18.2.2 Fire and the definition of demographic stages

Fire prevents tree invasion, as 40 years of protection from burning have demonstrated in Lamto [72, 73, 27]. This led to the conclusion that fire was the major factor explaining the existence of Lamto savannas in climatic conditions where rainforests could survive [54].

Other arguments support the idea that fire is a major driving force of tree communities in Lamto: the dominant savanna tree species all show a very high resistance to fire, based on a high resprouting ability for dicotyledon trees [39] and on a very good bud protection for *Borassus* palm trees [71].

Like forest trees, adult savanna trees can resist fire through adaptations like a thick bark [42, 61] and a high resprouting ability from belowground organs. But the regularity of savanna fire, the low flame height, and the short time of exposure to flame (Sect. 4.5) give adult trees other possibilities of resistance: leaf fall occuring during the dry season minimizes the chance of a crown fire; once a tree overgrows the grass stratum, most of its buds are located above flame height [32]. There is evidence that low intensity fires kill smaller stems, but not taller ones in fire-tolerant species [57]. Before young trees can reach the safe zone above the grass layer, they have to survive within the fuel bed where fire intensity is highest [24]. Therefore, fire objectively defines a recruitment stage, where a tree acquires a perennial, fire-resistant trunk instead of producing resprouts each year from belowground organs, like a grass tuft. Based on average flame height, Menaut and César [52] considered

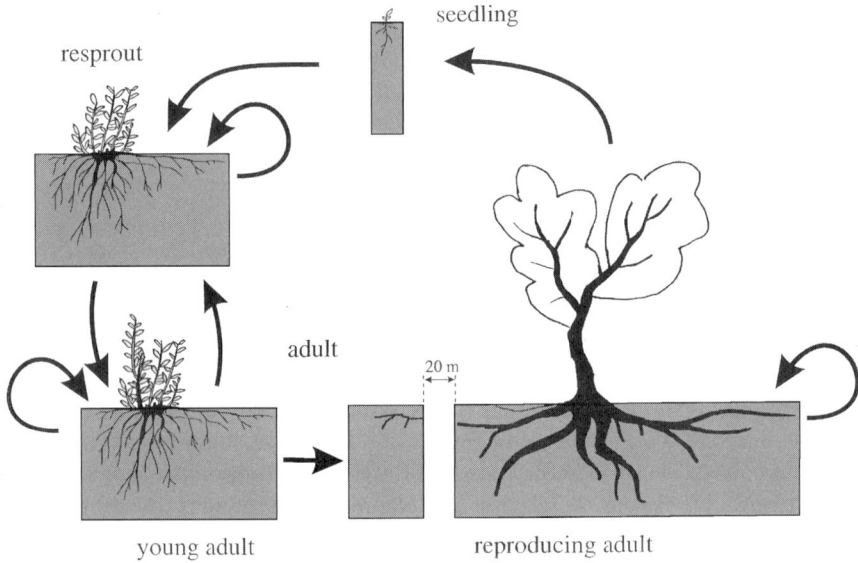

Fig. 18.2. Definition of demographic stages based on morphological traits (fire scars). Perennial parts of the trees are in black and annual parts are in white. Seedlings are identified by the presence of cotyledons or their small size together with the absence of record at their location at the previous census. Resprouts have a perennial belowground system and an annual aboveground systems, i.e., all stems or their tallest stem bear no fire scars. Adults (either young or reproducing) have a perennial aboveground system, i.e., their tallest stem bears fire scars.

a 2 m tall limit for defining the young and adult stages. Through a detailed study of fire resistance in *Crossopteryx febrifuga* and *Piliostigma thonningii*, Gignoux et al. [39] demonstrated that this limit actually varied among species: *Crossopteryx* can recruit as soon as it reaches 25 cm and will almost certainly build a trunk when taller than 65 cm, while *Piliostigma* cannot recruit before 100 cm and will almost certainly recruit when 225 cm tall.

It is possible to see on all individual trees (except palms) the scars left by fire (Fig. 18.2). This was used as a basis for classifying trees in meaningful demographic stages, since tree age cannot be assessed simply by ring counting (growth rings are irregular (unpublished data by Menaut)). Unburned stems can be distinguished from burned ones, since burned stems always have their killed top remaining during most of the wet season. This enables one to classify trees as adults when they have a perennial trunk (2 years old) or as resprouts when they only have annual stems; intermediate cases exist when both annual and older stems are present: they have been classified as resprouts or adults according to whether their tallest stem was an annual stem or an older stem.

Cussonia has a particular architecture (Leeuwenbergs model [43]) and does not branch before it reaches 2-3 m in height; but the leaf scars are visible on the stems and tend to be closer to each other at the end of the growing season, so that is is possible to count the number of growing seasons on young (1-5 years) stems and apply the same stage classification as for the other species.

18.3 Spatial patterns of tree species

Spatial pattern analysis relies on the hypothesis that is is possible to infer some conclusion about the processes leading to the observed patterns simply by analyzing these patterns [49]. Problems arise from the fact that it is usually difficult to find a single cause for a pattern: many processes can often explain the same pattern. However, Barot [11] argues that with a carefully chosen set of hypotheses and appropriate data, it is possible to draw some conclusions on the underlying processes (or at least to screen out processes) from the observed patterns. As for the grass population patterns (Sect. 17.2), we analyzed the spatial pattern of trees using Diggle's and Ripley's methods [29, 66] with a software developed by Gignoux [38]. Our set of hypotheses to interprete spatial patterns is as follows:

- Competition leads to regularity within a population and to spatial inhibition between populations [62, 3, 60].
- Plants should be aggregated and associated to nutrient-rich patches if they are nutrient-limited [58, 15].
- Different stages within a population have different requirements and can have different spatial patterns.
- Low dispersal distance generates aggregated patterns [45, 6].
- Due to lower fuel (grass phytomass) loads in tree clumps, fire sensitive species should tend to have a clumped pattern [53, 67].

The spatial data set comprises 7 plots of increasing tree density where all plants were mapped (Table 18.1). Other data sets (transects by Bonvallot et al. [17] where all trees, soil depths, and an indication of the grass cover, were mapped) were not included in this analysis because of the conspicuous gradient in tree density linked to topographic position observed. We expect fire intensity to be lower on the highest density plots, and competition for light and nutrients to increase with tree density.

18.3.1 Spatial distribution of tree species

The patterns presented here have been inferred from works by Gignoux et al. [37, 40] on *Crossopteryx febrifuga*, *Piliostigma thonningii*, *Cussonia arborea*, and *Bridelia ferruginea*, on a detailed study of *Borassus aethiopum* populations [11], and on original results (Tables 18.2 and 18.3).

340   Jacques Gignoux et al.

Table 18.1. Savanna plots used for spatial pattern analysis and demographic studies: "Tree demography" plots were set up in May 1969 and were censused in 1970, 1973, 1975, 1989, May 1991, Dec. 1991, May 1992, Dec. 1992, May 1993, Dec. 1993, 1994 and 1995. All trees > 2 m in height were mapped, tagged, and measured over the 1969 to 1989 period; all smaller individuals were added in the following censuses. "Spatial pattern" studies: All trees > 1 m were mapped on plots PL-PP; Only *Borassus* palms were mapped on plot PM. "Long term dynamics" plots: All trees (incl. seedlings) were counted every 3 years from 1965. Fire regimes: unb., unburned; l.f., late fires; exc., except.

Study	Year	Plot	Name/location	Type	Fire treatment	Size (m)	Surface (ha)
Tree	1969	A	Plateau	AS	yearly	50 × 50	0.25
(Menaut,	to	C	Savane Gruyère	AS	unb. 64-69	50 × 50	0.25
Dauget,	2003	G	Virage glissant	AS	yearly	50 × 50	0.25
Gignoux,		H	Maison du garde	SW	yearly	50 × 50	0.25
Lahoreau)		I	Seaux à Barbault	AS	yearly	50 × 50	0.25
Spatial	1994	PL/TS3	Plateau	AS	yearly	150 × 140	3.50
pattern		BA	Barrière	SW	yearly	150 × 150	2.25
(Barot)		PP/GS2	Pont de Paris	LS	yearly	200 × 250	5.00
Demography	1995	GS1	Savane du Rocher	LS	yearly	150 × 250	3.75
of *Borassus*	to	PP/GS2			as above		
(Barot)	1999	TS1	Savane du Rocher	AS	yearly	200 × 200	4.00
		TS2	Piste du Sud	AS	yearly	128 × 250	3.20
		PL/TS3			as above		
		BA			as above		
Long-term	1965	1	Mare Portères	LS	yearly	50 × 50	0.25
dynamics	to	2	Pont de Paris	AS	l.f. 64-66	50 × 50	0.25
(Vuattoux)	2002	3	Campement	AS	unb. 62-	50 × 50	0.25
		4	Piste de Zougoussi	AS	unb. 64-65	50 × 50	0.25
		5	Marigot salé	LS	unb. 64-69	50 × 50	0.25
		6	Non brûlé en brûlé	AS	unb. 66-	50 × 50	0.25
					exc. 85, 89	50 × 50	0.25
		7	Terres noires	AS	unb. 62-	50 × 50	0.25
					exc. 67	50 × 50	0.25
		S	Sismographe	SW	yearly	50 × 50	0.25
Tree/grass	1989	PM	Piste du Sud	AS	yearly	100 × 230	2.30
interactions							
(Mordelet)							

Results presented here concern plots of Andropogoneae savanna with different tree densities. Plots H and BA have a high tree density (savanna woodland) and plot G is a shrub savanna evolving toward a savanna woodland. Except for *Borassus aethiopum*, results concern only >2 m individuals.

The overall spatial pattern of adult trees (all species together) is aggregated on all savanna plots, even in savanna woodlands (plots H and BA). Three groups of species can be distinguished based on the overall spatial pattern of adult trees:

Group 1: Species with a random spatial pattern on all plots, whatever the density of trees: *Crossopteryx febrifuga* and *Borassus aethiopum*. For

Table 18.2. Number of the various mapped items (adult trees, termite mounds, and rocky outcrops) on the spatial pattern analysis plots. Spatial analyses were performed only when sample size was ≥9.

Mapped item	Plot						
	A	C	G	H	I	PL	BA
Annona senegalensis	3	1	1	6	21	5	23
Borassus aethiopum	6	12	3	3	5	25	20
Bridelia ferruginea	7	12	22	39	18	66	534
Crossopteryx febrifuga	17	17	23	13	29	163	124
Cussonia arborea	11	7	14	60	10	113	88
Piliostigma thonningii	2	9	3	90	1	9	129
Pterocarpus erinaceus	0	0	8	1	0	4	20
Terminalia shimperiana	1	1	1	61	2	12	248
All trees	47	59	75	310	88	404	1289
Termite mounds	—	—	—	—	—	48	18

Borassus, this result was confirmed from analysis for another plot of 100×230 m of shrub savanna (mapped in 1990 by P. Mordelet, 1993, and analyzed by Gignoux [37]).

Group 2: Species with an aggregated pattern at high tree density (in savanna woodlands) and a random pattern at low density: *Annona senegalensis*, *Piliostigma thonningii*, *Terminalia shimperiana*.

Group 3: Species with a random or aggregated pattern, without a clear relation between the change in pattern and tree density: *Cussonia arborea*.

Bridelia ferruginea could be classified either in group 1 or 2, since it has a random pattern on almost all plots except the BA plot, where it has an aggregated pattern and reaches its highest density; however, its pattern becomes random in the H plot, where tree density is maximal.

From our hypotheses, very fire-resistant species should have a random pattern, relatively independent from the presence of clumps, because they have no advantage in recruiting in such safe sites. We can therefore classify species of the group 1 (*Crossopteryx* and *Borassus*) as good candidates for being very fire-resistant. Since *Bridelia* also tends to have a random pattern in low tree density, e.g., most fire-prone plots, we can also suspect it of being a very resistant species; its tendency toward aggregation when fire conditions become milder is probably explained by another trait than fire resistance. All the other species would then be more fire-sensitive.

Competition for light is expected to cause an increase in regularity of the patterns from initially aggregated patterns to random and even regular patterns as time goes on. However, none of the species here shows such a change in pattern with increasing density, except maybe *Bridelia* when comparing plots BA and H: it is probable that on plot H, forest species are further in the process of outcompeting savanna species than on BA. *Bridelia* could then be classified as a relatively fire-resistant species sensitive to competition for light.

Table 18.3. Summary of spatial pattern and spatial association test results: For each item, three tests were run (based on Diggle's G and F functions and on Ripley's K function for spatial patterns, and on the $G_{1,2}$, the $G_{2,1}$, and the $K_{1,2}$ functions for spatial associations), and the conclusion of the tests (decision made as in [11]) is shown for each class of mapped items on each plot as: A for aggregation or spatial association; - for random pattern or spatial independence; R for regular pattern or spatial repulsion; empty cell when the tests could not be performed because of a too small sample size.

Mapped item	Plot						
	PL	A	C	G	I	BA	H
Spatial pattern analyses							
All tree species	A					A	
Annona senegalensis					-	A	
Borassus aethiopum	-		A			-	
Bridelia ferruginea	-		-	-	-	A	-
Crossopteryx febrifuga	-		-	-	-	-	-
Cussonia arborea	A	-		A	-	A	A
Piliostigma thonningii	-		A			A	A
Pterocarpus erinaceus					-	A	
Terminalia shimperiana	-					A	A
Termite mounds	R					-	
Spatial association analyses							
Annona × *Crossopteryx*						-	-
Annona × *Borassus*						-	
Annona × mounds						-	
Borassus × *Crossopteryx*	-		A			-	
Borassus × mounds	A					-	
Bridelia × *Crossopteryx*	A		-	-	A	-	-
Bridelia × *Borassus*	R		-				
Bridelia × mounds	A					-	
Crossopteryx × mounds	A					-	
Cussonia × *Crossopteryx*	A	A		A	A	-	A
Cussonia × *Borassus*	A					-	
Cussonia × mounds	A					A	
Piliostigma × *Crossopteryx*	-		-			-	-
Piliostigma × *Borassus*	-		R			-	
Piliostimga × mounds	A					A	
Terminalia × *Crossopteryx*	-					-	A
Terminalia × *Borassus*	-					-	
Terminalia × mounds	-					-	

For the other species, the aggregated pattern can be the result of grouping within nutrient-rich areas or to fire sensitivity favoring recruitment within tree clumps or to low dispersal distances combined with a relatively constant juvenile mortality. Spatial association analyses help to sort out among these hypotheses.

18.3.2 Association to environment heterogeneities

Spatial association tests [28] were used to distinguish between specific attraction of one species by another and average or symmetric association, i.e., association where no species can be pointed out as dependent from the other.

Nutrient-rich patches

Mounds, that are nutrient-rich patches, were mapped on two plots (PL and BA). Termite mounds have a regular spatial pattern in the low tree density savanna and a random pattern in the savanna woodland. Some tree species are associated to mounds, while others are not:

- *Cussonia* and *Piliostigma* are spatially attracted by mounds on both plots.
- *Bridelia* and *Borassus* are attracted by mounds in the open savanna, but not in the savanna woodland.
- *Terminalia* and *Crossopteryx* are independent from mounds on both plots. On the only plot where the analysis could be done, *Annona* also seems to be independent of mounds.

Since tree clumps and termite mounds are nutrient-rich patches, we can classify *Terminalia*, *Crossopteryx*, and maybe *Annona* as species not strongly nutrient demanding, *Cussonia* and *Piliostigma* as nutrient-demanding species (since they always need to grow close to a mound), and *Bridelia* and *Borassus* as intermediate species, since their need to be close to a mound in the open savanna vanishes when tree density, and, from our hypotheses, average nutrient richness, increase.

Fire-safe sites

We can confirm the supposed high fire resistance of *Cros-sopteryx febrifuga* and *Borassus aethiopum* by checking whether they constitute kernels of clumps for more sensitive species on the lower density plots (PL to I).

- *Crossopteryx* and *Borassus* are independent from each other, confirming their high fire resistance.
- *Annona*, *Piliostigma*, and *Terminalia* are independent of *Crossopteryx* or *Borassus*, in the plots PL to I where fire intensity is higher because of lower tree densities. This suggests that these species have an aggregated pattern not because of their fire sensitivity, but for some other reason.
- *Cussonia* is systematically associated to *Crossopteryx* and *Borassus*, suggesting that it needs protection from fire to recruit successfully.

Our results at this stage do not enable us to validate the clump formation mechanism proposed by Menaut et al. [53], partly because reality is more complex than model assumptions (i.e., fire is not the only cause of tree aggregation). We would require more detailed analysis based on tree size and grass presence to check whether tree recruitment is favored in grass-free areas.

Dispersal

Crossopteryx and *Terminalia* are anemochorous, *Borassus* and *Piliostigma* are barochorous (with a possible secondary dispersal by animals), and *Bridelia*, *Cussonia*, and *Annona* are apparently zoochorous (Menaut, personal observations). All these dispersal modes should generate clumped patterns of seedlings, but these might evolve later to less aggregated patterns.

The spatial pattern of seedlings and resprouts of *Bridelia*, *Crossopteryx*, *Cussonia*, and *Piliostigma* is aggregated on plots A, C, G, H, and I [37], suggesting very low dispersal distances for all these species. *Borassus* seeds are never dispersed further than 10 m away from the mother tree [12], even though Vuattoux observed in the 1960s secondary dispersal of *Borassus* seeds and transport on mounds or rocks by baboons, now extinct from the area (Vuattoux, personal observation).

The spatial patterns of seedlings is aggregated for all species, as expected for short dispersal distances. However, this initial pattern can be later modified by differential mortality or recruitment rates linked to local environment. In extreme cases like *Borassus*, those processes can lead to a random pattern of adults [11]. We should therefore expect that the pattern of seedlings resulting from seed dispersal shows little relation with the spatial pattern of adults.

18.3.3 Case study: *Borassus aethiopum*

A detailed study of the spatial pattern of *Borassus* has been performed by Barot et al. [11, 8]. This palm tree is dioecious and shows a marked senescence period where fecundity declines [9]. The spatial pattern of 3 different stages (seedlings, juveniles, adults, split into males and females) revealed the following:

- There is a strong competition between seedlings and between juveniles for recruitment; females have a negative (direct or indirect) influence on their own offsprings survival [12]; intraspecific competition increases when tree density increases.
- All classes of *Borassus* palms are nutrient demanding; adults are loosely associated to nutrient-rich patches (mounds and tree clumps) thanks to an efficient root foraging strategy [56], while juveniles and seedlings are closely associated to nutrient-rich patches; association distances decrease with demographic stage, adult females being sometimes independent of patches while most seedlings grow in nutrient-rich patches.
- There is a discrepancy between the spatial pattern and the location of adults and juveniles, which is apparently due to a complex set of interactions between at least four processes: seed dispersal, negative effect of females on juveniles, the need for juveniles to be close to a nutrient-rich patch, and density-dependent mortality of juveniles within a clump of juveniles.

- Male and female palms have different spatial patterns, which can be explained only by a difference in nutrient requirement associated to reproduction costs: from the results, females should have a higher nutrient demand and have developed a more efficient root foraging strategy than males. This results in females being independent of nutrient-rich patches while males are associated to them. The cause of this pattern could be differential survival of sexes or environment-induced sex.

From those results, a spatialized scenario of the life cycle of *Borassus* was proposed [8]: (1) short distance seed dispersal initiates clumps of seedlings around mother palms; (2) the recruitment and survival of seedlings are higher near nutrient-rich patches and far from the mother palm, leading to a weaker aggregation and association to female adults at older stages; (3) when the mother dies, juveniles from the periphery of the clumps have the highest chance to recruit as adults because they suffer less competition than juveniles from the center of the clump; (4) as time goes on, only one adult remains per former clump of juveniles, and the pattern of adults more or less reflects that of clumps (random), with a weak association to nutrient-rich patches thanks to the root foraging ability of adult *Borassus*.

18.3.4 Conclusion: Vital attributes of savanna trees inferred from their spatial patterns

The interpretation of spatial pattern and association analyses is not straightforward because of the inherent ambiguities of our set of assumptions. For example, the aggregated pattern of *Cussonia* and *Piliostigma* at high densities could be due to important nutrient requirements as well as low fire resistance. Their random pattern at low density would tend to indicate a good fire resistance, but the systematic association of *Cussonia* to *Crossopteryx* and *Borassus* would rather indicate sensitivity to fire. A parsimonious interpretation would be to consider *Piliostigma* as more resistant to fire than *Cussonia*, based on the overall set of results and not on a single particular result. Experiments on fire resistance as already performed for two other species [39] would enable one to decide between those two hypotheses.

From the simple analyses of spatial pattern and spatial association presented so far, we can infer major properties (i.e., some vital attributes as defined by Noble and Slatyer [59]) of savanna tree species, as follows:

Competition between adult trees apparently plays only a minor role in the savanna, since thinning is only observed for one species (*Bridelia*) in the highest density savanna woodland plot; all other species either show no change in spatial pattern with increasing tree density, or a change opposite to what the competition hypothesis predicts (more aggregation as density increases).

We can roughly classify tree species on a scale of fire resistance on the basis of their spatial pattern, with *Crossopteryx* and *Borassus* as the most fire resistant, then *Bridelia* as a less resistant species, and *Cussonia*, *Piliostigma*, *Annona*, and *Terminalia* as the most fire-sensitive species. An independent study of individual tree resistance to fire [39] effectively proved that *Crossopteryx* was much more fire resistant than *Piliostigma*.

We can also classify the major tree species on a scale of nutrient demand, with *Cussonia* and *Piliostigma* being the most nutrient demanding, *Bridelia* and *Borassus* nutrient demanding only in the open savanna, but not in the richer savanna woodland areas, and *Crossopteryx*, *Annona*, and *Terminalia* showing no evidence of nutrient limitation.

These results and the detailed study of *Borassus* illustrate the interest of spatial analyses for the integrated study of an ecosystem like Lamto, which has an obvious structure at all scales. This clearly showed that demography in Lamto savannas is influenced by and influences spatial pattern in a complex way that needs to be taken into account in modeling exercises addressing the question of the stability of the ecosystem.

18.4 Tree population dynamics

Three datasets consitute our main information on tree community dynamics in Lamto (Table 18.1):

Seven plots (labeled 1 to 7 and S) were set up in 1965 by Vuattoux in various savanna facies and under various fire regimes. Trees were counted by species and stage on these plots every 3 years until 1998, so that a detailed demographic analysis is not possible, but the long-term dynamics of the plot is documented.

Five plots (labeled A, C, G, H, I) were set up in 1969 by Menaut and César, where all trees have been mapped, tagged, and measured in 1969, 1970, 1973, 1975, 1989, and yearly since 1991 (twice a year in 1991-1993). These data can be used for the estimation of demographic parameters for adults on the 1969-1989 period, and also for smaller trees (including seedlings) afterward.

Five plots (labeled GS1, PP/GS2, TS1, TS2, PL/TS3, BA) were set up by Barot and S. Konaté and censused yearly between 1995 and 1999 for a detailed study of *Borassus aethiopum* [11].

18.4.1 Long-term dynamics

Comparison of treatments of the long-term dynamics study (Table 18.1 and Fig. 18.3) provides insights on the major forces driving the ecosystem [72, 73, 27]:

1. Under the normal fire regime, the tree community did not seem stable: tree density increased by 20-50% in the "standard" Andropogoneae savanna plots (plots A, G, and I), by 600% in the *Loudetia* savanna plot (plot 1), and by 50% in the high density plots (plots S and H). Only these

18 Structure and Dynamics of the Tree Community   347

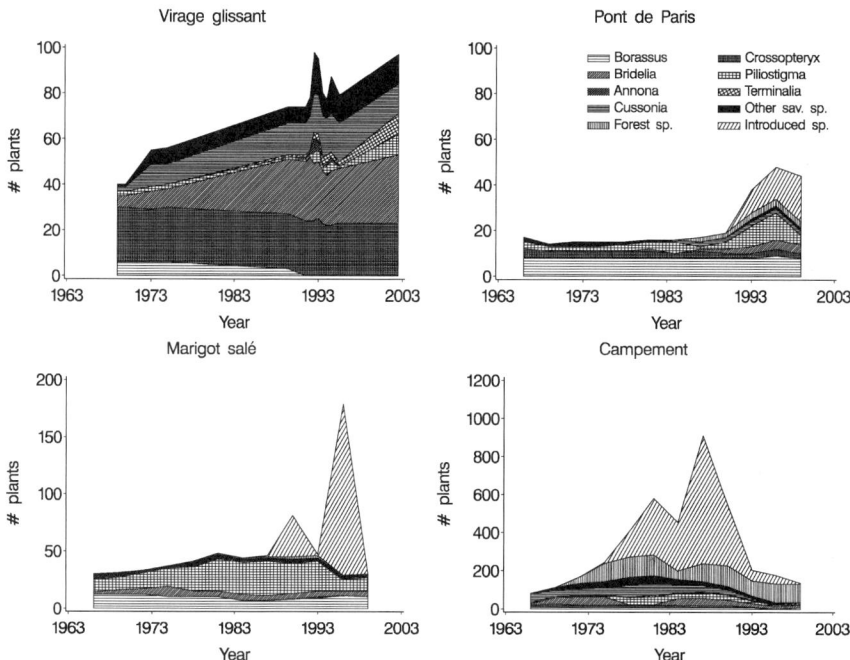

Fig. 18.3. Dynamics of tree numbers on four 0.25 ha plots (Table 18.1). Fire treatments: plot "Virage glissant," annually burned; plot "Pont de Paris," 3 years of late fires (1964-66); plot "Marigot salé," protected from fire during 4 years (1964-65, 1968-69); plot "Campement," protected from fire over 40 years (after [72, 73, 27] and unpublished data by R. Vuattoux).

latter plots were invaded by other species: fire-sensitive savanna species like *Terminalia shimperiana* and *Ficus sur*, the invading weed *Chromolaena odorata* [35], and forest species [25]. In the other plots, increase in density was mainly due to *Piliostigma thonningii*, *Cussonia arborea*, *Annona senegalensis*, and *Bridelia ferruginea*.

2. In savannas unburned for a few years (C, 4, and 5), the same conclusions hold. On plot 5, the increase in *Piliostigma* can be linked to fire protection. The later decrease of the population indicates a relatively short life span (< 30 years) for this species under the normal fire regime, since numbers are now equal to their initial values in 1965.

3. In the 3-year late burned plot (plot 2), there was a decrease in tree density just after the treatment for all species except *Borassus aethiopum* and *Crossopteryx febrifuga*; *Piliostigma thonningii* started to recover after 10-15 years; *Bridelia ferruginea* and *Cussonia arborea*, absent from the plot in 1965, started to invade it roughly at the same period. The time lag

between the end of the late fire treatement and the recruitment of new adults gives a duration of the juvenile/resprout stage of 10-15 years for the main savanna species.

4. Unburned savannas (Plots 3, 6, 7) had a qualitatively identical evolution, with slight quantitative differences. A succession started with fire protection, where savanna species already present first invaded the plot, *Bridelia ferruginea* and *Ficus sur* being the fastest invaders, followed by *Cussonia arborea* and *Piliostigma thonningii*; other more fire-sensitive savanna species like *Terminalia shimperiana* then took the advantage, followed by species growing in humid areas, forest-savanna edge species, and true forest species starting to invade 6-12 years after fire protection. After 15 years, *Borassus aethiopum* and *Crossopteryx febrifuga* (which both showed very little response to fire protection) started to decline. This overall pattern was disturbed by the introduced weed *Chromolaena odorata*, which apparently slowed down the invasion by forest species [35]. The timescale of this succession also depended on the distance of the plot to the source of forest species seeds (gallery forests). Accidental fires had little effect on the succession once started, whereas major droughts seemed to be important.

These long-term data enable one to infer savanna species habits compatible with that of the analysis of their spatial patterns: *Crossopteryx* and *Borassus* appear as the most fire-resistant species, able to grow in any type of savanna under any fire regime (including late fires), and decrease in density only when tree density increases (in savanna woodlands or unburned savanna). The other major savanna species (*Annona*, *Bridelia*, *Cussonia*, and *Piliostigma*) are able to increase in density even in the normal fire regime although they seem to have much more fluctuating numbers. They quickly take advantage of any reduction of fire severity or frequency, but they decrease under a late fire regime. Other rarer savanna species like *Terminalia shimperiana*, *Pterocarpus erinaceus*, and *Ficus sur* only invade when fire is excluded, either temporally or spatially. These patterns are consistent with the vital attributes inferred from the spatial pattern analyses.

The behavior of the main savanna species is also consistent with the predictions of Gignoux et al. [39]: (1) there should be a trade-off between competitive ability for resources and fire resistance and (2) fire-sensitive trees should recruit either in clumps or in cohorts. Indeed, the most fire-resistant *Crossopteryx* and *Borassus* never increase in numbers when fire is excluded, while the more fire-sensitive *Annona*, *Bridelia*, *Piliostigma*, and *Cussonia* all respond to fire protection by increasing in numbers; they are later excluded by the even more fire-sensitive savanna and forest species. This strongly supports the existence of a trade-off between individual growth rate (competitive ability) and fire resistance. Furthermore, *Annona*, *Bridelia*, *Piliostigma*, and *Cussonia* densities fluctuate under the normal yearly fire regime, while

Borassus and *Crossopteryx* are much more stable. These fluctuations can be interpreted as cohort recruitment events linked to less intense fires, as suggested in hypothesis 2.

Although there is no difference in the increase in tree density between Andropogoneae savannas and savanna woodlands under a normal fire regime (all plots except plot I increase by ca. 50% over 25 years), there is an important qualitative difference: fire-sensitive species (first *Terminalia shimperiana* and *Ficus sur*, then edge and forest species) appear in the densest parts of the plots, apparently initiating a succession like that observed in the unburned savanna plots. The mechanism explaining this evolution in dense plots is probably that hypothesized by Gignoux [36] and Menaut et al. [53]: when tree density locally increases, some dense tree clumps may appear at random (due to the initial pattern of the tree community); once a clump is established, it becomes a fire safe site where the recruitment of fire-sensitive species is possible. This idea seems confirmed by the comparison of plots G and I. These plots had different densities in 1969 (156 ha^{-1} trees for G and 300 for I). Plot I remained a typical open savanna, with a slight increase in density over 25 years (352 ha^{-1} trees in 1992), while plot G seems to evolve toward a savanna woodland, with the invasion of *Pterocarpus erinaceus* and *Terminalia shimperiana*, fire-sensitive savanna species. This difference can be explained only by the fortuitous coalescence of 2-3 smaller clumps on plot G, a random event that did not happen on plot I although the initial spatial patterns was quite similar (Fig. 18.4).

Gautier [34] estimated from the comparison of aerial photographs taken in 1962 and 1989 that the density of savanna trees had doubled in Lamto over this period. Possible explanations are (1) a change in the rainfall regime (a deficit in rainfall has been observed over the same period (Sect. 3.5)) or (2) a change in the fire regime due to the imposed date of fires at the heart of the dry season since the set up of the Lamto reserve in 1962. Hypothesis 1 would mean that the Guinea savannas are more sensitive to the water balance than expected given their high rainfall and average water availability, and that tree density increases when rainfall decreases. Hypothesis 2 would mean that Guinea savannas need some variations in the date of fires to be stable: trees normally invade the savanna, but are prevented from doing so by occasionnal late fires which, although usually less intense than mid-season fires, are more harmful to adult trees because of their phenological stage. Hypothesis 2 seems confirmed by the results of the plot 2. It also has the advantage of explaining the overall increase in density observed in the region of Lamto (not only in the reserve): the increase in human population is expected to reduce the occurrence of late fires. Since people tend to light fires for safety reasons (Sect. 4.5) early in the dry season, the probability that areas remain unburned until the end of the dry season decreases with human density.

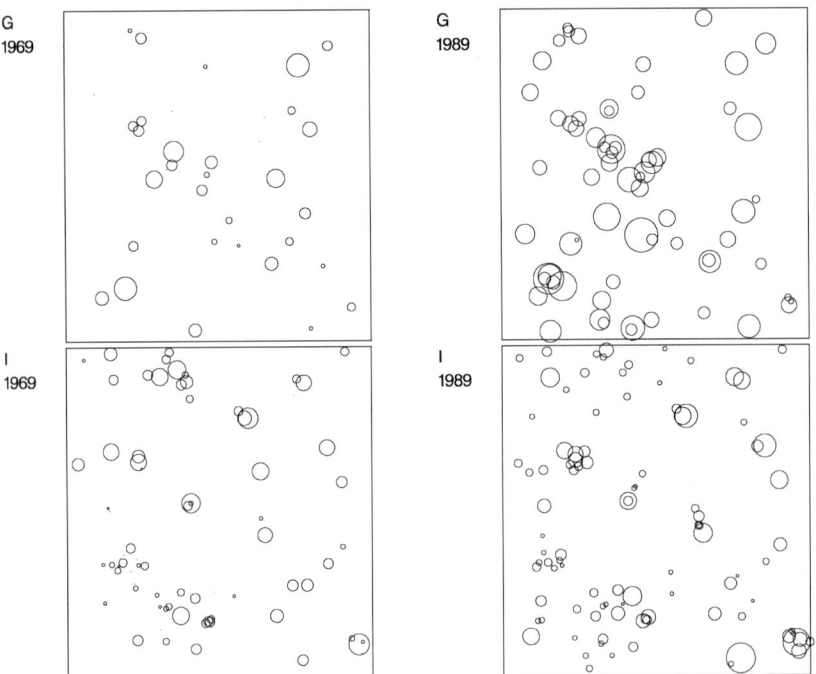

Fig. 18.4. Maps of two savanna plots censused in 1969 and 1989. Plots are 50×50 m. Circle surface equal to tree crown surface. Only >2 m high trees have been mapped (J.M. Dauget and J.C. Menaut, unpublished data).

18.4.2 Size structure of tree populations

Main tree species

The best measure of tree size in Lamto savannas is probably basal circonference for adult trees, since trunks are rarely straight and branch at a low height. The four main species (*Bridelia*, *Crossopteryx*, *Cussonia*, *Piliostigma*: Fig. 18.5) have very different population structures. *Bridelia* has a classical population structure, with many small individuals and few large ones. *Piliostigma* and *Cussonia* have similar structures, but the structure is not identical at all on most plots: the mode of the distribution moves toward higher classes across time on plot H (where the invasion by other species is important), but also on plots A, G, and I for *Cussonia*. This is consistent with the results of the previous section, suggesting that for the species *Cussonia*, *Piliostigma*, and *Bridelia*, recruitment is linked to temporal variations in fire severity and produces cohorts of relatively even-sized individuals. *Crossopteryx* population has a particular structure with very flat histograms with almost the same numbers of individuals in all classes. Since only >2 m individuals were considered to draw the histograms, this simply means that

18 Structure and Dynamics of the Tree Community   351

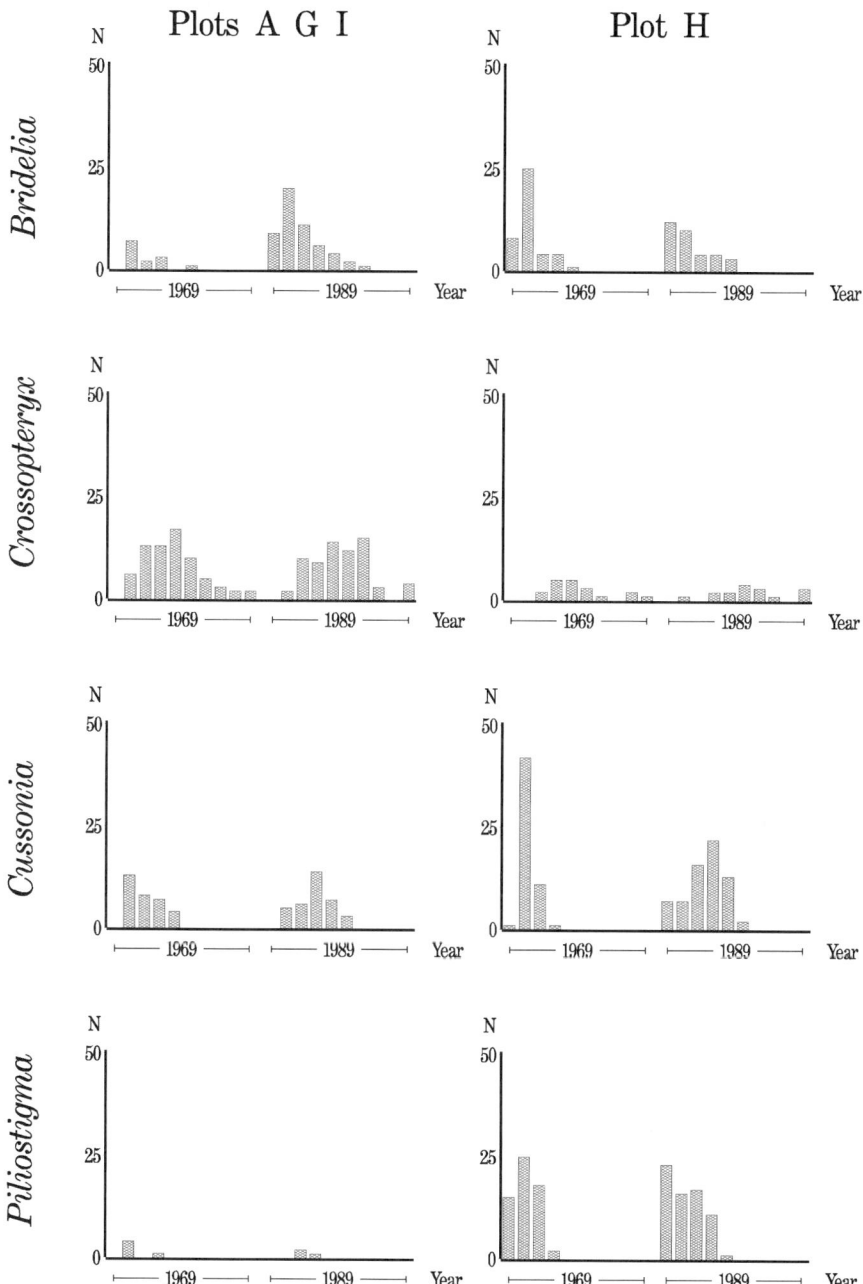

Fig. 18.5. Histograms of basal circumference for the four main species (*Bridelia, Crossopteryx, Cussonia,* and *Piliostigma*) over the 1969-1989 period on plots A, G, H, and I. Circumference in classes of 20 cm from 0-20 to 180-200 cm; frequency in numbers (J.C. Menaut and J.M. Dauget, unpublished data).

352 Jacques Gignoux et al.

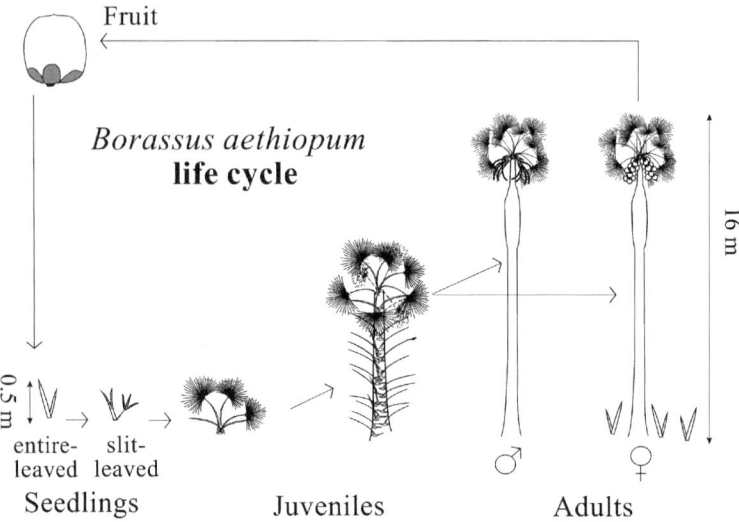

Fig. 18.6. Schematic diagram of *Borassus aethiopum* life cycle.

recruitment to the adult stage occurs long before 2 m for *Crossopteryx* (i.e., only the tail of the distribution is observed) and around 2 m for the other species. This again is fully compatible with the spatial pattern analyses of the previous section and the analysis of fire resistance of Gignoux et al. [39].

Case study: The population structure of *Borassus aethiopum*

Barot [9] defined four stages for *Borassus*: Entire-Leaved seedlings (EL-seedlings), Slit-Leaved seedlings (SL-seedlings), juveniles, and adults (Fig. 18.6). Germination is remote-tubular [71] so that seedlings are buried about 40 cm deep in the soil by the cotyledonary axis that extends downward. In both seedling stages and in the younger juveniles, the terminal bud is far below ground level: this defines the establishment phase. EL-seedlings have one or two elongated entire leaves. SL-seedlings have one or two leaves that are slitted a few times. Juveniles and adults have the same fan-shaped, induplicate and costapalmate leaves. Petioles of dead leaves remain on juveniles stems. In a few years, these petioles fall down, a swelling appears on the stem, and sexual maturity is reached.

Most seedlings die before reaching the adult stage (Fig. 18.7), as in most tree species [44]. The distribution of juveniles and adults classified according to their height is strongly bimodal (Fig. 18.8): there are very few juveniles in the 2-8 m height class. This could be due to (1) temporal variations of survival and recruitment rates (the population would be far from the stable stage distribution, and some decades ago, juveniles could have suffered from an exceptionally sharp mortality event due to some disturbance) or (2) size-dependent growth rate of juveniles (if juveniles of the intermediate height

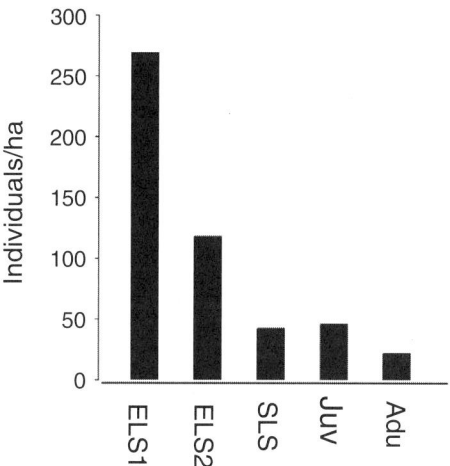

Fig. 18.7. Palm density by stage (ELS1, one leaf Entire Leaved Seedling; ELS2, two leaves Entire Leaved Seedling; SLS, Slitted Leaved Seedling; Juv, Juvenile; Adu, Adult). Individuals of 4 plots (17.5 ha all together) were pooled together (S. Barot, unpublished data).

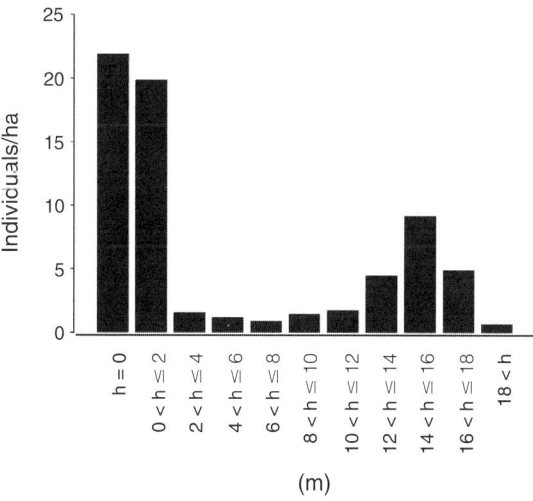

Fig. 18.8. Adult and juvenile palm density by height class. Individuals of 4 plots (17.5 ha all together) were pooled together (S. Barot, unpublished data).

classes grow faster than smaller juveniles, these classes will be less represented in the population) [46]. Since there is no hint of a past important disturbance and since it has been proved that the taller a juvenile is, the faster it grows [9], the second hypothesis seems to be the more likely. This is supported by the study of *Borassus* demography (see below): the palm population seems to be close to the stable stage distribution.

18.4.3 Demographic parameters

Although substantial data already exist for estimating the main demographic parameters of trees, no comprehensive analysis has been performed yet. We report here the few available estimates, plus a detailed case study of the palm tree *Borassus aethiopum*.

Main tree species

Survival - Seedling survival was estimated from the censuses of plots A-I and from experiments where seedlings were planted in an annually burned savanna (Table 18.4, from [37]). This experiment was designed to test the effect of tree cover on seedling growth and survival. Seedlings were planted in the field between grass tufts, under a tree clump or in the open; they were protected against rodent predation. There is a clear difference between species, and apparently no difference between treatments, suggesting that the relatively small tree clumps chosen were not big enough to cause a reduction in fire intensity. The survival of *Piliostigma* seedlings is strikingly high given their small size. At the seedling stage, *Piliostigma* has a higher survival than *Bridelia*, maybe because of a higher investment into belowground organs (Sect. 5.4).

Indications of adult and resprout stages life span is provided by the long-term dynamic data (Fig. 18.3): on plot 5, *Piliostigma* numbers start to increase 5 years after the end of fire exclusion, suggesting a duration of the resprout stage of at least 5 years, and adults recruited at this time are now slowly declining, suggesting a life span of the adult stage in this

Table 18.4. Experiment on seedling survival: numbers of alive seedlings. 20 seedlings of *Bridelia ferruginea* and *Piliostigma thonningii* were planted in the field in May 1992, under a tree clump and in the open, with a protection against rodent predation (after [37]).

		Number of seedlings alive			
Species	Treatment	Sept. 92	Oct. 93	Oct. 94	Oct. 95
Bridelia	Cover	20	18	6	5
ferruginea	Open	20	20	7	6
Piliostigma	Cover	20	20	20	19
thonningii	Open	20	20	20	20

species of 20-30 years, which is remarkably short. These conjectures would need further work to be confirmed.

Growth - Using data of two unburned plots measured by Menaut between 1969 and 1975 [51], Gignoux [36] estimated an empirical growth equation for adult trees of the shape:

$$\frac{dD^2H}{dt} = G D^2 \left(1 - \frac{DH}{DH_{max}}\right)(1 - \alpha N),$$

where D is basal diameter in cm, H is total height in cm, G is the intrinsic growth rate, DH_{max} is the maximal value of the DH product in cm^2, α is a competition factor, and N is the number of crown-overlapping neighbors (Table 18.5). This equation was derived from Botkin et al. [18] and enables one to estimate tree maximal height and basal diameter. The analysis was performed for all species mixed. The estimated value for the competition parameters leads to a reduction of growth by 55% when a tree has 5 crown-overlapping neighbors, explaining the poor competitive ability of savanna trees in shaded conditions.

Reproduction and germination - From preliminary analyses of the long-term demographic data, the average recruitment rate into the adult population for the four main species is around 4 ha^{-1} y^{-1}. Ponce de Leon Garcia [64] studied the germination ecophysiology of *Bridelia* and *Piliostigma*, germination of the other two main species being very difficult to obtain. These two species display tegumentary inhibition of germination (they need scarification to germinate). *Bridelia* seeds are still able to germinate after 5 years of storage, a relatively long life span compared to most tropical forest species. Seedlings of some species are apparently subject to heavy predation in their early stages, e.g., by rodents for *Piliostigma* and by crickets for *Cussonia* (unpublished field observations by Gignoux and Simioni).

Resistance to fire - Based on architectural descriptions of trees, César and Menaut [24] distinguished two main strategies enabling young trees to resist fire:

1. "Hide-and-resprout" strategy: As temperature rise is very low or negligible in the soil, young individual trees can survive by resprouting each year from belowground storage structures. To recruit into the adult population, such resprouts have to successfully establish a fire-resistant perennial trunk which will allow further growth in height the

Table 18.5. Estimates of adult growth equation parameters. Non-linear adjustment, all effects significant, $R^2 = 0.39$ (after [36]).

Parameter	Estimate	Standard error
G	53	4
DH_{max}	36000	2000
α	0.16	0.02

following year. This is achieved only when belowground structures are strong enough to produce between two successive fires (in some cases, this can be as short as one growing season only) a trunk (i) reaching a height where the terminal buds are able to resist the existing fire conditions and (ii) thick enough at its base to resist the high fire intensity existing in the fuel bed.
2. "Stay-and-resist" strategy: Young individuals can also survive by directly building an aerial fire-resistant structure (i.e., a thick trunk with a fire-protecting bark) enabling it to resist all fire conditions.

Gignoux et al. [39] have demonstrated that fire resistance resulted from the interaction of at least three properties defining a continuous set of strategies: (1) intrinsic resistance of stems, linked to bark properties; (2) specific trunk profile; (3) growth rate between two successive fires. Some species, like *Crossopteryx febrifuga*, are able to resist through a high intrinsic resistance and a thick trunk, while others like *Piliostigma thonningii* rely on a fast growth between fires. These differences should result in different spatio-temporal patterns of recruitment probabilities and probably affect tree dynamics through interactions with competition, which is consistent with the spatial patterns and long-term dynamics reported here.

Case study: The demography of *Borassus aethiopum*

Results presented here are based on a detailed demographic study based on a series Lefkovitch matrix population model presented in Sect. 19.2 [9, 14, 13, 10].

Demographic parameters - The model was based on the demographic stages defined in the previous section. Parameters were estimated for the populations of 2 plots (PP/GS2 and TS2: Table 18.1) for the period April 1995-April 1997 for SL-seedlings ($n = 369$), juveniles ($n = 423$), and adults ($n = 80$). Fecundity and EL-seedling survival and recruitment rates were estimated from the offspring of 6 females (999 EL-seedlings) during the period April 1996-April 1997 [12].

Death rate is U-shaped (respectively for the 4 stages 0.0930, 0.0100, 0.0065, 0.0400) as for probably most trees [74]. Juveniles seem to be the individuals the less likely to die. Recruitment rates are relatively high for seedlings (0.015 and 0.17) and much lower for juveniles (0.0035). Fecundity has been estimated as 28 seedlings produced per female per year.

Estimates of palm age - Mean age of palms (mean age of residence in the original paper), mean age at recruitment, and mean remaining life span were calculated [22] for each stage as well as their standard deviation (Table 18.6). The recruitment of juveniles into the adult population occurs very late, when juveniles are in average 116 years old. Adults then reproduce for a relatively short period since the mean remaining life span of adult is 25 years and since the mean age in the adult stage is 140

Table 18.6. Mean age, mean age of recruitment, and mean remaining life span for *Borassus* individuals of each stage. σ, standard deviation (after [14], with permission of Cambridge University Press).

Variable	EL-seedlings	SL-seedlings	Juveniles	Adults
y (mean age)	9.26	14.81	114.81	139.81
$\sigma(y)$	8.74	10.09	100.01	102.96
τ (mean age of recruitment)	-	10.26	15.81	115.81
$\sigma(\tau)$	-	8.74	10.09	100.01
Ω (mean remaining life span)	24.30	108.26	108.75	25.00
$\sigma(\Omega)$	53.97	101.63	101.25	24.49

years. The mean remaining life span is nearly the same for SL-seedlings (108 years) and juveniles (109 years). Individuals of these 2 stages behave quite in the same way: they both have very high survival rates, and the recruitment rate between the 2 stages is relatively high. Standard deviations values are close to the mean ages values. Thus, according to the model hypothesis, individuals can have very different temporal life histories and individuals in the same stages may have very different ages.

Senescence - Number of leaves, fecundity [9], and survival probability decrease with adult age, indicating a senescence period. The particularity of this senescence is that it seems to start just when sexual maturity is reached: adult number of leaves is maximum for the younger adults that have just recruited. The adult stage seems to be globally a senescent stage. Senescence is known for some tree species [74] and other palm trees [30, 21] but we do not know any other comparable senescence as strong as in *Borassus*, and beginning nearly as soon as individuals begin to reproduce.

Conclusion: The demographic strategy of *Borassus* - The demographic strategy of *Borassus* was compared to available results for other palm tree species: *Podococcus barteri* [20], *Astrocaryum mexicanum* [63], and *Rhopalostylis sapida* [31]. There was no proof of senescence in any of those (forest) species [9]. The estimates of stage ages computed for all the species reveal a much longer reproductive life for all the other species: they spend more than half of their life span reproducing, while *Borassus* only reproduces during a third of its life [14, 10]. The fact that *Borassus* is a savanna palm whereas the three others are forest palms could have played a role in the divergence of their life histories since the two kinds of ecosystem do not provide the same constraints. In forest, competition for light plays a very important role: it is an offspring-size-beneficial habitat [16]. In humid savannas, competition between trees is less intense, at least for adult palms that almost never grow close to each other [11], and competition, if any, is more likely to be for nutrients. Fire is also likely to have played an important evolutionary role in savannas and other fire-prone ecosystems [19, 68, 39].

Conclusion: Tree dynamics and demographic strategies

There is substantial evidence that the tree community in Lamto is unstable: trees are slowly but surely invading the savanna, both by an increase in savanna tree density [34] and a progression of gallery forest edges over the savanna [33]. The only estimate of a population asymptotic growth rate we have (for *Borassus*) is consistent with those descriptive results (population doubling in 20 years with the estimated value). There is enough data available to assess that the fluctuating numbers of some of the main species, probably linked to temporal fire variability, do not hide this long-term trend.

The scarce information we have so far on the demographic strategies of the main species show surprising facts: relatively short life spans for the main tree species (< 30 years for the adult stage and < 15 years for the resprout stage), and an original life cycle with a quick senescence for the palm tree *Borassus*. At the moment, we lack a comprehensive comparative study of the demographic strategies of all the major tree species of Lamto to propose any sound explanation for these particularities.

However, if demographic analysis results are consistent with the long-term trend, they do not explain it or provide any mechanism explaining it. Some results indicate that the long-term trend could be due either to (1) a change in the fire regime causing a change in tree recruitment patterns and (2) an effect of tree aggregation (i.e., comparison of the fate of plots I and G: Fig. 18.4). We explored this question through (1) analyses of the links between spatial structure of the ecosystem and population dynamics (next section) and through (2) modeling studies of the tree demography in relation to the major factor apparently constraining their recruitment, fire (Chap. 19).

18.5 Discussion: The interaction of demography and spatial patterns and its effect on savanna stability

The studies conducted so far on tree dynamics at Lamto illustrate once more the great complexity of this dynamics in savanna ecosystems, even without explicitly considering the problem of tree-grass interactions (Chap. 8). As demonstrated by fire exclusion experiments, fire effectively shapes Guinea savannas by selecting the species able to survive in these extreme conditions, but among the community of fire-resistant species, it becomes a secondary factor affecting their demography. Spatial pattern analyses showed that for two dominant species (*Crossopteryx* and *Borassus*), fire was not a problem, and for some other ones, nutrients were at least as important. This is supported in Lamto by the fact that only late fires are able to maintain tree populations, normal fire being unable to prevent tree invasion. For the less fire-resistant savanna species (*Piliostigma, Cussonia, Annona, Bridelia*), we observe fluctuating numbers in the long term, unstable population structures, and heterogeneous or clumped spatialpatterns, as in other savanna systems

[65, 69]. Although the exact reason for these fluctuations is not known, it is certainly not due to herbivores as in [65] given the low herbivore load in Lamto (Chap. 10), but more probably to fire, as previously hypothesized [39].

The second most important regulator of tree populations appears to be nutrients, which are characterized in Lamto by a very heterogeneous distribution: nutrients are concentrated in patches, which provides a key for analyzing spatial patterns. Nutrient patchiness (together with fire protection and short-distance seed dispersal) contributes to the formation of clumps associated to those patches, a cause often invoked to explain clumped patterns [67, 23] but never evidenced as here. We observe here these clumps for adult trees, but the studies of Barot clearly demonstrated that different stages in the same species could have different nutrient requirements, or different abilities to capture nutrients just because of their sizes: patterns within a species could complicate the picture. Furthermore, a detailed study of the fate of *Borassus* seedlings as a function of distance to their mother or to nutrient-rich patches [12] shows that the key assumption of matrix population models, that demographic parameters are homogeneous within a stage, is not fulfilled in Lamto because these parameters depend on the spatial pattern of the population.

Since water is probably not the main limiting factor in Lamto (Sects. 4.2, 3.2 and 18.2), the last factor affecting tree patterns and dynamics is competition for light. Little evidence of self-thinning has been found, except for young stages in the *Borassus* study [11], as in many other savanna systems [23, 70].

The key input of the studies of tree dynamics conducted at Lamto is the demonstration that ecosystem heterogeneity, tree demography, and spatial patterns are closely linked: over the initial patchiness of soil nutrients, demography, through the differences in seed dispersal, survival at different stages, and fire resistance of different species, adds a second layer of patchiness that eventually reinforces the initial patchiness, tree clumps becoming nutrient-rich and fire-protected patches. This feedback of tree demography on spatial pattern and environmental heterogeneity is apparently able to produce nursing effects comparable to those documented in other savannas [5, 4, 7] and lead to a local invasion by trees even under a normal fire regime as observed on plots H and G and modeled by Gignoux et al. [41]. In such a case, large clumps would constitute stable features of the landscape, even if they start randomly.

There are many unanswered questions left: there is still a dispersal problem in *Borassus* (how can the life cycle be accomplished when favorable sites for seedlings and adults are different?); the effect of fire on seedling survival and resprout recruitment has been studied at the individual level, but the response of the population has not been studied; seed dispersal is a key factor linking pattern and demography and is worth a detailed study; a comprehensive and comparative analysis of spatial pattern and demography of the major tree species, such as that performed for *Borassus*, is needed to understand the determinants of the demographic strategies of savanna tree species; linking functioning and dynamics is also necessary to understand the link between recruitment fluctuations and climate. Most of this work involves long-term processes and thus requires a substantial modeling effort to be achieved.

References

1. L. Abbadie, M. Lepage, and X. Le Roux. Soil fauna at the forest-savanna boundary: Role of termite mounds in nutrient cycling. In P.A. Furley, J. Proctor, and J.A. Ratter, editors, *Nature and dynamics of forest-savanna boundaries*, pages 473–484. Chapman & Hall, London, 1992.
2. L. Abbadie, A. Mariotti, and J.C. Menaut. Independence of savanna grasses from soil organic matter for their nitrogen supply. *Ecology*, 73(2):608–613, 1992.
3. J. Antonovics and D.A. Levin. The ecological and genetic consequences of density-dependent regulation in plants. *Annual Review of Ecology and Systematics*, 11:411–452, 1980.
4. S. Archer. Development and stability of grass/woody mosaics in a subtropical savanna parkland, Texas, USA. *Journal of Biogeography*, 17:34, 1990.
5. S. Archer, C. Scifres, C.R. Bassham, and R. Maggio. Autogenic succession in a subtropical savanna: Conversion of grassland to thorn woodland. *Ecological Monographs*, 58(2):111–127, 1988.
6. C.K. Augspurger. Seedling survival of tropical tree species: interactions of dispersal distance, light-gaps and pathogens. *Ecology*, 65(6):1705–1712, 1984.
7. P.W. Barnes and S. Archer. Influence of an overstorey tree (*Prosopis glandulosa*) on associated shrubs in a savanna parkland: Implications for patch dynamics. *Oecologia*, 105(4):493–500, 1996.
8. S. Barot. *Interactions entre répartition spatiale, hétérogénéité environnementale et démographie: Cas du palmier Rônier dans une savane humide de Côte d'Ivoire*. Ph.D. thesis, Université de Paris 6, Paris, 1999.
9. S. Barot and J. Gignoux. Population structure and life cycle of *Borassus aethiopum* Mart.: Evidence of senescence in a palm tree. *Biotropica*, 31(3):439–448, 1999.
10. S. Barot, J. Gignoux, and S. Legendre. Matrix models and age estimations in plants. *Oikos*, 96:56–61, 2000.
11. S. Barot, J. Gignoux, and J.C. Menaut. Demography of a savanna palm tree: Predictions from comprehensive spatial pattern analyses. *Ecology*, 80(6):1987–2005, 1999.
12. S. Barot, J. Gignoux, and J.C. Menaut. Seed shadows, survival and recruitment: How simple mechanisms lead to the dynamics of population recruitment curves. *Oikos*, 86:320–330, 1999.
13. S. Barot, J. Gignoux, and J.C. Menaut. Neighborhood analysis in a savanna palm: Interplay of intraspecific competition and soil patchiness. *Journal of Vegetation Science*, 14:79–88, 2000.
14. S. Barot, J. Gignoux, R. Vuattoux, and S. Legendre. Demography of a savanna palm tree in Ivory coast (Lamto): Population persistence and life history. *Journal of Tropical Ecology*, 16:637–655, 2000.
15. S.W. Beatty. Influence of microtopography and canopy species on spatial patterns of forest understory plants. *Ecology*, 65(5):1406–1419, 1984.
16. M. Begon, J.L. Harper, and C.R. Townsend. *Ecology - Individuals, populations and communities*. Blackwell Scientific Publications, Boston, 1986.

17. J. Bonvallot, M. Dugerdil, and D. Duviard. Recherches écologiques dans la savane de Lamto (côte d'ivoire): Répartition de la végétation dans la savana préforestière. *La Terre et la Vie*, 1:3–21, 1970.
18. D.B. Botkin, J.F. Janak, and J.R. Wallis. Some ecological consequences of a computer model of forest growth. *Journal of Ecology*, 60:849–871, 1972.
19. R.A. Bradstock and T.D. Auld. Soil temperatures during experimental bushfires in relation to fire intensity: Consequences for legume germination and fire management in south-eastern Australia. *Journal of Applied Ecology*, 32(1):76–84, 1995.
20. S.H. Bullock. Demography of an undergrowth palm in littoral Cameroon. *Biotropica*, 12(4):247–255, 1980.
21. R.L. Chazdon. Patterns of growth and reproduction of *Geonoma congesta*, a clustered understory palm. *Biotropica*, 24(1):43–51, 1992.
22. M.E. Cochran and S. Ellner. Simple methods for calculating age-based life history parameters for stage-structured populations. *Ecological Monographs*, 62(3):345–364, 1992.
23. P. Couteron and K. Kokou. Woody vegetation spatial patterns in a semi-arid savanna of Burkina Faso, West Africa. *Plant Ecology*, 132:211–227, 1997.
24. J. César and J.C. Menaut. Analyse d'un écosystème tropical humide: La savane de Lamto (Côte d'Ivoire). II. Le peuplement végétal. *Bulletin de Liaison des Chercheurs de Lamto*, S2:1–161, 1974.
25. J.M. Dauget and J.C. Menaut. Evolution sur vingt ans d'une parcelle de savane boisée non protégée du feu dans la réserve de Lamto (Côte d'Ivoire). *Candollea*, 47:621–630, 1992.
26. J. Delmas. Recherches écologiques dans la savane de Lamto (Côte d'Ivoire): Premier aperçu sur les sols et leur valeur agronomique. *La Terre et la Vie*, 21(3):216–227, 1967.
27. J.L. Devineau, C. Lecordier, and R. Vuattoux. Evolution de la diversité spécifique du peuplement ligneux dans une succession préforestière de colonisation d'une savane protégée des feux (Lamto, Côte d'Ivoire). *Candollea*, 39:103–134, 1984.
28. P.J. Diggle. On parameter estimation and goodness-of-fit testing for spatial point patterns. *Biometrics*, 35(0):87–101, 1979.
29. P.J. Diggle. *Statistical analysis of spatial point patterns*. Mathematics in Biology. Academic Press, London, 1983.
30. N.J. Enright. Age, reproduction and biomass allocation in *Rhopalostylis sapida* (Nikau Palm). *Australian Journal of Ecology*, 10:461–467, 1985.
31. N.J. Enright and A.D. Watson. Population dynamics of the Nikau palm, *Rhopalostylis sapida* (Wendl. et Drude), in a temperate forest remnant near Auckland, New Zealand. *New Zealand Journal of Botany*, 30:29–43, 1992.
32. P.G.H. Frost and F. Robertson. The ecological effects of fire in savannas. In B.H. Walker, editor, *Determinants of tropical savannas*, volume 3 of *Monograph series*, pages 93–140. International Council of Scientific Unions Press, Miami, FL, 1985.
33. L. Gautier. Contact forêt savane en Côte d'Ivoire centrale: Evolution de la surface forestière de la réserve de Lamto (sud du V baoulé). *Bulletin de la Société Botanique de France*, 136(3):85–92, 1989.
34. L. Gautier. Contact forêt-savane en Côte d'Ivoire centrale: Evolution du recouvrement ligneux des savanes de la réserve de Lamto (sud du V baoulé). *Candollea*, 45:627–641, 1990.

35. L. Gautier. *Contact forêt-savane en Côte d'Ivoire - Rôle de* Chromolaena odorata *(L) dans la dynamique de la végétation*. Doctorat, Université de Genève, Genève, 1992.
36. J. Gignoux. *Modélisation de la dynamique d'une population ligneuse - Application à l'étude d'une savane africaine*. M.Sc. thesis, Institut National Agronomique Paris-Grignon, Paris, 1988.
37. J. Gignoux. *Modélisation de la coexistence herbes/arbres en savane*. Ph.D. thesis, Institut National Agronomique Paris-Grignon, Paris, 1994.
38. J. Gignoux. *SPASTAT: Un logiciel pour l'analyse de répartitions spatiales par les méthodes de Diggle et Ripley*. Ecole Normale Supérieure, Paris, 1994.
39. J. Gignoux, J. Clobert, and J.C. Menaut. Alternative fire resistance strategies in savanna trees. *Oecologia*, 110(4):576–583, 1997.
40. J. Gignoux, C. Duby, and S. Barot. Comparing the performances of Diggle's tests of spatial randomness for small samples with and without edge-effect correction: Application to ecological data. *Biometrics*, 55(1):156–164, 1999.
41. J. Gignoux, J.C. Menaut, I.R. Noble, and I.D. Davies. A spatial model of savanna function and dynamics: model description and preliminary results. In D.M. Newbery, H.H.T. Prins, and N.D. Brown, editors, *Dynamics of tropical communities*, volume 37 of *Annual symposium of the BES*, pages 361–383. Blackwell Scientific Publications, Cambridge, 1998.
42. A.M. Gill and D.H. Ashton. The role of bark type in relative tolerance to fire of three central Victorian eucalypts. *Australian Journal of Botany*, 16:491–498, 1968.
43. F. Hallé and R.A.A. Oldeman. *Essai sur l'architecture et la dynamique de croissance des arbres tropicaux*. Masson, Paris, 1970.
44. P.A. Harcombe. Tree life tables. *BioScience*, 37(8):557–568, 1987.
45. H.F. Howe and J. Smallwood. Ecology of seed dispersal. *Annual Review of Ecology and Systematics*, 13:201–228, 1982.
46. M.A. Huston and D.L. DeAngelis. Size bimodality in monospecific populations: a critical review of potential mechanisms. *The American Naturalist*, 129(5):678–707, 1987.
47. S. Konaté, X. Le Roux, D. Tessier, and M. Lepage. Influence of large termitaria on soil characteristics, soil water regime, and tree leaf shedding pattern in a West African savanna. *Plant and Soil*, 206:47–60, 1999.
48. X. Le Roux and T. Bariac. Seasonal variation in soil, grass and shrub water status in a West African humid savanna. *Oecologia*, 113:456–466, 1998.
49. S. A. Levin. The problem of pattern and scale in ecology. *Ecology*, 73(6):1943–1967, 1992.
50. J.C. Menaut. Chutes de feuilles et apport au sol de litière par les ligneux dans une savane préforestière de Côte d'Ivoire. *Bulletin d'Ecologie*, 5:27–39, 1974.
51. J.C. Menaut. Evolution of plots protected from fire since 13 years in a Guinea savanna of Ivory coast. In *Actas Del IV Symposium Internacional De Ecologia Tropical*, pages 541–558, Panama, 1977.
52. J.C. Menaut and J. César. Structure and primary productivity of Lamto savannas, Ivory Coast. *Ecology*, 60(6):1197–1210, 1979.
53. J.C. Menaut, J. Gignoux, C. Prado, and J. Clobert. Tree community dynamics in a humid savanna of the Côte d'Ivoire: Modelling the effects of fire and competition with grass and neighbours. *Journal of Biogeography*, 17:471–481, 1990.

54. Y. Monnier. *Les effets des feux de brousse sur une savane préforestière de Côte d'Ivoire*, volume 9 of *Etudes Eburnéennes*. Ministère de l'Education Nationale de Côte d'Ivoire, Abidjan, 1968.
55. P. Mordelet, L. Abbadie, and J.C. Menaut. Effects of tree clumps on soil characteristics in a humid savanna of West Africa (Lamto, Côte d'Ivoire). *Plant and Soil*, 153:103–111, 1993.
56. P. Mordelet, S. Barot, and L. Abbadie. Root foraging strategies and soil patchiness in a humid savanna. *Plant and Soil*, 182:171–176, 1996.
57. D.A. Morrison. Some effects of low-intensity fires on populations of co-occurring small trees in the Sydney region. *Proceedings of the Linnean Society of New South Wales*, 115:109–119, 1995.
58. J.J. Mott and A.J. McComb. Patterns in annual vegetation and soil microrelief in an arid region of Western Australia. *Journal of Ecology*, 62:115–125, 1974.
59. I.R. Noble and R.O. Slatyer. The use of vital attributes to predict successional changes in plant communities subject to recurrent disturbances. *Vegetatio*, 43:5–21, 1980.
60. D.L. Phillips and J.A. MacMahon. Competition and spacing patterns in desert shrubs. *Journal of Ecology*, 69:95–115, 1981.
61. M.A. Pinard and J. Huffman. Fire resistance and bark properties of trees in a seasonally dry forest in eastern Bolivia. *Journal of Tropical Ecology*, 13:727–740, 1997.
62. E.C. Piélou. The use of plant-to-plant distances for the detection of competition. *Journal of Ecology*, 50:357–367, 1962.
63. D. Piñero, M. Martínez-Ramos, and J. Sarukhán. A population model of *Astrocaryum mexicanum* and a sensitivity analysis of its finite rate of increase. *Journal of Ecology*, 72:977–991, 1984.
64. L. Ponce de Leon Garcia. *L'écophysiologie de la germination d'espèces forestières et de savane, en rapport avec la dynamique de la végétation en Côte d'Ivoire*, volume 1 of *Travaux des chercheurs de Lamto*. Ecole Normale Supérieure, Paris, 1982.
65. H.H.T. Prins and H.K. Van der Jeugd. Herbivore population crashes and woodland structure in East Africa. *Journal of Ecology*, 81:305–314, 1993.
66. B.D. Ripley. *Spatial statistics*. John Wiley & Sons, New York, 1981.
67. J.J. San José, M.R. Fariñas, and J. Rosales. Spatial patterns of trees and structuring factors in a *Trachypogon* savanna of the Orinoco Llanos. *Biotropica*, 23(2):114–123, 1991.
68. A.L. Schutte, J.H.J. Vlok, and B.E. Vanwyk. Fire-survival strategy - A character of taxonomic, ecological and evolutionary importance in fynbos legumes. *Plant Systematics and Evolution*, 195(3-4):243–259, 1995.
69. C.M. Shackleton. Demography and dynamics of the dominant woody species in a communal and protected area of the Eastern Transvaal Lowveld. *South African Journal of Botany*, 59(6):569–574, 1993.
70. C. Skarpe. Spatial patterns and dynamics of woody vegetation in an arid savanna. *Journal of Vegetation Science*, 2:565–572, 1991.
71. P.B. Tomlinson and E.C. Jeffrey. *The structural biology of palms*. Clarendon Press, Oxford, 1990.
72. R. Vuattoux. Observations sur l'évolution des strates arborée et arbustive dans la savane de Lamto (Côte d'Ivoire). *Annales de l'Université d'Abidjan, Série E*, 3(1):285–315, 1970.

73. R. Vuattoux. Contribution à l'étude de l'évolution des strates arborée et arbustive dans la savane de Lamto (Côte d'Ivoire). Deuxième note. *Annales de l'Université d'Abidjan, Série C*, 7(1):35–63, 1976.
74. A. Watkinson. Plant senescence. *Trends in Ecology and Evolution*, 7(12):417–420, 1992.

19

Modeling Tree and Grass Dynamics: From Demographic to Spatially Explicit Models

Jacques Gignoux and Sébastien Barot

19.1 Introduction

The question of the stability of Guinea savannas is still unresolved [23]: rainforests grow at the latitude of Lamto on similar soils. Fire is usually invoked to explain the stability of the forest-savanna boundary [20]. Without fire, trees are expected to outcompete grass. Because of the high production of the grass layer, the considerable fuel load present at the beginning of the long dry season results in the most severe surface fires observed on the West African climatic gradient. This would strongly limit tree recruitment, preventing their invasion in the area in spite of favorable climatic conditions.

This issue has been addressed in Lamto through experimental studies (Sect. 18.4), the main one being the 40-year fire exclusion experiment (Fig. 18.3) and Vuattoux's long term succession study on plots subject to different fire treatments [34, 35, 11]. Results of this experiment (Sect. 18.5) demonstrate that fire excludes forest species, but does not prevent savanna tree species from invading except when fires occur late in the dry season (i.e., when trees have started shedding their leaves). The question of savanna stability is thus open for further investigation through modeling of the long term community dynamics.

Classical population dynamics studies have examined the persistence of grass [15] and tree [4] populations under current and experimental fire regimes. These studies enable one to assess whether particular species are able to maintain themselves or invade, but they do not point out the causes of specific success or failure to survive in a recurrently burned vegetation. More mechanistic approaches focusing on spatial pattern effects have been developed [21, 24, 19] in order to analyze the mechanism by which fire prevents tree invasion. This chapter presents these results and demonstrates how spatial pattern emerged as a determinant feature of savanna ecosystem dynamics.

19.2 Persistence of savanna species under annual burning: Analysis through matrix population models

Two studies have focused on the demography of savanna species through classical methods (stage transition or Lefkovitch matrix models [7]) at Lamto: one examined the demography of one of the dominant grass species, *Hyparrhenia diplandra*, in relation to fire [15]; the other looked at the demography of the palm tree *Borassus aethiopum*, in order to determine whether the protected population of Lamto was stable in spite of the lack of formerly present animal dispersers [4].

19.2.1 Effect of fire on grass demography and persistence

Garnier followed for 3 years two populations of *Hyparrhenia diplandra* submitted to two fire treatments, yearly burned and unburned [15]. A total of 110 reproducing individuals plus 2867 dispersed seeds were tagged and followed during the experiment, enabling one to estimate the main demographic parameters of a 4 size-class stage-transition matric model. Two separate models, one for each treatment, were constructed (a fire survival parameter as included in the model for the burned treatment). Model analysis was performed with the ULM software [22, 13], which yields the population asymptotic growth rate (dominant eigenvalue of the transition matrix) and its sensitivity to parameters (e.g., sensitivities and elasticities).

Fire exclusion increased transition rates between size classes. From the analysis of the model, we have the following:

- The asymptotic growth rate was higher in the unburned plot than in the burned plot (1.23 instead of 1.07). Altering parameters to simulate a more severe fire resulted in a smaller value of 0.99. In the unburned plot, simulating litter accumulation by reducing both germination and recruitment rates resulted in population growth rates close to that of the burned plot (1.08).
- The transition to the reproducing adult stage made the largest contribution to the population growth rate. In burned plots, the transition to the previous (non-reproducing adults) stage was almost as important.
- Sensitivity to the germination rate was higher in the unburned plot.

Annual burning has a strong effect on grass population, even if they seem well adapted to this environment (through a tussock life-form insulating buds from excessive heat and an hygrometric awn on seeds facilitating their burying into the soil [16]). Fire significantly reduces the population growth rate and affects life history traits. Similar studies conducted in other tropical savannas on *Andropogoneae* species [31, 32, 25] showed a different pattern, with a higher population growth rate in the burned treatment. This was interpreted as a negative effect of litter accumulation, not observed over the duration of the experiment in Lamto, but likely in the long term according to model results.

Even an apparently well adapted, fire-prone species like the dominant grass species *Hyparrhenia diplandra* showed a positive response to fire exclusion. However, it is not clear whether the long term positive effect of fire exclusion on survival and recruitment balances the negative effect of litter accumulation. *Hyparrhenia diplandra* could well show no demographic response to fire exclusion in the long term (compare the 1.07 growth rate under the annual fire regime to the 1.08 value under fire exclusion + litter accumulation scenario), which is a clear disadvantage relative to trees: after 35 years of fire exclusion, grasses are outcompeted by trees through competition for light (Sect. 8.2).

19.2.2 Tree persistence and reproductive strategy: The case study of *Borassus aethiopum*

Barot designed a one-sex female based Lefkovitch matrix population model with five stages [4]: seeds, EL-seedlings, SL-seedlings, juveniles and adults (Sect. 18.4). For each stage, a probability of survival and a probability of recruitment were defined. Retrogression to a previous stage was allowed for non-reproductive stages. Parameter estimation was based on the census of four plots (two grass savanna and two shrub savanna plots), yearly censused between 1996 and 1998. An average sex ratio of 0.5 was used since there was no significant difference in sex ratio among plots. Fecundity and germination rate were estimated from specific studies (Barot, unpublished data). Matrix models were analyzed with the ULM software [22]. Estimations of ages of the different stages were computed using a method based on the implicit presence of age/time in stage-classified models [9]. Five model parameterizations were used, one for each savanna plot and an average model. No significant difference between demographic parameters according to year and plot was detected.

From the analysis of the model, we have the following:

- The average (averaged over all plots and years) asymptotic population growth rate was 1.009 with a standard deviation of 0.010. Single-plot models yielded growth rates between 0.993 (s.d. 0.019) and 1.007 (s.d. 0.009). Assuming a Normal distribution, none of these rates would be significantly different from 1.
- The observed stable stage distribution was significantly different from the predicted stable stage distribution in all models. The high sensitivity of the stable stage distribution to the juvenile-to-adult recruitment rate and the uncertainty in the estimation of this parameter could explain this discrepancy.
- Elasticities and sensitivity analyses demonstrate that the most sensitive stage is the juvenile stage (Fig. 19.1).
- The total conditional life span of adults is 114.9 years (s.d. 176.0), with a surprisingly low remaining life span of 22.7 years (s.d. 103.7). Juveniles have the highest remaining life span (79.1 years, s.d. 248.0).

The population of *Borassus aethiopum* seems very close to equilibrium (growth rate not significantly different from 1 and no significant difference between

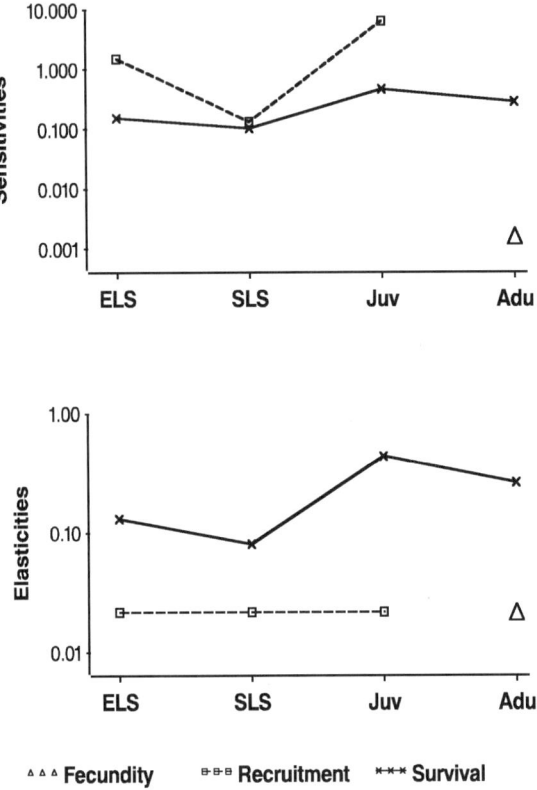

Fig. 19.1. Sensitivities and elasticities of the asymptotic growth rate to the matrix parameters for the *Borassus* population. Logarithmic scales were used as recommended in [6].

parameters across plots), although the stable stage distribution is not reached. This might be due to past disturbances (exceptionally fierce or late fires) or to the disparition in the Lamto area of seed dispersers (baboons and elephants), which would cause a change in demographic parameters. As previously suspected [1], *Borassus aethiopum* is very long lived, with a delayed reproduction and a short reproductive period, two unexpected features. Other studies of forest palm tree species show longer reproductive periods ([5, 12, 28, 26, 27] and Fig. 19.2).

Borassus aethiopum is able to persist in Lamto in the current conditions, and its population is very close to the demographic equilibrium. However, spatial pattern analyses (Sect. 18.3 and [1]) and analyses of seed dispersal [2] demonstrate that local variations of demographic parameters are significant in this species. We should therefore expect important changes in the demography

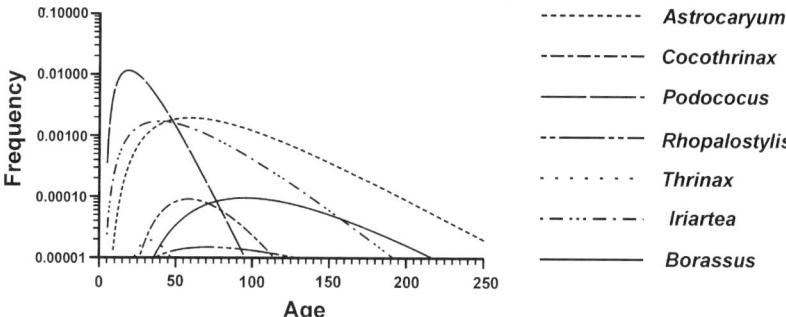

Fig. 19.2. Comparison of *Borassus aethiopum* stable age distribution estimated with matrix population models with other tropical forest palm tree species.

of *Borassus aethiopum* if those local parameters were to change, an hypothesis that the classical demographic approach used here is unable to test.

19.2.3 Population persistence, fire, and demographic strategies of savanna plants

The two previous modeling studies are the only attempts so far to look at the demography of the main plant species in Lamto through classical methods. We lack a larger sampling of species to really generalize these results, but two main facts arise form these studies.

First, the populations of two of the dominant species of Lamto are close to the demographic equilibrium, reinforcing the idea that fire is not a major problem for the persistence of these species in Guinea savannas. However, ambiguities remain and the matrix model approach used here does not allow a mechanistic analysis of the cause of this stability.

Second, the tree and grass species studied here have remarkably similar life spans: 120 years for the palm tree and 150 years for *Hyparrhenia diplandra*. These estimates are highly sensitive to the adult survival rates, which are difficult to estimate in short term studies. But even as order of magnitude, these estimates would look very similar. To generalize the conclusion of these studies, demographic analysis of the other dominant tree and grass species is required.

If dominant trees and grasses have similar life spans, can we find a reasonable selective pressure for this convergence, or is it due to pure chance? Given the competitive advantage of trees in the area, there should be a strong selective pressure on grass species to promote fire. This does not require them to be long lived (see, for example, the Northern Australian humid savannas, dominated by annual grasses [36]). Nutrient scarcity could select for this trait, since concentrating roots under grass tufts appears as an efficient way of recycling nitrogen (Sect. 15.5). Again, the comparison to North Australian savannas does not confirm this hypothesis, since nutrient levels are comparable in these annual-dominated savannas [10]). At this stage and without further studies on other species, there is no apparent reason for such a convergence in life spans.

19.3 Spatialized demographic models

Two articles have studied the effect of fire on tree spatial patterns and population dynamics through spatially explicit models, using very different formalisms [21, 24].

19.3.1 The role of spatial pattern and fire in savanna dynamics

The spatial pattern of the tree community in Lamto is complex (Sect. 18.3) and as for *Borassus aethiopum*, spatial variation in demographic parameters is expected [2, 3]. This seriously limits the use of classical matrix population models for understanding the stability of this community.

Fire has a spatial component: the grass layer constitutes the fuel, and fuel load is reduced under tree clumps because of competition for light in favor of trees in these areas (Chap. 8). Different from most spatial effects examined in the literature (e.g., [14, 8]), this effect is not due to preexisting environment heterogeneities, (like e.g., nutrient-rich patches due to soil heterogeneity independent of vegetation dynamics), but is generated by the dynamics of the tree populations, through their dispersal strategies and autogenerated spatial variations in recruitment rates. This led to the development of two spatial models of tree demography in relation to fire resistance.

19.3.2 Continuous spatial model

FRENCH [17, 24, 19] is a simulation model developed to explore the causal mechanisms of vegetation structure in savanna communities. The principal aim is to analyze how a spatially explicit formulation can account for the role of fire in savannas and to what extent the individual performance of trees is responsible for the observed spatial pattern.

Assumptions and model structure

The model is based on three life history stages (seedlings, juveniles, and reproducing adults) and is characterized by the explicit treatment of the spatial structure of tree stands. Computations are thus performed at the individual level to study local interactions and neighborhood relationships [29, 30]. Trees, located by their coordinates on the map, are distributed in a continuous space with wrap-around margins to avoid edge effects [33].

Emphasis is given to the tree layer. Grasses are treated as a heterogeneous, tree- and fire-dependent pattern of constraints, which affects tree recruitment. The grass cover (spatial pattern of biomass) is a function of tree pattern and density, decreasing down to total disappearance under tree clumps comprising more than 6 individuals. Depending on rainfall, the grass cover can be irregularly burned (intensity and pattern of fire). This is modeled by attributing a variance to the mean value of tree seedling survival in the grass layer; survival values are then randomly computed within the range defined by this variance.

Species survival depends on the time-space constraints that affect the performance of individuals throughout their life cycle. Survival constraints are represented by two factors: fire and neighborhood competition functions. Resources are not explicitly treated. Species are characterized by seven parameters: reproductive output (expressed as seedlings), dispersal, reproductive lifetime, life span (age-specific survival process), individual growth rate, maximum diameter, and height. In the original version of the model, only one "synthetic" species was considered. The model has now been extended to enable many different tree species.

The inputs of the model, running on a yearly time step, are as follows:

- An initial distribution of trees (spatial pattern and size) over a 50×50 m savanna plot. It is based either on actual field data or on a random generator of plant distributions and positions (according to a random, regular or clumped spatial pattern).
- Four life history attributes: age-specific survival rate, reproductive lifetime, seedling production, and seedling dispersal.
- allometric relationships among tree basal diameter, height, and crown surface.
- Growth rules for seedlings, resprouts, and mature trees.
- Height of the grass cover and a function of grass biomass decrease in relation to tree density. Grass biomass is not introduced as such, but through its effects on tree growth and survival.
- Fire intensity, introduced as a function of grass biomass and fire regime: basically once a year, with the possibility to test different regimes (once every n years, or total exclusion).

The outputs of the model are the following:

- a map of the plot with the positions and sizes of each individual, giving density and type of pattern; and
- the dimension and age of each individual, enabling calculation of the size and age class structure of the tree community.

All the details of the equations and parameters used in this model are presented in [24].

Results

Fire parameters appear to strongly determine the dynamics of the tree population [24]: in particular, with a homogeneous fire burning the whole surface, no recruitment of isolated trees (i.e., outside tree clumps) is possible; the intrinsic fire resistance of a tree species determines its ability to survive: low resistance species are quickly eliminated.

The fraction of the plot surface left unburned by chance is a key driver of tree dynamics [19] (Fig. 19.3): below a threshold surface left unburned (around 15%), trees go extinct; above this threshold and below a second one (25%), the tree population stabilizes on average to typical densities of 200 ha^{-1} observed

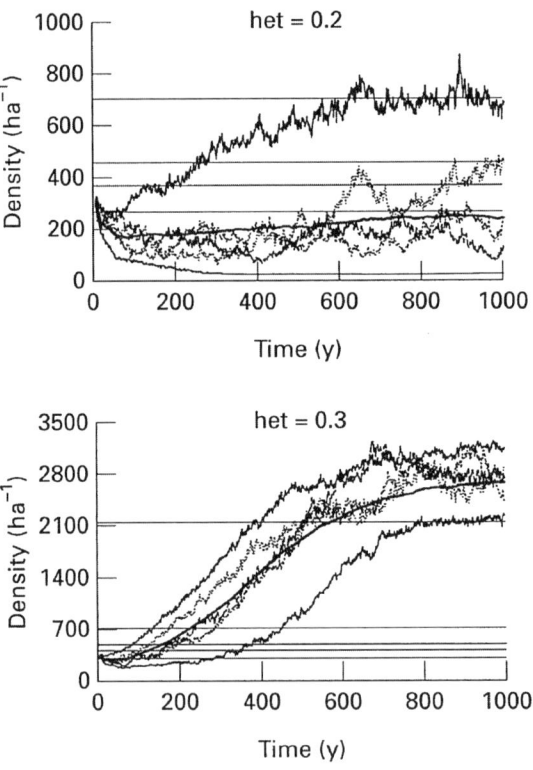

Fig. 19.3. Simulation of tree population dynamics with the FRENCH model. Parameter het measures the probability of escaping fire through pure chance for young tree. The system switches to fire-regulated, savanna-like (top), to competition-regulated, forest-like (bottom) for het between 0.2 and 0.3. Solid line, average of 100 simulations; top and bottom lines, confidence interval (envelope of 100 simulations); horizontal lines, tree densities measured on real savanna plots for comparison (after [19], with permission of Blackwell Publishing).

in the field; above the second threshold, the tree population stabilizes at large densities around 2500 ha^{-1}, only observed in the field in dense areas where grass has been excluded. This parameter thus separates a domain where tree existence is impossible from domains where the tree population is stabilized by fire or by intraspecific competition.

Spatial patterns generated through simulation were compatible with patterns observed in the field (97% of simulations had a clumped spatial pattern while all field plots exhibited such patterns).

19.3.3 Cellular automaton model

Assumptions and model structure

Hochberg et al. [21] used a minimally complicated algorithm to highlight the basic processes acting to determine spatial patterning in a single tree population. This approach corresponds closely to the category of cellular automaton models, where the habitat is broken up into a grid of spatially distinct, interacting cells. These cells can be either uninhabited or inhabited by a single dynamic entity such as an individual plant or a group of plants.

The system is modeled as a $n \times n$ grid of square cells, the whole grid being surrounded by grassland. The equations governing the changes in cell occupation by trees are iterated once per calendar year, and at the beginning of a given interaction, each cell can be in one of three states with respect to tree occupation: (i) unoccupied, (ii) occupied by reproductively immature trees, or (iii) occupied by mature trees. Demographic events can occur either purely within a cell (i.e., mortality not induced by fire), or in interaction with the eight immediately adjacent cells (i.e., mortality from fire, recruitment), or at longer distances (i.e., recruitment of dispersed seeds). The size of a given cell is 1m × 1 m, which is approximately that permitting the growth of, at most, a single maximally reproductive individual. Intraspecific competition is not explicitly considered in the model; when cells are occupied by more than one reproductively mature individual, the largest is assumed to rapidly displace its competitors. Tree-tree competition between cells is of the pre-emptive type, such that the first individual arriving in a cell cannot be directly displaced by another. In our model, interspecific effects occur via fire, such that the (constant) grass population has a negative impact on tree seedlings. Thus, the absence of trees implies the presence of (fire conducting) grasses.

Details of the algorithm and parameter estimations can be found in [21].

Results

Model simulations predict that, in the presence of yearly fires, the doubling time of an initially randomly dispersed tree population is ca. 20-30 years. This estimation holds true until approximately 40% of the system is initially occupied by trees, beyond which (due to limits in the number of cells in the system) the doubling time rapidly approaches infinity. When mature trees are initially in the center of the system, the doubling time is substantially increased. In accordance with the doubling time of 30-40 years, the tree population reaches an equilibrium of ca. 90% cell occupancy by mature trees in 1200 years following the introduction of a single mature tree into the system. In the absence of fire, the doubling time decreases to about 6 years.

Clumping indices were computed to study the dynamics of the spatial pattern. They both show that fire reinforces tree aggregation: higher values of the clumpiness index are reached with fire, and clumpiness increases even when the initial population is random (Fig. 19.4).

19.3.4 Fire and the stability of Guinea savannas as mixed life-form systems

The models have complementary properties: while the FRENCH model enables one to compare simulated spatial patterns to field data, the cellular automaton model enables one to quantify aggregation in a more convenient way. Their results are remarkably similar in spite of an apparent contradiction: the FRENCH model demonstrates that fire can stabilize the tree population in certain conditions, but not the cellular automaton model (trees always invade, although more or less slowly). However, results of the FRENCH model show that the savannas of Lamto should be unstable at the landscape level. In this model, transition is possible from a fire-stabilized system to a

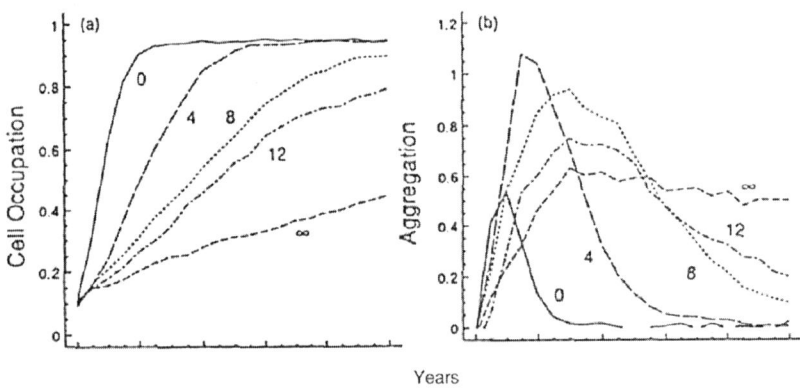

Fig. 19.4. Sensitivity of the cellular automata model to changes in burn threshold (number of encircling mature trees at or beyond which fire mortality is 0). (a) Fraction of cells occupied by mature trees. (b) Aggregation index as the number of couplings between mature trees divided by the expected number of couplings if trees were distributed randomly (from [21], with permission of Blackwell Publishing).

competition stabilized system, but the reverse transition is impossible because grass has been eliminated in a competition-stabilized system. Therefore, if for any reason a tree clump appears and initiates a competition stabilized kernel, nothing will stop the slow invasion of trees from clump margins, exactly what the cellular automaton model predicts.

If invasion is not impossible, it is slow. The cellular automaton model predicts that, for a 0.25 ha plot as those studied in the field, competition-stabilized equilibrium should be reached in 1200 years in a burned system instead of 350 years in an unburned system (this prediction is made in the absence of long distance seed dispersal, i.e., only adjacent cells can be colonized by trees. These slow rates of invasion leave the possibility for rare catastrophic events–fierce fire, diseases, storms, exceptional droughts–to bring the tree population down to lower levels, ensuring long term stability of the tree-grass system.

Field results (Chap. 18) tend to confirm these conclusions since a significant increase in tree density has been observed in Lamto over 30 years. For example, the dynamics of the G and I plot between 1969 and 1989 (Fig. 18.4) is particularly interesting: in 1969, the two plots roughly had the same density; 20 years later, the I plot remains stable while a large tree clump has started to increase on the G plot, where grass is almost excluded. Only a slight difference in initial spatial pattern is visible on these plots.

Which factor should be invoked if fire does not stabilize the Guinea savannas in the long term? From the present knowledge, two hypotheses can be proposed:

- Different tree species have different fire resistances, and the interaction of such species could stabilize the tree community. A field experiment [18] demonstrated that fire resistance strategies were different among species and that a trade-off between resistance and competitive ability (individual growth rate) existed. This trade-off could probably lead to a 3-functional groups stable savanna: fire resistant trees would sometimes initiate clumps where they would be outcompeted by more fire sensitive trees; but because those trees recruit as a cohort of even aged trees, their synchronous death in a few years would lead to a collapse of their own population due to the coming back of grass and fire (Fig. 18.3). Field evidence supports the first part of this explanation, i.e., the invasion of dense tree clumps by less fire resistance species is common. But the collapse of such a population has never been recorded so far.
- Late fires, not represented in any of the two models, are able to kill adult trees (Sect. 18.2). They could be the rare, catastrophic events previously proposed: to reach stability, late fire (i.e., fires occurring at a sensitive phenological stage for trees) would have to occur from time to time. The continuous increase in tree density observed at Lamto and in the surrounding area could be due to the disparition of such late fires, either due to the planification of fire dates in the Lamto reserve or to the increase in human

population outside the reserve (which statistically reduces the probability of an area reaching the end of the dry season without being burned).

The existing models can relatively easily be modified and parameterized to test these two hypotheses. If they remain inconclusive, other factors would have to be included to explain the stability of the Guinea savannas.

19.4 Conclusion: The dynamics of plant populations and spatial patterns

The major input of the works summarized here to the understanding of savannas is the evidence that population and community dynamics in mixed tree-grass systems are closely linked to their spatial patterns and to the dynamics of these patterns. Static spatial pattern studies (Sects. 17.2 and 18.3) demonstrate a clear link between population dynamics and preexisting soil heterogeneities like the nutrient-rich patches constituted by old termite mounds. Classical demographic studies do not allow one to infer more than the ability of a population to stay in the system. Spatialized demographic modeling enables one to show the link between self-generated spatial patterns and long term plant dynamics: the long term unstability observed in the FRENCH model is due to the spatial patterns generated by fire; in a non-spatial model, the conclusion would probably have been that Guinea savannas were stable in the long term. These works were the impetus for further studies focusing on spatial issues in Lamto, like studies on mechanisms of dispersal and distance-dependent mortality and recruitment of seedlings [2].

References

1. S. Barot, J. Gignoux, and J.C. Menaut. Demography of a savanna palm tree: Predictions from comprehensive spatial pattern analyses. *Ecology*, 80(6):1987–2005, 1999.
2. S. Barot, J. Gignoux, and J.C. Menaut. Seed shadows, survival and recruitment: How simple mechanisms lead to the dynamics of population recruitment curves. *Oikos*, 86:320–330, 1999.
3. S. Barot, J. Gignoux, and J.C. Menaut. Neighborhood analysis in a savanna palm: Interplay of intraspecific competition and soil patchiness. *Journal of Vegetation Science*, 14:79–88, 2000.
4. S. Barot, J. Gignoux, R. Vuattoux, and S. Legendre. Demography of a savanna palm tree in Ivory coast (Lamto): Population persistence and life history. *Journal of Tropical Ecology*, 16:637–655, 2000.
5. S.H. Bullock. Demography of an undergrowth palm in littoral Cameroon. *Biotropica*, 12(4):247–255, 1980.
6. H. Caswell. Life cycle models for plants. *Lectures on Mathematics in the Life Sciences*, 18:171–233, 1986.

7. H. Caswell. *Matrix population models*. Sinauer, Sunderland, MA, 1989.
8. P.L. Chesson. Coexistence of competitors in spatially and temporally varying environments: A look at the combined effects of different sorts of variability. *Theoretical Population Biology*, 28(3):263–287, 1984.
9. M.E. Cochran and S. Ellner. Simple methods for calculating age-based life history parameters for stage-structured populations. *Ecological Monographs*, 62(3):345–364, 1992.
10. G.D. Cook and M.H. Andrew. The nutrient capital of indigenous *Sorghum* species and other understorey components of savannas in North-Western Australia. *Australian Journal of Ecology*, 16:375–384, 1991.
11. J.L. Devineau, C. Lecordier, and R. Vuattoux. Evolution de la diversité spécifique du peuplement ligneux dans une succession préforestière de colonisation d'une savane protégée des feux (Lamto, Côte d'Ivoire). *Candollea*, 39:103–134, 1984.
12. N.J. Enright and A.D. Watson. Population dynamics of the Nikau palm, *Rhopalostylis sapida* (Wendl. et Drude), in a temperate forest remnant near Auckland, New Zealand. *New Zealand Journal of Botany*, 30:29–43, 1992.
13. R. Ferrière, F. Sarrazin, S. Legendre, and J.P. Baron. Matrix population models applied to viability analysis and conservation: Theory and practice using ULM software. *Acta Oecologica*, 17:629–656, 1996.
14. R.H. Gardner and R.V. O'neill. Pattern, process and predictability: The use of neutral models for landscape analysis. In M.G. Turner and H. Gardner, editors, *Quantitative methods in landscape ecology*, volume 82 of *Ecological studies*, pages 289–307. Springer-Verlag, Berlin, 1991.
15. L.K. Garnier and I. Dajoz. The influence of fire on the demography of a dominant grass species of West African savannas, *Hyparrhenia diplandra*. *Journal of Ecology*, 89:200–208, 2001.
16. L.K.M. Garnier and I. Dajoz. Evolutionary significance of diaspore polymorphism in a clonal grass of fire prone savannas. *Ecology*, 82:1720–1733, 2001.
17. J. Gignoux. *Modélisation de la dynamique d'une population ligneuse - Application à l'étude d'une savane africaine*. M.Sc. thesis, Institut National Agronomique Paris-Grignon, Paris, 1988.
18. J. Gignoux, J. Clobert, and J.C. Menaut. Alternative fire resistance strategies in savanna trees. *Oecologia*, 110(4):576–583, 1997.
19. J. Gignoux, J.C. Menaut, I.R. Noble, and I.D. Davies. A spatial model of savanna function and dynamics: model description and preliminary results. In D.M. Newbery, H.H.T. Prins, and N.D. Brown, editors, *Dynamics of tropical communities*, volume 37 of *Annual symposium of the BES*, pages 361–383. Blackwell Scientific Publications, Cambridge, 1998.
20. D. Gillon. The fire problem in tropical savannas. In F. Bourlière, editor, *Tropical savannas*, volume 13 of *Ecosystems of the world*, pages 617–642. Elsevier, Amsterdam, 1983.
21. M.E. Hochberg, J.C. Menaut, and J. Gignoux. The influences of tree biology and fire in the spatial structure of the West African savanna. *Journal of Ecology*, 82(2):217–226, 1994.
22. S. Legendre and J. Clobert. ULM, a software for conservation and evolutionary biologists. *Journal of Applied Statistics*, 22:817–834, 1995.
23. J.C. Menaut. The vegetation of african savannas. In F. Bourlière, editor, *Tropical Savannas*, volume 13, pages 109–149. Elsevier, 1983.

24. J.C. Menaut, J. Gignoux, C. Prado, and J. Clobert. Tree community dynamics in a humid savanna of the Côte d'Ivoire: Modelling the effects of fire and competition with grass and neighbours. *Journal of Biogeography*, 17:471–481, 1990.
25. J.J. Mott and A.J. McComb. Patterns in annual vegetation and soil microrelief in an arid region of Western Australia. *Journal of Ecology*, 62:115–125, 1974.
26. I. Olmsted and E.R. Alvarez-Buylla. Sustainable harvesting of tropical trees: Demography and matrix models of two palm species in Mexico. *Ecological Applications*, 5(2):484–500, 1995.
27. M. Piñard. Impacts of stem harvesting on populations of *Iriartea deltoidea* (Palmae) in extractive reserve in Acre, Brazil. *Biotropica*, 25:2–14, 1993.
28. D. Piñero, M. Martínez-Ramos, and J. Sarukhán. A population model of *Astrocaryum mexicanum* and a sensitivity analysis of its finite rate of increase. *Journal of Ecology*, 72:977–991, 1984.
29. C. Prado. *Un modèle de succession végétale: Rôle des traits biologiques des espèces et des contraintes spatiales*. Ph.D. thesis, Université de Paris 6, Paris, 1988.
30. C. Prado. Plant community dynamics and species growth rules: A simulation study based on a cellular automata formalism. In A. Pavé and G.C. Vansteekiste, editors, *Artificial intelligence in numerical and symbolic simulation*, pages 75–87. ALEAS, Paris, 1991.
31. J.F. Silva, J. Raventos, and H. Caswell. Fire and fire exclusion effects on the growth and survival of two savanna grasses. *Acta Oecologica*, 11(6):783–800, 1990.
32. J.F. Silva, J. Raventos, H. Caswell, and M.C. Trevisan. Population responses to fire in a tropical savanna grass, *Andropogon semiberbis* - a matrix model approach. *Journal of Ecology*, 79(2):345–356, 1991.
33. O. Van Tongeren and I.C. Prentice. A spatial simulation model for vegetation dynamics. *Vegetatio*, 65:163–173, 1986.
34. R. Vuattoux. Observations sur l'évolution des strates arborée et arbustive dans la savane de Lamto (Côte d'Ivoire). *Annales de l'Université d'Abidjan, Série E*, 3(1):285–315, 1970.
35. R. Vuattoux. Contribution à l'étude de l'évolution des strates arborée et arbustive dans la savane de Lamto (Côte d'Ivoire). Deuxième note. *Annales de l'Université d'Abidjan, Série C*, 7(1):35–63, 1976.
36. A.R. Watkinson, W.M. Lonsdale, and M.H. Andrew. Modelling the population dynamics of an annual plant sorghum intrans in the wet-dry tropics. *Journal of Ecology*, 77:162–181, 1989.

Part VI

Toward an Integration of Savanna Structure, Functioning, and Dynamics

20

A Synthetic Overview of Lamto Savanna Ecology: Importance of Structure-Functioning-Dynamics Relationships

Xavier Le Roux, Luc Abbadie, and Jacques Gignoux

20.1 Introduction

A tremendous amount of work has been carried on to describe and understand ecosystem structure, diversity, functioning and dynamics at Lamto for more than 40 years. This book did not aim at presenting all the knowledge gained from ecological studies performed during these four decades. For instance, the description of biodiversity and systematics of organisms encountered at Lamto (see [29] for bird communities; [4] for reptile communities; [7] for rodents; [8] for fishes; [13, 3, 11, 18] for invertebrates), the mutualistic interactions between specific organisms (e.g., *Blastophaga*/figs: [14]), and animal behavior (e.g., [15] for termites) were beyond the scope of this synthesis. Our objective was to show how much of the ecological work performed at Lamto has allowed emergence of a novel, comprehensive view of savanna functioning and dynamics. Four decades ago, the prevailing representation of Lamto savanna ecosystem was a compartmental one, focusing on biomass values of each compartment and fluxes transferred from one compartment to the next (Fig. 20.1). At this time, the ecosystem was viewed as relatively stable in time and ecosystem heterogeneity was largely disregarded or overviewed by classification into savanna types (see Sect. 5.2).

Our view of Lamto savanna has now moved toward an explicit account of savanna heterogeneity and unstability. In particular, it is the belief of the authors that a key for understanding and predicting savanna ecosystem functioning and dynamics is not only to account for savanna heterogeneity and unstability, but rather to put structure-functioning-dynamics relationships at the heart of modern savanna ecology studies. This concluding chapter presents rationales supporting this new approach of savanna ecology, and presents emblematic examples showing the importance of expliciting structure-functioning-dynamics relationships to understand the behavior of Lamto savanna.

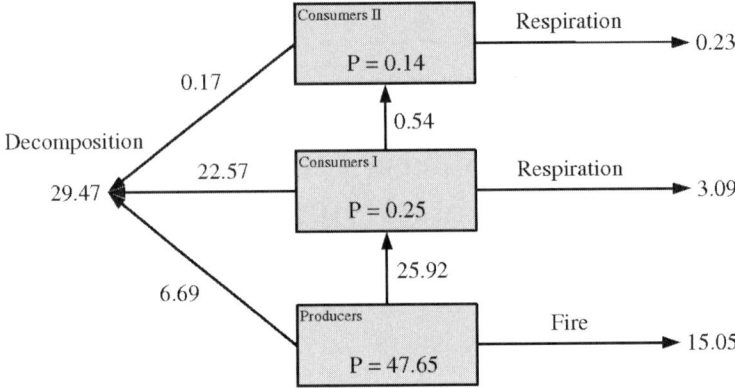

Fig. 20.1. Annual energy fluxes in Lamto savanna ($J\,m^{-2}, y^{-1}$) (data from [16]).

20.2 Rationales for the "structure-functioning-dynamics relationships" approach

The effect of ecosystems on energy and mass exchanges at large (landscape, regional and global) scales, the responses of ecosystems to disturbances (resistance and/or resilience of biodiversity and functions) and sustainable use of ecosystems to provide social and economical goods to human populations are fundamental problems of modern ecosystem ecology. For instance, in a global change context, there is a need to be able to predict the response of ecosystems to changes in natural or anthropic disturbance regimes (e.g., shifts in species composition, carbon storage capacity, gas and water fluxes to the atmosphere, reorganizations of food webs). Addressing these issues requires one to integrate many processes and, moreover, the interactions between these processes.

However, processes interacting to produce ecosystem properties rely on completely different logics (ecophysiology, micrometeorology, thermodynamics, demography, evolution, etc.), implying different methodologies (from flux measurements to capture-recapture methods) and related barriers between schools of thought. Crudely, ecosystem studies belong to three major groups: studies of ecosystem structure and/or diversity, ecosystem functioning and ecosystem dynamics, generally conducted with weak connections between them. For instance, the response of vegetation to climatic change has often been studied only through the impacts on plant physiology [30], and few studies consider impacts on competition, species composition and spatial patterns in a demographic context (reproduction, survival, fecundity, etc.) [6]. More generally, studies explicitly coupling functional and demographical responses of plant species to changed environmental conditions are scarce.

The few available studies coupling spatial structure to ecosystem dynamics have yielded extremely original results, showing, for example, that the effect of biodiversity on primary production in forests, or their response to elevated CO_2, were not properly predicted when detailed spatial structure effects were not considered [24].

The 40-year long Lamto experiment has clearly showed that facing the real world often implies connecting the three dimensions of the ecosystem: structure, functioning and dynamics (Fig. 20.2). At Lamto, several studies have provided original contributions to make the links between either ecosystem structure and functioning (e.g., [2] and Chap. 15 for N dynamics, and [27] and Chap. 6 for water and carbon fluxes), or ecosystem structure and dynamics explicit ([21, 5, 12] and Chaps. 18 and 19). The link between ecosystem

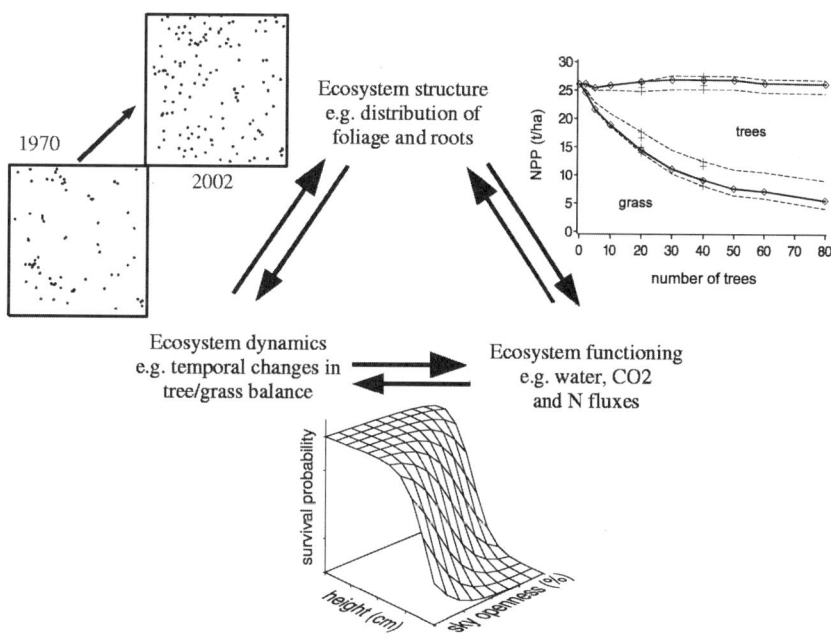

Fig. 20.2. Simple representation of the biological links between ecosystem structure, functioning and dynamics for Lamto savanna. Top left inset represents the evolution of the spatial structure of a 0.25 ha savanna plot between 1970 and 2002, with dots representing adult trees (J.C. Menaut, J. Gignoux and G. Lahoreau, unpublished data); top right inset shows the response of grass, tree and total primary productions to tree density and tree aggregation: top line is for aggregated patterns, solid line for random patterns and bottom line for regular patterns (from [27], with permission of the Ecological Society of America); bottom inset represents the relationship between tree mortality and local light level and tree height for *Bridelia ferruginea* (Lahoreau et al., unpublished data, see Fig. 20.4).

functioning and dynamics is currently studied for the tree layer at Lamto (importance of the local environment and ecophysiological traits of tree species for their growth, survival and reproductive success: Lahoreau, Gignoux and Le Roux, unpublished). Emblematic examples showing the paramount importance of expliciting structure-functioning-dynamics relationships to understand the behavior of Lamto savanna are presented below.

20.3 Structure-functioning relationships as a key to understanding the Lamto productivity paradox

Lamto savanna soils exhibit very poor nutrient availability as measured by classical fertility indices (low soil C and N concentrations; see Chap. 2). Because of such low soil organic matter content, of the low biodegradability of this organic matter (Chap. 11) and of the annual loss of nutrients through fire (Chap. 4), a low production of Lamto savannas could be expected because of nutrient shortage. On the contrary, Lamto savannas are among the most productive ecosystems in the world (Chap. 7). This paradox is partly explained by physiological mechanisms:

1. The particularly high photosynthetic nitrogen use efficiency of both C_4 grasses and C_3 tree species dominating Lamto vegetation ([17] and Chap. 6) is an important feature allowing high primary production in those nitrogen poor ecosystem.
2. The quasi-absence of nitrification, advantageous since nitrates are more leachable than ammonium, and of denitrification, which reduces gaseous losses, considerably reduce nitrogen losses from the soil (Chap. 15). The actual fertility of the soil therefore largely depends on the nature and activity of the biological communities in the ecosystem (microbes, plants, soil fauna).
3. The very efficient recycling of nitrogen of senescent leaves within plants and the atmospheric nitrogen fixation occurring in the rhizosphere further limit nitrogen losses (Chaps. 14 and 15).

However, the spatial organization of the soil/plant system and the importance of structure-functioning relationships for nutrient recycling is the major biological characteristic of Lamto savanna explaining the apparent high production/low nitrogen availability paradox.

Indeed, because of the clumped structure of the belowground part of the grass stratum (Sect. 15.5), mineralization of dead roots and associated soil fauna are located very close to the living roots: the nitrogen produced by this mineralization is very likely to be uptaken by the living roots. The efficiency of vegetation to recycle nutrients from root litter is thus high. This has been confirmed by measurements of ^{15}N composition of grass roots as a function

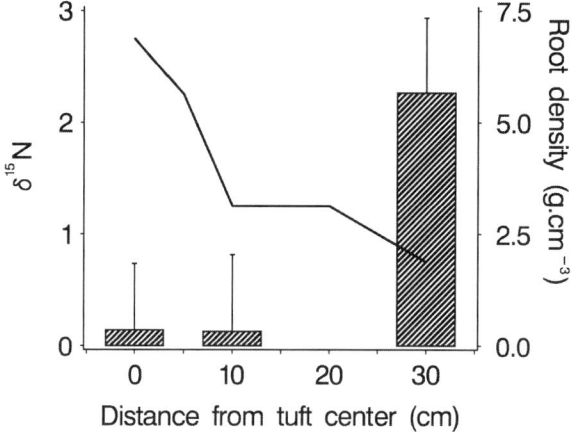

Fig. 20.3. Variations in root δ^{15}N values (left axis and bars) and root density (right axis and solid line) as a function of distance to grass tuft center. Mean tuft radius is 10 cm. According to the scheme developed by Abbadie et al. [2], low values indicate a high N recycling from root mineralization and a large independance from soil N supply (L. Abbadie and A. Mariotti, unpublished data).

of distance to tuft center (Fig. 20.3): δ^{15}N values are low at the center of tufts, indicating a high N recycling where root density is maximal, while δ^{15}N values increase out of the tuft, indicating that N uptake is more dependent on soil N supply where root density is low. More generally, the production of mineral nitrogen through decomposition of litter does not occur uniformly within the soil, but is concentrated in some activation sites like the wall of fungus growing comb chambers built by termites (Sect. 13.4 and 16.3), more generally in areas processed by soil fauna, usually concentrated close to grass tufts. These activation sites are likely long lived: more than 6 months for fungus-comb chambers, decades for grass tufts.

These results suggest that the tufted architecture and spatial aggregation of roots maximize nutrient intake for grasses in poor soil ecosystems. Similarly, the existence of localized hot spots of higher nutrient availability (termite mounds, earthworm galleries, tree clumps) is of major importance for the actual soil fertility experienced by savanna plants [1, 22]. Thus, the evaluation of soil fertility through mean stock and flux measurements as proposed by the classical compartmental approach only may lead to an underestimation of potential productivity. The soil fertility in Lamto savannas depends less on the pools of nitrogen and on the rhythm at which they are made available to plants than on the spatial distribution of nitrogen sources and sinks. This example clearly shows that understanding and modeling savanna functioning often requires an explicit accounting for structure-functioning relationships.

20.4 Structure-dynamics relationships as a key to understanding changes in tree/grass balance

Compared to other savanna sites in Africa [25], fire is a major constraint shaping Lamto savannas in their species composition and long term structure and physiognomy (Chap. 4). Fire deeply affects tree growth and has selected species able to survive to yearly, intense surface fires. However, fires are highly variable in intensity in space and time (Sect. 4.5), and assuming homogeneous fire intensity in space and time invariably leads to the elimination of trees [21, 12].

There is evidence that tree clumping [21] and cohort recruitment [9] are linked to spatial and temporal variations in fire intensity, respectively. Most fire sensitive species tend to form clumps and be associated to more fire resistant species, while rare savanna tree species are almost always found on rocks and other fire-safe sites (Chap. 18). Tree clumps apparently constitute fire-safe (or fire-safer) sites favorable to the recruitment of fire sensitive species. Years with less intense or no fire tend to produce cohorts of recruits in fire sensitive species.

In simulations [10], it is possible to reach a stable tree population if tree demographic parameters carefully balance their fire sensitivity. However, this equilibrium is unstable: once a tree clump starts to establish, it can only go on growing without limit, since fire is eliminated from the clump (Fig. 19.3). Since tree clumps can appear randomly at any time and since fire promotes tree clumping once a clump "kernel" is formed, we should expect in the very long term the savanna to be totally invaded by trees. This conclusion would not have been reached if we had not considered the spatial pattern of tree populations: fire exclusion experiments simply demonstrate that fire prevents tree invasion, but the stabilizing mechanism is not straightforward (see list of potential mechanisms in Chap. 19): one has to find a process able to cause the simultaneous death of all trees within a tree clump. This example illustrates the great importance of structure-dynamics relationships to understand and adequately predict changes in tree/grass balance.

20.5 Current approaches for studying tree functioning-dynamics relationships in Lamto savanna

Although different studies have made the structure-functioning and structure-dynamics relationships explicit at Lamto, the link between ecosystem functioning and dynamics has rarely been studied (most studies on functional aspects at best weakly assessed the importance of functioning for plant survival, reproduction and recruitment). During the last years, a project has been launched to study the importance of the local environment (at individual tree scale) and ecophysiological traits of tree species (leaf characteristics for water and CO_2 exchanges, allocation patterns, etc.) for their growth, survival

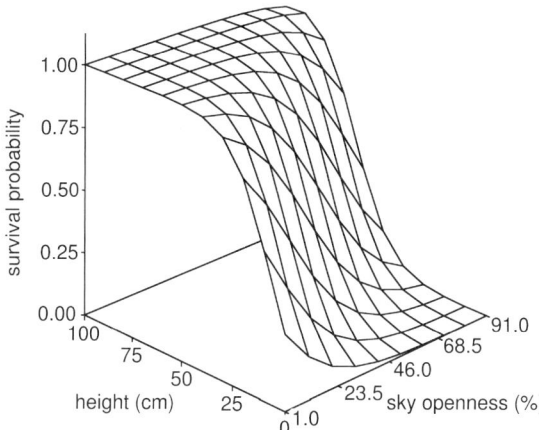

Fig. 20.4. Relationship between individual tree mortality and tree height and local light environment (sky openness) for *Bridelia ferruginea* (G. Lahoreau, J. Gignoux and X. Le Roux, unpublished data).

and reproductive success (Lahoreau, Gignoux and Le Roux, unpublished). A first step was to relate individual tree growth/survival recorded on different savanna plots (Chap. 18) to the light, water and nutrient availabilities experienced by the individuals along several years. The first results of this study showed that, for some species, treeindexmortality!tree is strongly correlated to the local light environment experienced by individuals (Fig. 20.4) whereas changes in water and nutrient availabilities play a minor role. This calls for a better understanding of how shading affects tree mortality through effects on tree functioning and growth (e.g., carbon gain and allocation to tree parts). This also calls for tools allowing one to synthesize the complex interactions among tree structure (namely tree density and spatial distribution), tree functioning and growth (particularly carbon balance) and tree dynamics (demographic traits). The importance of seedling carbon gain and allocation strategy for their survival, and the potential negative effect of investment of resources into reproduction (seeds/fruits) for mature tree growth need to be studied in this context.

20.6 Modeling as a synthesis tool for studying structure-functioning-dynamics relationships

When trying to analyze the interactions among structure, dynamics and functioning in ecosystems, at least two major difficulties arise. First, coupling different processes following different rules encompasses an enormous range of scientific expertise. At the moment, only simulation models are able to synthesize such knowledge. Second, the problem is not to represent (descriptive approach) spatial patterns at a given time or prescribe their temporal variations, but rather to let them emerge as a result of the functioning and dynamics of ecosystem components. For vegetation, a relevant level of analysis is clearly plant individuals, because functional and demographic traits operate at the individual scale in response to the local environment sensed by individuals. A good way to cope with spatial structures is thus to use spatially explicit, individual-based models. Complementary to analytical models (systems of equations that can be solved with mathematical techniques, but can rarely be tested with real data except for specific cases), simulation models enable one to couple many processes, can directly incorporate field data if properly built up can incorporate theoretical knowledge, and can represent spatial structures if they are spatially explicit. For these reasons, they appear as an attractive synthesis tool to study interactions between ecosystem structure, functioning and dynamics. They can be used as "virtual ecosystems," i.e., models (in the original sense of "imitation of reality at an accessible scale") of real ecosystems. They benefit from specific advantages compared to real systems, giving access to measures practically impossible in the field (like meaningful estimates of competition, access to long-term "experiments", test of complex scenarios of hypotheses, etc.). Of course, this approach requires that the virtual ecosystem be a sufficiently good representation of its real counterpart: it relies on an enormous effort in data acquisition and model validation. An example of this use of simulation models and of the heavy data acquisition effort they require is given by Pacala et al. [23]. The strength of this approach relies on this possible link with reality, which cannot be as tight with other approaches. It is also its limitation, since the field task remains the heavy one, compared to the modeling task, which tends to be facilitated by the recent development of generic simulation modeling framework [10].

Although analytical modeling approaches have been used to understand Lamto savanna ecology (e.g., [20]), most modeling studies have been based on simulation approaches (Chap. 9). Such modeling approaches allowed deeper understanding of key aspects of Lamto savanna functioning and dynamics: determinants of the outcome of tree/grass interactions [27, 26], key processes controlling plant growth response to grazing intensity [19] and determinants of tree dynamics [21, 12]. A significant achievement in modeling structure-functioning-dynamics relationships was the MUSE model [10], a generic ecosystem modeling environment for spatial individual-based models. In this model, most assumptions concern the spatial relationships between geometric objects representing plants of different types (e.g., grasses and trees): plants are represented by stratified piles of disks for their canopy and root system, and their competitive interactions are based on the overlap between the disks of neighboring plants (Fig. 20.5). Physiological, biological and ecological functions can be adapted to suit the needs of any particular modeling exercise. More than a model, MUSE is a synthesis and an operational formalization of the structure-functioning-dynamics interactions. It reached its limit with the

Fig. 20.5. Representation of savanna structure by the MUSE ecosystem modeling environment (after [10], with permission of Blackwell Publishing).

TREEGRASS approach, where different spatial representations of the same system coexist to suit the needs of very different process representations [28].

20.7 Conclusion

Development of new modeling approaches designed to correctly account for structure-functioning-dynamics relationships is obviously a priority for future ecological research on the Lamto savanna, both in the specialized direction (models adapted to treat the problems of savanna ecosystems) and in the generalized direction (i.e., generalizing the concepts to broaden the spectrum of possible ecosystem representations). Such studies should analyze how spatial organization influences exchanges between the ecosystem and the atmosphere, and how spatial organization can be a major ecosystem characteristic to resist disturbances.

Indeed, spatial organization is a key component for the functioning and dynamics of Lamto savannas: through their space occupation strategy at two scales (individual: plant architecture; and populational: distribution of individuals), trees and grasses modify the local environment that they experience, concentrate nutrient resources, create facilitation areas for their own group and create spots of risk to other groups. The most important lesson gained from ecological studies performed at Lamto during the last four decades is thus that heterogeneity is not a "complexity" or "difficulty" that we have to cope with when studying savanna ecology (i.e., heterogeneity as a constraint to represent and understand ecosystem behavior). Actually, spatial organization, rather than heterogeneity, is a key characteristic of the ecosystem that has to be fully taken into account as resulting from and influencing biological processes and ecosystem behavior (i.e., control of space occupation as a major adaptive mechanism to environment unpredictability).

References

1. L. Abbadie and M. Lepage. The role of subterranean fungus comb chambers (isoptera, macrotermitinae) in soil nitrogen cycling in a preforest savanna (Côte d'Ivoire). *Soil Biology and Biochemistry*, 21:1067–1071, 1989.
2. L. Abbadie, A. Mariotti, and J.C. Menaut. Independence of savanna grasses from soil organic matter for their nitrogen supply. *Ecology*, 73:608–613, 1992.
3. F. Athias. *Etude quantitative du peuplement en Microarthropodes du sol d'une savane de Côte d'Ivoire*. Ph.D. thesis, Faculté des Sciences, Université Pierre et Marie Curie, Paris, 1973.
4. R. Barbault. *Structure et dynamique d'un peuplement de Lézards: Les Scincidés de la savane de Lamto (Côte d'Ivoire)*. Ph.D. thesis, Université Pierre et Marie Curie, Paris, 1973.
5. S. Barot, J. Gignoux, and J.-C. Menaut. Demography of a savanna palm tree: Predictions from comprehensive spatial pattern analyses. *Ecology*, 80(6):1987–2005, 1999.
6. F.A. Bazzaz, S.L. Bassow, G.M. Berntson, and S.C. Thomas. Elevated CO_2 and terrestrial vegetation: Implications for and beyond the global carbon budget. In B. Walker and W. Steffen, editors, *Global change and terrestrial ecosystems*, volume 2, pages 43–76. Cambridge University Press, Cambridge, 1996.
7. L. Bellier. Le peuplement de rongeurs de la savane de Lamto. *Bulletin de Liaison des Chercheurs de Lamto*, S4:69–91, 1974.
8. J. Daget and P. Planquette. Sur quelques poissons de Côte d'Ivoire avec la description d'une espèce nouvelle, *Clarias lamottei* n. sp. (Pisces, Siluriformes, Clariidae). *Bulletin du Muséum National d'Histoire Naturelle, 2ème Série*, 39:278–281, 1967.
9. J.L. Devineau, C. Lecordier, and R. Vuattoux. Evolution de la diversité spécifique du peuplement ligneux dans une succession préforestière de colonisation d'une savane protégée des feux (Lamto, Côte d'Ivoire). *Candollea*, 39:103–134, 1984.
10. J. Gignoux, J.C. Menaut, I.R. Noble, and I.D. Davies. A spatial model of savanna function and dynamics: model description and preliminary results. In D.M. Newbery, H.H.T. Prins, and N.D. Brown, editors, *Dynamics of tropical communities*, volume 37 of *Annual symposium of the British Ecological Society*, pages 361–383. Blackwell Scientific Publications, Cambridge, 1998.
11. Y. Gillon. *Etude écologique quantitative d'un peuplement acridien en milieu herbacé tropical*. Ph.D. thesis, Université Pierre et Marie Curie, Paris, 1973.

12. M.E. Hochberg, J.C. Menaut, and J. Gignoux. The influences of tree biology and fire in the spatial structure of the West African savanna. *Journal of Ecology*, 82(2):217–226, 1994.
13. G. Josens. *Etudes biologiques et écologiques des Termites (Isoptera) de la savane de Lamto-Pakobo (Côte d'Ivoire)*. Ph.D. thesis, Faculté des Sciences de Bruxelles, 1972.
14. C. Kerdelhué and J.Y. Rasplus. Non-pollinating afrotropical fig wasps affect the fig-pollinator mutualism in *Ficus* within the subgenus *Sycamorus*. *Oikos*, 75(3):3–14, 1996.
15. S. Konaté. *Structure, dynamique et rôle des buttes termitiques dans le fonctionnement d'une savane préforestière (Lamto, Côte d'Ivoire): Le termite champignonniste Odontotermes comme ingénieur de l'écosystème*. Ph.D. thesis, Université Pierre et Marie Curie, Paris, 1998.
16. M. Lamotte. Les contraintes physico-chimiques et biologiques de la zone intertropicale. In *Enjeux de la tropicalité*. Masson , coll. Recherches en Géographie, Paris, 1989.
17. X. Le Roux and P. Mordelet. Leaf and canopy CO_2 assimilation in a West African humid savanna during the early growing season. *Journal of Tropical Ecology*, 11, 1995.
18. C. Lecordier. *Les peuplements de Carabiques (Coléoptères) dans la savane de Lamto (Côte d'Ivoire)*. Ph.D. thesis, Université Pierre et Marie Curie, Paris, 1975.
19. H. Leriche, X. Le Roux, J. Gignoux, A. Tuzet, H. Fritz, L. Abbadie, and M. Loreau. Which functional processes control the short-term effect of grazing on net primary production in grassland? *Oecologia*, 129:114–124, 2001.
20. C. De Mazancourt, M. Loreau, and L. Abbadie. Grazing optimization and nutrient cycling: Potential impact of large herbivores in a savanna system. *Ecological Applications*, 9(3):784–797, 1999.
21. J.C. Menaut, J. Gignoux, C. Prado, and J. Clobert. Tree community dynamics in a humid savanna of Côte d'Ivoire: Modelling the effects of fire and competition with grass and neighbours. *Journal of Biogeography*, 17:471–481, 1990.
22. P. Mordelet, L. Abbadie, and J.C. Menaut. Effects of tree clumps on soil characteristics in a humid savanna of West Africa (Lamto, Côte d'Ivoire). *Plant and Soil*, 153.103–111, 1993.
23. S.W. Pacala, C.D. Canham, J. Saponara, J.A. Silander, Jr., R.K. Kobe, and E. Ribbens. Forest models defined by field measurements: estimation, error analysis and dynamics. *Ecological Monographs*, 66(1):1–43, 1996.
24. S.W. Pacala and D.H. Deutschman. Details that matter: the spatial distribution of individual trees maintains ecosystem function. *Oikos*, 74:357–365, 1995.
25. M. Sankaran, N.P. Hanan, R.J. Scholes, J. Ratnam, D.J. Augustine, B.S. Cade, J. Gignoux, S.I. Higgins, X. Le Roux, F. Ludwig, J. Ardo, F. Banyikwa, A. Bronn, G. Bucini, K.K. Caylor, M.B. Coughenour, A. Diouf, W. Ekaya, C.J. Feral, E.C. February, P.G.H. Frost, P. Hiernaux, H. Hrabar, K.L. Metzger, H.H.T. Prins, S. Ringrose, W. Sea, J. Tews, J. Worden, and N. Zambatis. Determinants of woody cover in African savannas: A continental scale analysis. *Nature*, 438:846–849, 2005.
26. G. Simioni. *Importance de la structure spatiale de la strate arborée sur les fonctionnements carboné et hydrique des écosystèmes herbes-arbres*. Ph.D. thesis, Université Pierre et Marie Curie, Paris, 2001.

27. G. Simioni, J. Gignoux, and X. Le Roux. Tree layer spatial structure can affect savanna production and water budget: Results of a 3D model. *Ecology*, 84(7):1879–1894, 2003.
28. G. Simioni, X. Le Roux, J. Gignoux, and H. Sinoquet. TREEGRASS: a 3D, process-based model for simulating plant interactions in tree-grass ecosystems. *Ecological Modelling*, 131:47–63, 2000.
29. J.M. Thiollay. Le peuplement avien d'une savane préforestière (Lamto, Côte d'Ivoire). Thèse de spécialité, Université d'Abidjan, 1970.
30. R.T. Watson, M.C. Zinyowera, and R.H. Moss. *Climate change 1995 - Impacts, adaptations and mitigation of climate change: scientific-technical analyses*. Cambridge University Press, Cambridge, 1996.

21

Perspectives: From the Lamto Experience to Critical Issues for Savanna Ecology Research

Jacques Gignoux, Xavier Le Roux, and Luc Abbadie

21.1 Introduction

Studies performed at Lamto for 40 years helped to recognize the importance of the structure-functioning relationships in savanna ecology and, more generally, in ecosystem science [34, 6, 2, 51]. The major challenge for the future is to develop strategies and tools to better explore these relationships.

Savanna models can be classified along spatial and temporal resolution axes (Fig. 21.1), according to the scale at which ecological mechanisms are represented within them and the scale of model outputs. We can summarize these scales naming them according to the scale of process representation rather than model outputs:

The "ecophysiology" scale: In models operating at this scale, interest is in energy, carbon, water and nitrogen fluxes, plant growth and architectural development, over a small plot of 0.1 to a few hectares, typically in response to climate forcing over one to a few seasonal cycles. Most ecophysiological processes are represented by so-called mechanistic models using short time steps (day or shorter). The spatial resolution is usually coarser, but in savannas, it has to be detailed (infra individual for trees, square meter for the grass layer) because of the complex physics of energy and matter exchanges in those highly structured media. The TREEGRASS model [52] is a typical example of such fine scale representations of Lamto savanna functioning (Chap. 9).

The "demography" scale: In this case, interest is in long term stability, interannual variability, spatial structuration of plant communities, typically over small areas (same plots as before), but over longer time periods. Processes involved are demographic (reproduction, seed dispersal, germination, survival, response to disturbances) and can allow model running with longer time steps (month, year). Detailed spatial representation is still required by the numerous local interactions between plants (local competition, fire protection by large trees, attraction to nutrient rich patches, short range seed dispersal) and the strong spatial structure of

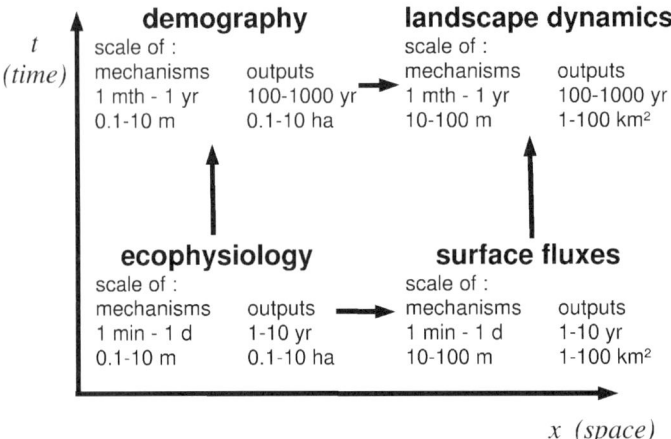

Fig. 21.1. Temporal and spatial scales of models of savanna functioning and dynamics.

savannas (Chap. 18). Different modeling approaches have been used to study tree dynamics at Lamto [21, 34].

The "surface flux" scale: At this scale, interest is usually in establishing carbon or water budgets over areas relevant to the physics of air masses or to hydrology, typically small water catchments or "landscapes" of 1-100 km^2, and over short periods of time (a few seasonal cycles). Modeling these processes at landscape scale (i.e., accounting for topography, heterogeneity in land cover and human land use) requires models using short time steps (day or shorter) and relatively coarse spatial resolution (e.g., 20 × 20 m pixels or larger) [40], because computing power generally prevents the representation of ecophysiological processes at the scale they really happen (the individual) for such large areas. In the last years, recent modeling efforts have been devoted to represent such processes for Lamto area [10].

The "landscape dynamics" scale: Interest now switches to examining the landscape dynamics in the long term. New processes such as human behavior, management decisions, animal population dynamics and migrations, large scale disturbances (fires and storms) arise [14]. As before, models are limited by computing power and must represent processes at relatively coarse spatial and temporal scales. Such a landscape dynamics approach has never been developed for Lamto savanna.

Should these different spatial and temporal scales be reconciled or coupled, and how could that be done? Splitting processes simply according to the spatial and temporal scales of their representations often breaks important feedbacks of ecological systems. For example, vegetation spatial pattern influences fluxes in the short term, but in the long term, the spatial pattern largely results from the demography of the plant species. Most demographic models are currently not linked to the underlying ecophysiological processes. For example, we do not know how a change in photosynthesis translates into population parameters like survival or recruitment. Similarly, savanna functioning models like TREEGRASS clearly showed that processes are deeply interacting with fine scale spatial structure of the vegetation. However, producing landscape models able to tackle infra-pixel spatial structure effects on energy and mass fluxes still represents a major challenge. Future tree/grass dynamics models should be able to run over long periods (centuries) but also to capture key features of structure-functioning-dynamics relationships.

The strong ecological emphasis on dominant processes lead to a great simplification of biodiversity in Lamto savanna models (trees vs. grass at the extreme), and the few studies conducted comparing different species within life-form types or functional groups demonstrated that species of trees or grasses are not exchangeable for most of their ecophysiological traits [25, 29, 16, 53, 28]. Not surprisingly, this also holds for other organisms like termites [26]. Most field and modeling studies have poorly analyzed the role of within functional group biodiversity on Lamto savanna functioning so far.

Summarizing savanna as a pure tree/grass system led to consider other ecological agents as animals and man at best as disturbances (e.g., [32, 31]). This simplification, sustained by the low density and impact of large herbivores (Chap. 10) and the low levels of human activities other than fire in Lamto savanna (Chap. 4), is too strong to apply to most other savanna types. In particular, grazing and browsing are major driving forces in other savannas [43, 33], which implies considering the whole trophic web in the ecosystem analysis. This question has just recently started to be addressed in Lamto (Chap. 10).

Representing in more detail the trophic and functional diversity of savanna ecosystems is a different question from just scaling in space and time. Besides the spatial and temporal scales appears a scale of "ecological complexity", or "functional biodiversity" that must be considered in the future. The challenge is thus now to integrate these views into comprehensive models/studies of savanna functioning, which, given the particularities of savannas, could lead to generic approaches for the analysis of these ecosystems. The question is particulary challenging because the three axes of scaling (time, space and complexity) are of course not independent. In this context, three difficulties arise: bridging the gap between short and long terms studies, between small and large spatial scales and integrating system complexity.

21.2 Scaling across time

21.2.1 Linking physiology and demography

A key reason for integrating physiology and demography when representing savanna behavior is the feedback of demography on spatial structure: the spatial pattern of vegetation influences the fluxes of energy, water, carbon and nitrogen *in the short term* [51]. At this time scale (Fig. 21.1), the change in vegetation spatial pattern through time can be neglected. But in the longer term, plants may die, grow or recruit and spatial pattern will change accordingly [7, 8]. Although measured separately and often considered a separate research field, demographic traits are not fixed in marble: they display "environmental" or "interannual" long term fluctuations. Ultimately, demographic parameters such as survival probability and per capita fecundity depend on the condition/health of individual plants, which depend on the resources they have access to and the way they uptake them: in the long term, primary production and all related short term processes must translate into demographic parameters. Suprisingly, although this long term feedback of ecophysiology on demography has been coded into most models of vegetation dynamics (e.g., [9] and its numerous descents [49, 48, 47]), it has been studied in the field only once [38].

Another reason for coupling physiology and demography is the increasing importance this coupling takes when predicting ecosystem response to disturbances for increasing time scales. For instance, the growth response of individual plants to CO_2 enrichment can be largely buffered at the community scale due to competitive interactions between individuals [27]. On a longer term, the buffering can increase both with plant mitigation to CO_2 and with changes in nutrient availability [5]. On an even longer one, ecosystem response may imply changes in vegetation composition [42] and alterations in reproduction physiology. Thus, understanding and predicting ecosystem behavior on the long term (e.g., effects of climate change and human-induced disturbances) requires one to couple functional studies (Chaps. 6 to 9) and studies focusing on plant dynamics (Chaps. 17 to 19).

Relating plant demographic traits (germination, survival, growth and allocation to reproduction) to the local environmental conditions experienced by plant individuals is currently under progress for Lamto savanna trees (Chap. 20). This should allow development of tree/grass dynamics models where demographic traits of plant individuals will actually depend on local environmental conditions according to biologically sound rules. Such models will be valuable tools to predict long term responses of savanna ecosystems to future climatic or anthropic scenarios.

21.2.2 Linking vegetation dynamics to soil organic matter decomposition/sequestration

Another process linking short and long time scales is the dynamics of soil organic matter: the slow accumulation of plant litter produces in the long term stabilized soil organic matter, and in the short term, pulses of mineral nitrogen very important in such nutrient depleted systems as Lamto. Models and experiments rarely reconcile these two approaches: most models developed to simulate long term carbon sequestration in soils are unable to simulate short term nutrient or carbon dynamics [54]; in contrast, plant nutrition models are not designed to simulate long term organic matter accumulation. The organic matter-decomposing microorganisms system poses at a different spatial scale (but at a similar temporal one) the same questions as the vegetation. Microbial physiology and demography interact with the seasonal rhythm of litter inputs and soil microclimate to produce complex dynamics that can only be understood through complex decomposition models such as SOMKO (Chap. 12 and [18]). The interaction of this subsystem with vegetation is a key issue that needs further development. Such an interaction involves the slow process of litter production by plants and the feedback of the organic matter system to the mineral nutrient pool and plant nutrition.

21.3 Scaling across space: From plot to landscape and region

21.3.1 From plot to landscape

Most physical processes involved in plant ecophysiology naturally integrate at the scale of the landscape (a few km^2). For instance, lateral water flows (runoff and runon) become particularly important at the scale of km^2, especially in dry areas. Satisfactory methodologies exist to scale short term processes up to landscape in pure tree or pure grass system. This does not hold for tree/grass systems: in savannas, the spatial pattern of the tree community (the terms "layer" and "stratum" are really inappropriate in Lamto) affects fluxes even at the plot scale [51], making the scaling up to the landscape not straightforward. Many landscape or regional scale models for tree/grass systems have been developed ignoring this constraint or assuming it can be averaged empirically [46, 55, 39, 45] Only one explicitly considers tree clumping within pixels as a factor potentially affecting fluxes [14]. Results from the TREEGRASS model (Chap. 9) could be used as a basis for a mechanistic parameterization of within-pixel spatial structure effects in coarse spatial resolution models. Work on this subject is currently in progress for humid and dry savannas of West Africa (Fig. 21.2) and should provide the basis to scale up from plot to landscape.

Such upscaling approaches need also to be carried on experimentally: at the moment, we can estimate landscape level fluxes in Lamto only through modelling. Carbon and water flux measurements over large areas are a way to explore in the near future, for example in the context of research programs on West African monsoon [3]. These data would be of great value in Lamto, where the knowledge of the finer resolution processes, i.e., the causal mechanisms of fluxes, is well advanced.

21.3.2 From landscape to region

A key interest of regional (region meaning here a significant portion of continent, in our case West Africa) modeling is in understanding the feedback of

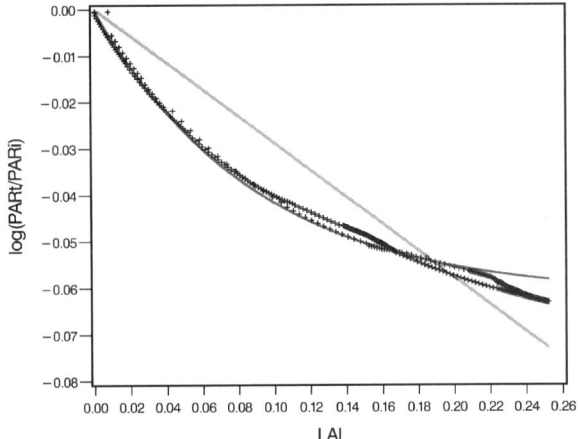

Fig. 21.2. Comparison of relationships between photosynthetically active radiation transmitted by the tree layer as a function of tree leaf area index, predicted by (light grey line) the classical Beer's law, (+) the TREEGRASS model, and (dark grey line) a modified Beer's law accounting for tree spatial pattern (after Boulain et al., unpublished).

vegetation on climate, a particularly burning issue in West Africa where local climatology is suspected to depend for a significant part on this local feedback (Chap. 3).

For ecologists, one of the challenges of the upscaling to region problem is the genericity of the models developed on a specific site ant their portability to different environmental (climate and soil) conditions. Such a genericity can only be achieved through a mechanistic analysis of ecosystem processes. Besides the use of the Lamto site as a representative well known reference site for West Africa, the deliberately mechanistic approach of ecosystem functioning taken at Lamto and the models and concepts developed on this paradigm constitute useful tools for such a spatial integration ([50] and Chap. 9).

Lamto has been one of the main sites of the SALT West African savanna transect (Chap. 1). More recently, Lamto has been included as a major site of the "Zones atelier de recherche sur l'environnement" French initiative, which deals with establishing long term regional scale research on environmental issues—a program comparable to the US LTER (long term ecological research) program, and of the West African Monsoon research project aiming at understanding the mechanisms responsible for the dynamics of the regional climate. This inserts Lamto into a network of reference sites, within sub-regions of a few hundred square kilometers where remote sensing efforts, hydrological and meteorological equipement will be concentrated [3]. An important modeling effort in climatology, atmospheric chemistry, and "surface processes" will be conducted during this program.

21.4 Scaling across system complexity

Most studies presented in this book have considered the savanna as a dynamic equilibrium between trees and grasses. Although this is more complex than many 'single group' ecosystem models (e.g., forest models [9, 23, 38, 41, 47]), it is a crude simplification of the savanna species richness and variety of processes [4].

21.4.1 Savanna biodiversity and functioning

In many studies conducted at Lamto, biodiversity effects have started to pull down the simplification of a savanna as two life-form groups, grasses and trees:

- The study of the long term stability of Lamto savannas (Chap. 18) raises the question of the minimal functional biodiversity needed to name an ecosystem a savanna: since a simple grass and trees system seems unstable, we have to propose various mechanisms to stabilize it, among which the existence of response groups to fire. It is not easy to clearly state the minimal number of groups needed to be considered in a savanna model to yield a realistic prediction of its functioning, and results are probably not portable from a site to another.
- While grass species seem to behave in a remarkably similar way for many functions, tree species do not have the same properties with respect to photosynthesis [53], water uptake strategy [25, 29] and fire resistance strategy [16].
- Within the same grass species, *Hyparrhenia diplandra*, two ecotypes with a very different behavior with respect to the nitrogen cycle (Chap. 15) coexist in Lamto [28].
- Some preliminary data tend to show that the interaction between termites and trees depends on the pairs of species involved and may profoundly influence the ecosystem spatial structure [6].
- Recent data documented the genetic diversity of nitrite oxidizers in Lamto soils (Fig. 21.3), underlining the potential role of microbial diversity for soil N dynamics.

Whether these results should be implemented in a savanna model depends on their significance in the long term patterns (e.g., stability, primary production) studied at the ecosystem level, but without including them, their significance is not testable.

21.4.2 Savanna as a trophic web

So far, all modeling experience and the general structure-dynamics-functioning approach (Chap. 20) have focused on resource uptake and competitive interactions between plants, with the noticeable exception of the SOMKO model [18].

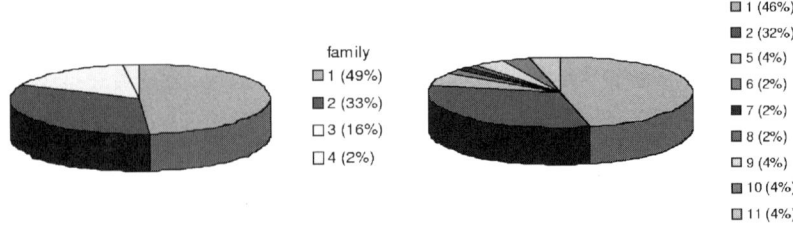

Fig. 21.3. Genetic diversity of the nitrite oxidizing bacterial community in Lamto soils. Left, nitrifying soils; right, non-nitrifying soils. Community structure was assessed using primers targeting the functional norA gene, encoding the catalytic subunit of the nitrite oxidoreductase of nitrifers (after Degrange and Poly, personal communication).

Considering savanna as a trophic network with competition within trophic levels and physical constraints acting on the whole system would lead to a completely new family of experimental approaches and models. The challenge will be to implement the trophic network within the already existing spatially explicit, individual based frameworks designed to tackle the structure-functioning-dynamics problems. Recent experimental and modeling work on herbivores [32, 15, 31] in Lamto provide some background to tackle this issue.

21.4.3 Savanna as a managed system

A striking feature of Lamto savannas is that a few species seem to shape the ecosystem structure and functioning (Table 21.1).

Examples include the following:

- The action of termites on soils (clay moving and local enrichment in mounds, recycling of plant litter) creates spots of nutrient richness which are exploited by trees [1, 25], potentially influencing the functioning and dynamics of the whole system (Chaps. 13 and 16 and Sect. 18.3) and *providing termites an abundant source of protein-rich food* in the form of tree litter.
- An ecotype of the grass *Hyparrhenia diplandra* is able to control nitrification in its rhizosphere [28], preventing nitrate production and hence *nitrogen loss through leaching* (Sect. 15.6), with probably a deep influence on the nitrogen cycle which consequences have not been examined at the whole system level so far.
- The tufted structure of the grass layer and the resulting concentration of roots not only enables grasses to better exploit the existing nitrogen, but *causes the capitalization of nitrogen* in their rhizosphere [2] (Chap. 15). Grasses thus create islands of fertility at their own scale and for their own use.

Table 21.1. Suggested biological traits and main species involved in the ecosystem organization and functioning in Lamto.

Resource/constraint	Traits	Organism
Light	Tall	Trees
	Small	Grasses, herbs and shrubs
	Shade tolerant	Forest tree species, *Imperata*
	Shade intolerant	Savanna tree species, grasses?
Carbon	C_4	Grasses
	C_3	Trees, herbs and shrubs
Water	Shallow rooted	Grasses, *Crossopteryx*
	Deep rooted	*Cussonia*, other trees?
Hydromorphy	Tolerant	*Loudetia, Annona*?
	Intolerant	Trees
Nitrogen	Large nutrient patch Maker (and user)	Trees, *Odontotermes*
	Small nutrient patch Maker (and user)	Andropogoneae grasses, fungus Growing termites, earthworms
	Nutrient patch user	*Borassus*, other trees?, grasses?
	Nitrogen wasters	Trees, non-Andropogoneae grasses?
	Nitrogen keepers	*Hyparrhenia dipl.* Non-nitrifying ecotype (other species?)
Fire	Resistant	Grasses, *Crossopteryx, Borassus, Bridelia? Annona*?
	Sensitive	*Piliostigma, Cussonia*, other species?
	Non-resistant	Forest tree species

More than ecosystem engineers [22] that modify their environment to better exploit it, we face here "ecosystem managers" that are able to capitalize scarce nutrient resources or induce the production of resources by other actors. We hypothesize that resource capitalization should improve resource use efficiency at the ecosystem level: if an ecosystem engineer simply better exploits resources, we should expect no improvement of resource use efficiency at the ecosystem level; if it is able to capitalize or induce the production of new resource by others, then we expect a better overall resource use efficiency.

This concept opens new perspectives for management of ecosystems through the manipulation of keystone "ecosystem manager" species. A challenge of future research in Lamto will be to propose alternative methods of natural resources management (e.g., management of soil fertility, fire sensitivity of vegetation, vegetation quality for cattle nutrition, etc.) based on the control of the physical environment by biological agents.

21.5 The whole picture: Modeling a spatially organized trophic web and its physical environment

Including the couplings presented in the previous sections, which all deal with some scaling, in time, space and complexity, will not be reached by simply adding up new functions to existing models, but rather implies the development of a new modeling framework designed from the very beginning to handle these scalings. A first attempt was made with the MUSE modeling framework [19]. This model, or modeling environment, has been designed to handle a large variety of spatially explicit individual-based representations of ecosystems, together with any kind of biological/ecological functions representing process (Table 21.2).

The work is currently being carried on beyond MUSE limitations using modern computing tools [13, 35] within the 3Worlds project [44, 17], based on the following principles:

Hierarchy principle: An ecosystem can be organized into nested sub-systems at many different levels.

Coexistence of physics and biology: An ecosystem obeys to different, often conflicting, logics belonging to different scientific fields, the main ones being physics and biology. Physics tends to represent an ecosystem as a complex medium carrying fluxes of quantities represented by state variables: things are smooth and continuous and nicely follow the rules of algebra. Biology represents an ecosystem as a population or community of interacting individuals that have a birth and a death and make decisions during their life: things are discrete and follow the rules of statistics, probabilities and catastrophe theory. The same ecosystem can have two different representations in these two different worlds.

Process-based spatial representations: A possible solution to make different hierarchical levels, and different representation backgrounds such

Table 21.2. List of published models that have been re-implemented within the MUSE modeling environment, demonstrating the generality of its structure (after [19], with permission of Blackwell Publishing).

Model family	Name	Ecosystem	Source
Spatial models	"French"	Savanna	[34]
	Sortie	Temperate forest	[37]
	Selva	Tropical forest	[20]
	Spatial Forska	Boreal forest	Gignoux and Davies, unpubl. (after [41], [42])
Gap models (non-spatial, individual-based)	Jabowa	Temperate forest	[9]
	Foret	Temperate forest	[49]
	Kiambram	Tropical forest	[47]
	Brind	Dry temperate forest	[36]
	Forska 1	Boreal forest	[41]
	Forska 2	Boreal forest	[42]
Non-spatial models	Competitive lottery	2-species community	[12]
	Pepsee-grass	Tropical grassland	[30]
		Temperate rainforest	[24]

as physics and biology coexist in the same model is to build up a spatially explicit representation of an ecosystem, compatible with all the different processes. Ultimately, all interactions in an ecosystem are located in space and time. Location in space and time is often the only way to associate a physical state with a biological unit, hence to measure the interaction between processes and to model it.

Based on these principles, we think that a general ecosystem model applicable to savanna ecosystems can be built and will be able to address the issues raised in the previous sections.

21.6 Conclusion

Ecological research in Lamto evolved from the primary production, energetics, "trophic chain" paradigm of the pioneer period (1962-1979), synthesized in Bourlière [11] to the "structure-functioning-dynamics" paradigm (1980-2000), whose major work is synthetized in this book. The current science is now moving to (1) considering spatial patterns as organization rather than heterogeneity, i.e., as a part of the response of the plant and animal communities to their environmental constraints, (2) developing the concept of ecosystem manager and its implications for ecosystem functioning and relations to biodiversity and (3) including the recent developments of trophic network and stoichiometric ecology to link functional ecology and community ecology.

We believe that Lamto is an ideal place to conduct such a program, because of (1) the striking spatial organization of all its components at many different scales, (2) the extremely intense macrophysical constraints that make processes more visible than in temperate systems, and enable one to easily relate plant and animal traits to their environment and (3) the low number of dominant species which helps one find the rules for hierarchizing processes and reduces the number of interactions to consider.

References

1. L. Abbadie, M. Lepage, and X. Le Roux. Soil fauna at the forest-savanna boundary: Role of termite mounds in nutrient cycling. In P.A. Furley, J. Proctor, and J.A. Ratter, editors, *Nature and dynamics of forest-savanna boundaries*, pages 473–484. Chapman & Hall, London, 1992.
2. L. Abbadie, A. Mariotti, and J.C. Menaut. Independence of savanna grasses from soil organic matter for their nitrogen supply. *Ecology*, 73(2):608–613, 1992.
3. AMMA community. French white book on African monsoon and its various components. Research project, CNRS/INSU, CNES, IRD, Météo-France, Paris, 2001.
4. S. Archer, M. Coughenour, C. Dall'Aglio, C. W. Fernandez, J. Hay, W. Hoffmann, C. Klink, J. F. Silva, and O. T. Solbrig. Savanna biodiversity and ecosystem properties. In O. T. Solbrig, E. Medina, and J. F. Silva, editors, *Biodiversity and savanna ecosystem properties*, pages 207–215. Springer-Verlag, Berlin, 1996.

5. R. Barnard, L. Barthes, X. Le Roux, H. Harmens, R. Raschi, J.F. Soussana, B. Winkler, and P.W. Leadley. Atmospheric CO_2 elevation has little effect on nitrifying and denitrifying enzyme activity in European grasslands. *Global Change Ecology*, 10:488–497, 2003.
6. S. Barot, J. Gignoux, and J.C. Menaut. Demography of a savanna palm tree: Predictions from comprehensive spatial pattern analyses. *Ecology*, 80(6):1987–2005, 1999.
7. B. Bolker and S.W. Pacala. Using moment equations to understand stochastically driven spatial pattern formation in ecological systems. *Theoretical Population Biology*, 52:179–197, 1997.
8. B.M. Bolker and S.W. Pacala. Spatial moment equations for plant competition: Understanding spatial strategies and the advantages of short dispersal. *American Naturalist*, 153(6):575–602, 1999.
9. D.B. Botkin, J.F. Janak, and J.R. Wallis. Some ecological consequences of a computer model of forest growth. *Journal of Ecology*, 60:849–871, 1972.
10. N. Boulain. *Modélisation de l'interaction végétation-hydrologie de surface en zone sahélienne: Effets du régime des précipitations et de l'occupation des sols.* M.Sc. thesis, Université de Paris 6, Paris, 2000.
11. F. Bourlière. *Tropical savannas*, volume 13. Elsevier, Amsterdam, 1983.
12. P.L. Chesson and R.R. Warner. Environmental variability promotes coexistence in lottery competitive systems. *American Naturalist*, 117(6):923–943, 1981.
13. P. Coquillard and D.R.C. Hill. *Modélisation et simulation d'écosystèmes. Des modèles déterministes aux simulations à événements discrets.* Recherche en écologie. Masson, Paris, 1997.
14. M.B. Coughenour. Savanna - landscape and regional ecosystem model. Documentation, Colorado State University, Fort Collins, CO, 1994.
15. C. de Mazancourt, M. Loreau, and L. Abbadie. Grazing optimization and nutrient cycling: Potential impact of large herbivores in a savanna system. *Ecological Applications*, 9(3):784–797, 1999.
16. J. Gignoux, J. Clobert, and J.C. Menaut. Alternative fire resistance strategies in savanna trees. *Oecologia*, 110(4):576–583, 1997.
17. J. Gignoux, I.D. Davies, and D.R.C. Hill. 3Worlds: A new platform for simulating ecological systems. In D.R.C. Hill, V. Barra, and M. Traoré, editors, *First open international conference on modelling and simulation*, volume in press, pages 1–15, Clermont-Ferrand, 2005. Blaise Pascal University Press.
18. J. Gignoux, D. Hall, J. House, D. Masse, H.N. Nacro, and L. Abbadie. Design and test of a generic cohort model of soil organic matter decomposition: The SOMKO model. *Global Ecology and Biogeography*, 10:639–660, 2001.
19. J. Gignoux, J.C. Menaut, I.R. Noble, and I.D. Davies. A spatial model of savanna function and dynamics: model description and preliminary results. In D.M. Newbery, H.H.T. Prins, and N.D. Brown, editors, *Dynamics of tropical communities*, volume 37 of *Annual symposium of the BES*, pages 361–383. Blackwell Scientific Publications, Cambridge, 1998.
20. S. Gourlet-Fleury. *Modélisation individuelle spatialement explicite de la dynamique d'un peuplement de forêt dense tropicale humide (dispositif de Paracou, Guyane Française).* Ph.D. thesis, Université Claude Bernard-Lyon I, Lyon, 1997.
21. M.E. Hochberg, J.C. Menaut, and J. Gignoux. The influences of tree biology and fire in the spatial structure of the West African savanna. *Journal of Ecology*, 82(2):217–226, 1994.

22. C.G. Jones, J.H. Lawton, and M. Shachak. Organisms as ecosystem engineers. *Oikos*, 69:373–386, 1994.
23. S. Kellomäki and H. Väisänen. Application of a gap model for the simulation of forest ground vegetation in boreal conditions. *Forest Ecology and Management*, 42:35–47, 1991.
24. T. Kohyama. Density-size dynamics of trees simulated by a one- sided competition multi-species model of rain forest stands. *Annals of Botany*, 70(5):451–460, 1992.
25. S. Konaté, X. Le Roux, D. Tessier, and M. Lepage. Influence of large termitaria on soil characteristics, soil water regime, and tree leaf shedding pattern in a West African savanna. *Plant and Soil*, 206:47–60, 1999.
26. S. Konaté, X. Le Roux, B. Verdier, and M. Lepage. Effect of underground fungus-growing termites on carbon dioxide emission from soils at the chamber- and landscape-scales in an African savanna. *Functional Ecology*, 17:305–314, 2003.
27. C. Körner. The response of complex multispecies systems to elevated CO_2. In B. Walker and W. Steffen, editors, *Global change and terrestrial ecosystems*, volume 2, pages 20–42. Cambridge University Press, Cambridge, 1996.
28. J.C. Lata, J. Durand, R. Lensi, and L. Abbadie. Stable coexistence of contrasted nitrification statuses in a wet tropical savanna ecosystem. *Functional Ecology*, 13:762–768, 1999.
29. X. Le Roux and T. Bariac. Seasonal variation in soil, grass and shrub water status in a West African humid savanna. *Oecologia*, 113:456–466, 1998.
30. X. Le Roux, J. Gignoux, J.C. Menaut, A. Tuzet, A. Perrier, O. Zurfluh, and B. Monteny. Modelling phenology, primary production and water balance in tropical humid savannas. *Annales Geophysicae*, 13 (IIs):477, 1995.
31. H. Leriche, X. Le Roux, F. Desnoyers, D. Benest, G. Simioni, and L. Abbadie. Response of grass dry-matter- and nitrogen- yields to clipping in an African savanna: An experimental test of the herbivory optimization hypothesis. *Ecological Applications*, 13:1346–1354, 2003.
32. H. Leriche, X. Le Roux, J. Gignoux, A. Tuzet, H. Fritz, L. Abbadie, and M. Loreau. Which functional processes control the short-term effect of grazing on net primary production in grassland? *Oecologia*, 129:114–124, 2001.
33. S.J. McNaughton. Evolutionary ecology of large tropical herbivores. In P.W. Price, T.M. Lewinsohn, Wilson Fernandes, G., and W.W. Benson, editors, *Plant-animal interactions: Evolutionary ecology in tropical and temperate regions*, pages 509–522. John Wiley & Sons, New York, 1991.
34. J.C. Menaut, J. Gignoux, C. Prado, and J. Clobert. Tree community dynamics in a humid savanna of the Côte d'Ivoire: Modelling the effects of fire and competition with grass and neighbours. *Journal of Biogeography*, 17:471–481, 1990.
35. P.A. Muller and N. Gaertner. *Modélisation objet avec UML*. Eyrolles, Paris, 2000.
36. I.R. Noble, H.H. Shugart, and J.S. Schauer. A description of brind, a computer model of succession and fire response of the high altitude eucalyptus forests of the brindabella range, australian capital territory. software documentation, USDA, 1980.
37. S.W. Pacala, C.D. Canham, J. Saponara, J.A. Silander, Jr, R.K. Kobe, and E. Ribbens. Forest models defined by field measurements: estimation, error analysis and dynamics. *Ecological Monographs*, 66(1):1–43, 1996.

38. S.W. Pacala, C.D. Canham, and J.A. Silander. Forest models defined by field measurements: I. the design of a northeastern forest simulator. *Canadian Journal of Forest Research*, 23:1980–1988, 1993.
39. W.J. Parton, M.B. Coughenour, J.M.O. Scurlock, D.S. Ojima, T.G. Gilmanov, R.J. Scholes, D.S. Schimel, T.B. Kirchner, J.C. Menaut, T.R. Seastedt, E. Garcia Moya, A. Kamnalrut, J.I. Kinyamario, and D.O. Hall. Global grassland ecosystem modelling: development and test of ecosystem models for grassland systems. In A.I. Breymeyer, D.O. Hall, J.M. Melillo, and G.I. Ågren, editors, *Global change: Effects on coniferous forests and grasslands*, volume 56, pages 229–270. John Wiley & Sons, Chichester, 1996.
40. C. Peugeot, B. Cappelaere, B. Vieux, L. Séguis, and A. Maia. Hydrologic process simulation of a semiraid, endoreic catchment in Sahelian West Niger: 1. Model-aided data analysis and screening. *Journal of Hydrology*, 279:224–243, 2003.
41. I.C. Prentice and R. Leemans. Pattern and process and the dynamics of forest structure: A simulation approach. *Journal of Ecology*, 78:340–355, 1990.
42. I.C. Prentice, M.T. Sykes, and W. Cramer. A simulation model for the transient effects of climate change on forest landscapes. *Ecological Modelling*, 65:51–70, 1993.
43. H.H.T. Prins and H.K. Van der Jeugd. Herbivore population crashes and woodland structure in East Africa. *Journal of Ecology*, 81:305–314, 1993.
44. A. Roche. *Techniques de simulations et simulateurs d'écosystèmes*. 2nd year thesis, ISIMA, Clermont-Ferrand, 2002.
45. J.C. Scanlan. A model of woody-herbaceous biomass relationships in eucalypt and mesquite communities. *Journal of Range Management*, 45:75–80, 1992.
46. P.J. Sellers, R.E. Dickinson, D.A. Randall, A.K. Betts, F.G. Hall, J.A. Berry, G.J. Collatz, A.S. Denning, H.A. Mooney, C.A. Nobre, N. Sato, C.B. Field, and A. Henderson-Sellers. Modeling the exchanges of energy, water, and carbon between continents and the atmosphere. *Science*, 275:502–509, 1997.
47. H.H. Shugart, M.S. Hopkins, I.P. Burgess, and A.T. Mortlock. The development of a succession model for subtropical rain forest and its application to assess the effects of timber harvest at wiangaree state forest, new south wales. *Journal of Environmental Management*, 11:243–265, 1980.
48. H.H. Shugart and I.R. Noble. A computer model of succession and fire response of the high-altitude Eucalyptus forest of the Brindabella range, Australian Capital Territory. *Australian Journal of Ecology*, 6:149–164, 1981.
49. H.H. Shugart and D.C. West. Development of an Appalachian deciduous forest succession model and its application to assessment of the impact of the chestnut blight. *Journal of Environmental Management*, 5:161–179, 1977.
50. G. Simioni. *Importance de la structure spatiale de la strate arborée sur les fonctionnements carboné et hydrique des écosystèmes herbes-arbres*. Ph.D. thesis, Université Pierre et Marie Curie, Paris, 2001.
51. G. Simioni, J. Gignoux, and X. Le Roux. Tree layer spatial structure can affect savanna production and water budget: Results of a 3D model. *Ecology*, 84(7):1879–1894, 2003.
52. G. Simioni, X. Le Roux, J. Gignoux, and H. Sinoquet. TREEGRASS: A 3D, process-based model for simulating plant interactions in tree-grass ecosystems. *Ecological Modelling*, 131:47–63, 2000.
53. G. Simioni, A. Walcroft, X. Le Roux, and J. Gignoux. Leaf gas exchange characteristics and water- and nitrogen-use efficiencies of dominant grass and tree species in a West African savanna. *Plant Ecology*, 173:233–246, 2004.

54. P. Smith, J.U. Smith, D.S. Powlson, W.B. McGill, J.R.M. Arah, O.G. Chertov, K. Coleman, U. Franko, S. Frolking, D.S. Jenkinson, L.S. Jensen, R.H. Kelly, H. Klein-Gunnewiek, A.S. Komarov, J.A.E. Molina, T. Mueller, W.J. Parton, J.H.M. Thornley, and A.P. Whitmore. A comparison of the performance of nine soil organic matter models using datasets from seven long-term experiments. *Geoderma*, 81:153–225, 1997.
55. A. Verhoef, H.A.R. De Bruin, and B.J.J.M. Van den Hurk. Some practical notes on the parameter kb-1 for sparse vegetation. *Journal of Applied Meteorology*, 36:560–572, 1997.

Index

aerosol, 27–29, 31, 36, 37, 41, 80
aging, *see* senescence
albedo, *see* radiation (reflectance)
allocation, *see* carbon (allocation), nitrogen (allocation)
allometry, *see* plant architecture (allometry)
ammonification, *see* nitrogen (ammonification)
ammonium, *see* nitrogen (ammonium)
Ancistrotermes, 237, 239, 242–245, 300, 304
Andropogon, 68, 72, 78, 84, 168, 169
  ascinodis, 66, 67, 316, 318, 321
  canaliculatus, 66, 68, 281, 316, 320
  schirensis, 67, 68, 99, 101, 202, 210, 223–225, 278–280, 282, 316, 321
architecture, *see* plant architecture
atmosphere
  chemistry, 27, 36, 398
  transmission, 31, 50, 87

bacteria, 204, 208, 210, 219, 220, 255, 258, 259, 262, 264, 265, 271, 277, 289, 290, 294, 400
bedrock, 15, 18, 20, 22, 23, 255
biodiversity, *see* diversity
biogenic structure, 236, 238, 242, 245, 299, 303, 306, 308, 309
  earthworm cast, *see* earthworm (cast)
  fungus-comb chamber, *see* termite (fungus-comb chamber)
  termitaria, *see* termite (termitaria)
  termite mound, *see* termite (mound)

biomass
  animal, 45, 186, 235, 236, 299, 303, 308, 381
  burning, *see* fire (fuel load)
  earthworm, *see* earthworm (biomass)
  herbivore, *see* herbivory (herbivore biomass)
  microbial, 219, 220, 223, 224, 228–232, 305–307
  plant, *see* phytomass
  termite, *see* termite (biomass)
Borassus aethiopum, 1, 63, 66, 119, 125, 149, 241, 335, 337, 339–349, 352–354, 356–359, 366–370, 401
Bridelia ferruginea, 64, 66, 67, 70, 72, 78, 135, 202, 223–225, 335, 339, 341–348, 350, 351, 354, 355, 358, 383, 387, 401
buffalo, *see Syncerus caffer*

C_3/C_4, *see* photosynthesis (photosynthetic pathway)
calcium, 49, 50, 264
carbon
  allocation, 77, 104, 129, 134, 135, 164, 175–177, 192, 194, 386, 387, 396
  CO_2, *see* CO_2
  mineralization, *see* respiration
  sequestration, 201, 303, 396
  soil, 143, 204–206, 214, 219, 224, 235, 245
carbon dioxide, *see* CO_2
Chromolaena odorata, 347, 348
CO_2, 230

atmospheric, 141–143
photosynthesis, *see* photosynthesis
respiration, *see* respiration
competition, 16, 104, 321, 335, 337, 339, 345, 355, 356, 371, 372, 375, 382, 388, 393, 400
  intra/interspecific, 321–323, 344, 345, 357, 373
  root:shoot ratio, *see* root:shoot ratio (effect of competition on -)
  tree-grass, *see* tree-grass interactions
Crossopteryx febrifuga, 64, 66, 69, 70, 72, 78, 96–98, 101, 140, 149, 202, 223–225, 335–358, 401
Cussonia arborea, 64, 66, 70, 78, 96–98, 101, 140, 149, 202, 223–225, 335, 336, 339, 341–348, 350, 351, 355, 358, 401
cyanobacteria, 258, 259, 271

decomposition, 203, 208, 228–230, 236, 245, 279, 294, 382, 397
  litter, 48, 55, 124, 129, 130, 132, 143, 153, 167, 207, 214, 223–226, 231, 232, 279, 285, 290, 293, 385
  soil organic matter, 201, 202, 205, 207, 214, 222, 225–230, 241, 242, 245, 295, 396
demography, *see* population dynamics
denitrification, *see* nitrogen (denitrification)
deposition, 27, 270, 272, 290, 294
  dry, 255–258, 271, 283
  nitrogen, 105, 106, 255–258, 263, 269, 271, 305
  ozone, 104, 105
  wet, 255–258, 271, 283
detritivory, 186, 235, 240, 270
deuterium, *see* isotope (^{2}H)
dew, *see* water (dew)
diversity, 5, 18, 23, 57, 149, 325, 381–383, 395, 399, 403
  grasses, 325, 327–331
  mammals, 56, 57, 196
  soil micro-organisms, 192, 219, 399, 400
  termites, 236
drainage, *see* water (drainage)

drought, *see* water (soil - potential, soil - content)

earthworm, 186, 194, 235–246, 283, 299–309, 401
  biomass, 57, 235, 236, 301, 305, 307, 308
  carbon flux, *see* respiration (soil)
  cast, 237, 238, 241, 245, 246, 266, 305–307
  nitrogen flux, *see* nitrogen
  organic matter, *see* soil (organic matter quality, organic matter decomposition)
  soil structure, 143, 385
energy budget
  soil-vegetation-atmosphere, 33, 78, 79, 86–88, 91, 92, 164, 166, 167, 393, 395, 396
  trophic network, 6, 210, 229, 235, 237, 262, 290, 382
evaporation, *see* water (soil evaporation)
evapotranspiration, *see* water (evapotranspiration)

fecundity, *see* population dynamics (fecundity)
fire
  ashes, 49, 84, 268–270
  effect on nitrogen budget, 49, 80, 194, 204, 267–272, 278, 294, 295, 309, 384
  effect on vegetation composition, 45, 52, 69–70, 259, 335–360, 365, 374, 375, 386, 399
  effect on vegetation dynamics, 52, 53, 57, 67, 72, 122, 124, 135, 178, 261, 316, 324, 335–360, 365–367, 369, 370, 372–375, 386
  exclusion, 1, 3, 9, 52, 53, 55, 70, 95, 104, 118, 124, 125, 219, 227, 228, 236, 259–261, 281, 287, 288, 324, 340, 347–349, 354, 355, 358, 365–367, 375, 386
  frequency, 51, 185, 348
  fuel load, 36, 37, 48, 52, 55, 79, 106, 268, 269, 272, 309, 337, 339, 356, 365, 370, 371
  gas emission, 36, 79, 106

root:shoot ratio, *see* root:shoot ratio (effect of fire on -)
severity, 53–55, 63, 68, 92, 268, 335, 339, 343, 354, 365, 371, 386
temperature, 52–56, 355
tolerance/resistance, 63, 70, 135, 149, 338, 341, 343, 345, 348, 352, 356, 359, 370, 372, 375, 386, 399, 401
Flag, VII
functional group, 192, 219, 239, 375, 395
fungi, 204, 208, 210, 219, 236, 238–240, 242–244, 246, 279, 290, 299, 300, 302–304, 385, 401

GCM, *see* modeling (general circulation -)
genetic structure, 315, 324, 325, 329, 331
grazing, *see* herbivory
greenhouse trace gas
  CO_2, *see* CO_2
  NO, 79, 80, 104–107, 266, 267, 269, 271, 272, 306
  NO_2, 27, 79, 80, 104–106
  O_3, *see* O_3

herbivory, 45, 56–57, 181, 185–196, 395
  effect on primary production, 185, 188–196
  effect on root:shoot ratio, *see* root:shoot ratio (effect of herbivory on -)
  effect on soil processes, 185, 192, 193
  herbivore biomass, 56, 57, 185, 186, 188, 189, 194
  herbivore consumption, 45, 57, 164, 270, 388
  insects, 185–186
  large mammals, 5, 51, 187–189, 272
  modeling, *see* modeling (herbivory)
Hyparrhenia, 68, 72, 78, 92, 168, 169, 192, 193, 219, 267, 291–293
  diplandra, 63, 66, 97, 99–102, 202, 206, 210, 213, 219, 221, 262, 263, 265, 267, 269, 271, 278–280, 282, 286, 287, 290–292, 295, 316, 318, 320, 321, 323–326, 330, 331, 366, 367, 369, 399–401

smithiana, 66, 131, 202, 210, 265, 278–280, 282, 283, 316, 318, 320, 321

Imperata cylindrica, 63, 66, 68, 70, 72, 281, 401
isotope
  ^{13}C, 117, 134, 146, 202, 203, 227, 241, 245
  ^{15}N, 207–208, 213, 283–286, 307, 384, 385
  ^{18}O, 79, 89, 96, 147
  ^{2}H, 79, 89, 96, 147

Kobus kob, 56, 187, 270

leaching
  nutrient, 50, 210, 255, 263–264, 270, 271, 279, 295
  water, *see* water (drainage)
leaf
  area index, 46, 79, 82, 83, 89, 93, 103, 140, 141, 173, 176, 177
  biochemical composition, 223
  demography, 130, 131
  nitrogen content, *see* nitrogen (leaf)
  photosynthesis, *see* photosynthesis (leaf)
  specific leaf area, 77–79
  stomatal conductance, *see* resistance (stomatal)
  stomatal density, 100
  transpiration, *see* water (transpiration)
legume, 63, 69, 156, 223, 259–262, 271, 278–283
life span, *see* mortality
light, *see* radiation
long-term dynamics, 5, 7–9, 28, 46, 52, 78, 181, 192, 196, 201, 240, 245, 246, 268, 306, 335, 340, 346, 354, 358, 360, 365, 386, 388, 393–396, 398, 399
Loudetia simplex, 41, 63, 66, 67, 69, 92, 93, 95, 106, 118, 121, 123, 127, 128, 130, 131, 134, 169, 202, 204, 205, 208–212, 219, 220, 223, 224, 226, 238, 244, 258–260, 262, 263, 265, 267–269, 278–282, 286,

287, 290, 291, 301, 309, 316, 318, 320–322, 336, 346, 401

Macrotermes, 300
magnesium, 49, 50, 264
Microtermes, 237, 239
Millsonia, 236, 241–243, 246, 301, 305
modeling
  general circulation -, 26, 106, 174, 178–180, 393–403
  herbivory, 189, 192–195, 388, 393–403
  population dynamics, 52, 356–359, 365–376, 393–403
  soil decomposition, 223, 226, 228–231, 305, 307, 393–403
  spatially explicit -, 154, 156, 165–180, 343, 346, 359, 370–376, 388, 389, 393–403
  vegetation functioning, 8, 9, 32, 40, 46, 78, 80, 83, 86, 91, 92, 98, 99, 101, 104, 106, 107, 115, 129, 136, 146, 163–181, 385, 389, 393–403
mortality, 92, 230
  grass, 323, 369
  leaf, 129, 132–134, 153, 167
  root, 203
  tree, 342, 344, 347, 352, 354–358, 367, 369, 371, 373, 374, 376, 383, 387

nitrate, *see* nitrogen (nitrate)
nitric oxide, *see* greenhouse trace gas (NO_2)
nitrification, *see* nitrogen (nitrification)
nitrogen
  ammonification, 288
  ammonium, 37, 50, 192, 221, 222, 255, 256, 283, 286–295, 305, 306
  denitrification, 106, 192, 194, 224, 255, 256, 264–384
  deposition, *see* deposition (nitrogen)
  effect of fire on -, *see* fire (effect on nitrogen budget)
  fixation, non-symbiotic, 255, 258, 259, 262, 263, 271, 283, 284, 294, 384
  fixation, symbiotic, 259, 262, 271, 283
  greenhouse gas emission, *see* greenhouse trace gas
  immobilization, 192, 221, 224, 229, 230, 264, 269

  in ashes, *see* fire (ashes)
  leaf, 101, 102, 116, 255, 278, 279, 281, 303
  mineralization, 143, 192, 194, 220, 222, 229, 230, 270, 283, 284, 287–289, 295, 303–307
  nitrate, 192, 221, 222, 255–258, 263, 264, 267, 269, 271, 283, 286–288, 290–293, 295, 384, 400
  nitrification, 106, 192–194, 264, 266, 267, 288, 290–295, 306, 330, 384, 400
  root, 116, 223, 277, 278, 280–285, 290, 295
  soil fauna, 300, 301
  soil organic -, 48–50, 116, 143, 206, 221, 222, 230, 231, 272, 277, 283, 286–287, 289, 291, 294, 300, 304, 306
  use efficiency, 101, 103, 149, 384
nitrogen monoxide, *see* greenhouse trace gas (NO)
NPP, *see* primary production
NUE, *see* nitrogen (use efficiency)
nutrient
  calcium, *see* calcium
  competition for -, *see* tree-grass interactions (nutrients)
  magnesium, *see* magnesium
  nitrogen, *see* nitrogen
  phosphorus, *see* phosphorus
  rich patch, *see* spatial structure (nutrient-rich patches)

O_3, 105
  deposition, *see* deposition (ozone)
  emission, 79, 80, 104, 105
Odontotermes, 237–240, 244, 245, 401
ozone, *see* O_3

PAR, *see* radiation (photosynthetically active -)
partitioning, *see* allocation
phenology, *see* vegetation (phenology)
phosphorus, 22, 49, 50, 143, 201, 264
photosynthesis, 101, 104, 135, 154, 165, 262, 384, 395
  canopy, 32, 50, 51, 103, 104, 140, 142, 153, 192, 399

leaf, 99–104, 151, 167, 255
photosynthetic pathway, 99, 146, 202, 203, 227, 279
quantum yield, 99–101, 103
phytomass, 5, 6, 9, 45, 55, 71, 77, 115–136, 141, 153, 154, 168–171, 173, 174, 189, 202, 231, 232, 261, 268, 277, 281, 282, 284, 293–295, 320, 371, 381
aboveground biomass, 63, 68, 71, 72, 78, 115–136, 152, 153, 168–170, 186, 193–195, 224, 255, 259, 260, 279–282, 295, 308
aboveground necromass, 55, 115–136, 153, 168–170, 282
belowground -, 18, 20, 22, 68, 69, 116, 119, 124–136, 147, 148, 150, 153, 189, 202, 203, 259, 260, 282, 285, 286, 294, 308
fuel load, see fire (fuel load)
Piliostigma thonningii, 64, 66, 67, 70, 72, 78, 135, 202, 223–225, 335, 338, 339, 341–348, 350, 351, 354–356, 358, 401
plant architecture, 92, 389
allometry, 116, 118
grass, 232, 295, 385
root, see root (architecture)
tree, 70, 339
popotte, VII, 2
population dynamics, 9, 181, 185, 223, 307, 315, 323, 324, 335, 337, 338, 340, 344, 346–360, 365–370, 373, 376, 382, 386, 393, 395–397
fecundity, 337, 344, 356, 357, 367, 382, 396
life history, 370, 371, 387, 388, 396
matrix population model, see modeling (population dynamics)
mortality, see mortality
parameter estimation, 346, 354–359, 366–368, 370, 373, 396
seed dispersal, 320, 321, 325, 329–331, 335, 339, 342, 344–345, 359, 368, 370, 371, 375, 376, 393
size structure, 350–354
precipitation, 25–27, 30–31, 37, 45, 46, 48, 49, 92, 98, 132, 144, 167
acidity, 27

amount, 23, 26–31, 37–41, 45–47, 49, 57, 92–94, 96, 98, 132–134, 136, 145, 146, 156, 167, 171, 173, 174, 189, 202, 231, 256, 263, 264, 302, 335, 349, 371
chemistry, 50, 255–258, 264, 271, 290, 294
frequency, 25, 31, 37, 38, 40, 71, 156, 231
primary production
aboveground, 127–136, 152, 153, 156, 172, 188, 195, 203, 214, 244, 270, 294, 320
belowground, 129–130, 132–133, 203, 240, 295
effect of grazing on -, see herbivory (effect on primary production)
modeling, see modeling (vegetation functioning)
total, 45, 49, 70–72, 99, 103, 116, 118, 119, 127–133, 135, 136, 139, 140, 150, 151, 163, 164, 171, 175, 176, 180, 186, 188, 195, 201, 202, 204, 215, 240, 255, 268, 270, 277, 282, 286, 287, 294, 295, 299, 315, 332, 383, 384, 396, 399, 403

quantum yield, see photosynthesis (quantum yield)

radiation
budget, 79–81, 83, 91
incoming longwave -, 33, 80–82, 87
incoming solar -, 27, 29, 31, 32, 36, 40, 41, 46, 50, 51, 80, 82, 141, 178, 180
interception by vegetation, 80, 83, 140, 141, 153, 383
light as a resource, 45, 50, 68, 83, 99, 100, 139–141, 145, 146, 149, 153–156, 164, 165, 321, 337, 339, 341, 357, 359, 387, 401
light use efficiency, 129, 167
net -, 80–82, 86–90, 152
photosynthetically active -, 32, 50, 51, 79, 83, 141, 142, 167, 176, 398
reflectance, 79–85, 269

shading, 83, 139, 140, 144, 149, 154, 165, 171, 337, 341, 357, 359, 367, 370, 387
rainfall, *see* precipitation
reflectance, *see* radiation (reflectance)
remote sensing, 29, 78, 84, 86, 180, 398
  data assimilation, 174–178, 180
  reflectance, *see* radiation (reflectance)
  vegetation indices, 84–86
resistance
  aerodynamic, 90, 91
  stomatal, 91, 92, 100
  surface, 90, 91
respiration
  soil, 45, 92, 102, 104, 142, 186, 194, 201, 202, 207, 208, 210, 212, 213, 215, 219–232, 240–246, 262, 264, 266, 269, 270, 277, 300, 382
  vegetation, VII, 226
root
  architecture, 68–70, 72, 119, 125, 146, 149, 163, 166, 167, 232, 241, 277, 284–286, 295, 344, 345, 369, 385, 388, 400
  biochemical composition, 203, 223–225, 232
  effect on nitrification, 192, 292, 293
  effect on soil organic matter mineralization, 225, 226, 241, 288
  nitrogen content, *see* nitrogen (root)
  phytomass, *see* phytomass (belowground)
  profile, 46, 48, 49, 69, 79, 91, 92, 98, 125–127, 132, 146, 147, 150, 166, 214, 290, 335
  rhizosphere, 258, 262, 264, 265
root:shoot ratio, 167, 175, 177, 307, 308
  effect of competition on -, 134, 153–155, 164
  effect of fire on -, 135, 176
  effect of herbivory on -, 194, 195
runoff, *see* water (runoff)

satellite, *see* remote sensing
scaling, 245, 395–401
seed dispersal, *see* population dynamics (seed dispersal)
senescence, 86, 99, 100, 279
  leaf, 85, 384

plant, 315, 344, 357, 358
shading, *see* radiation (shading)
soil
  ammonium, *see* nitrogen (ammonium)
  bulk density, 143
  carbon, *see* carbon (soil)
  evaporation, *see* water (soil evaporation)
  heat flux, 86–88, 270
  nitrate, *see* nitrogen (nitrate)
  organic matter decomposition, *see* decomposition
  organic matter quality, 144, 201–215, 228–230, 232, 241, 265, 306
  organic nitrogen, *see* nitrogen (soil organic -)
  respiration, *see* respiration (soil)
  temperature, 28, 140, 164, 207, 208, 220, 228, 229, 264, 269, 270, 287–290
  texture, 20–22, 46, 142, 207, 214, 229, 237, 246, 286, 294, 308
  type, 18–23, 48, 49, 93–95, 265, 286, 287, 294
  water status, *see* water (soil - content, soil - potential)
SOM, *see* soil (organic matter quality)
spatial structure, 180, 181, 306, 318, 376, 381, 382, 388, 390, 394, 396, 403
  environment heterogeneity, 80, 100, 105, 107, 147, 181, 246, 277, 288, 320, 322, 325, 337, 359, 370, 371
  grass tufts, 68, 115, 315, 316, 318–324, 326, 328, 330, 332, 366
  nutrient-rich patches, 149, 150, 270, 299, 337, 339, 343–345, 359, 370, 376, 393
  roots, 68, 285, 286, 291, 385
  tree individuals, 339–346, 348, 349, 352, 356, 358, 359, 365, 368, 370, 371, 373–375, 383, 386, 397, 398
specific leaf area, *see* leaf (specific leaf area)
stomate
  stomatal conductance, *see* resistance (stomatal)

stomatal density, *see* leaf (stomatal density)
Syncerus caffer, 56, 187, 188, 270

temperature
 air, 27–30, 33–34, 36, 37, 39–41, 45, 46, 55, 87, 99, 152, 202, 210, 226, 232, 294
 fire, *see* fire (temperature)
 sea surface -, 26, 39, 40
 soil, *see* soil (temperature)
termite, 48, 143, 194, 235–246, 266, 283, 299–309, 381, 395, 399–401
 biomass, 57, 186, 236, 237, 244, 299–302
 carbon flux, *see* respiration
 fungus-comb chamber, 238, 239, 242–245, 300, 303, 304, 306–308, 337, 385
 mound, 17, 23, 48, 63, 69, 72, 105, 143, 145, 146, 181, 203, 238, 240–246, 267, 303, 304, 306, 308, 336, 337, 341–344, 376, 385, 400
 nitrogen flux, *see* nitrogen
 organic matter, *see* soil (organic matter quality, organic matter decomposition)
 termitaria, 238, 267, 303
topography, 17, 18, 41, 48, 63, 72, 115, 203, 265, 268, 339, 394
transpiration, *see* water (transpiration)
tree-grass interactions, 9, 92, 117, 120, 139–156, 163, 171, 185, 196, 340, 358, 376, 386, 388, 395, 396
 nutrients, 146, 268, 357, 399
 shading, *see* radiation (shading)
 soil water uptake, *see* water (uptake by plants)
Trinervitermes, 237, 238, 240, 300, 302

unburned savanna, *see* fire (exclusion)

vegetation, 382
 ecotone, 41, 52, 95, 204, 205, 227, 231, 365
 life-forms, 69–70, 135, 136, 139, 146, 147, 366, 374, 395, 399

modeling, *see* modeling (vegetation functioning, population dynamics)
 phenology, 55, 71–72, 91, 103, 107, 149, 164, 167, 178, 329, 349, 375
 species composition, grass, 68, 115, 202, 204, 277, 282, 289, 292, 316, 382
 species composition, tree, 202, 204, 277, 382
 types, 41, 65, 78, 98, 104, 106, 107, 118, 121, 130, 133, 134, 220, 322, 346

water
 air vapor pressure, 27, 29, 33, 34, 41, 87, 89, 90, 99–101
 balance, 46, 91–99, 139, 144, 164, 168, 169, 171, 194, 349
 dew, 30, 31, 34, 89, 92
 drainage, 18, 20–23, 92, 95, 96, 98, 167, 264
 evapotranspiration, 33, 87–92, 98, 140, 144, 164, 171, 173, 180
 interception, 92, 93, 98, 263
 modeling - balance, *see* modeling (vegetation functioning)
 plant - potential, 96–98, 149, 151, 152
 precipitation, *see* precipitation
 rain - recycling by vegetation, 25
 runoff, 92–94, 98, 99, 167, 397
 soil - content, 41, 46–48, 63, 72, 79, 84, 85, 87, 89, 91 98, 103, 104, 106, 132–134, 140, 144–149, 152–154, 164, 167, 168, 171, 173, 177, 180, 193, 194, 220, 225, 228, 241, 264, 280, 289, 290, 348, 375
 soil - potential, 89
 soil evaporation, 28, 36, 46, 47, 87, 90, 167, 175
 transpiration, 89, 91, 92, 144, 145, 152, 165, 167, 171, 173, 175
 uptake by plants, 8, 79, 96, 98, 146–150, 164, 399
 use efficiency, 100, 101
WUE, *see* water (use efficiency)

Ecological Studies

Volumes published since 2001

Volume 143
Global Climate Change and Human Impacts on Forest Ecosystems: Postglacial Development, Present Situation and Future Trends in Central Europe (2001)
P. Joachim and U. Bernhard

Volume 144
Coastal Marine Ecosystems of Latin America (2001)
U. Seeliger and B. Kjerfve (Eds.)

Volume 145
Ecology and Evolution of the Freshwater Mussels Unionoida (2001)
B. Gerhard and K. Wächtler (Eds.)

Volume 146
Inselbergs: Biotic Diversity of Isolated Rock Outcrops in Tropical and Temperate Regions (2001)
S. Porembski and W. Barthlott (Eds.)

Volume 147
Ecosystem Approaches to Landscape Management in Central Europe (2001)
J.D. Tenhunen, R. Lenz, and R. Hantschel (Eds.)

Volume 148
A Systems Analysis of the Baltic Sea (2001)
F.V. Wulff, L.A. Rahm, and P. Larsson (Eds.)

Volume 149
Banded Vegetation Patterning in Arid and Semiarid Environments: Ecological Processes and Consequences for Management (2001)
D.J. Tongway, C. Valentin, and J. Seghieri (Eds.)

Volume 150
Biological Soil Crusts: Structure, Function and Management (2001)
J. Belnap and O.L. Lange (Eds.)

Volume 151
Ecological Comparisons of Sedimentary Shores (2001)
K. Reise (Ed.)

Volume 152
Future Scenarios of Global Biodiversity (2001)
F.S. Chapin III, O.E. Sala, and E. Huber-Sannwald (Eds.)

Volume 153
UV Radiation and Arctic Ecosystems (2002)
D.O. Hessen (Ed.)

Volume 154
Geoecology of Antartic Ice-Free Costal (2002)
L. Beyer and M. Bolter (Eds.)

Volume 155
Conserving Biodiversity in East African Forests: A Study of the Eastern Arc Mountains (2002)
W.D. Newmark (Eds.)

Volume 156
Urban Air Pollution and Forests: Resources at Risk in the Mexico City Air Basin (2002)
M.E. Fenn, L.I. de Bauer, and T. Hernández-Tejeda (Eds.)

Volume 157
Mycorrhizal Ecology (2002, 2003)
M.G.A. van der Heijden and I.R. Sanders

Volume 158
Diversity and Interaction in a Temperate Forest Community: Ogawa Forest Reserve of Japan (2002)
T. Nakashizuka and Y. Matsumoto

Volume 159
Big-Leaf Mahogany: Genetics, Ecology, and Management (2003)
A.E. Lugo, J.C. Figueroa Colón, and M. Alayon (Eds.)

Volume 160
Fire and Climatic Change in Temperate Ecosystems of the Western Americas (2003)
T.T. Veblen, W.L. Baker, G. Montenegro, and T.W. Swetnam (Eds.)

Volume 161
Competition and Coexistence (2003)
U. Sommer and B. Worm (Eds.)

Volume 162
How Landscapes Change (2003)
G.A. Bradshaw and P.A. Marquet (Eds.)

Volume 163
Fluxes of Carbon, Water and Energy of European Forests (2003)
R. Valentini (Ed.)

Volume 164
Herbivory of Leaf-Cutting Ants (2003)
R. Wirth, H. Herz, R.J. Ryel, W. Beyschlag, and B. Hölldobler (Eds.)

Volume 165
Population Viability in Plants (2003)
C.A. Brigham and M.W. Schwartz (Eds.)

Volume 166
North American Temperate Deciduous Forest Response to Changing Precipitation Regimes (2003)
P. Hanson and S.D. Wullschleger (Eds.)

Volume 167
Alpine Biodiversity in Europe (2003)
L. Nagy, G. Grabherr, C. Körner, and D.B.A. Thompson (Eds.)

Volume 168
Root Ecology (2003)
H. de Kroon and E.J.W. Visser (Eds.)

Volume 169
Fire in Tropical Savannas: The Kapalga Experiment (2003)
A.N. Andersen, G.D. Cook, and R.J. Williams (Eds.)

Volume 170
Molecular Ecotoxicology of Plants (2004)
H. Sandermann (Ed.)

Volume 171
Coastal Dunes: Ecology and Conservation (2004)
M.L. Martínez and N. Psuty (Eds.)

Volume 172
Biogeochemistry of Forested Catchments in a Changing Environment: A German Case Study (2004)
E. Matzner (Ed.)

Volume 173
Insects and Ecosystem Function (2004)
W.W. Weisser and E. Siemann (Eds.)

Volume 174
Pollination Ecology and the Rain Forest: Sarawak Studies (2005)
D.W. Roubik, S. Sakai, and A.A. Hamid Karim (Eds.)

Volume 175
Antartic Ecosystems: Environmental Contamination, Climate Change, and Human Impact (2005)
R. Bargagli (Ed.)

Volume 176
Forest Diversity and Function: Temperate and Boreal Systems (2005)
M. Scherer-Lorenzen, C. Körner, and E.-D. Schulze (Eds.)

Volume 177
A History of Atmospheric CO_2 and Its Effects on Plants, Animals, and Ecosystems (2005)
J.R. Ehleringer, T.E. Cerling, and M.D. Dearing (Eds.)

Volume 178
Photosynthetic Adaptation: Chloroplast to Landscape (2004)
W.K. Smith, T.C. Vogelmann, and C. Critchley (Eds.)

Volume 179
Lamto: Structure, Functioning, and Dynamics of a Savanna Ecosystem (2006)
L. Abbadie, J. Gignoux, X. Le Roux, and M. Lepage (Eds.)

Volume 180
Plant Ecology, Herbivory, and Human Impact in Nordic Mountain Birch Forests (2005)
F.E. Wielgolaski (Ed.) and P.S. Karlsson, S. Neuvonen, and D. Thannheiser (Ed. Board)

Volume 181
Nutrient Acquisition by Plants: An Ecological Perspective (2005)
H. BassiriRad (Ed.)

Volume 182
Human Ecology: Biocultural Adaptations in Human Communities (2006)
H. Schutkowski

Volume 183
Growth Dynamics of Conifer Tree Rings: Images of Past and Future Environments
(2006)
E.A. Vaganov, M.K. Hughes, and
A.V. Shashkin

Volume 184
Reindeer Management in Northernmost Europe: Linking Practical and Scientific Knowledge in Social-Ecological Systems
(2006)
B.C. Forbes, M. Bölter, L. Müller-Wille,
J. Hukkinen, F. Müller, N. Gunslay, and
Y. Konstantinov (Eds.)